For Product Safety Concerns and Information please contact our EU representative GPSR@taylorandfrancis.com Taylor & Francis Verlag GmbH, Kaufingerstraße 24, 80331 München, Germany

DC/AC Electrical Fundamentals

RIVER PUBLISHERS SERIES IN ELECTRONIC MATERIALS, CIRCUITS AND DEVICES

Series Editors:

The "River Publishers Series in Electronic Materials, Circuits and Devices" is a series of comprehensive academic and professional books which focus on theory and applications of advanced electronic materials, circuits and devices. This includes analog and digital integrated circuits, memory technologies, system-on-chip and processor design. Also theory and modeling of devices, performance and reliability of electron and ion integrated circuit devices and interconnects, insulators, metals, organic materials, micro-plasmas, semiconductors, quantum-effect structures, vacuum devices, and emerging materials. The series also includes books on electronic design automation and design methodology, as well as computer aided design tools.

Books published in the series include research monographs, edited volumes, handbooks and textbooks. The books provide professionals, researchers, educators, and advanced students in the field with an invaluable insight into the latest research and developments.

Topics covered in this series include:-

- Analog Integrated Circuits
- Data Converters
- Digital Integrated Circuits
- Electronic Design Automation
- Insulators
- Integrated circuit devices
- Interconnects
- Memory Design
- MEMS
- Nanoelectronics
- Organic materials
- Power ICs
- Processor Architectures
- Quantum-effect structures
- Semiconductors
- Sensors and actuators
- System-on-Chip
- Vacuum devices

For a list of other books in this series, visit www.riverpublishers.com

DC/AC Electrical Fundamentals

Dale R. Patrick
USA

Stephen W. Fardo
Eastern Kentucky University, USA

Ray E. Richardson
Eastern Kentucky University, USA

Vigyan (Vigs) Chandra
Eastern Kentucky University, USA

River Publishers

Routledge
Taylor & Francis Group
LONDON AND NEW YORK

Published 2022 by River Publishers
River Publishers
Alsbjergvej 10, 9260 Gistrup, Denmark
www.riverpublishers.com

Distributed exclusively by Routledge
4 Park Square, Milton Park, Abingdon, Oxon OX14 4RN
605 Third Avenue, New York, NY 10017, USA

DC/AC Electrical Fundamentals / by Dale R. Patrick, Stephen W. Fardo, Ray E. Richardson, Vigyan (Vigs) Chandra.

Routledge is an imprint of the Taylor & Francis Group, an informa business

ISBN 978-87-7022-740-7 (print)
ISBN 978-10-0085-177-9 (online)
ISBN 978-10-0337-726-9 (ebook master)

Contents

Preface

DC/AC Electrical Fundamentals by Patrick, Fardo, Richardson, and Chandra explores many essential topics in a basic and easy-to-understand manner. This book, and the accompanying **Electronic Devices and Circuit Fundamentals** by Patrick, Fardo, Richardson, and Chandra, has been modified with significant updates in content. The books are developed using a classic textbook – **Electricity and Electronics: A Survey (5th Edition)** by Patrick and Fardo – as a framework. Both new books have been structured using a similar sequence and organization as previous editions. The previous edition of **Electricity and Electronics: A Survey** contained 18 chapters – 8 in the **Electricity** section and 10 in the **Electronics** section.

DC/AC Electrical Fundamentals has been expanded to include 18 chapters and 6 informative appendices, further simplifying content and providing a more comprehensive coverage of content. The content has been continually updated and revised through new editions and by reviewers throughout the years. Additional quality checks to ensure technical accuracy, clarity, and coverage of content have always been an area of focus.

Each edition of the text has been improved through the following features:

1. Improved and updated text content
2. Improved usage of illustrations and photos
3. Use of color to add emphasis and clarify content

Organization of the Book

The two separate books, **DC/AC Electrical Fundamentals** and **Electronic Devices and Circuit Fundamentals**, now provide an even better comprehensive reference for the following electrical engineering/technology topics:

- Survey of Electrical and Electronic Engineering Fundamentals
- Direct Current (DC) Circuit Fundamentals
- Alternating Current (AC) Circuit Fundamentals
- Electronic Device Fundamentals
- Electronic Circuit Fundamentals

The expanded **DC/AC Electrical Fundamentals** is a basic introductory text with comprehensive coverage of fundamental electrical topics. Key concepts in the textbook are presented using the "big picture" or "systems" approach that greatly enhances learning. Many applications, testing procedures, and operational aspects of electrical devices and circuits are discussed through specific applications and illustrations. The text is divided into two sections: 1) **Direct Current (DC)** and 2) **Alternating Current (AC)**. The chapters are organized as follows:

- Introduction
- Learning Objectives
- Chapter Outline
- Major Content Discussions
- Self-examinations/Review Questions
- Summary
- Formulas/Problems
- Answers to Self-examinations

Key terms that are introduced in each chapter are defined at the end of the chapter. ***The Self-examinations*** included in each chapter allow a check of understanding of major topics covered. Practical applications and problem solving are emphasized in each chapter. The basic design of the book is to be easy to understand. The ***Problems*** included in most chapters are specifically focused on important topics that are discussed.

Acknowledgments

The authors would like to thank the many companies who have provided photographs and technical information for the book.

Dale R. Patrick
Stephen W. Fardo
Ray E. Richardson
Vigyan (Vigs) Chandra

List of Figures

List of Tables

List of Abbreviations

AC	Alternating current
ADC	Analog to digital converter
AWG	American Wire Gauge
BW	Bandwidth
CCW	Counter-clockwise
CEMF	Counter electromotive force
CRT	Cathode-ray tube
DC	Direct current
DCWV	DC working voltage
DMM	Digital multimeter
DPDT	Double-pole double-throw
DPST	Double-pole single-throw
EMF	Electromotive force
ESG	Electro static discharge
EV	Electric vehicle
GFCI	Ground-fault circuit interrupter
KCL	Kirchhoff's current law
KVL	Kirchhoff's voltage law
LCD	Liquid crystal display
LED	Light emitting diode
MFD	Microfarad
MMF	Magnetomotive force
MMFD	Micro-microfarad
MSDS	Material Safety Data Sheets

NC	Normally closed
NiMH	Nickel-metal-hydride
NO	Normally open
OL	Overload protection
PCB	Printed circuit board
PF	Power factor
PRF	Pulse repetition frequency
PRR	Pulse repetition rate
PRT	Pulse repetition time
psi	Per square inch
RF	Radio frequency
RMS	Root mean square
SI	Systems International
SPST	Single-pole single-throw
UPS	Uninterruptible power supply
VA	Volt-ampere
VAR	Volt-amperes reactive
VOM	Volt-ohm-milliammeter/volt-ohm-meter

Part 1

Direct Current (DC)

1

DC (Direct Current) Electrical Fundamentals

Electrical science is fascinating in many different ways. Advancements have provided us with the many devices that make our lives easier and more enjoyable. For example, devices such as cellular phones enable us to communicate from long distances with family and friends and to conduct business from home. Common household appliances enable us to heat and cool our homes and to keep our foods at an appropriate temperature so that they do not spoil. Televisions, radios, stereophonic equipment, and "smart" devices, such as iPods, provide us with entertainment. There are so many other applications that we use each day that it would be difficult to list all of them here.

In this chapter, you will learn about the basic principles of electrical theory and some practical applications. For example, you will learn what electrical current is and how it flows through a circuit to energize devices and equipment we use. You will also learn about the basic parts and characteristics of electrical circuits and learn to describe a complete electrical system. This information will serve as a foundation on which to build your knowledge as you proceed through this textbook and study fundamental electrical topics. Chapter 1 discusses basic topics in the study of electrical theory. These topics include basic electrical systems, energy and power, the structure of matter, electrical charges, static electricity, electrical current, voltage, resistance, and power.

Objectives

Upon the completion of this chapter, you should be able to:

1. Explain the composition of matter
2. Explain the laws of electrical charges
3. Explain current flow in electrical circuits
4. Define the terms insulator, conductor, and semiconductor
5. Define the term voltage
6. Define the term current

7. Define the term resistance
8. Define the relationship between energy, work, and power
9. Explain the parts of an electrical system

Chapter Outline

1.1 Structure of Matter

In the study of electricity and electronics, it is necessary to understand why electrical energy exists before you can understand its nature. To gain this understanding, let us look first at how certainnatural materials are made.

1.1 Determine if an atom is stable or unstable.

In order to achieve objective 1.1, you should be able to:

- Explain the composition of matter.
- Identify the parts of an atom.
- Define *matter*, *element*, *compound*, *molecule*, *atom*, *electrons*, *protons*, *neutrons*, *nucleus*, *atomic number*, *valence shell*, and *valence electrons*.

We begin with some basic scientific terms. These terms are often used in the study of chemistry. They are also important in the study of electrical theory. First, we say that **matter** is anything that occupies space and has weight. Matter can exist in a **solid**, **liquid**, or **gaseous** state. Solid matter includes such things as metal and wood; liquid matter is exemplified by water or gasoline; and gaseous matter includes such things as oxygen and hydrogen. Solids can be converted into liquids, and liquids can be made into gases. For example, water can be a solid in the form of ice. Water can also be a gas in

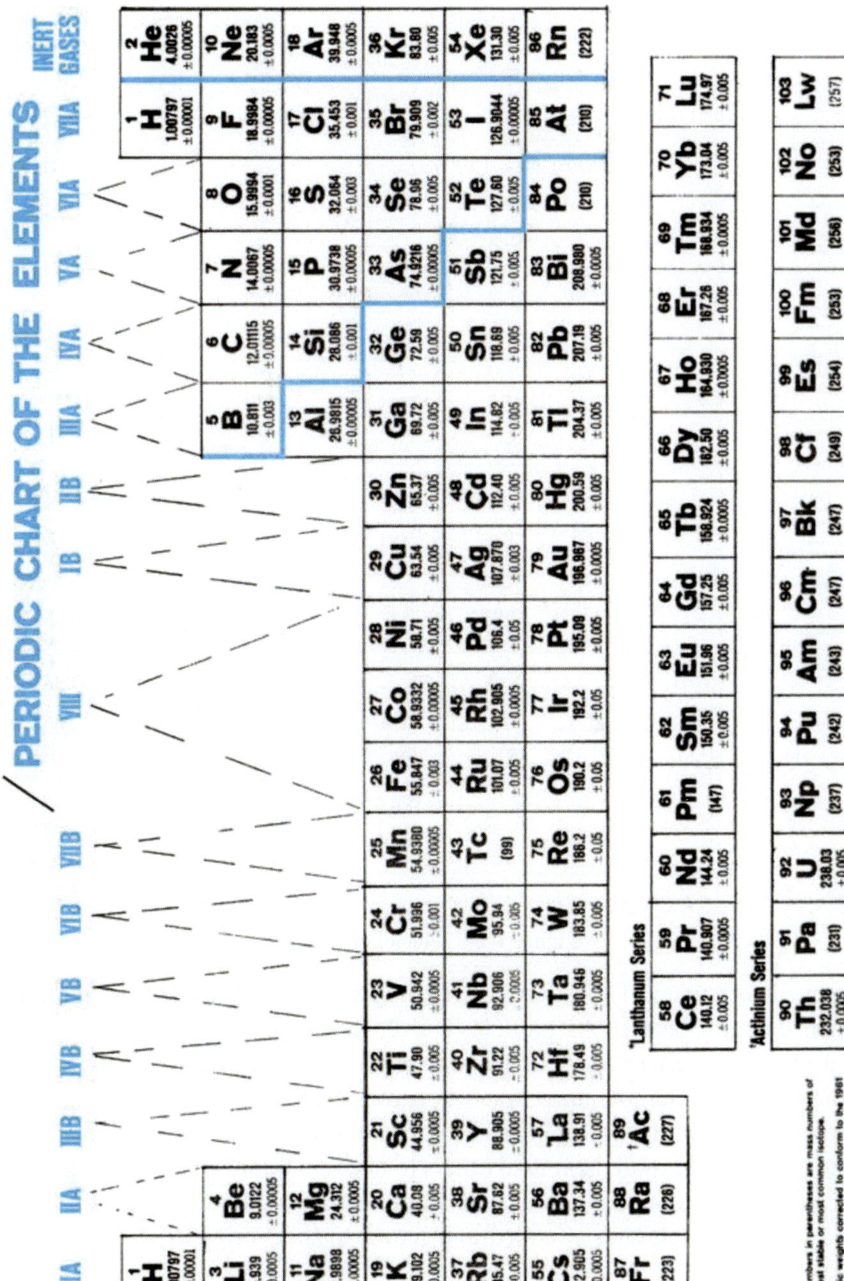
PERIODIC CHART OF THE ELEMENTS
IA IIA IIIB IVB VB VIB VIIB VIII IB IIB IIIA IVA VA VIA VIIA INERT GASES
*Lanthanum Series
†Actinium Series
( ) Numbers in parentheses are mass numbers of most stable or most common isotope.
Atomic weights corrected to conform to the 1961 values of the Commission on Atomic Weights.

An Alphabetical List of the Elements

Element	*Symbolic*	*Atomic No.*	*Atomic Weight*	*Element*	*Symbolic*	*Atomic No.*	*Atomic Weight*
Actinium	Ac	89	277*	Molybdenum	Mo	42	95.95
Aluminum	Al	13	26.97	Neodymium	Nd	60	144.27
Americium	Am	95	243*	Neon	Ne	10	20.183
Antimony	Sb	51	121.76	Neptunium	Np	93	237*
Argon	Ar	18	39.944	Nickel	Ni	28	58.69
Arsenic	As	33	74.91	Niobium	Nb	41	92.91
Astatine	At	85	210*	Nitrogen	N	7	14.008
Barium	Ba	56	137.36	Nobelium	No	102	253
Berkelium	Bk	97	247*	Osmium	Os	76	190.2
Beryllium	Be	4	9.013	Oxygen	O	8	16.000
Bismuth	Bi	83	209.00	Palladium	Pd	46	106.7
Boron	B	5	10.82	Phosphorus	P	15	30.975
Bromine	Br	35	79.916	Platinum	Pt	78	195.23
Cadmium	Cd	48	112.41	Plutonium	Pu	94	244
Calcium	Ca	20	40.08	Polonium	Po	84	210
Californium	Cf	98	251*	Potassium	K	19	39.100
Carbon	C	6	12.01	Praseodymium	Pr	59	140.92
Cerium	Ce	58	140.13	Promethium	Pm	61	145*
Cesium	Cs	55	132.91	Protactinium	Pa	91	231*
Chlorine	Cl	17	35.457	Radium	Ra	88	226.05
Chromium	Cr	24	52.01	Radon	Rn	86	222
Cobalt	Co	27	58.94	Rhenium	Re	75	186.31
Copper	Cu	29	63.54	Rhodium	Rh	45	102.91
Curium	Cm	96	247	Rubidium	Rb	37	85.48
Dysprosium	Dy	66	162.46	Ruthenium	Ru	44	101.1
Einsteinium	E	99	254*	Rutherfordium	Rf	104	260*
Erbium	Er	68	167.2	or	or		
Europium	Eu	63	152.0	Kurchatonium	Ku		
Fermium	Fm	100	255*	Samarium	Sm	62	150.43
Fluorine	F	9	19.00	Scandium	Sc	21	44.96
Francium	Fr	87	233*	Selenium	Se	34	78.96
Gadolinium	Gd	64	156.9	Silicon	Si	14	28.09
Gallium	Ga	31	69.72	Silver	Ag	47	107.880
Germanium	Ge	32	72.60	Sodium	Na	11	22.997
Gold	Au	79	197.0	Strontium	Sr	38	87.63
Hafnium	Hf	72	178.6	Sulfur	S	16	32.066
Hahnium	Ha	105	262*	Tantalum	Ta	73	180.95
Helium	He	2	4.003	Technetium	Tc	43	97*
Holmium	Ho	67	164.94	Tellurium	Te	52	127.61
Hydrogen	H	1	1.0080	Terbium	Tb	65	158.93
Indium	In	49	114.76	Thallium	Tl	81	204.39
Iodine	I	53	126.91	Thorium	Th	90	232.12
Iridium	Ir	77	192.2	Thulium	Tm	69	168.94
Iron	Fe	26	55.85	Tin	Sn	50	118.70
Krypton	Kr	36	83.8	Titanium	Ti	22	47.90
Lanthanum	La	57	138.92	Tungsten	W	74	183.92
Lawrencium	Lr	103	257*	Uranium	U	92	238.07
Lead	Pb	82	207.21	Vanadium	V	23	50.95
Lithium	Li	3	6.940	Xenon	Xe	54	131.3
Lutetium	Lu	71	174.99	Ytterbium	Yb	70	173.04
Magnesium	Mg	12	24.32	Yttrium	Y	39	88.92
Manganese	Mn	25	54.94	Zinc	Zn	30	65.38
Mendelevium	Mv	101	256*	Zirconium	Zr	40	91.22
Mercury	Hg	80	200.61				

*Mass number of the longest-lived of the known available forms of the element.

Figure 1-1 Table of elements.

the form of steam. The difference is the increased movement of particles when they are heated. As particles move, they strike one another, causing them to move farther apart. Ice is converted into a liquid by adding heat. If heated to a high temperature, water becomes a gas. All forms of matter exist in their most familiar forms because of the amount of heat they contain. Some materials require more heat than others to become liquids or gases. However, all materials can be made to change from a solid to a liquid or from a liquid to a gas if enough heat is added and return to their original state if heat is taken away.

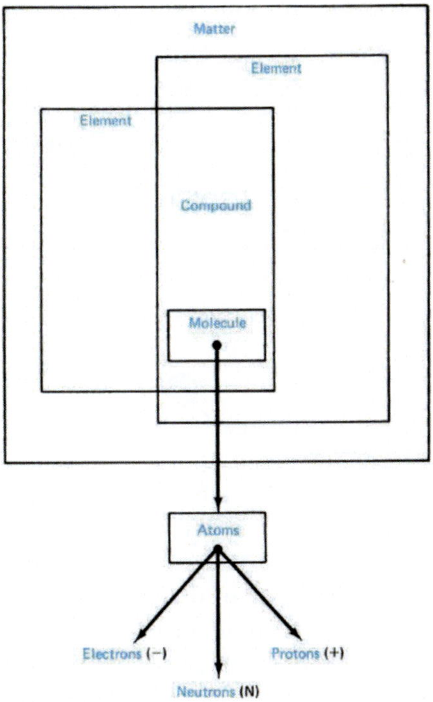

Figure 1-2 The relationship between matter, elements, compounds, molecules, atoms, electrons, protons, and neutrons.

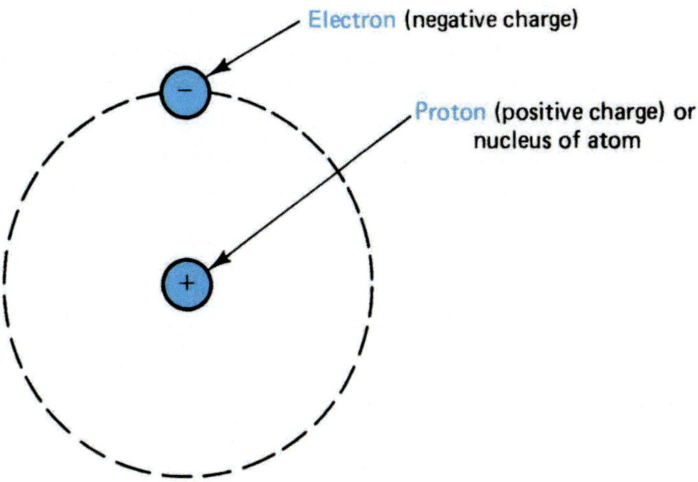

Figure 1-3 Hydrogen atom.

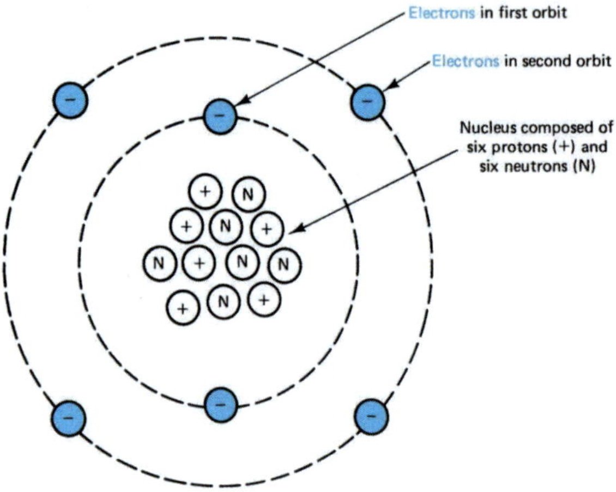

Figure 1-4 Carbon atom.

Elements

The next important term in the study of the structure of matter is **element**. An element is considered to be the basic material of which matter is made. Materials such as hydrogen, aluminum, copper, iron, and iodine are a few of the over 100 elements known to exist. A table of elements is shown in **Figure 1-1**. Some elements exist in nature, and some are manufactured. Everything around us is made of elements.

There are many more materials in our world than there are elements. Materials are made by combining elements. A combination of two or more elements is called a **compound**. For example, water is a compound made by the elements hydrogen and oxygen. Salt is made from sodium and chloride.

Another important term is **molecule**. A molecule is the smallest particle to which a compound can be reduced before breaking down into its basic elements. For example, one molecule of water has two hydrogen atoms and one oxygen atom.

Atoms

An even deeper look into the structure of matter shows particles called atoms. Within these atoms are the forces that cause electrical energy to exist. An **atom** is considered to be the smallest particle to which an element can be reduced and still have the properties of that element. If an atom were broken down any further, the element would no longer exist.

The particles found in all atoms are called **electrons**, **protons**, and **neutrons**. Elements differ from one another on the basis of the amounts of these particles found in their atoms. The relationship between matter, elements, compounds, molecules, atoms, electrons, protons, and neutrons is shown in **Figure 1-2**.

The simplest atom, hydrogen, is shown in **Figure 1-3**. The **hydrogen atom** has a center part called a **nucleus**, which has one proton. A **proton** is a particle which is said to have a positive (+) charge. The hydrogen atom has one **electron**, which orbits the nucleus of the atom. The electron is said to have a negative (–) charge. Most atoms also have neutrons in the nucleus. A neutron has neither a positive nor a negative charge and is considered neutral.

A **carbon atom** is shown in **Figure 1-4**. A carbon atom has six protons (+), six neutrons (N), and six electrons (–). The protons and the neutrons are in the nucleus, and the electrons orbit the nucleus. The carbon atom has two orbits. In the first orbit, there are two electrons. The other four electrons are in the second orbit. Electrons move easily in their orbits around the nucleus of an atom. It is the movement of electrons that causes electrical energy to exist.

The number of protons that each atom has is called an **atomic number**. Look at the atomic number for elements in the table in **Figure 1-1**. Note that each element has a different atomic number and, thus, a different number of **protons** in the nucleus of each atom. This causes each element to be different. For example, hydrogen has 1 proton, carbon has 6, oxygen has 8, and lead has 82. Since neutrons have no charge and protons have positive charge, the nucleus of an atom has net positive charge. Although science does not agree on the exact size of protons, the mass or weight of a proton is thought to be over 1800 times more than that of an electron.

Orbitals

Early models of atoms show electrons orbiting the nucleus like planets orbit the sun. This model is inconsistent with much modern experimental evidence. Atomic orbitals are very different from the orbits of satellites. Atoms consist of a dense, positively charged nucleus surrounded by a cloud or series of clouds of electrons that occupy energy levels called *shells*. The outer shell of an atom, which has the highest energy, is known as the **valence shell**, and the electrons in it are known as **valence electrons**. Electrons behave as both particles and waves; so descriptions of them always refer to their probability of being in a certain region around the nucleus. Representations or orbitals are boundary surfaces enclosing the probable areas in which the

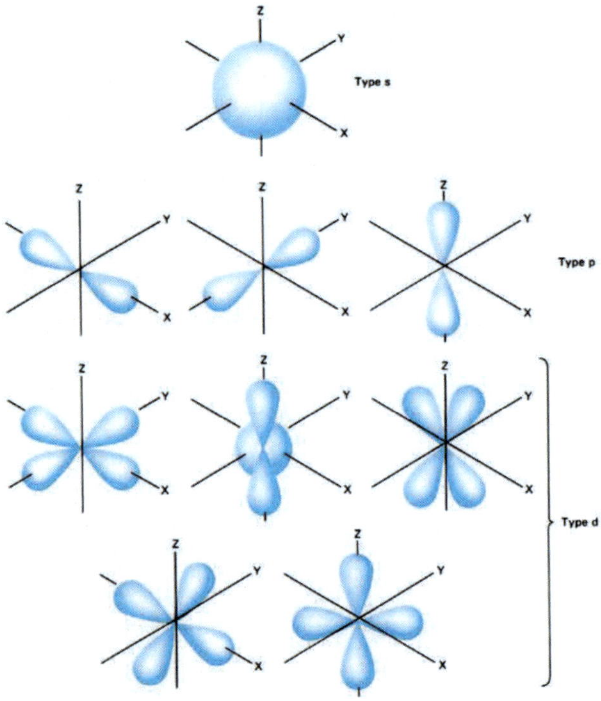

Figure 1-5 Orbitals.

electrons are found. All s **orbitals** are spherical, p orbitals are egg shaped, d orbitals are dumbbell shaped, and f orbitals are double-dumbbell shaped (see **Figure 1-5**).

An exact pattern is thought to be followed in the placement an atom's electrons (see **Figure 1-6**). Note that the first **shell** (K) or orbital contains up to two electrons. The next shell (L) contains up to eight electrons. The third (M) shell contains up to 18 electrons, which is the largest quantity this shell can contain. New shells are started as soon as shells nearer the nucleus have been filled with the maximum number of electrons.

Atoms with an incomplete **valence shell** are very active. When two unlike atoms with incomplete valence shells come together, they try to share their valence electrons. When their combined valence electrons are enough to create one complete shell, stable atoms are formed. For example, **oxygen** has eight electrons: two in the first shell (K) and six in its outer shell (L). There is room for a total of eight electrons in the outer shell. **Hydrogen**

Shells	Energy Level	Maximum Number of Electrons
K	s	2
L	s	2
	p	6
M	s	2
	p	6
	d	10
N	s	2
	p	6
	d	10
	f	14
O	s	2
	p	6
	d	10
	f	14
	g	18
P	s	2
	p	6
	d	10
	f	14
	g	18
	h	22
Q	s	2

Figure 1-6 Placement of atom's electrons.

has one electron in its outer shell. When two hydrogen atoms come near, oxygen combines with the hydrogen atoms by sharing the electrons of the two hydrogen atoms. Water is formed, as shown in **Figure 1-7**. All the electrons are then bound tightly together, and a very stable **water molecule** is formed. The electrons in the incomplete valence shell of an atom are the only electrons that will combine with other atoms to form compounds. They are also the only electrons used to cause electric current. For this reason, it is necessary to understand the structure of matter.

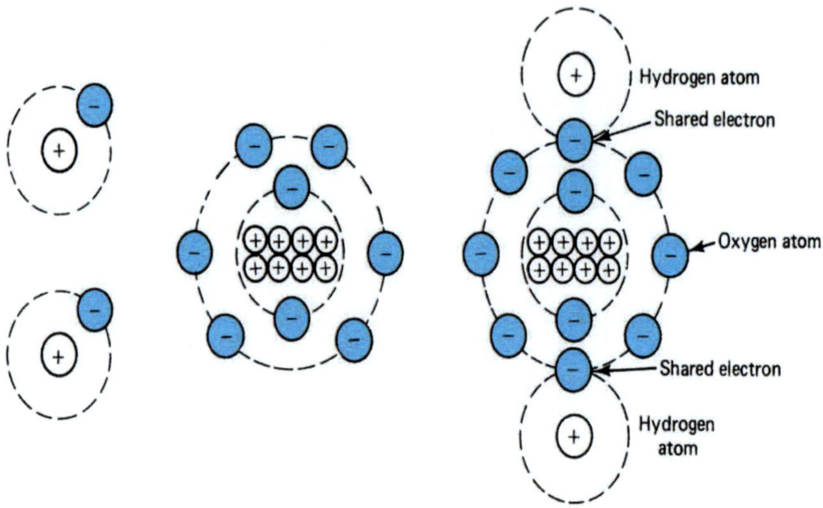

Figure 1-7 Two hydrogen atoms and one oxygen atom share electrons to form a stable water molecule.

1.2 Electric Charges

Section 1.1 discussed the positive and negative charges of particles called *protons* and *electrons*.
Recall that in every type of atom, there are an equal number of protons and electrons. Therefore, the positive and negative charges of an atom cancel one another, producing an atom with a neutral charge. However, atoms of some materials can be made to gain or lose electrons, thus giving the material a net positive or net negative charge. In this section, you will learn more about the nature of these charges and how these charges affect one another.

1.2 Determine the electric charge of a material.
In order to achieve objective 1.2, you should be able to:

- Explain the law of electric CHARGES.
- *Define electric charge, static electricity, coulomb, electrostatic field, and lines of force.*

Some materials can be made to gain or lose electrons by contacting one type of material with another. For example, if you rub a glass rod with a piece of silk cloth, the glass rod loses electrons (–); so it now has a net positive (+) charge. The silk cloth pulls away electrons (–) from the glass. Because the silk cloth gains new electrons, it now has a net negative (–) charge. Some materials are charged when they are brought close to another charged object. For example, if a charged rubber rod is touched against another material, the other material may become charged.

When an atom gains or loses electrons, it then becomes charged. These charges are called electric charges. Therefore, we can say that an electric charge is due to a deficiency or an excess of electrons in an atom or material. When there is a deficiency of electrons in an atom or material, the atom or material exhibits a net positive electrical charge. An atom or material that has an excess of electrons exhibits a net negative electric charge. The deficiency or excess of electrons causes an imbalance in the electrical charge of a material's atoms. This imbalance is called **static electricity**.

The Coulomb

The unit of measurement of an electric charge is the **coulomb**. It is estimated that 1 coulomb is equal to 6,250,000,000,000,000,000 electrons. This is expressed as 6.25×10^{18} in scientific notation form. You will learn about scientific notation in Chapter 2.

Law of Electric Charges

Each charged particle (proton and electron) of an atom is surrounded by an **electrostatic field** or **e-field**. This field is the space or area around the charged particle in which an electric charge is experienced. **Figure 1-8** shows how

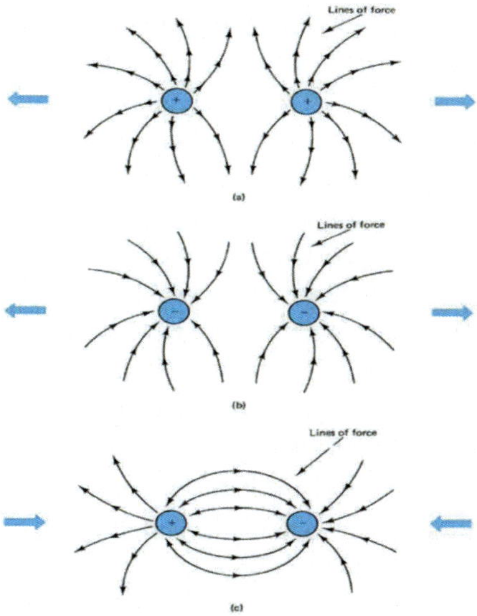

Figure 1-8 Electrostatic charges (a) Positive charges repel (b) Negative charges repel (c) Positive and negative charges attract.

electric charges affect each other. Note that the electric charges either repel (move away) or attract (come together) each other. This action is as follows:

- Positive charges repel each other (**Figure 1-8(A)**).
- Negative charges repel each other (**Figure 1-8(B)**).
- Positive and negative charges attract each other (**Figure 1-8(C)**).

The arrowed lines shown in **Figure 1-8** represent **lines of force**. These lines are imaginary; however, they represent a force that is exerted in all directions around a charged material. This force is similar to the force of gravity around the earth, called a *gravitational field.*

1.3 Electrical Current

1.3 Calculate the amount of electric current for a circuit.
In order to achieve objective 1.3, you should be able to:

- Explain electric current.
- Explain the difference between electron flow and conventional current.
- Define *electric current, free electron, ampere, circuit, open circuit,* and *short circuit.*

Electrical current is the controlled movement of electric charges through a material called a **conductor**. It is produced when electrons (–) are removed from their atoms. Some electrons in the valence shells of atoms or certain elements are easy to remove. A force applied to a material causes electrons to be removed. To understand how electric current takes place, it is necessary to know about the atoms of conductors. Conductors, such as copper, have atoms that are loosely held together. Copper is said to have atoms connected by metallic bonding. A copper atom has one valence electron which is loosely held to the atom. These atoms are so close together that their outer orbits overlap. These electrons can, therefore, move easily from one atom to another.

In any conductor, because of its atomic structure, the valence electrons continually move in a random manner from one atom to another. However, the random movement of electrons does not result in current. Electrons must move in the same direction to cause electric current. If electric charges are placed on each end of a conductor, the free electrons move in one direction.

The free electrons have a negative (–) charge; so they are repelled by the negative (–) charge applied to one end of the conductor. The **free electrons**

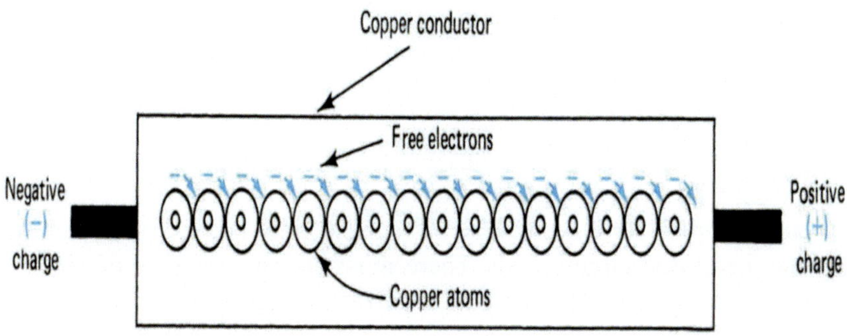

Figure 1-9 Movement of electrons through a conductor.

are attracted to the positive (+) charge that is applied to the other end of the conductor. The free electrons move from one atom to another toward the positive charge. If the electric charges on each end of the conductor are increased, more free electrons move. This increased movement of free electrons causes more electric current. The energy released by these electrons as they move allows work to be done. As more electrons move along a conductor, more energy is released. This will be discussed in more detail later in this chapter. The movement of electrons through a conductor is shown in **Figure 1-9**.

Amount of Electric Current

The amount of electric current that occurs in a circuit is measured by the number of electrons, or electric charges, that pass a point in a certain time. Recall that the coulomb is a unit of measurement of electric charge. It is estimated that 1 **coulomb** is 6,250,000,000,000,000,000 electrons (6.25×10^{18}). When 1 coulomb passes a point on a conductor in 1 second, 1 A of electric charge flows in the circuit. This is shown in the following formula:

$$1 \text{ ampere (A)} = [1 \text{ coulomb of electric charge}]/[1 \text{ second}] \quad (1\text{-}1) \quad \text{or}$$

$$I = Q/t \quad (1\text{-}2)$$

where I stands for current, Q stands for the unit of charge, and t for time in seconds.

The unit is named after Andre-Marie Ampere, an eighteenth-century scientist who studied electricity. Electric current is commonly measured in units called milliamperes and microamperes. These are smaller units of electric current than the ampere. A milliampere is one thousandth (1/1000) of an ampere and a microampere is one millionth (1/1,000,000) of an ampere. Electrical units of measurement such as these are discussed in detail in Chapter 2.

Circuits

A **circuit** is a path for current. Electric current only occurs when there is a complete, or closed-circuit, path. There must also be a source of electrical energy, such as a battery, to cause the electric charges to flow along a closed path.

Figure 1-10 shows a battery used as an energy source to cause electric charges to flow through a light bulb. Note that the path or circuit is complete. Light is given off by the light bulb due to the work done as electric charges flow through a closed circuit. Electric energy produced by the battery is changed to light energy in this circuit. You will learn more about work and energy later in this chapter.

Electric charges cannot flow if a circuit is open. An **open circuit** does not provide a complete path for current. If the circuit of **Figure 1-10** became open, no electric charges would flow. The light bulb would not glow. Free electrons of the conductor would no longer move from one atom to another.

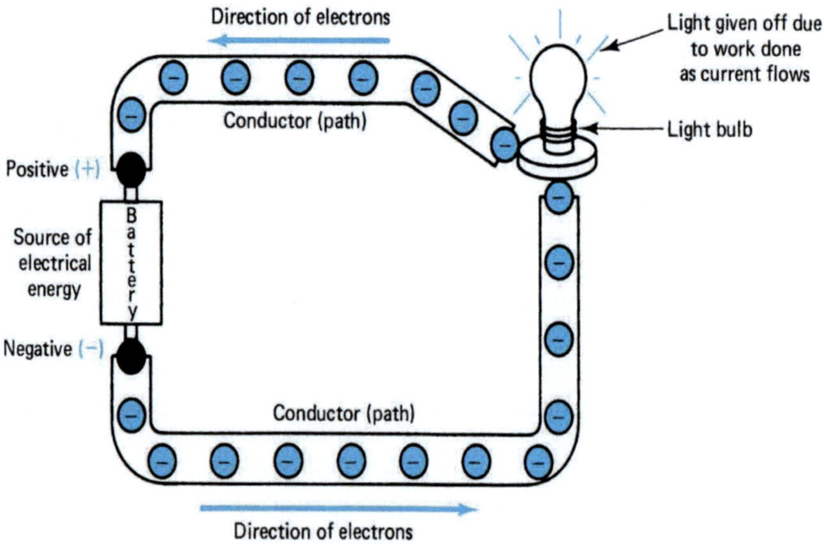

Figure 1-10 Current flow in a closed circuit.

An example of an open circuit is when a light bulb "burns out." Actually, the filament (part that produces light) has become open. The open filament of a light bulb stops electric charge flow. This causes the light bulb to stop producing light.

Another common circuit term is **short circuit**. In electrical work, a short circuit can be very harmful. It occurs when a conductor connects directly across the terminals of an electrical energy source. For example, if a wire is placed across a **battery**, a short circuit occurs. For safety purposes, a short circuit should never happen. Short circuits cause too much current. The battery would probably be destroyed, and the wire could get hot or possibly melt due to a short circuit.

Direction of Current

As you know, electrons are negative charges. Negative charges are attracted to positive charges and repelled by other negative charges. Electrons move from the negative terminal of a battery to the positive terminal. This is called **electron flow**.

Another way to look at the direction of current is in terms of charges. Electric charge movement is from an area of high charge to an area of low charge. A high charge can be considered positive and a low charge, negative. Using this method, an electric charge is considered to move from a high charge to a low charge. This is called **conventional current**. Electron flow and conventional current should not be confusing. They are just two different ways of looking at the direction of current. One deals with electron movement and the other deals with charge movement. For most applications, the assumed direction does not matter. In this book, **electron flow** is used.

1.4 Electric Potential

Circuits need a source of electrical pressure to move the free electrons through a material. In this section, you will learn about the characteristics of electric pressure, also called voltage, and how it affects current.

1.4 Determine the output voltage of an electrical energy source.

In order to achieve objective 1.4, you should be able to:

- Describe the relationship between voltage and current.
- Describe the characteristics of voltage.
- Define voltage, volt, and load.

Recall that electric current is the movement of free electrons in a conductor. Electric current is the rate of flow. Just as water pressure is needed to force water along a pipe, electrical pressure is needed to force current along a conductor. Pressure in water is produced by a pump and causes water to flow through pipes. A **voltage source** such as a battery or generator causes current to flow through a circuit. Water pressure is usually measured in pounds per square inch (psi). Electrical pressure is measured in volts and is referred to as voltage. The unit of measure for electrical force is the **volt**. The term voltage can be understood by looking at a flashlight battery. The battery is a source of voltage that is commonly used to supply electrical pressure to circuits. The battery has positive (+) and negative (–) terminals.

An electrical circuit with a battery is shown in **Figure 1-10**. The battery is a source of electrical pressure or voltage. The conductor is a path to allow the electrical current to pass to the load. The load is the part of an electrical system that converts electrical energy into another form of energy, such as an electric motor. An electric motor converts electrical energy into mechanical energy. Thus, the lamp is a load, because it changes electrical energy into light energy.

When a circuit is closed, current flows because of the electrical pressure produced by the battery. A material will not release electrons until enough force is applied. As voltage is increased, the amount of current in a circuit is also increased. As voltage is decreased, the amount of current in a circuit is also decreased. For example, a lamp rated at 120 volts requires 120 volts of voltage applied to the lamp to force the proper amount of current through it. More pressure would increase the current flow and burn out the lamp. The lamp would not operate properly with too much or too little voltage. Less pressure would not force enough current to flow and the lamp would illuminate dimly or not illuminate at all. The amount of voltage in a circuit is the work done by units of charge expressed in **coulombs** of charge. The unit of **work** in electronics is defined as the **joule (J)**. The amount of voltage is the amount of work divided by the unit charge as the following equations show:

$$\text{voltage} = [\text{work}]/[\text{charge}]$$
$$1\ \text{volt} = [1\ \text{joule}]/[1\ \text{coulomb}]$$

Voltage is also called potential, potential difference, or **electromotive force (EMF)**. The term electromotive force is largely responsible for the usage of E as a symbol for voltage. With the development of solid-state and computer electronics, E has other meanings. To avoid confusion, the letter V is now commonly used as the symbol for voltage.

1.5 Resistance

Recall that atoms of some materials give up their valence electrons easily. These materials offer low opposition to current. Other materials hold their valence electrons and offer high opposition to current. In this section, you will learn more about this opposition and how it affects current.

1.5 Determine the resistance of a circuit.

In order to achieve objective 1.5, you should be able to:

- Describe the relationship between resistance and current.
- Explain the phenomenon of resistance at the molecular level.
- State the formula for resistance and conductance.
- Define *resistance*.

The opposition to current in electrical circuits is called **resistance**. Even a very good conductor has some resistance that limits the flow of free electrons. As you have learned, the material of which an object is made affects its degree of resistance. The ease with which different materials give up their valence electrons is very important in determining resistance. For example, silver is an excellent conductor of electricity. Copper, aluminum, and iron have more resistance but are more commonly used as they are less expensive. All materials conduct an electric current to some extent, even though some (insulators) have extremely high resistance.

With a constant amount of electric potential (voltage) and a large opposition (resistance) to current, the number of electrons moving (current) through the material is small. With constant voltage, current can be increased by decreasing resistance. Therefore, by increasing or decreasing the amount of resistance in a circuit, the amount of current flow can be changed.

1.6 Conductors and Insulators

Conductors

As you have learned, the ability of a free electron to flow down a path or circuit depends greatly on the material through which the electrons will travel. In electricity, materials are classified based on their ability to allow (or restrict) the movement or conduction of the electrons. This section discusses four types of materials through which electricity may use as a path: conductors, insulators, semiconductors, and superconductors.

1.6 Explain how valence electrons affect a material's ability to conduct electricity.

In order to achieve objective 1.6, you should be able to:

- Name some examples of conductors, insulators, semiconductors, and superconductors.
- Define *conductor*, *insulator*, *semiconductor*, and *superconductor*.

A material through which current flows easily is called a **conductor**. Copper and aluminum wire are commonly used as conductors. Conductors are said to have low resistance to electric current.

Conductors usually have three or fewer electrons in the valence shells of their atoms.

Many metals are conductors. Each metal has a different ability to conduct electric current. For example, silver is a better conductor than copper, but silver is too expensive to use in large amounts. Aluminum does not conduct electric current as well as copper, but its use is common. It is less expensive and lighter than other conductors. Copper is used more than any other conductor. Materials with only one valence electron (gold, silver, copper, etc.) are the best conductors. Note that these elements are located in column IB of the periodic table **(Figure 1-1**).

Insulators

A material through which electric current does not flow easily is called an **insulator**. The electrons of materials that are insulators are difficult to release. Some insulators have their valence shells filled with eight electrons. Others have valence shells that are over half filled with electrons. The atoms of materials that are insulators are said to be stable. Insulators have high resistance to the movement of electric current. Some examples of insulators are plastic and rubber. **Figure 1-11** shows some types of insulators.

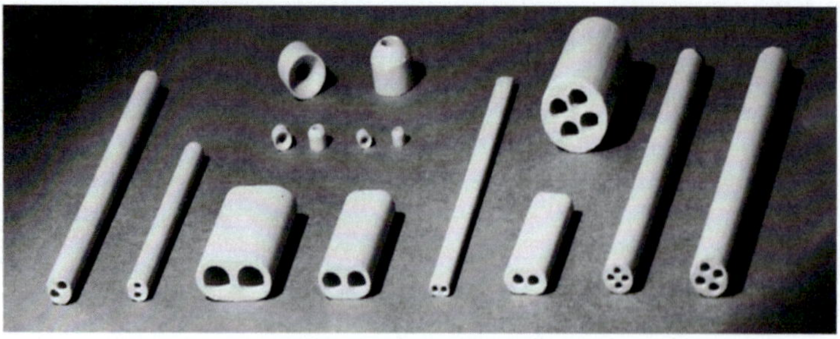

Figure 1-11. Insulators.

Semiconductors

Semiconductors have become extremely important in electronics. A **semiconductor** is a material that does not conduct electric current easily; however, they are not good insulators. Semiconductors have four electrons in their valence shells. Remember that conductors have valence shells less than half full, and insulators ordinarily have valence shells more than half full.

Some common types of semiconductor materials are silicon, germanium, and selenium. Note that these elements are located in column IVA of the periodic table shown in **Figure 1-1**. **Figure 1-12** compares conductors, insulators, and semiconductors.

Superconductors

Much effort is being put forth in superconductor research. A **superconductor** is a conductor that has no resistance to electric current. This phenomenon was first observed in the early 1900s by a Dutch scientist, Kamerlingh Onnes. Onnes discovered that if a mercury crystal was cooled to a temperature just

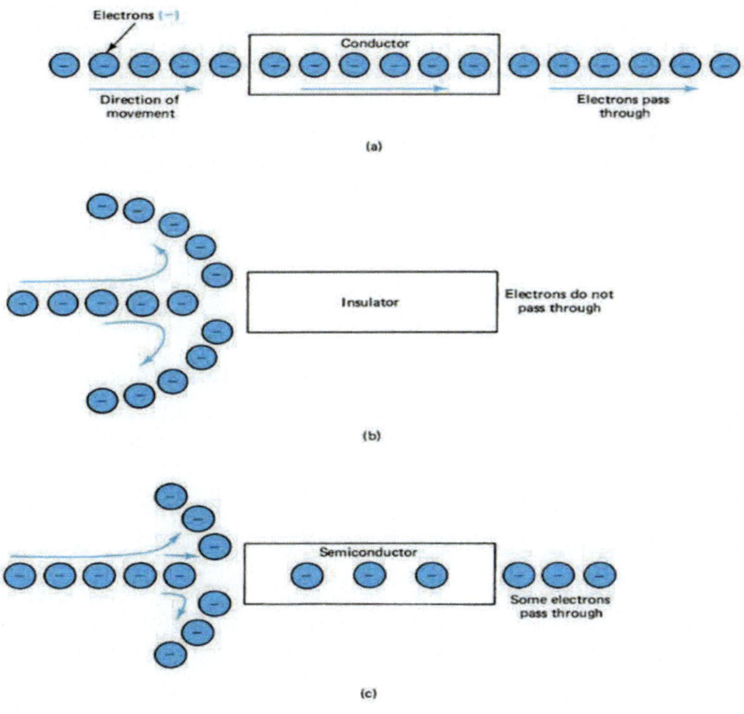

Figure 1-12 Comparison of (a) conductors, (b) insulators, and (c) semiconductors.

above absolute zero (–459.67°F or –273.15°C), it lost all resistance to the flow free electrons. Only recently has this process begun to be understood. The basic idea of superconductors is simple. A normal conductor has resistance to electric current. This resistance converts energy to heat. If current continues, a source of energy must be present to replace the energy lost as heat. A superconductor has no resistance and, therefore, does not generate heat. If free electrons flow in a superconductor, they will flow without ceasing.

In a normal conductor, **free electrons** are continually colliding with atoms that make up the conductor. Each collision causes energy to be lost and heat to be generated. In a superconductor, each passing electron causes a small vibration in the conductor. This action clears a path for more electrons. As the temperature rises, the motion of the conductor's atoms increases. Eventually, the bond between electrons is broken and superconduction stops. For this reason, superconduction requires an extremely low operating temperature.

Superconduction has been observed in certain metals when they were cooled to temperatures near absolute zero and in some ceramic compounds. Research scientists would like to create a room temperature superconductor. The idea of room temperature superconductors now seems to be a real possibility. Superconductive power lines could save enormous amounts of energy in the transmission of electric power. Typically, 15%–20% of the energy produced by electric power plants is lost in transmission. Superconducting transmission lines would lower the cost of producing electricity, conserve natural resources, and reduce pollution. Other applications have also been proposed for superconductors, such as high-speed trains levitated by magnetic fields created around superconductive rails.

1.7 Energy, Work, and Power

An understanding of the terms energy, work, and power is necessary in the study of electricity and electronics. These terms are used to describe mechanical and electrical energies.

1.7 Calculate the rate at which work in a circuit is done.

In order to achieve objective 1.7, you should be able to:

- Calculate the amount of energy expended in an electric circuit.
- Explain the relationship between energy, work, and power.
- Define energy, kinetic energy, potential energy, work, power, and watt.

The first term, **energy**, means the capacity to do work. For example, the capacity to light a light bulb, to heat a home, or to move something requires energy. Energy exists in many forms, such as electrical, mechanical, chemical, and heat energy. If energy exists because of the movement of some item, such as a ball rolling down a hill, it is called kinetic energy. If it exists because of the position of something, such as a ball at the top of the hill but not yet rolling, it is called potential energy. A second important term is **work**. Work is the transferring or transforming of energy. In mechanical terms, work is done when a force is exerted to move something over a certain distance against opposition. It is represented by the following formula:

work = force × distance.

Work is done when a chair is moved from one side of a room to the other. An electrical motor used to drive a machine performs work. When force is applied to open a door, work is performed. Work is also done each time energy changes from one form into another.

A third important term is **power**. Power is the rate at which work is done. It considers not only the work that is performed but also the amount of time in which the work is done. For instance, mechanical power is the rate at which work is done as an object is moved against opposition over a certain distance. It is represented by the following formula:

power = [work]/[time]

or

power = [force × distance]/[time].

Electric power is the rate at which work (energy transferred from one form to another) is done. It is represented by the following formula:

power = voltage × current.

A **watt** is the unit of measurement of electrical power. You will learn more about power in the following chapters.

1. Name the two types of energy.
2. Force and distance are needed to define ____________.
3. The rate of work is known as ________________.

1.8 Electrical Systems

The concept of "electrical systems" allows discussion of some complex things in a simplified manner. This method is used to present much of the material in this book in order to make it easier to understand. The systems concept serves as a "big picture" in the study of electricity and electronics. The role played by each part then becomes clearer, and it is easy to understand the operation of a complete electrical system.

1.8 Explain the parts of an electrical system.

In order to achieve objective 1.8, you should be able to:

- Name some examples of sources of electricity, paths, controls, loads, and indicators.
- Identify the energy source, path, control, load, and indicator of an electrical system.

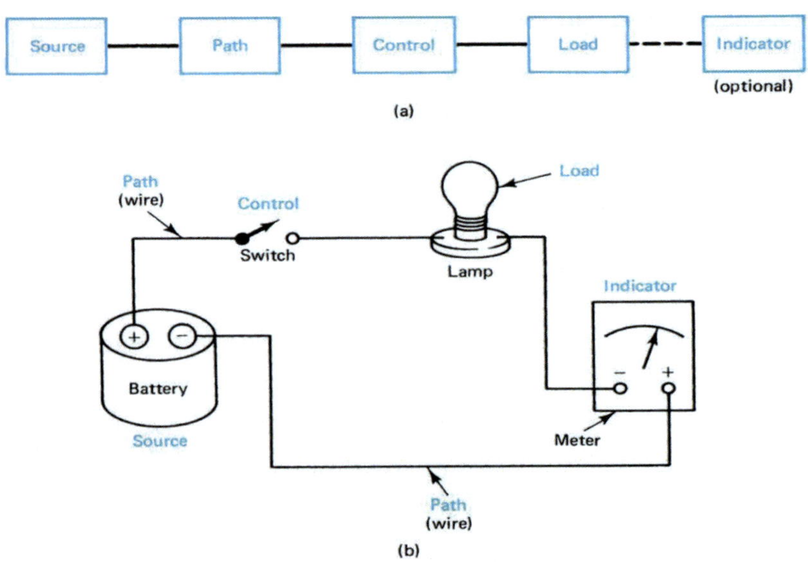

Figure 1-13 Electrical system. (a) Block diagram. (b) Pictorial diagram.

The parts of an **electrical system** are the **energy source**, **path**, **control**, **load, and indicator**. A simple electrical system block diagram and pictorial diagram are shown in **Figure 1-13**. Using a block diagram allows a better

(a)

(b)

(c)

Figure 1-14 Sources of electrical energy. (a) Batteries – convert chemical energy into electrical energy. (h) Fossil-fueled steam plant in Paradise, Kentucky – converts mechanical energy and heat energy into electrical energy. (c) Solar cells – convert light energy into electrical energy [(a) Courtesy of Union Carbide Corp.; (b) courtesy of Tennessee Valley Authority; (c) courtesy of International Rectifier].

understanding of electrical equipment and provides a simple way to "fit pieces together." The system block diagram can be used to simplify many types of electrical circuits and equipment. Each block of an electrical system has an important role to play in the operation of the system. Hundreds and even thousands of components are sometimes needed to form an electrical system. Regardless of the complexity of the system, each block must achieve its function when the system operates.

Electrical System Parts

The **energy source** of an electrical system provides electric energy for the system. Heat, light, chemical, and mechanical energy may be used as sources of electrical energy. **Figure 1-14** shows some sources of electrical energy.

The path of an electrical system is simple compared to other system parts. This part of the system provides a path for the transfer of electrical energy. It starts with the energy source and continues through the load. In some systems, this path is an electrical wire. In other systems, a complex supply line is placed between the source and the load, and a return line from the load to the source is used. There are usually many paths within a complete electrical system. **Figure 1-15** shows the distribution path of electrical power from its source to where it is used.

The control section of an electrical system is the most complex part of the system. In its simplest form, control is achieved when a system is

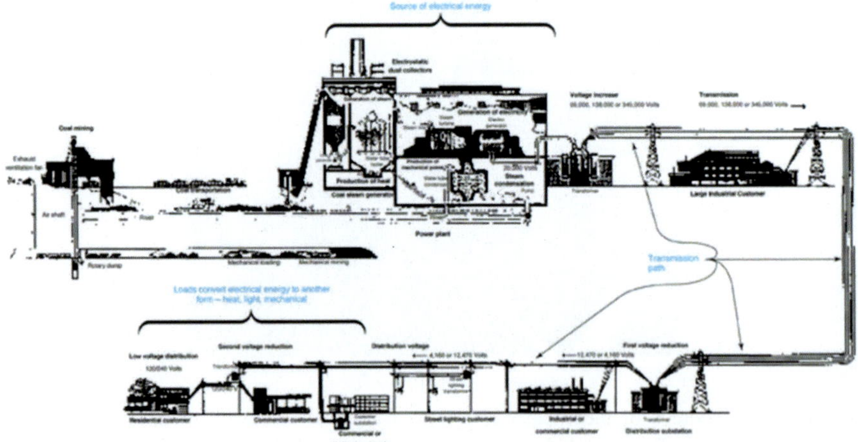

Figure 1-15 Distribution path for electrical power from its source to where it is used (Courtesy of Kentucky Utilities Co.).

Basics
Limit
Pendant
Pushbutton
Rocker
Rotary
Slide
Special Purpose
Toggle
Transducers
Trim
Hardware and Accessories

(a)

OFF
ON

(b)

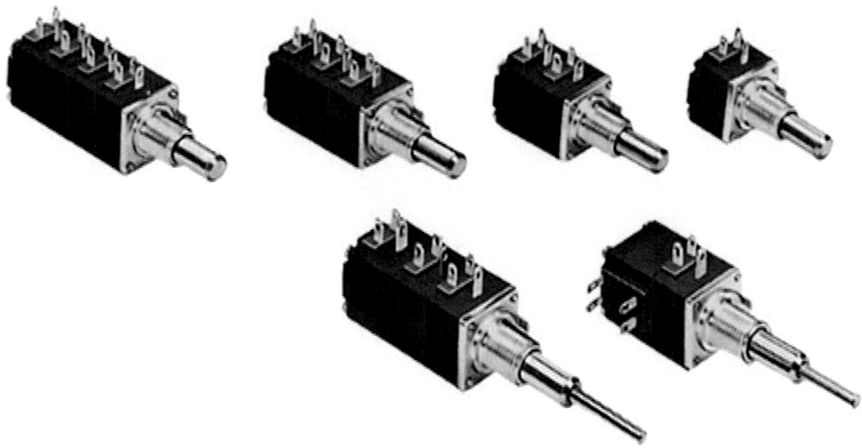

(c)

Figure 1-16 Common control devices. (a) Miscellaneous types of switches. (b) Potentiometers – partial control [(a) Courtesy of Eaton Corp., Cutler-Hammer Products; (b) Courtesy of Allen-Bradley Co.].

turned on or off. Control of this type may take place anywhere between the source and the load. The term **full control** is used to describe this operation. A system may also use some type of partial control. **Partial control** causes some type of operational change in the system other than turning it on or off. A change in the amount of electric current is a type of change that is achieved by partial control. Some common control devices are shown in **Figure 1-16**.

The **load** of an electrical system is the part or group of parts that do some type of work. Work occurs when energy goes through a transformation or change. Heat, light, and mechanical motions are forms of work produced by loads. Much of the energy produced by the source is changed to another type by the load. The load is usually the most obvious part of the system because of the work it does. An example is a light bulb which produces light. Some common loads are shown in **Figure 1-17**.

The **indicator** of an electrical system displays a particular operating condition. In some systems, the indicator is an optional part that is not really needed; in other systems, it is necessary for proper operation. In some cases, adjustments are made by using indicators; in other cases, an indicator is attached temporarily to the system to make measurements. Test lights, panel meters, oscilloscopes, and chart recorders are common indicators used in electrical systems. Electrical indicators are shown in **Figure 1-18**.

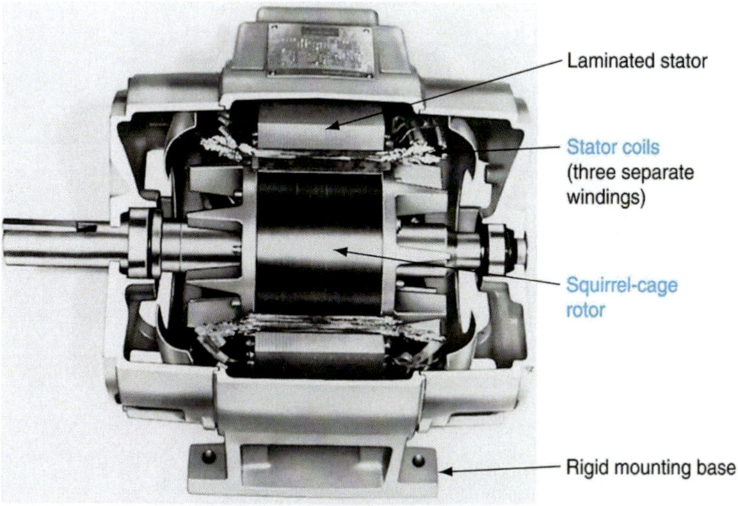

(a)

(b)

Figure 1-17 Common electrical loads. (a) Light bulb – converts electric energy into light energy. (b) Heat pump home heating system – converts electrical energy into heat energy. (c) Electric motor – converts electrical energy into mechanical energy [(a) Courtesy of Philips Lighting Co.; (b) courtesy of Williamson Co.; (c) courtesy of Delco Products Division—General Motors Corp.].

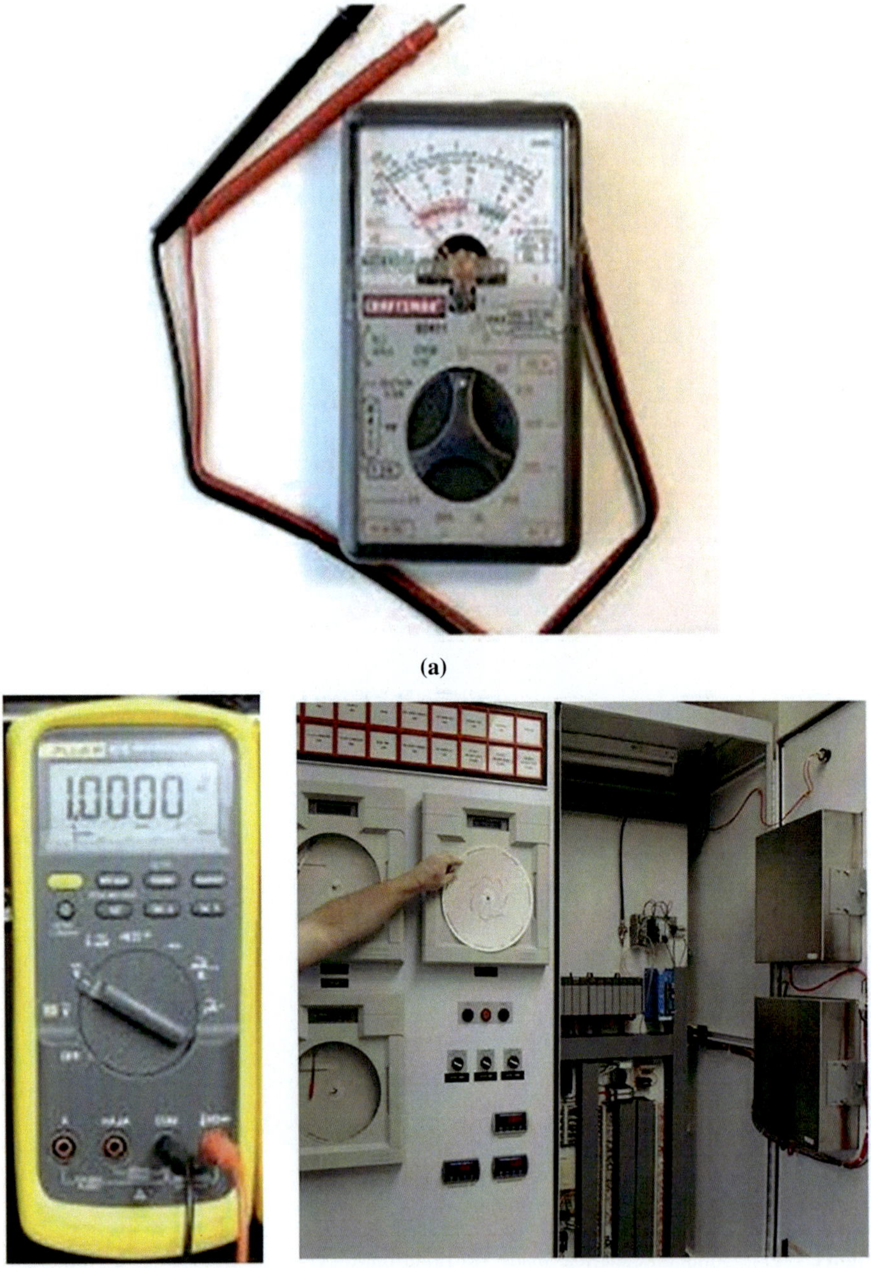

(a)

(b) **(c)**

Figure 1-18 Electrical indicators. (a) Analog meter. (b) Digital multimeter. (c) Chart recorder that makes a permanent record of some quantity [(a) Courtesy of Craftsman Co.; (b) courtesy of Fluke Instrument Corp.; (c) courtesy of Gould Inc., Instruments Division].

Electrical System Examples

Nearly everyone has used a flashlight. This device is designed to serve as a light source.

Flashlights are a simple type of electrical system. **Figure 1-19** is a cutaway drawing of a flashlight with each part shown.

- The battery of a flashlight serves as the **energy source** of the system. Chemical energy in the battery is changed into electrical energy to cause the system to operate. The energy source of a flashlight – the battery may be thrown away. Batteries are replaced periodically when they lose their ability to produce energy.
- The **path** of a flashlight is a metal case or a small metal strip. Copper, brass, or plated steel are used as paths.
- The **control** of electrical energy in a flashlight is achieved by a slide switch or a pushbutton switch. This type of control closes or opens the path between the source and the load device. Flashlights have only a means of full control, which is operated manually by a person.
- The **load** of a flashlight is a small lamp. When electrical energy from the source passes through the lamp, the lamp produces a bright glow. Electrical energy is then changed into light energy. A certain amount of work is done by the lamp when this energy change takes place.
- Flashlights do not use an **indicator** as part of the system. Operation is indicated, however, when the lamp produces light. The load of this system also acts as an indicator.

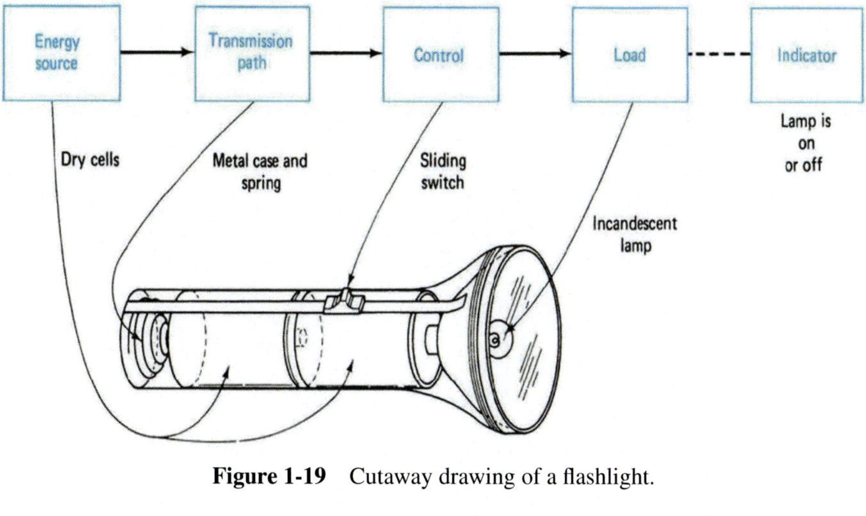

Figure 1-19 Cutaway drawing of a flashlight.

A more complex example of a system is the electrical power system that supplies energy to buildings, such as our homes. **Figure 1-20** shows a sketch of a simple electrical power system.

The **energy source** of an electrical power system is much more complex than that of a flashlight. The source of energy may be coal, natural gas, atomic fuel, or moving water. This type of energy is needed to produce mechanical energy. The mechanical energy develops the motion needed to turn a turbine. Large generators are then rotated by the turbine to produce electrical energy. The energy conversion of this system is quite complex from start to finish. The functions of the parts of the system remain the same regardless of complexity.

In an electrical power system, the **path** consists of many electrical conductors. Copper wire and aluminum wire are ordinarily used as conductors. Metal, water, the earth, and the human body can all be made paths for electrical energy transfer.

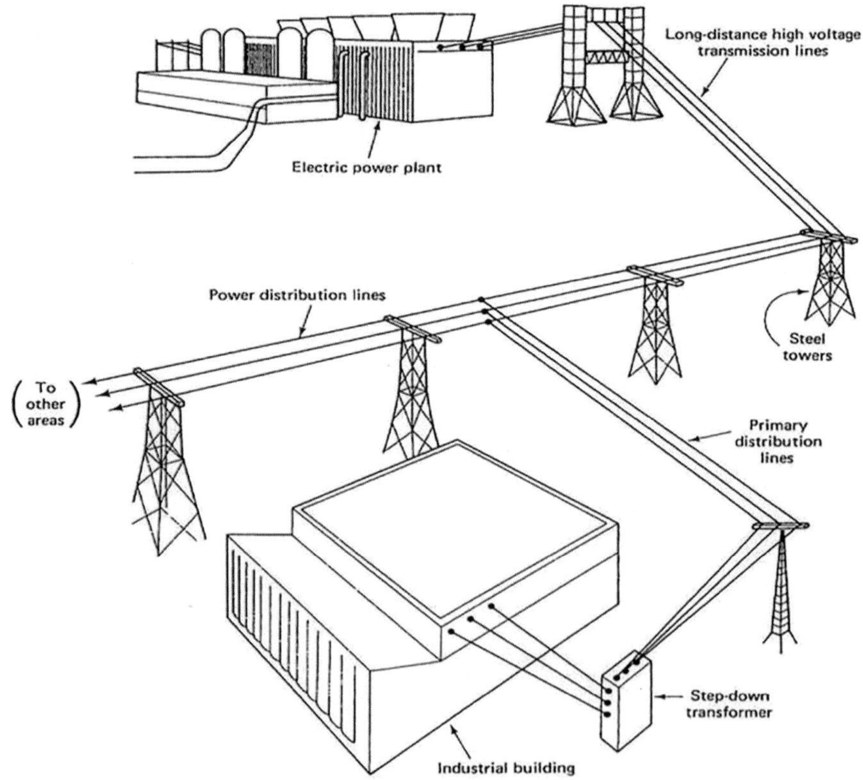

Figure 1-20 Simplified electrical power system.

The **control** function of an electrical power system is performed in many different ways. **Full-control** devices include switches, circuit breakers, and fuses. **Partial control** of an electrical power system is achieved in many ways; for example, transformers are used throughout the system. These partial-control devices are designed to change the amount of voltage in the system.

The **load** of an electrical power system includes everything that uses electrical energy from the source. The total load of an electrical power system changes continually. The load is the part of the system that actually does work. Motors, lamps, electrical ovens, welders, and power tools are some common load devices. Loads are classified according to the type of work they produce, such as light, heat, and mechanical.

The **indicator** of an electrical power system is designed to show the presence of electrical energy. It may also be used to measure electrical quantities. Panel-mounted meters, oscilloscopes, and chart recording instruments are some of the indicators used in this type of system. Indicators of this type provide information about the operation of the system.

The **systems concept** is a method that may be used to study electrical engineering, electricity, and electronics. This method provides a common organizational plan that applies to most electrical systems. An understanding of a basic electrical system helps to overcome some of the problems involved in understanding complex systems.

Summary

- Matter is anything that occupies space and has weight.
- An element is the basic material of which matter is made.
- Electrons (negative charge), protons (positive charge), and neutrons (neutral charge) are the particles found in atoms.
- Atoms have the same amount of electrons as they have protons and are therefore electrically neutral.
- Electrons orbit the nucleus of an atom in various energy levels called shells.
- The outermost shell of an atom is called a valence shell and the electrons in it are called valence electrons.
- Atoms with incomplete valence shells are very active.
- A material becomes electrically charged when it loses or gains valence electrons.

- The basic unit of electric charge is the coulomb and is represented by the letter Q.
- One coulomb is equal to 6,250,000,000,000,000,000 electrons.
- Charged particles and materials are surrounded by an electrostatic field.
- Opposite charges attract, like charges repel.
- Electric current is the controlled movement of electric charges through a conductor.
- One coulomb of electric charges past a point is equal to 1 ampere or amp.
- A circuit is a closed path for current to flow.
- A closed, or complete, circuit provides a complete path for electricity, and an open circuit does not.
- Resistance is the opposition to current.
- A conductor easily passes electrons, and insulators do not.
- A semiconductor is a material that does not conduct electric current easily; however, they are not good insulators.
- Electrical systems consist of a source, load, and path.
- Additional components in an electrical system may include controls and indicators.

Self-examination/Answers

Answer the following questions.

1. The ____ is the basic material of which matter is made.
2. A combination of two or more elements is called a(n) _________.
3. The smallest particle to which an element can be reduced and still have the properties of that element is called a(n) ______________.
4. The three primary parts of an atom are the _____, _____, and _____.
5. Electrons are arranged in ________.
6. Atoms with an equal number of electrons and protons are called _____________.
7. Charges are expressed as ________ and ________.
8. Like charges _______ each other.
9. Unlike charges ______ each other.
10. Electrons that make up current flow are called _____ electrons.
11. A path for current flow is called a(n) ___________.
12. Electron movement from negative to positive through a circuit is called ___________ current flow.

13. Charge movement from positive to negative through a circuit is called ____________current flow.
14. One coulomb of charge per second is called a(n) ____________.
15. Electrical force is expressed as ________.
16. The unit of measure for voltage is the __________.
17. A(n) _________ is commonly used to supply electrical pressure to a circuit.
18. In a circuit, the current follows the path made by ______________.
19. A decrease in circuit voltage will cause an (increase, decrease) in current flow.
20. Opposition to current flow refers to __________.
21. Insulators have a (high/low) resistance.
22. If circuit resistance increases, current flow (increases, decreases).
23. Name the two types of energy.
24. Force and distance are needed to define ____________.
25. The rate of work is known as ________________.
26. Name three materials that make the best conductors.
27. Name two examples of insulators.
28. Name three semiconductor materials.
29. Name the five primary components of an electrical system.
30. The load in a flashlight would be the ____________.

Answers

1. element
2. compound
3. atom
4. proton, neutron, electron
5. shells
6. stable
7. positive and negative
8. repel
9. attract
10. free
11. circuit
12. electron
13. conventional
14. ampere or amp
15. voltage

16. volts
17. battery
18. conductors
19. decrease
20. conductors
21. insulators
22. semiconductors
23. kinetic and potential
24. work
25. power
26. gold, silver, and copper
27. plastic and rubber
28. silicon, germanium, selenium
29. source, path, control, control, load, indicator
30. lamp

Problems

Answer the following.

1. Silicon has the atomic number 14. How many electrons does this atom contain?
2. Silicon has the atomic number 14. How many shells of electrons does this atom contain?
3. What is the abbreviation for the element iron? What is the atomic number for iron? How many electrons are found in an atom of iron?
4. A movement of current is found to contain 21,980,000,000,000,000,000 electrons per second. What is the amount of current?
5. Six coulombs of electrons move past a point in 1 second. What is the current?
6. Six coulombs of electrons move past a point in 2 seconds. What is the current?

Glossary

Matter
Any material that occupies space and has weight. It can be a solid, a liquid, or a gas.

Element
The basic material that makes up all other materials. They exist by themselves (such as copper, hydrogen, and carbon) or in combination with other elements (water is a combination of the elements hydrogen and oxygen).

Compound
The chemical combination of two or more elements to make an entirely different material.

Molecule
The smallest particle to which a compound can be reduced before being broken down into its basic elements.

Atom
The smallest particle to which an element can be reduced and still retain its characteristics.

Electrons
Atomic particles that have a negative (–) charge. They cause the transfer of electrical energy from one place to another.

Protons
Particles in the center of an atom which have a positive (+) electrical charge.

Neutrons
Particles in the nucleus (center) of an atom which have no electrical charge or is neutral.

Nucleus
The core or center part of an atom, which contains protons having a positive charge and neutrons having no electrical charge.

Atomic number
The number of particles called protons in the nucleus (center) of an atom.

Valence shell
The outer shell of an atom, which has the highest energy.

Valence electrons
Electrons in the outer shell, or energy level, of an atom.

Electrostatic field
The space or area around a charged body in which the influence of an electrical charge is experienced.

Lines of force
A charge that is exerted in all directions around a charged particle.

Static electricity
Electricity with an excess or deficiency of positive and negative electrical charges.

Free electrons
Electrons located in the outer orbit of an atom which are easily removed, resulting in electrical current flow.

Circuit
A path for current flow.

Open circuit
A circuit that has a broken path so that no electrical current can flow through it.

Short circuit
A circuit that forms a direct path across a voltage source or component so that a very high and possibly unsafe electrical current flows.

Coulomb
A unit of electrical charge that represents a large number of electrons (about 6,250,000,000,000,000,000).

Ampere
The electrical charge movement which is the basic unit of measurement for current flow in an electrical circuit. One coulomb of electrical charge passing a point in 1 second.

Voltage
The electrical force or pressure that causes current to flow in a circuit.

Volt
The unit of measurement of electrical potential.

Load
The part of an electrical system that converts electrical energy into another form of energy. An example is an electric motor, which converts electrical energy into mechanical energy.

Conductor
A material that allows electrical current to flow through it easily.

Insulator
A material that offers high resistance to electrical current flow.

Semiconductor
A material that has a value of electrical resistance between that of a conductor and an insulator.

Superconductor
A material that has no resistance to current flow.

Resistance
Opposition to the flow of current in an electrical circuit; its unit of measurement is the ohm (Ω).

Energy
Something that is capable of producing work, such as heat, light, chemical, and mechanical action.

Kinetic energy
Energy due to motion.

Potential energy
Energy due to position.

Work
The transforming or transferring of energy.

Power
The rate of doing work, or transferring energy, in electrical circuits. Found by using the equation $P = I \times V$.

Watt
The unit of measurement of electrical power; the amount of power converted when 1 A of current flows under an electrical pressure of 1 volt.

2

Electrical Components and Diagrams

Most electrical equipment is made of several parts or components that work together. It would be almost impossible to explain how electrical equipment operates without using symbols and diagrams. Electrical diagrams show how the component parts of equipment fit together. Common electrical components are easy to identify. It is also easy to learn the symbols used to represent electrical components. The components of electrical equipment work together to form a functional electrical system.

Objectives

Upon completion of this chapter, you will be able to:

1. Diagram a simple electrical circuit.
2. Identify schematic electrical symbols.
3. Convert electrical quantities from metric units to English units and English units to metric units.
4. Use scientific notation to express numbers.
5. Identify different types of resistors.
6. Identify resistor value by color code and size.
7. Explain the operation of potentiometers (variable resistors).
8. Construct basic electrical circuits using a diagram.

Chapter Outline

2.1 Components, Symbols, and Diagrams

Anyone who studies electricity and electronics should be able to identify the components used in simple electrical circuits. *Components* are represented by symbols. **Symbols** are used to make diagrams. A **diagram** shows how the components are connected in a circuit. For example, it is easier to show symbols for a battery connected to a lamp than to draw a pictorial diagram of the battery and the lamp connected together. There are several symbols that are important to recognize. These symbols are used in many electrical diagrams. Diagrams are used for installing, troubleshooting, and repairing electrical equipment. Using symbols makes it easy to draw diagrams and to understand the purpose of each circuit. Common electrical symbols are listed in **Appendix A**.

Conductors Electrical circuits must have some means of connecting various electrical devices together; so the conductance as described in Chapter 1 can take place. Typically, this is performed by wires or traces of a **conductive material** on a printed circuit board. These "paths" of conductance can be made from many materials, but copper is the most popular. Whenever you handle a wire to "plug in" a device such as a fan or connect a USB cable to your computer, you are working with conductors.

Because conductors allow electrons from their outer shells of orbit to the adjoining orbit, wires and other devices to connect electrical circuits are made of conductors. Typically, wire conductors are made of copper, but some applications require aluminum or gold as a conductor. Typically, conductors are specified based on their gauge. This number or gauges is based on a standard known as the **American Wire Gauge**, or **AWG**. In this system, the larger the gauge number, the smaller in cross-sectional area the wire becomes. For example, a larger 6 gauge wire may be used for an auto battery, a 12 gauge wire for a home, and a small 22 gauge wire for portable circuits.

Wire conductors are also identified by physical characteristics other than the cross-sectional area, or **gauge**. The wire may be solid or made of multiple strands of single wires twisted or braided together, simply referred to as stranded wire. The voltage rating of wires actually has nothing to do with the conductor size but the value of the insulator that surrounds the wire. Wires assembled in a particular configuration are referred to as **cables**. This includes the three-wire cable for most household appliances, the four-wire cable for a traditional telephone, the eight-wire conductor for a CAT-5 computer networking cable, or the multiple conductors found in the ribbon cable that connects the motherboard of a computer to the various drives inside a computer.

Switches

In addition to providing a path for the movement of electrons through conductors such as cables, there is typically a need to control the electricity. The most common type of control is the **switch**. Switches come in a variety of types and styles and are identified by function, as well as voltage and current ratings. In general, any device capable of connecting, disconnecting, or changing the conduction in a circuit can be considered a switch.

Switches generally are identified in two distinct groups: manually operated and mechanically operated. **Manually operated switches** are those controlled "manually" or by a person. **Mechanically operated switches** are those switches that are operated by something from their environment, such as striking an object, temperature, or pressure. Manually operated switches have many familiar forms. Familiar types are toggle, push-button, slide, key lock, tactile, rotary selectors, and many other configurations. Mechanically operated switches come in several types as well. Limit, snap-action, tilt, temperature, overload, proximity, pressure, and light-activated switches are all part of this category. **Relays** can also be considered mechanical switches and will be discussed in the upcoming chapter regarding magnetism.

Whether mechanically or manually operated, **switches** are classified or rated on characteristics such as poles, throws, electrical ratings of the contacts, and any physical characteristics such as size or special hook-up or mounting configurations.

Poles and **throws** refer to the capabilities of the switch. A pole refers to the number of contacts that can be used to control the paths of conduction. The throw refers to the number of positions a switch can be actuated. The typical **toggle switch** used to control the lighting in one room such as a bedroom is typically a single pole single throw switch. Switches can come with any number of poles, but single and double are the most popular. Similarly, with throws, rotary switches have multiple throws, but single and double are very popular, specifically in toggle switch configurations.

Contacts are the actual pieces of metal or conductors inside the switch that control the connection or disconnection. Typically, switches are issued a **rating** based on the maximum voltage and current the contacts are capable of. For example, a typical toggle switch might be rated at 3 A, 120 VAC. This means the switch can control up to 3 A at 120 VAC. Lesser currents and voltages are acceptable, but the switch would be dangerous to operate if either rating was exceeded.

Switches in electrical circuits are not drawn as they appear in the device. Toggle switches may appear the same as a slide switch in symbol form. The symbol used for various switches depicts how they are actuated (toggle, pushbutton, etc.) and the number of poles and throws.

Common symbols, such as switches, are included in **Appendix A**.

Over-current Protective Devices

Circuits are designed or engineered within certain specifications or tolerance. Should any of these specifications be exceeded, it can be dangerous for the components in the circuit. Most damaging is a device that draws too much current, often generating heat and causing catastrophic failure of an electrical device. To prevent **over-current** situations, protection devices such as **fuses** and **circuit breakers** have been developed. These devices use various methods to detect a current greater than anticipated and open or turn off a circuit that may be malfunctioning. The cause for an over-current situation might include:

- a short circuit caused by two conductors (wire)s touching;
- too much equipment on the same circuit;
- worn insulation allowing bare wires to touch grounded objects.

Fuses and circuit breakers are used for the protection of circuits and **not** individuals; there are other procedures and equipment used to protect people, and fuses and circuit breakers will not protect them. Always remember to use common sense when working with electrical equipment or circuits. Follow safe practices in the electrical lab and in the home. Detailed safety information is available from the **National Safety Council** and other organizations. **It is always wise to be safe**.

Fuses are electrical devices that open an electrical circuit when excess current flows by melting away an internal conductor. Typical packages for fuses are called **cartridges**. Cartridges are round cylinders of varying size and material that contain an element or filament that melts when the specified current rating is exceeded. These cartridges may be small glass or plastic cylinders less than 5 mm in diameter and 20 mm in length for consumer devices to cartridges many inches in diameter used in industrial applications. These cartridges may have flat ends, pointed ends, or bladed ends so that they fit properly in their associated fuse holder. While there are many sizes and types of fuses, they typically are lumped into two categories: fast-acting or slow-blow. Once a fuse has detected an over-current condition and has been actuated, or "blown," it must be replaced with a fuse of the same rating before the circuit can be used again.

Another type of fuse is the **resettable fuse**. Resettable fuses react differently to an over-current condition than a standard fuse because it resets itself and does not need replacement after excessive current. Also, rather than opening a circuit, it offers a very high resistance to the circuit, closely simulating an open circuit.

Circuit breakers offer a distinct advantage over fuses in the fact that they can be reset after an over-current condition, rather than replacement as required by a traditional cartridge fuses. When encountering an over-current condition, the circuit breaker "trips" to create the open circuit, and once the problem has been corrected, the circuit breaker can be reset to resume normal circuit operation. Generally, circuit breakers are classified by the means they use to detect the over-current situation. Thermal circuit breakers contain a bimetallic strip that bends as it heats up due to increased current. When the given current rating is exceeded, the bimetallic strip bends far enough for the contacts to become open, thereby disconnecting the circuit. As it cools, the bimetallic strip straightens, allowing the breaker to be reset (manually or automatically). A **magnetic circuit breaker** uses the principle of an electromagnet to trip or open the circuit. In a magnetic breaker, an electromagnet is included in the circuit. When a current rating is exceeded, the electromagnet is strong enough to open the contacts in the circuit breaker, thereby disconnecting the circuit. Large circuit breakers may incorporate both thermal and magnetic mechanisms.Fuses and circuit breakers typically have only two electrical **ratings**: current and voltage. Of these two, current is the most critical. **Current ratings** are in amps and can range from less than ¼ amps (250 mA) to over a hundred amps in industrial settings. The **voltage rating** is the maximum voltage at which a fuse is designed to operate. Resettable fuses are rated a bit differently; they have both a hold and trip current ratings. When specifying a replacement fuse, applications that are less than the rated voltage are acceptable, but the circuit voltages should never exceed the fuse voltage rating.

Slow-blow (also called time-lag or **time-delay fuses**) are rated at the amount of time they can be operated in an over-current condition. Typically, this time is in fractions of a second, but very large applications may have time lags measured in seconds.

Other ratings for fuse and breakers refer to their physical packaging, size, or the type of panel or holder they are to be connected to. As with switches, the symbol used to represent a fuse or circuit breaker only depicts the presence of a device but does not tell in detail the physical description of the device. Fuse symbols are shown in **Appendix A**. The circuit designer may choose to include the current rating of the fuse as a label, but whether the fuse is of a cartridge or blade design is not depicted in the schematic. The same is true with circuit breakers. While the current rating of the breaker may be included as a label, details such as magnetic or thermal action is not depicted on the schematic symbol.

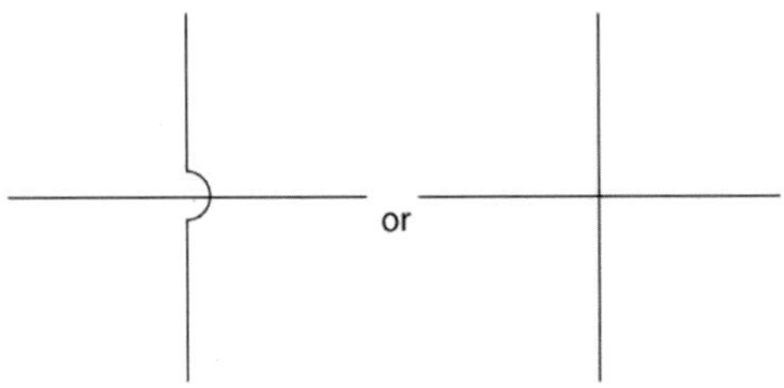

(a)

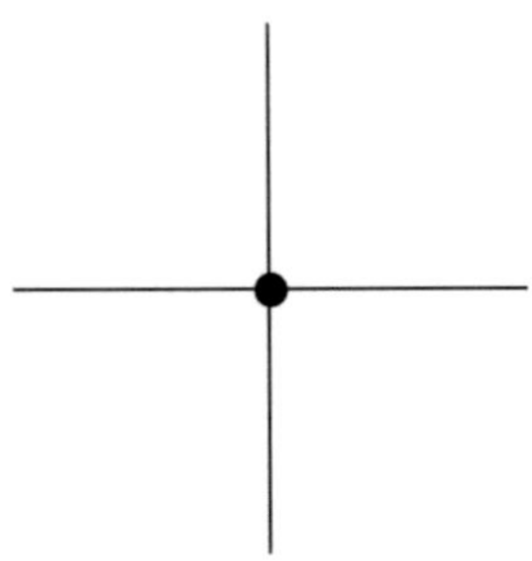

(b)

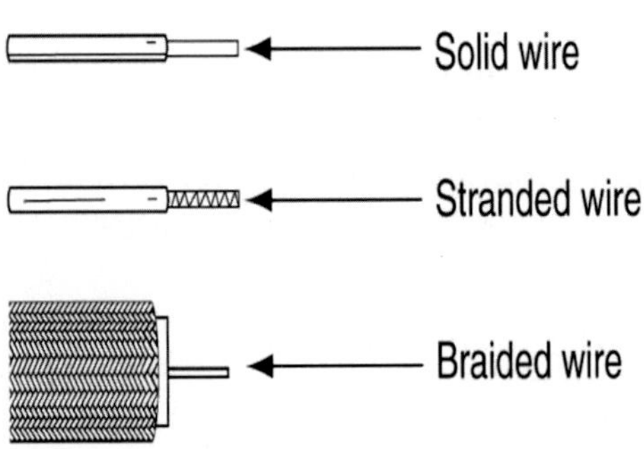

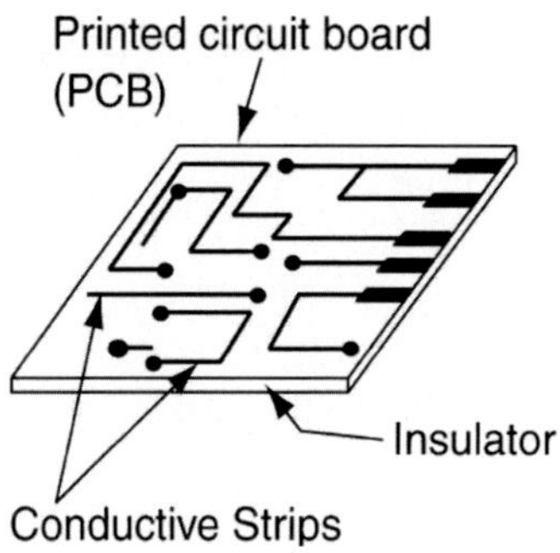
Printed circuit board
(PCB)
Insulator
Conductive Strips

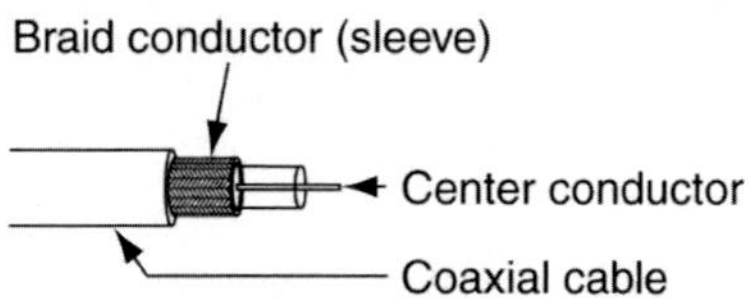
Braid conductor (sleeve)
Center conductor
Coaxial cable

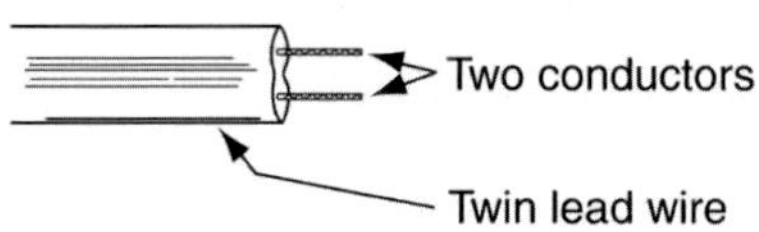
Two conductors
Twin lead wire

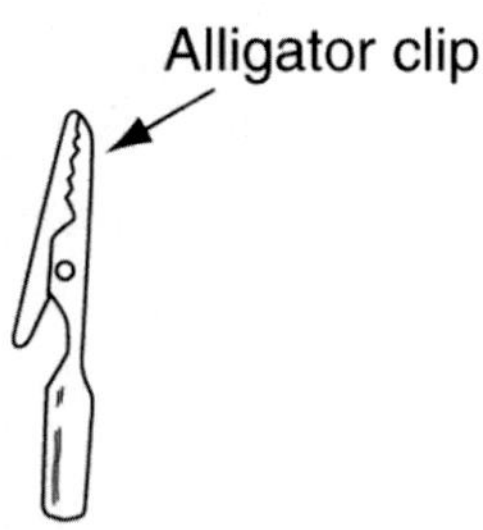
Alligator clip

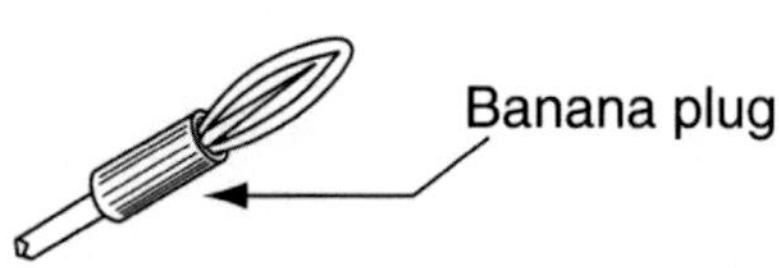
Banana plug

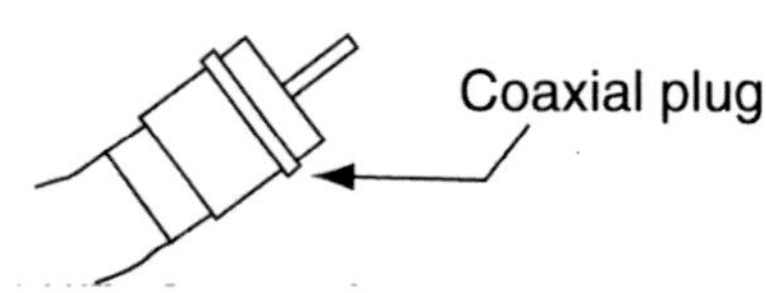
Coaxial plug

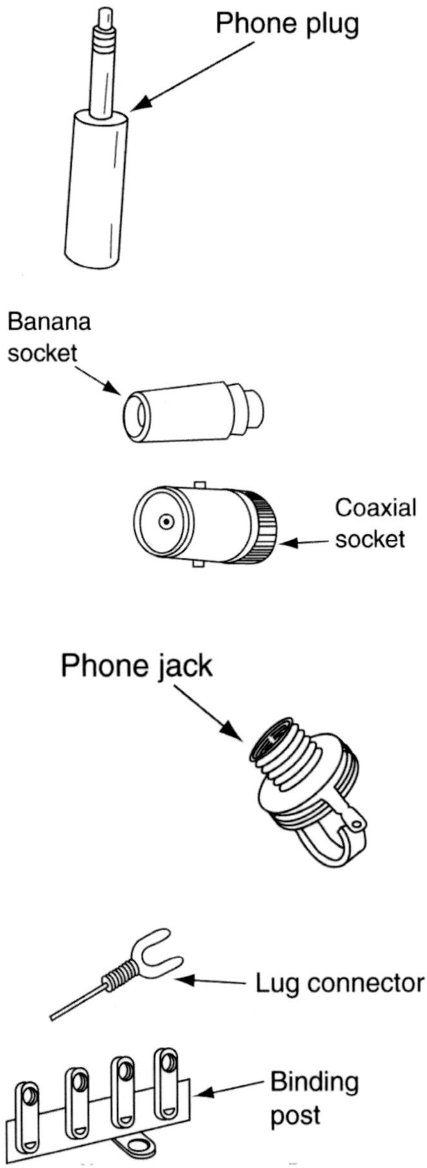

Figure 2-1 Symbols for electrical conductors: (a) conductors crossing; (b) conductors connected; (c) common types of conductors and connectors.

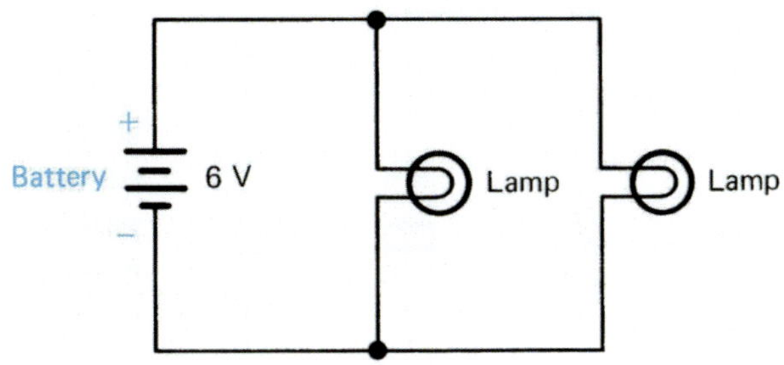

Figure 2-2 Symbols for a battery connected across two lamps.

Figure 2-3 Illustration of a single-pole single-throw (SPST) switch.

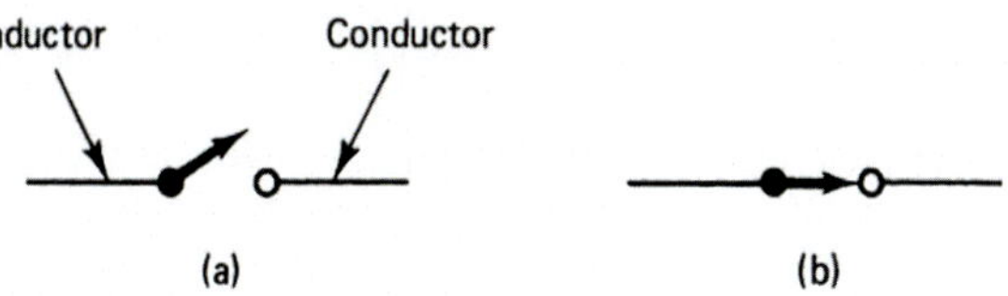

Figure 2-4 Symbol for a single-pole single-throw (SPST) switch: (a) off or open conditions; (b) on or closed condition.

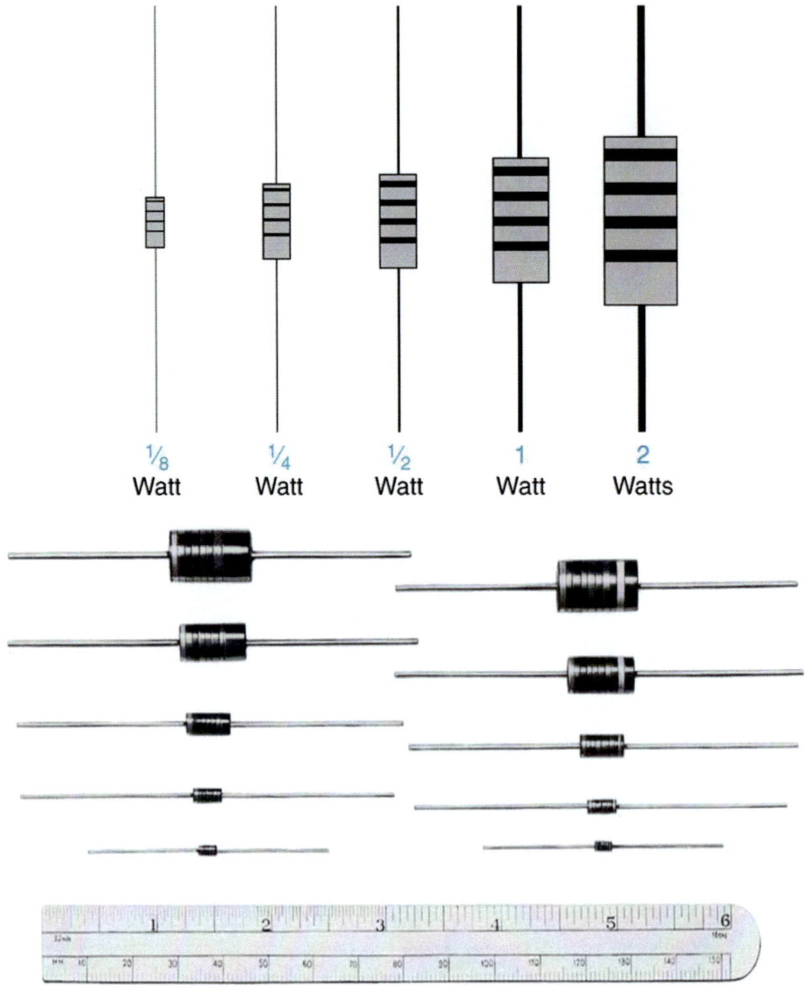

Figure 2-5 Resistors: (a) Resistor wattage rating – larger resistors are used for higher wattages. (b) Various sizes of resistors (Courtesy of Allen-Bradley Co.).

Most electrical equipment uses wires (**conductors**) to connect its components or parts together. The symbol for a conductor is a narrow line. If two conductors cross one another on a diagram, they may be shown by using symbols. **Figure 2-1(a)** shows two conductors crossing one another. If two conductors are connected together, they may also be identified by symbols, as shown in **Figure 2-1(b)**. **Figure 2-1(c)** shows some common types of conductors and connectors that are used to secure them to a circuit.

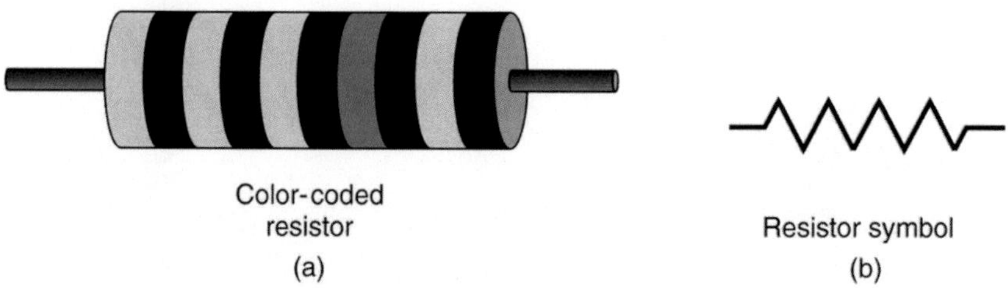

Figure 2-6 Color-coded resistor.

The symbols for two lamps connected across a battery are shown in **Figure 2-2** using symbols for the battery and lamps. Notice the part of the **diagram** where the conductors are connected together. A common electrical component is a **switch** such as the toggle switch shown in **Figure 2-3**. The simplest switch is a single-pole single-throw (SPST) switch. This switch turns a circuit on or off.

Figure 2-4. **(a)** shows the symbol for a switch in the off or open position. There is no path for current to flow from the battery to the lamp. The lamp will be off when the switch is open. **Figure 2-4(b)** shows a switch in the on or closed position. This switch position completes the circuit and allows current to flow.

Many electrical circuits use a component called a resistor. **Resistors** are usually small, cylinder-shaped components such as those shown in **Figure 2-5**. They are used to control the flow of electrical current.

A typical **color-coded resistor** and its symbol are shown in **Figure 2-6**. The most common type of resistor uses color coding to mark its value. Resistor value is always in **ohms**. For instance, a resistor might have a value of 100 Ohms. The symbol for ohms is the Greek capital letter omega (Ω). Each color on the resistor represents a specific number. Resistor color-coded values are easy to learn.

Another type of resistor is called a **potentiometer** or "pot." A pot is a variable resistor whose value can be changed by adjusting a rotary shaft. For example, a 1000-Ohm pot can be adjusted to any value from 0 to 1000 Ohms by rotating the shaft. The pictorial and symbol of this component are shown in **Figure 2-7(a)** and **(b)**. In the example shown in **Figure 2-7(c)**, potentiometer 1 is adjusted so that the resistance between points A and B is zero. The resistance between points B and C is 1000 W. By turning the shaft as far in the opposite direction as it will go, the resistance between

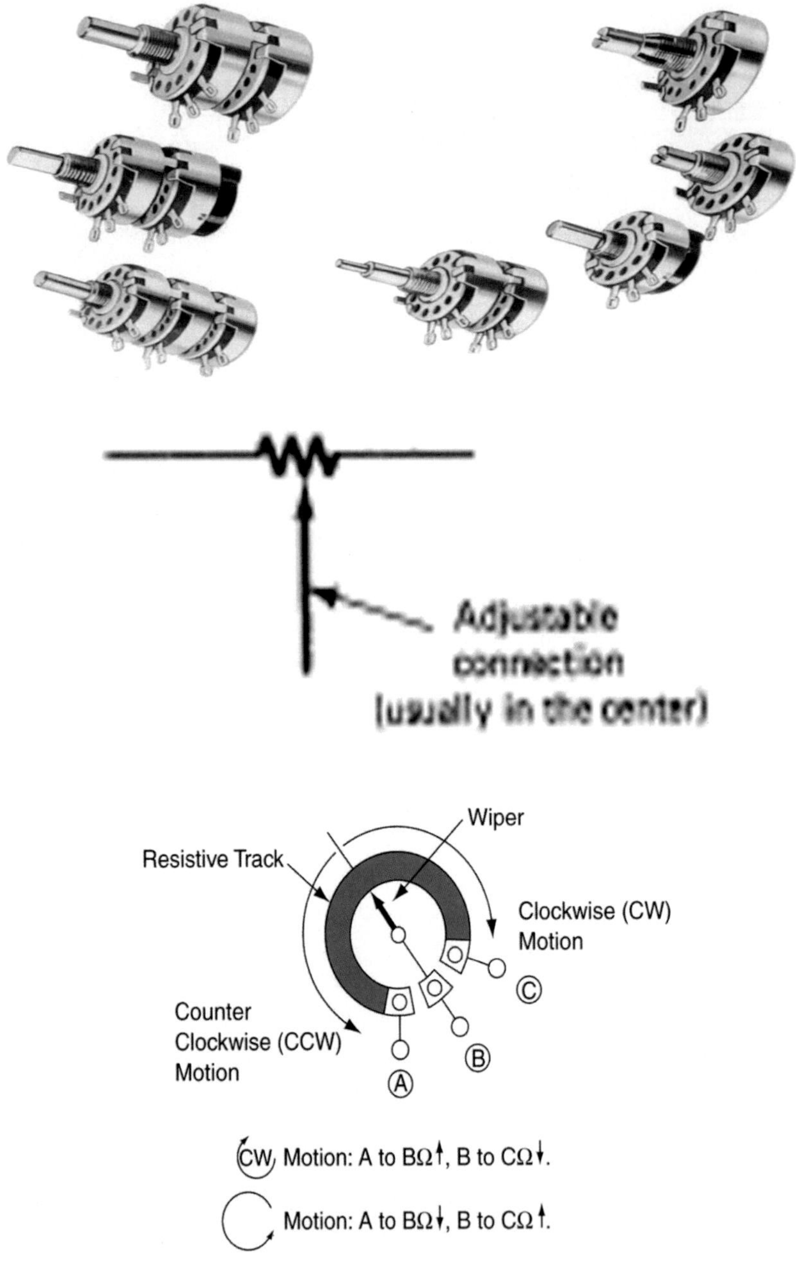

Figure 2-7 Potentiometers: (a) pictorials; (b) symbol; (c) examples [(a) Courtesy of Allen-Bradley Co.].

Figure 2-8 Common types of batteries (Courtesy of Union Carbide Corp.).

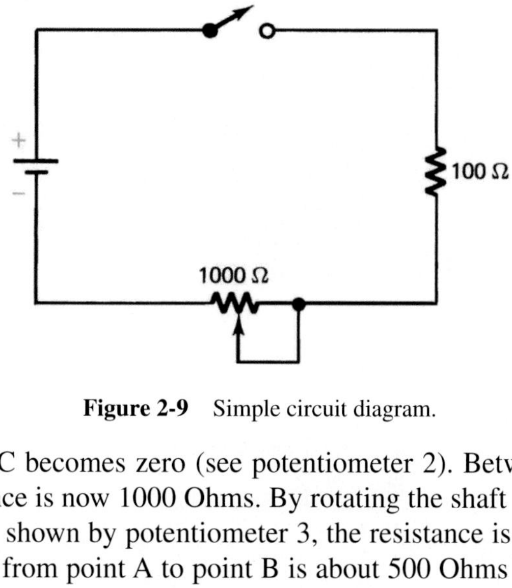

Figure 2-9 Simple circuit diagram.

points B and C becomes zero (see potentiometer 2). Between points A and B, the resistance is now 1000 Ohms. By rotating the shaft to the center of its movement, as shown by potentiometer 3, the resistance is split in half. Now the resistance from point A to point B is about 500 Ohms and the resistance from point B to point C is about 500 Ohms.

The symbol for a **battery** is shown in **Figure 2-2**. The symbol for any battery over 1.5 volts (V) is indicated by two sets of lines. A 1.5-V battery or cell is shown with one set of lines. The voltage of a battery is marked near its symbol. The long line in the symbol is always the positive (+) and the short line is the negative (–) side of the battery. Some common types of batteries are shown in **Figure 2-8**.

A simple **circuit diagram** using symbols is shown in **Figure 2-9**. This diagram shows a 1.5-V battery connected to an SPST switch, a 100-Ohm resistor, and a 1000-Ohm potentiometer. Because symbols are used, no words have to be written beside them. Anyone using this diagram should recognize the components represented and how they fit together to form a circuit.

2.2 Resistors

There is some resistance in all electrical circuits. Resistance is added to a circuit to control current flow. Devices that are used to cause proper resistance in a circuit are called **resistors**. A wide variety of resistors are used; some have a fixed value, and others are variable. Resistors are made of resistance wire, special carbon material, or metal film. Wire-wound resistors are ordinarily used to control large currents, and carbon resistors control currents which are smaller. Some types of resistors are shown in **Figure 2-10**.

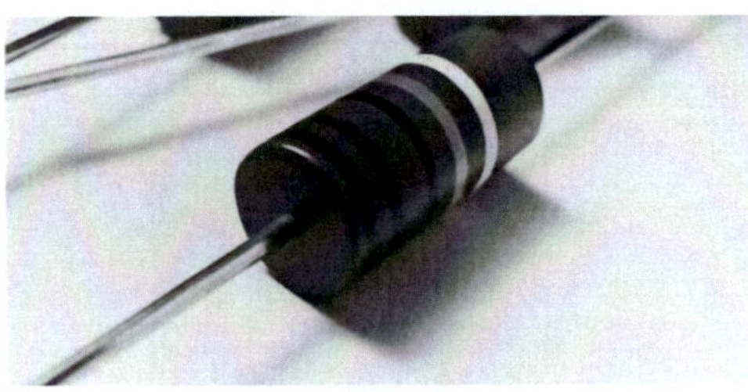

(a)

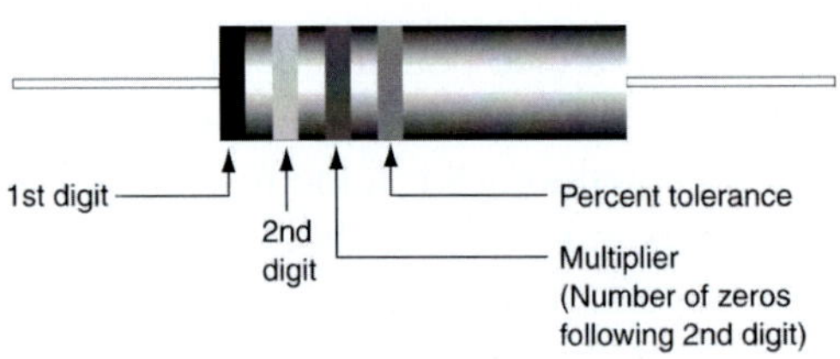

Color coding

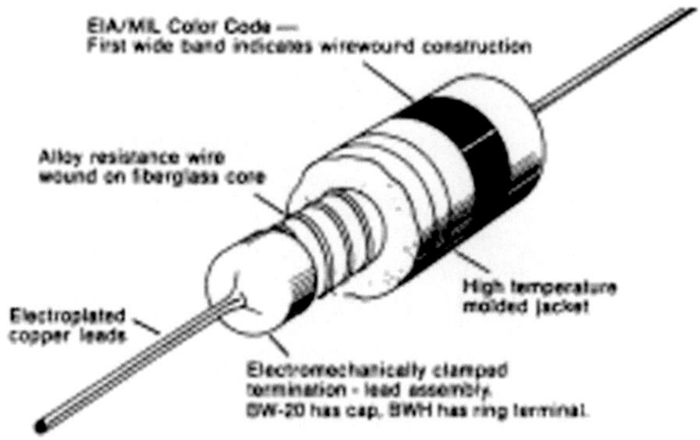
EIA/MIL Color Code —
First wide band indicates wirewound construction
Alloy resistance wire
wound on fiberglass core
High temperature
molded jacket
Electroplated
copper leads
Electromechanically clamped
termination - lead assembly.
BW-20 has cap, BWH has ring terminal.

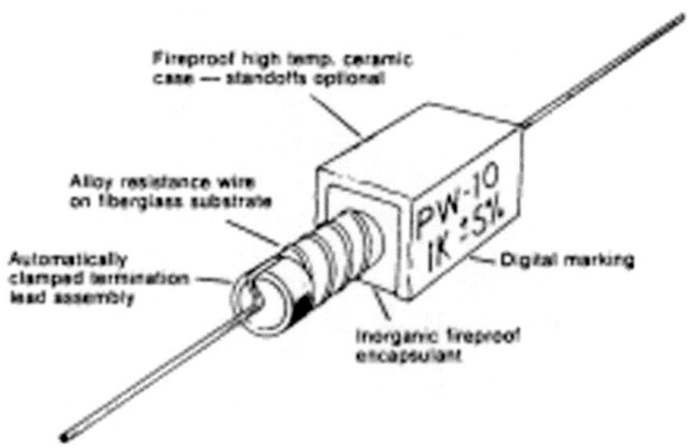
Fireproof high temp. ceramic
case — standoffs optional
Alloy resistance wire
on fiberglass substrate
PW-10
1K ±5%
Automatically
clamped termination
lead assembly
Digital marking
Inorganic fireproof
encapsulant

(d)

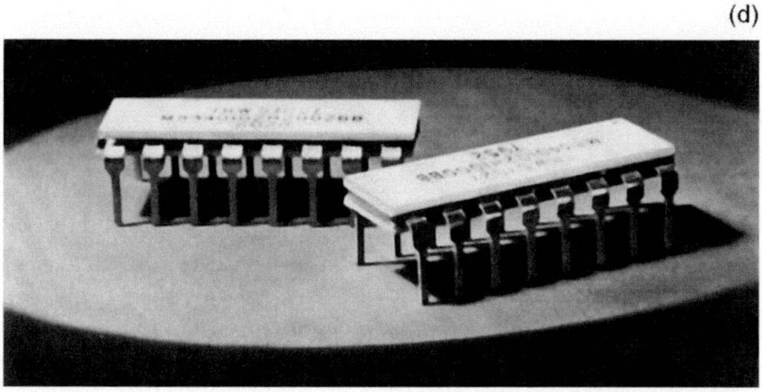

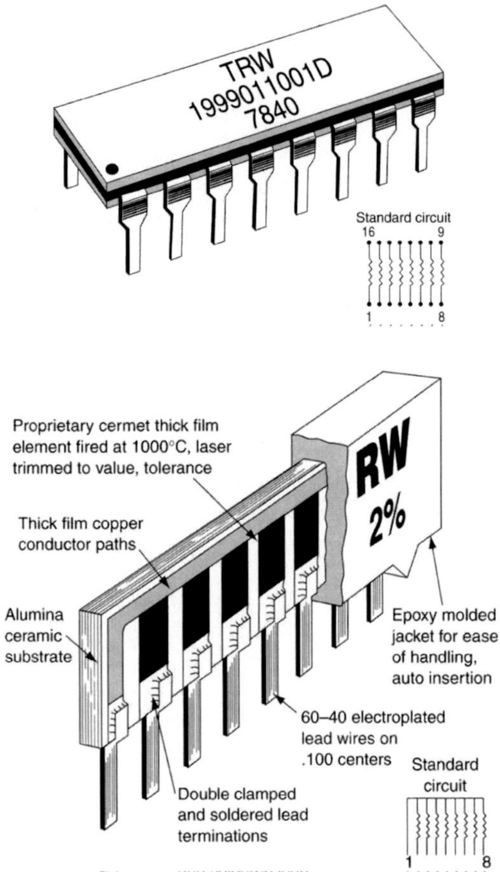

Figure 2-10 Types of resistors: (a) carbon composition resistor; (b) molded wire-wound resistor (all courtesy of TRW/UTC Resistors).

Wire-wound resistors are constructed by winding resistance wire on an insulating material. The wire ends are attached to metal terminals. An enamel coating is used to protect the wire and to conduct heat away from it. Wire-wound resistors may have fixed taps which can be used to change the resistance value in steps. They may also have sliders which can be adjusted to change the resistance to any fraction of their total resistance. **Precision-wound resistors** are used where the resistance value must be accurate, such as in measuring instruments.

Carbon resistors are constructed of a small cylinder of compressed material. Wires are attached to each end of the cylinder. The cylinder is then

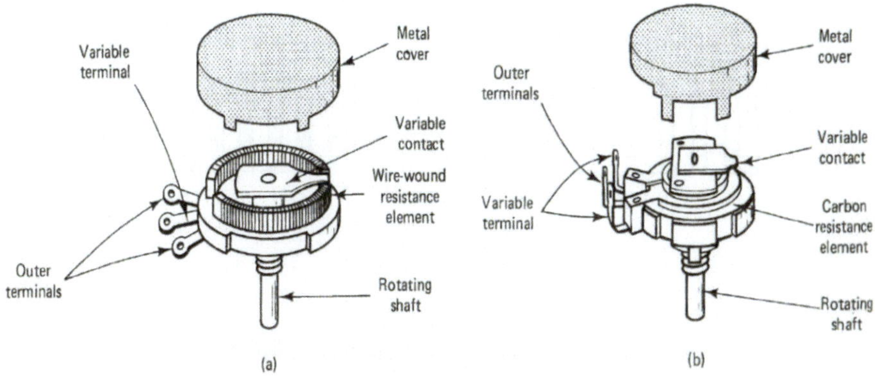

Figure 2-11 Variable-resistor construction: (a) wire-wound variable resistor; (b) carbon variable resistor.

covered with an insulating coating. **Variable resistors** are used to change resistance while equipment is in operation. They are called **potentiometers** or **rheostats**. Both carbon and wire-wound *variable* resistors are made by winding on a circular form [see **Figure 2-11(a)**]. A contact arm is attached to the form to make contact with the wire. The contact arm can be adjusted to any position on the circular form by a rotating shaft. A wire connected to the movable contact is used to vary the resistance from the contact arm to either of the two outer wires of the variable resistor. For controlling smaller currents, carbon variable resistors are made by using a carbon compound mounted on a fiber disk [see **Figure 2-11(b)**]. A contact on a movable arm varies the resistance as the arm is turned by rotating a metal shaft.

Resistor Color Codes

It is usually easy to find the value of a resistor by its color code or marked value. Most wire-wound resistors have resistance values (in ohms) printed on the resistor. If they are not marked in this way, an ohmmeter must be used to measure the value. Most carbon resistors use color bands to identify their value. Resistors are of two common types: **four-color band** and **five-color band**, as shown in **Figure 2-12**.

Resistors are color coded with an **end-to-center** color band system of marking. In the color-coding system, colors are used to indicate the resistance value in ohms. A color band also is used to indicate the tolerance of the resistor. The colors are read in the correct order from the end of a resistor. Numbers from the resistor color code, as shown in **Figure 2-13**, are

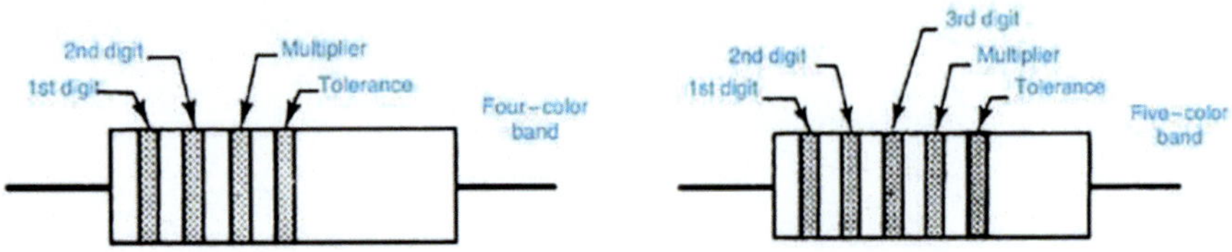

(a) (b)

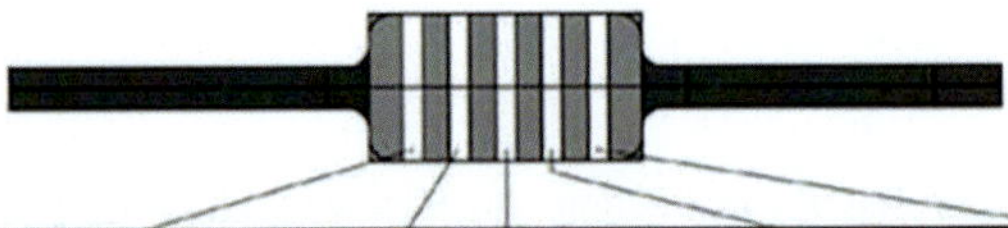

Color	1st Band	2nd Band	*3rd Band	Multiplier	Tolerance (%)
Black	0	0	0	1Ω	
Brown	1	1	1	10Ω	± 1%
Red	2	2	2	100 Ω	± 2%
Orange	3	3	3	1K Ω	
Yellow	4	4	4	10K Ω	
Green	5	5	5	100K Ω	± 0.5%
Blue	6	6	6	1M Ω	± 0.25%
Violet	7	7	7	10M Ω	± 0.10%
Grey	8	8	8		
White	9	9	9		
Gold				0.1 Ω	± 5%
Silver				0.01 Ω	± 10%

*Note: For carbon film resistors, exclude this column.

(c)

Figure 2-12 Carbon resistors: (a) four-color band; (b) five-color band; (c) resistor color-coded bands.

substituted for the colors. Through practice using the resistor color code, the value of a resistor may be determined at a glance.

It is difficult to manufacture a resistor to the exact value required. For many uses, the actual resistance value may be as much as 20% higher or lower than the value marked on the resistor without causing any problem. In most uses, the actual resistance does not need to be any closer than 10% higher or lower than the marked value. This percentage of variation between the marked color code value and the actual value is called **tolerance**. A resistor with a 5% tolerance should be no more than 5% higher or lower than the marked value. Resistors with tolerances of lower than 5% are called precision resistors. Resistors are marked with color bands starting at one

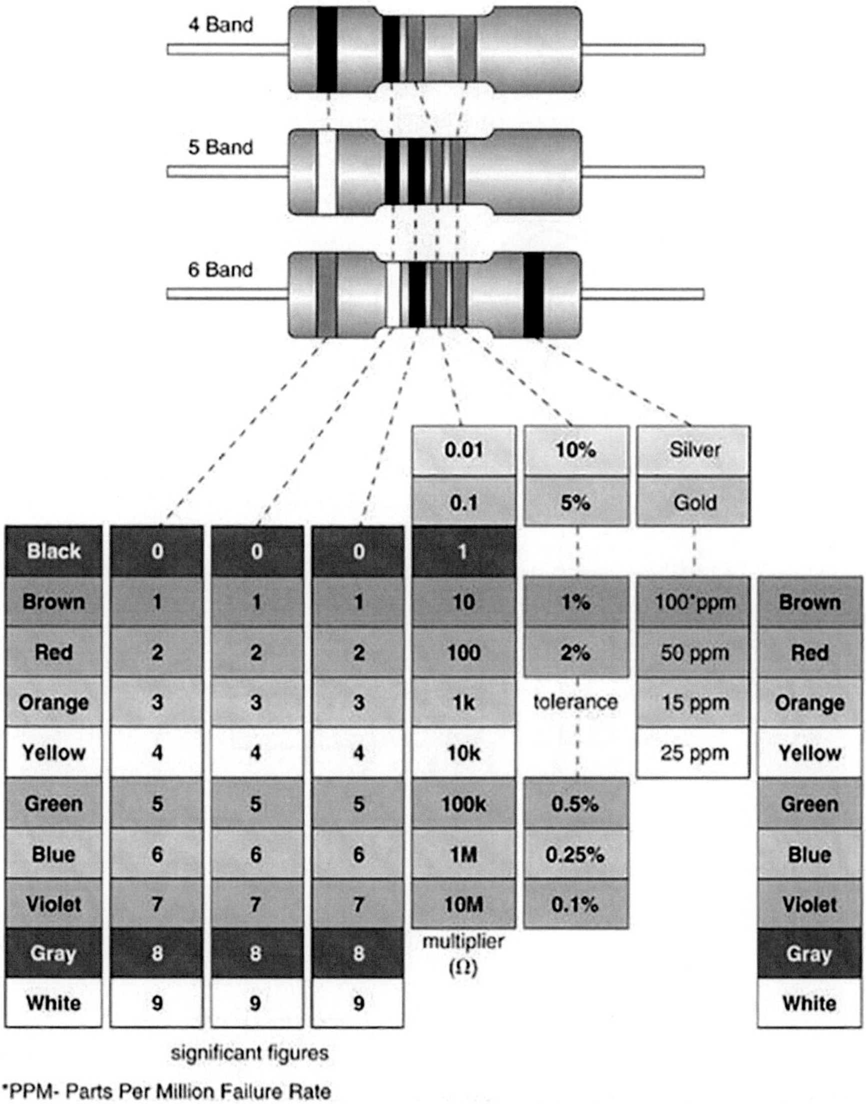

Figure 2-13 Resistor color code (four-band, five-band, and six-band).

end of the resistor. For example, a carbon resistor that has three color bands (yellow, violet, and brown) at one end has the color bands read from the end toward the center, as shown in **Figure 2-14**. The resistance value is 470 Ohms. Remember that black as the third color means that no zeros are to be added to the digits. A resistor with a green band, a red band, and black band has a value of 52 Ohms.

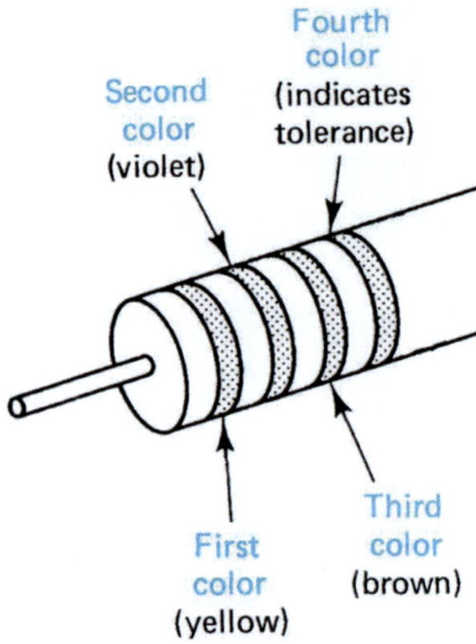

Figure 2-14 End-to-center system for carbon resistors – resistor value = 470 Ohms.

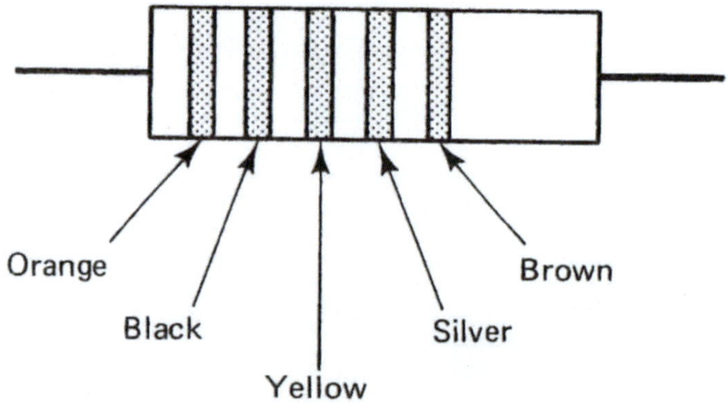

Figure 2-15 Resistor color-code example (five-band).

An example of a resistor with the five-band color-coded system is shown in **Figure 2-15**. Read the colors from left to right from the end of the resistor where the bands begin. Use the resistor color chart of **Figure 2-13** to determine the value of the resistor in ohms, and its tolerance. Note that a six-band resistor is included. These are not used as often.

The first color of the resistor in **Figure 2-15** is orange. The first digit in the value of the resistor is 3. The second color is black; so the second digit in the value of the resistor is 0. The next color is yellow; so the third digit is 4. The first three digits of the resistor value are 304. The fourth color is

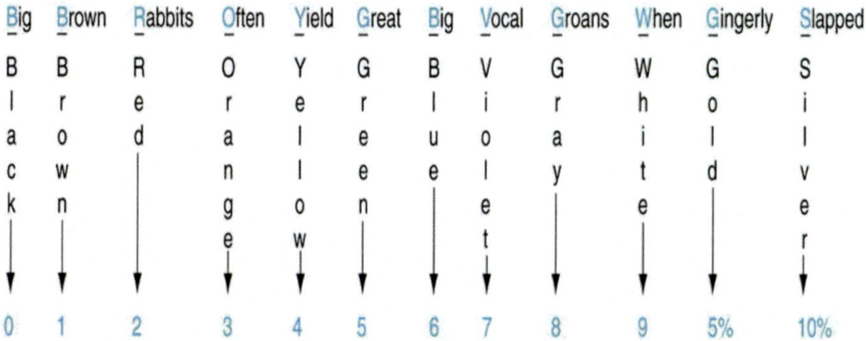

Figure 2-16 Mnemonic device used to memorize resistor color code.

silver, which indicates the number by which the first three digits are to be multiplied. The color silver is a multiplier of 0.01. Multiply 304 × 0.01 to obtain the value of the resistor (304 × 0.01 = 3.04).

The fifth color is brown. The tolerance of the resistor is given with this band. The tolerance shows how close the actual value of the resistor should be to the color-coded value. **Tolerance** is a percentage of the actual value. In this example, brown shows a tolerance of ± 1%, which means that the resistor should be within 1% of 3.04 Ohms in either direction.

Sometimes, there are only three colors on the body of the resistor. Then the tolerance is 20%. Using the color code, it is easy to list the ohms value and tolerance of resistors.

Another way to determine the value of color-coded resistors is to remember the following mnemonic statement:

Big **B**rown **R**abbits **O**ften **Y**ell **G**reat **B**ig **V**ocal **G**roans **W**hen **G**ingerly **S**urprised.

The first letter of each word in the mnemonic statement is the same as the first letter in each of the colors used in the color code. The words of the device are counted (beginning with zero) to find the word corresponding to each digit or the multiplier (refer to **Figure 2-16**). Use a method such as this mnemonic statement to remember the color code.

Power Rating of Resistors

The size of a resistor helps to determine its power rating. Larger resistors are used in circuits that have high power ratings. Small resistors will become damaged if they are put in high-power circuits. The power rating of a resistor indicates its ability to give off or dissipate heat. Common power (wattage) ratings of color-coded resistors are 1/8, 1/4, 1/2, 1, and 2 watts (W). Note the size differences of the resistors in **Figure 2-5**. Resistors that are larger in physical size will give off more heat and have higher power ratings.

2.3 Units of Measurement

An important activity in the study of electricity and electronics is **measurement**. Because many units measured in electricity and electronics may be quite large, larger than a million millions, or incredibly small, smaller than one billionth, special notations are used to better express these extreme quantities. This unit covers these units of measure, as well as information about how these units are expressed in a manner where it can be read by other professionals in electronics.

All quantities can be measured. The distance between two points may be measured in meters, kilometers, inches, feet, or miles. The weight of an object may be measured in ounces, pounds, grams, kilograms, or many other values. **Table 2-1** shows some of the base units of the SI (Systems International) system or metric system of units.

Electrical quantities may also be measured. The more common electrical units of measurement are discussed in this section and summarized in Appendix B. There are four common units of electrical measurement. **Voltage** is used to indicate the force that causes electron movement. **Current** is a measure of the amount of electron movement. **Resistance** is the opposition to electron movement. The amount of work done or energy used in the movement of electrons in a given period is called **power**. **Table 2-2** shows some common electrical units.

Table 2-1 Base units of the SI system.

Measurement quantity	*Unit*	*Symbol*
Length	Meter	m
Mass	Kilogram	kg
Time	Second	s
Electric current	Ampere	A
Temperature	Kelvin	K
Luminous intensity	Candela	cd
Amount of substance	Mole	mol

Table 2-2 Common electrical units.

Measurement quantity	*Unit*	*Symbol*
Electric capacitance	Farad	F
Electric charge	Coulomb	C
Electric conductance	Siemens	S
Electric potential	Volt	V
Electric resistance	Ohm	Q
Energy	Joule	J
Force	Newton	N
Frequency	Hertz	Hz
Illumination	Lux	lx
Inductance	Henry	H
Luminous intensity	Lumen	lm
Magnetic flux	Weber	Wb
Magnetic flux density	Tesla	T
Power	Watt	W
Pressure	Pascal	Pa

Small Units

The electrical unit used to measure a certain value is often less than a whole unit (less than 1). Examples of this are 0.6 V, 0.025 A, and 0.0550 W. When this occurs, **prefixes** are used. Some common prefixes are shown in **Table 2-3**.

For example, a millivolt is 1/1000 of a volt and a microampere is 1/1,000,000 of an ampere. The prefixes in **Table 2-3** may be used with any electrical unit of measurement. The unit is divided by the fractional part of the unit. For example, if 0.6 V is to be changed to millivolts, 0.6 V is divided by the fractional part of the unit. So, 0.6 V equals 600 millivolts (mV) or 0.6 divided by 0.001 = 600 mV. If 0.0005 A is changed to microamperes, 0.0005 A is equal to 500 microamperes (*m*A) or 0.0005 divided by 0.000001 = 500 *m*A. When changing a basic electrical unit to a prefix unit, move the decimal point of the unit to the right by the same number

of places in the fractional prefix. To change 0.8 V to millivolts, the decimal point of 0.8 V is moved three places to the right (800) since the prefix *milli* has three decimal places. So, 0.8 V equals 800 mV. The same method is used for converting any electrical unit to a unit with a smaller prefix.

Table 2-3 Prefixes of units smaller and larger than 1.

Prefix	*Symbol*	*Factor by which the unit is multiplied*
exa	E	$1{,}000{,}000{,}000{,}000{,}000{,}000 = 10^{18}$
peta	P	$1{,}000{,}000{,}000{,}000{,}000 = 10^{15}$
tera	T	$1{,}000{,}000{,}000{,}000 = 10^{12}$
giga[a]	G	$1{,}000{,}000{,}000 = 10^{9}$
mega	M	$1{,}000{,}000 = 10^{6}$
kilo	k	$1{,}000 = 10^{3}$
hecto	h	$100 = 10^{2}$
deka	da	$10 = 10^{1}$
deci	d	$0.1 = 10^{-1}$
centi	c	$0.01 = 10^{-2}$
milli	m	$0.001 = 10^{-3}$
micro	μ	$0.000001 = 10^{-6}$
nano	n	$0.000000001 = 10^{-9}$
pico	P	$0.000000000001 = 10^{-12}$
femto	f	$0.000000000000001 = 10^{-15}$
atto	a	$0.000000000000000001 = 10^{-18}$

Often an electrical unit with a prefix is **converted** back to the basic unit. For example, milliamperes may be converted back to amperes. Microvolt, sometimes, is converted back to volts. When a unit with a prefix is converted back to a basic unit, the prefix must be multiplied by the fractional part of the whole unit of the prefix. For example, 68 mV converted to volts is equal to 0.068 V. When 68 mV is multiplied by the fractional part of the whole unit (0.001 for the prefix **milli**), this equals 0.068 V (68 mV × 0.001 = 0.068 V).

When changing a fractional prefix unit into a basic electrical unit, move the decimal in the prefix unit to the left the same number of places of the prefix. To change 225 mV to volts, move the decimal point in 225 three places to the left (0.225) since the prefix **milli** has three decimal places. So, 225 mV equals 0.225 V. This same method is used when changing any fractional prefix unit back to the original electrical unit. **Figure 2**-17 shows a simple conversion scale for large and small numbers.

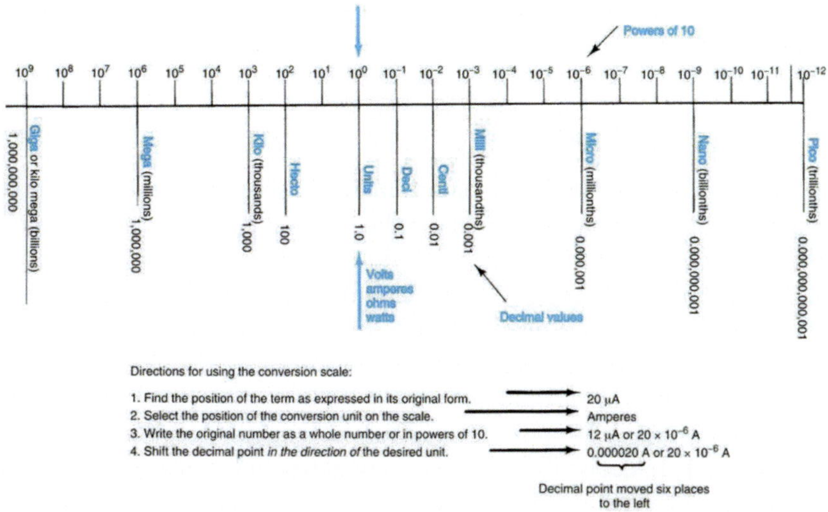

Figure 2-17 Simple conversion scale for large and small numbers.

Large Units

Sometimes electrical units of measurement are quite large, such as 20,000,000 W, 50,000 W, or 38,000 V. When this occurs, prefixes are needed to make these large numbers easier to use. Some prefixes used for large electrical values are shown in **Table 2-4**. To change a large value to a smaller unit, divide the large value by the number of the prefix. For example, 48,000,000 Ohms is changed to 48 mega ohms (MW) by dividing (48,000,000 divided by 1,000,000 = 48 MW). To convert 7000 V to 7 kilovolts (kV), divide 7000 by 1000 (7000 divided by 1000 = 7 kV). To change a large value to a smaller value, move the decimal point in the large value to the **left** by the number of zeros of the prefix; thus, 3600 V equals 3.6 kV (3600).

To convert a prefix unit back to a large number, move the decimal point to the right by the same number of places in the unit. Also, the number may be multiplied by the number of the prefix. If 90 MW is converted to ohms, move the decimal point six places to the right (90,000,000). The 90-MW value may also be multiplied by the number of the prefixes, which is 1,000,000. Thus, 90 MW × 1,000,000 = 90,000,000 W.

The simple **conversion scale** shown in **Figure 2-17** is useful when converting either large or small units to units of measurement with prefixes. This scale uses either powers of 10 or decimals to express the units. Refer to the calculator procedures for conversion of electrical units in the section on scientific calculators.

2.4 Scientific Notation

Using **powers of 10** or **scientific notation** greatly simplifies math operations. A number that has many zeros to the right or to the left of the decimal point is made simpler by putting it in the form of scientific notation (powers of 10). For example, 0.0000035 × 0.000025 is difficult to multiply. It can be put in the form $(3.5 \times 10^{-6}) \times (2.5 \times 10^{-5})$. Notice the number of places that the decimal point is moved in each number.

Table 2-4 lists some of the powers of 10. In a whole number, the power to which the number is raised is **positive**. It equals the number of zeros following the 1. In decimals, the power is **negative** and equals the number of places the decimal point is moved to the left of the 1. Easy powers of 10 to remember are $10^2 = 100$ (10 × 10) and $10^3 = 1000$ (10 × 10 × 10).

Any number written as a multiple of a power of 10 and a number between 1 and 10 is said to be expressed in **scientific notation**. For example,

$$81{,}000{,}000 = 8.1 \times 10{,}000{,}000 \text{ or}$$
$$8.1 \times 10^7$$

Table 2-4 Powers of 10 (scientific notation).

	Number	*Power of 10*
	1,000,000	10^6
	100,000	10^5
Whole	10,000	10^4
numbers	1000	10^3
	100	10^2
	10	10^1
	1.0	10^0
	0.1	10^{-1}
	0.01	10^{-2}
	0.001	10^{-3}
Decimals	0.0001	10^{-4}
	0.00001	10^{-5}
	0.000001	10^{-6}

$500{,}000{,}000 = 5 \times 100{,}000{,}000$ or

5×10^8

$0.0000000004 = 4 \times 0.0000000001$ or

4×10^{-10}.

Scientific notation simplifies multiplying and dividing large numbers of small decimals. For example,

$$4800 \times 0.000045 \times 800 \times 0.0058$$
$$= (4.8 \times 10^3) \times (4.5 \times 10^{-5}) \times (8 \times 10^2) \times (5.8 \times 10^3)$$
$$= (4.8 \times 4.5 \times 8 \times 5.8) \times (10^{3-5+2-3})$$
$$= 1002.24 \times 10^{-3}$$
$$= 1.00224$$

$$95{,}000 \div 0.0008$$
$$= 9.5 \times 10^4/8 \times 10^{-4}$$
$$= 9.5 \times 10^{4-(-4)}/8$$
$$= 9.5 \times 10^8/8 = 1.1875 \times 10^8$$
$$= 118{,}750{,}000.$$

With some practice, the use of scientific notation becomes easy.

2.5 Scientific Calculator

Many of the concepts in electronics are mathematically based and can be represented by mathematical models. Because of this close relationship between mathematics and electrical engineering technology, it is important to be proficient in any tool needed to perform mathematical functions. This may include various pieces of software packages such as simulation, online resources, or specialized laboratory equipment. Perhaps, the most universally used tool is the **scientific calculator**. By using a scientific calculator to express the notations used in electronics (such as milli, kilo, or scientific notation such as 8×10^2), this tool can greatly simplify many electronics-related calculations. This unit of study is intended to some practical examples of how to use a scientific calculator to make calculations related to the study of electricity and electronics.

When selecting a scientific calculator, be certain the calculator has the minimum functions expected, sufficient memory, and any necessary graphing or connectivity to a computer.

Entering Numbers

Figure 2-17 depicts a typical **scientific calculator**. You may notice that it shares many of the same features as common "pocket" calculator such as numerals, addition, subtraction, division, and multiplication. However, there are many other keys available. In order to be proficient in using the additional features on a scientific calculator, you will need to learn the function of a few of these additional keys.

Perhaps the most often used feature of a scientific calculator in electrical problem solving is the entry of numbers in scientific notation format. For example, suppose you needed to enter the number 8.1×10^7 into the scientific calculator in order to perform a calculation. In order to do this, you must find the **exponent** key. On the calculator depicted in **Figure 2-18**, the key is labeled "Exp." Other calculators label it as "EE." The proper format is to enter the base number (8.1 in the example) and follow with the exponent key and the actual exponent (7 in our example). Typically, this would be entered as

8.1 – Exp – 7.

Note: On some hand-held scientific calculators, keys may be assigned multiple functions. On such units, the exponent key is often assigned as a second function to an existing key. Typically, this is accessed through a shift

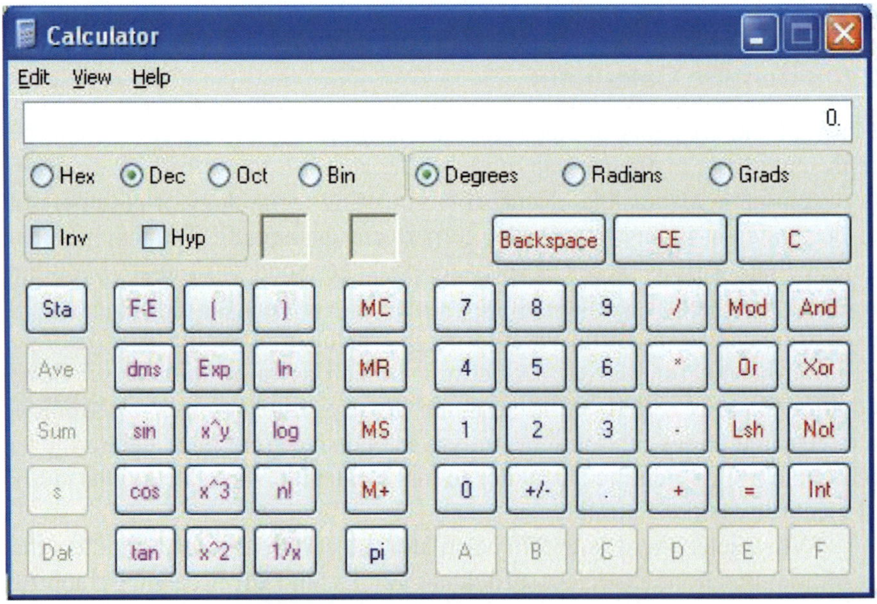

Figure 2-18 A typical scientific calculator.

or function key. Consult the manual for the calculator you are using for exact information.

If the number is very small and has a **negative exponent** such as 9.2×10^{-5}, one additional key must be identified, the button allowing the sign of a number to be changed. Just as before, the number is entered first (9.2 in our example), followed by the exponent key (Exp in our example), followed by the actual exponent (5) and the sign change key (+/– in our example). Typically, this would be entered as

9.2 – Exp – 5 – +/–.

On rare occasions, you will find the need to enter a small negative number with negative exponent. Such a number might be -8.75×10^{-5}. Again, the sign change key is used to change the sign of the digit as well as the exponent. As before, the number is entered first (8.75 in our example), the sign change key (+/– in our example), followed by the exponent key (Exp in our example), followed by the actual exponent (5) and the sign change key (+/– in our example). Typically, this would be entered as

8.75 – +/– – Exp – 5 – +/–.

After you become skilled in entering numbers in scientific notation into the scientific calculator, the methods of adding, subtracting, dividing, or multiplying these numbers in scientific notation is much like using a "pocket" calculator. As an illustration, we will multiply the first two numbers in scientific notation from the previous examples together. This would look like

$8.1 \times 10^{7} * 9.2 \times 10^{-5}$.

This would be entered as: 8.1 – Exp – 7 – * (or whatever multiplication sign is represented on the calculator used) – 9.2 – Exp – 5 – +/– – and the = (equals sign). Your calculator should display 7452 or 7.452×10^{3}. Depending on the make and model of the calculator, the number may remain in scientific notation on the display of the calculator, or the calculator may drop the notation if the number can be fit into the display area.

Even if the calculator you are using is not the same as shown in this book, the method for entering numbers in scientific notation should be similar and discussed in the **instruction manual** for the scientific calculator.

2.6 Electrical Diagrams

Schematic diagrams are used to represent the parts of electrical equipment or circuits. They show how the components or parts of each circuit fit together. Schematic diagrams are used to show the details of the electrical connections

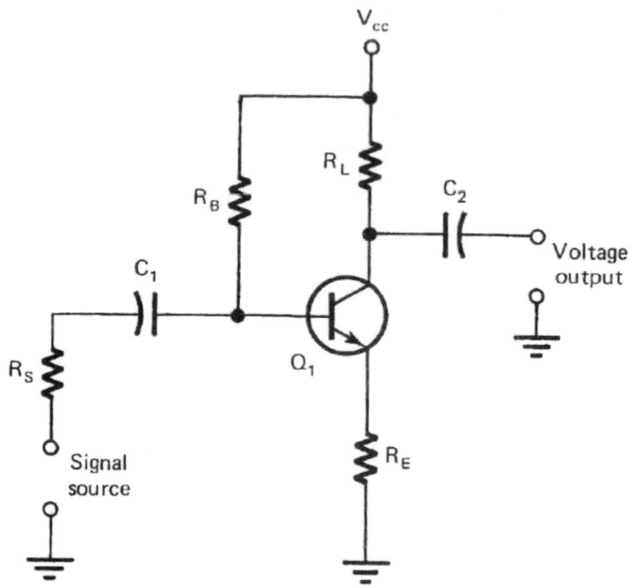

Figure 2-19 Schematic diagram of a transistor amplifier.

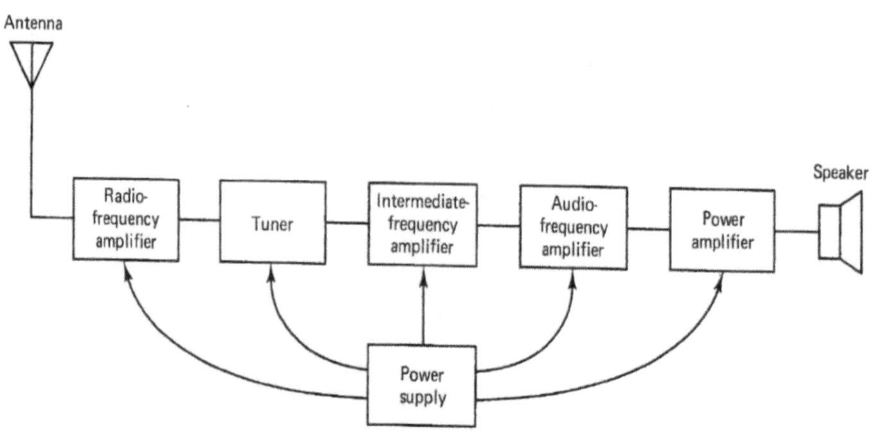

Figure 2-20 Block diagrams that represent electrical circuits and systems of a radio.

of any type of circuit or system. Schematics are used by manufacturers of electrical equipment showing operation and as an aid in servicing the equipment. A typical schematic diagram is shown in **Figure 2-19**. Note the symbols that are used. **Symbols** are used to represent electrical components in schematic diagrams. Standard electrical symbols are used by all equipment

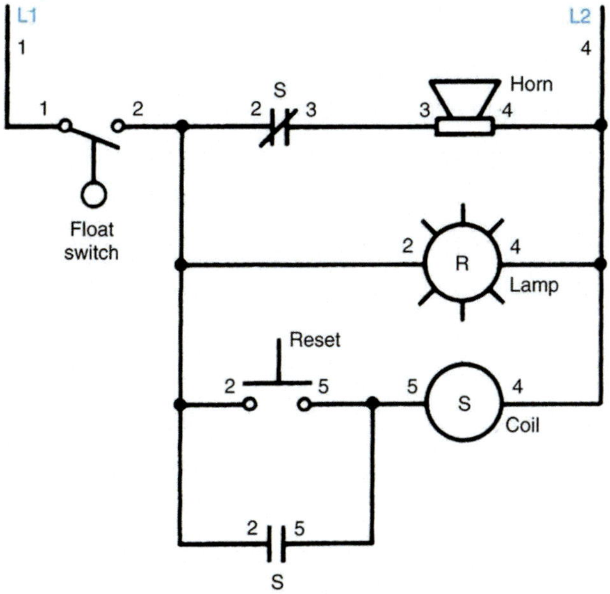

Figure 2-21 Simple wiring diagram.

manufacturers. Some common basic electrical symbols are shown in **Appendix A**. These symbols should be memorized.

Another way to show how electrical equipment operates is to use **block diagrams**. Block diagrams show the functions of the subparts of any electrical system. A block diagram of an electrical system was shown in **Figure 1-19** in the previous chapter. The same type of diagram is used to show the parts of a radio in **Figure 2-20**.

Inside the blocks, symbols or words are used to describe the function of the block. Block diagrams usually show the operation of the whole system. They provide an idea of how a system operates; however, they do not show details like a schematic diagram. It is easy to see the major subparts of a system by looking at a block diagram. **Figure 2-19(b)** shows block diagrams that represent electrical circuits and systems.

Another type of electrical diagram is called a wiring diagram (sometimes called a cabling diagram). Wiring diagrams show the actual location of parts and wires on equipment. The details of each connection are shown on a wiring diagram. Schematic and block diagrams show only how parts fit together electrically. Wiring diagrams show the details of actual connections. A simple wiring diagram is shown in **Figure 2-21**.

Problems

Electrical Unit Conversions

Answer each of the following electrical unit problems.

1.	0.65 A	=	____________	milliamperes
2.	0.12 microfarad	=	____________	picofarads
3.	0.215 mV	=	____________	V
4.	0.0000005 farad	=	____________	microfarads
5.	255 *m*A	=	____________	A
6.	45,000 Ohms	=	____________	MW
7.	0.85 MW	=	____________	Ohms
8.	6500 W	=	____________	kilowatts
9.	68,000 V	=	____________	kV
10.	9200 W	=	____________	megawatts

Scientific Notation

Write the following numbers as powers of 10 on a sheet of paper.

1. 0.00001
2. 0.00000001
3. 10,000,000
4. 1000
5. 10
6. 0.01
7. 10,000
8. 0.0001
9. 1.0
10. 1,000,000

Write the following numbers in scientific notation (as a number between 1 and 10 times a power of 10).

11. 0.00128
12. 1520
13. 0.000632
14. 0.0030
15. 28.2

16. 7,300,000,000
17. 52.30
18. 8,800,000
19. 0.051
20. 0.000006

Metric Conversions (See Appendix C) Answer each of the following metric conversion problems on a sheet of paper.

1 meter	=	____________	centimeters
1 centimeter	=	____________	millimeters
5000 grams	=	____________	kilograms
2 liters	=	____________	milliliters
1 meter	=	____________	inches
1 centimeter	=	____________	inches
1 mile	=	____________	kilometers
1 gram	=	____________	ounces
30 ounces	=	____________	grams
1 kilogram	=	____________	pounds
3 liters	=	____________	quarts
1 gallon	=	____________	liters
10 cups	=	____________	liters
50 miles	=	____________	kilometers
2 cubic centimeters (cm^3)	=	____________	cubic inches ($in.^3$)
4 cubic feet (ft^3)	=	____________	cubic meters (m^3)

Resistor Color Code

1 Using **Figure 2-13** for four-color band resistors, write the resistance value and tolerance of each problem as ohms; tolerance.

a.	(1) Violet	d. (1) Gray	f. (1) Green
	(2) Green	(2) Red	(2) Blue
	(3) Orange	(3) Black	(3) Black

(4) Gold	(4) Silver	(4) Gold
b. (1) Yellow	e. (1) White	g. (1) Brown
(2) Violet	(2) Brown	(2) Black
(3) Green	(3) Orange	(3) Black
c. (1) Green	(4) Gold	h. (1) Orange
(2) Blue		(2) White
(3) Red		(3) Black

Summary

- Controls in electrical circuits are typically switches
- Switches are identified by how they are actuated and by poles and throws
- Over-current devices are typically switches and circuit breakers
- Circuit breakers are resettable
- Resistance is the opposition to current flow
- Resistors are electronic devices that offer resistance to the flow of current
- Resistors are measured in Ohms, which is abbreviated by the Greek symbol Omega (Ω)
- The values of resistors are determined by color-coded bands that depict their values, multipliers, and tolerance
- Variable resistors are referred to as potentiometers (pots) or rheostats
- All electrical components are abbreviated by symbols
- A group of electrical symbols identifying a circuit is referred to as a schematic diagram
- There are four common units of electrical measurement
- Measurements in electronics frequently involve very large or very small numbers
- Large or small numbers may be expressed with prefixes
- Prefixes may be abbreviated with a letter or symbol
- It is convenient to express extremely large or small numbers in scientific notation

- The base number in scientific notation is between 1 and 10
- The exponent in scientific notation follows the 10
- Large numbers in scientific notation have a positive exponent
- Small numbers in scientific notation have a small exponent
- Scientific calculators greatly aid in engineering technology calculations
- Exponent and sign change keys are important on a scientific calculator
- Scientific calculators can mix standard decimal numbers and scientific notation
- Electrical diagrams, such as schematic, block, and wiring diagrams, are used to represent electrical circuits.

Self-examination

1. What prefix means the same as 10^3?
2. How many kilograms is 2000 grams?
3. How many zeros would be needed to express a 2.2-MΩ resistor in decimal?
4. Express 25 microamps as amps.
5. Why is scientific notation used?
6. Why is it necessary to convert large or small electrical units by using prefixes?
7. What is 314,000 in scientific notation?
8. What is 0.000044 in scientific notation?
9. Add the numbers 2×10^2 and 8×10^4 on a scientific calculator.
10. Find one-third of 12×10^6
11. What does AWG stand for?
12. Which wire is physically larger, a 10 gauge or 20 gauge?
13. What is the wire with the conductors twisted together called?
14. ___ are frequently used to control electrical circuits.
15. A DPDT switch refers to what kind of switch?
16. Toggle switches and slide switches may appear the same on a schematic.
17. How are the switch contacts rated?
18. What are the two most common devices use for over-current protection?
19. Circuit breakers differ from common fuses because they are ____.
20. The most important rating on a fuse is the ____ rating.

Answers

1. Kilo
2. 2 kilograms
3. 6 zeros
4. 0.0000025 amps
5. Easier to handle very large or very small numbers
6. Ease of use
7. What is 3.14×10^5
8. 4.4×10^{-5}
9. 8.02×10^4 or 80200
10. 4×10^6
11. American Wire Gauge
12. 10 gauge
13. Stranded
14. Switches
15. Double-pole double-throw
16. True
17. Maximum voltage and maximum current
18. Circuit breakers and fuses
19. Resettable
20. Maximum current

Glossary

Ammeter
A meter used to measure current flow.

Battery
An electrical energy source consisting of two or more cells connected together.

Block diagram
A diagram used to show how the parts of a system fit together.

Cell
An electrical energy source that converts chemical energy into electrical energy.

Component
An electrical device used in a circuit.

Conductor
A material that allows electrical current to flow through it easily.

Continuity check
A test to see if a circuit is an open or closed path.

Lamp
An electrical load device that converts electrical energy to light energy.

Multifunction meter
A meter that measures two or more electrical quantities, such as a volt-ohm-milliammeter (VOM), which measures voltage, resistance, and current. Such meters are also commonly called **multimeters**.

Multirange meter
A meter that has two or more ranges to measure an electrical quantity.

Ohmmeter
A meter used to measure resistance.

Polarity
The direction of an electrical potential (– or +) or a magnetic charge (north or south).

Potentiometer
A variable-resistance component used as a control device in electrical circuits.

Precision resistor
A resistor used when a high degree of accuracy is needed.

Prefix
An attachment to the beginning of a word to alter the meaning.

Resistor
A component used to control either the amount of current flow or the voltage distribution in a circuit.

Schematic diagram
A diagram used to show how the components of electrical circuits are wired together.

Scientific notation
The use of "powers of 10" to simplify large and small numbers.

Switch
A control device used to turn a circuit on or off.

Symbol
Used as a simple way to represent a component on a diagram or an electrical quantity in a formula.

Voltage drop
The electrical potential (voltage) that exists across two points of an electrical circuit.

Voltmeter
A meter used to measure voltage.

Volt-ohm-milliammeter (VOM)
A multifunction, multirange meter which is usually designed to measure voltage, current, and resistance (also called a **multimeter**).

Wiring diagram
A diagram that shows how wires are connected by showing the point-to-point wiring and the path followed by each wire.

3

Meters and Measurements

Another important activity in the study of electricity and electronics is **measurement**. Measurements are made in many types of electrical circuits. Special instruments are used to measure the quantities discussed in the first chapter. Students of electricity and electronics technology need to be proficient in the operation of analog and digital instruments as well as the proper manner to connect these instruments to a circuit. Learning the proper ways of measuring resistance, voltage, and current is important as these are the three most commonly measured quantities.

Objectives

Upon the completion of this chapter, you should be able to:

1. Identify the differences between analog and digital meters
2. Identify the similarities between analog and digital meters
3. Demonstrate how the voltmeter, ammeter, and ohmmeter are configured and connected in a circuit
4. Demonstrate safety while using electrical measurements

Chapter Outline

3.1 Meters

Perhaps, the most popular of all instruments used to measure various electrical quantities are **digital**-based instruments. Not long ago, the field of electronic instrumentation was nearly all **analog**. However, as true in many areas of electronics, digital devices are favored over traditional analog devices; so the change has been made in electronics instrumentation. Instruments such as digital counters, digital meters, and other digital instrumentation are

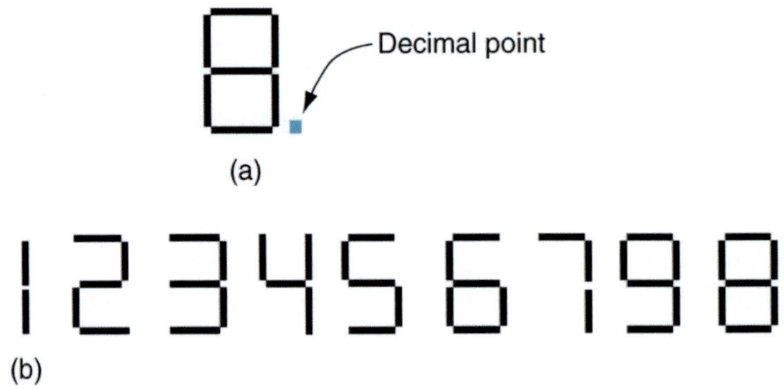

Figure 3-1 Seven-segment displays.

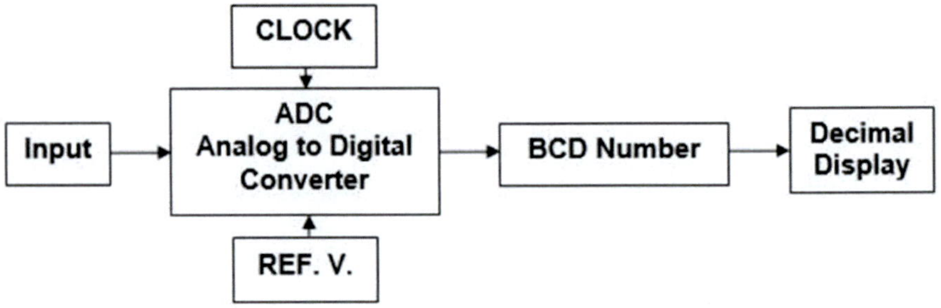

Figure 3-2 DMM block diagram.

commonly used. They employ numerical readouts to simplify the measurement process and to make more accurate measurements. Perhaps, the most striking advantage is the **numeric display**, which, unlike an analog meter, can be directly read by the user, requiring no interpretation, as sometimes required with analog instrumentation.

Many meters use a liquid crystal display (LCD) screen, rather than an analog needle moving across a scale such as that found in their analog counterparts. In addition, digital meters often have automatic ranging (auto-ranging) and some include automatic function settings, while analog instruments often require ranges and functions to be selected manually.

Digital multimeters (DMMs) have numerical readouts which display the measured quantity on seven-segment displays. **Figure 3-1(a)** shows the seven-segment display and **Figure 3-1(b)** indicates the parts of the display

Figure 3-3 Digital meters.

which are illuminated when each number is displayed. The **resolution** of a digital multimeter is the smallest change of a quantity that the meter can measure. The smaller the change a meter can detect, the better the resolution of the meter. The resolution of a meter is determined by the number of digits in the display. Many meters have 3½ digits in the display. A 3½-digit multimeter has three digit positions that can indicate from 0 through 9 and one digit position that can indicate only a value of 1. This latter digit is called the half-digit and is always the digit on the left of the display. For example, if the reading of 0.999 V, increased by 0.001–1 V, the display shows 1.000 V. A change of 0.001 V is the resolution of the 3-1/2-digit multimeter.

To operate, the **readout** of a digital meter is designed to transform electrical signals into numerical data. While this may appear complicated, a modern digital multimeter is quite simplistic in its construction, consisting of only one integrated circuit, a LCD display, a selector switch, and test leads. Inside the heart of the meter, the integrated circuit contains a built-in oscillator (clock), a reference voltage generator, an analog to digital converter (ADC), and related circuitry to perform all the functions and ranges. A block diagram of a typical digital multimeter is shown in **Figure 3-2**.

Central to this integrated circuit is the ADC. The **analog to digital converter** simply takes a continuously varying analog input signal from the sample being tested and transforms it to a digital equivalent. The output of this single integrated circuit is a numeric representation of the sample suitable

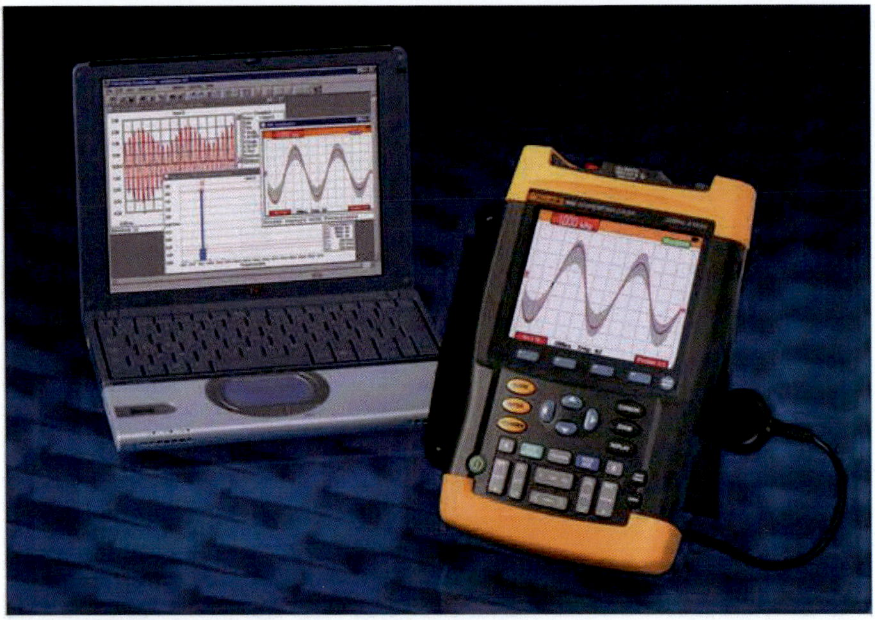

Figure 3-4 Scope meter with graphical display (courtesy of Fluke Co.).

Figure 3-5 Some types of digital meters.

for driving a **3½ digit LCD displays**, the most common visual indicator for handheld digital. Because of the half-digit in the readout, it is important to be able to interpret the reading properly. For example, if the reading is 999, as shown in **Figure 3-3**, is increased by 0.001–1 V, the display shows 1.000 V.

A change of 0.001 V is the resolution of the meter shown in **Figure 3-3(b)**. The number of digits in the display is known as the **resolution** of the meter.

Digital Meter Features

Some additional features may include the ability to hold a measurement to the display, meters that are waterproof, shockproof, and drop-proof, and meters that automatically power down after a period of time without use to preserve battery power. In addition, some meters have graphical displays, such as that shown in **Figure 3-4**, which display the quantity being measured in a graphical format.

Several types of meters offer computer compatible interfaces that allow the meter to perform data acquisition, which allows the technician to observe an electrical quantity as it changes over time and use that data on a computer, typically with a spreadsheet, to further aid in diagnosis. The technology available in digital instrumentation can be equal to or higher than instruments only found on bench tops only a few years ago. **Figure 3-4** is of a **digital scopemeter**. This meter offers oscilloscope functions with memory in a hand-held battery powered package. Combined with the software, the instrument is capable of taking the data from the measurements and analyzing, comparing, or archiving the results, or simply allowing the data from the testing to be included into a word processor for writing reports.

Some additional types of **digital meters** are shown in **Figure 3-5**. It is easy to see why digital meters have nearly replaced analog meters in the electricity and electronics technology fields. With advantages such as the elimination of human error in display reading, removal of the sensitivity to polarity an often-increased accuracy (analog 1%–2%; digital 0.1% or better), as well as a more rugged design, digital instruments are often preferred. However, there are a couple of important changes in measuring semiconductors and capacitors. Unlike their analog counterparts, digital meters need special settings to make measurements in these functions. Also, a digital meter has to have a good battery to make any measurements; analog meters can measure voltage and current without battery power.

Analog Meters

Before the rise in popularity of digital instrumentation in the 1980s, the electronics industry relied on analog instruments for the measurement of electrical quantities. Instruments that rely on the motion of a hand or pointer are called **analog instruments** (see **Figure 3-6**).

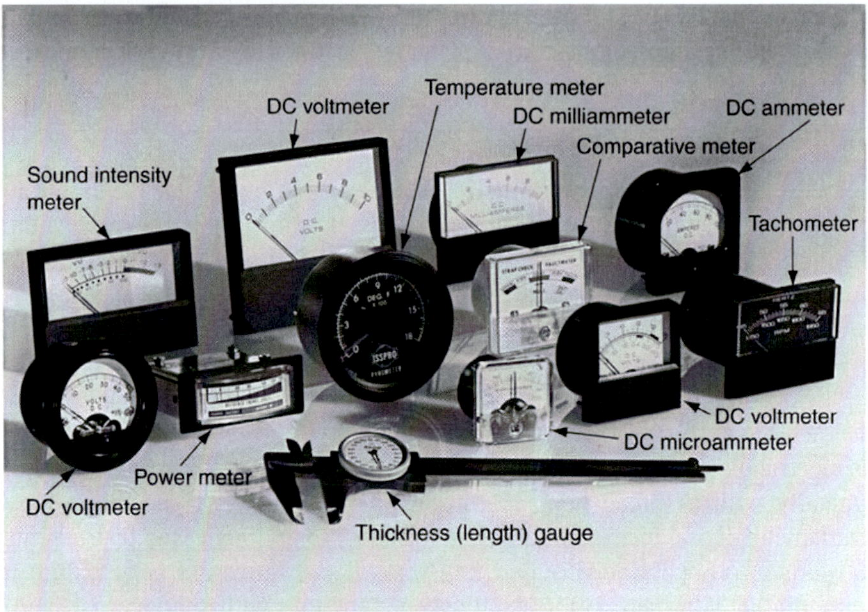

Figure 3-6 Single-function analog instruments (courtesy of Hout Electrical Instruments).

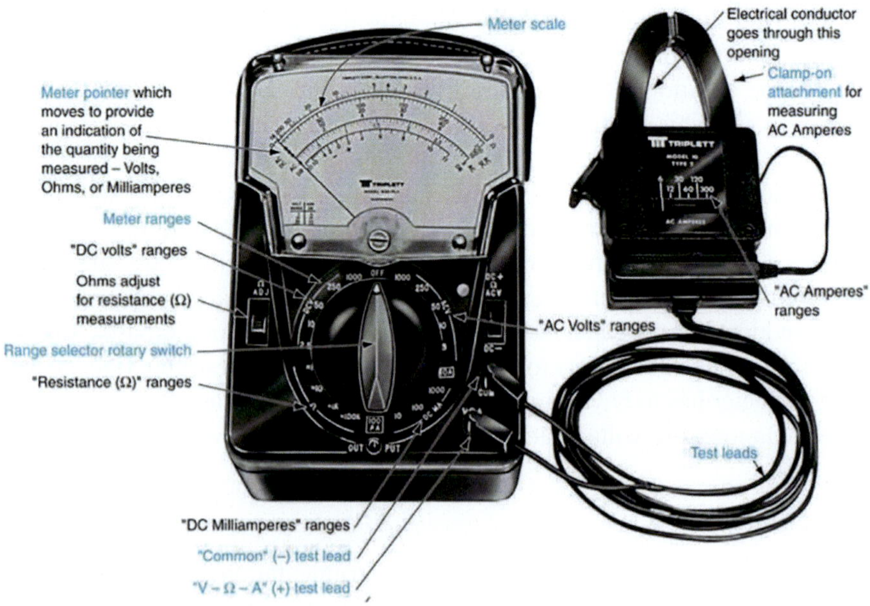

Figure 3-7 Volt-ohm-milliammeter (VOM) analog meter (courtesy of Triplett Corp.).

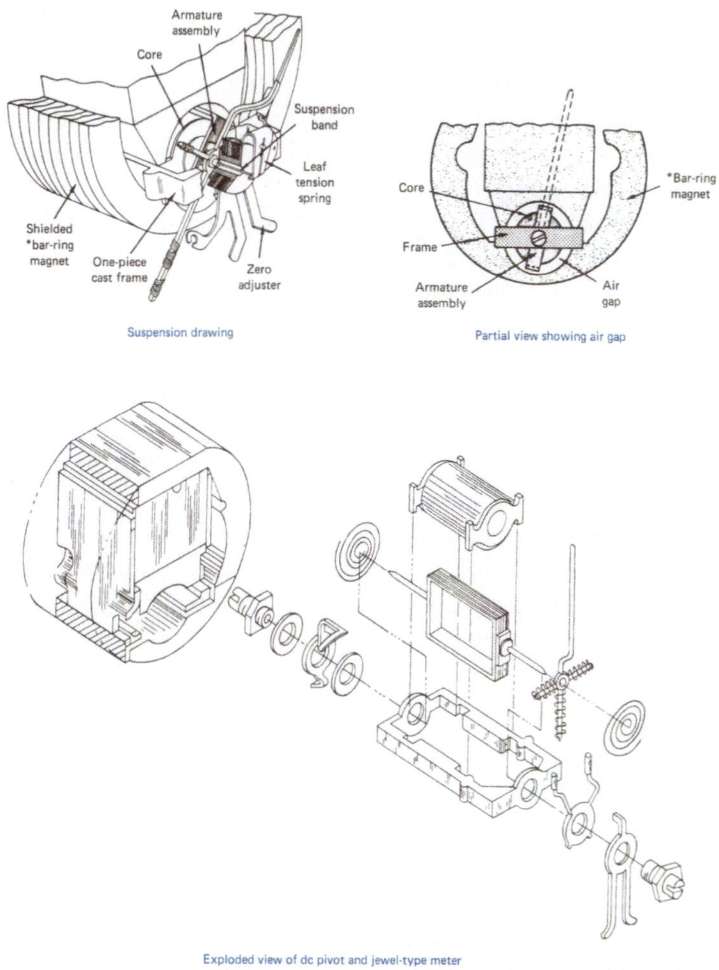

Figure 3-8 d'Arsonval movement.

Although these instruments are not used very frequently today, the interpretation of analog scales and meter design is a fundamental electrical engineering/technology competency.

The **volt-ohm-milliammeter (VOM)** is one type of analog instrument. The VOM or multimeter is an instrument used for measuring several electrical quantities. Single-function analog meters, such as those shown in **Figure 3-7**, are also used to measure electrical quantities. They measure only one quantity.

The basic part of an analog meter is called a **meter movement**. Physical quantities such as airflow or fluid pressure are also measured by analog meters. The movement of the hand or pointer over a calibrated scale indicates the quantity being measured. Many meters use the **d'Arsonval** or moving-coil type of meter movement. The construction details of this meter movement are shown in **Figure 3-8**.

The hand or pointer of the movement stays on the left side of the calibrated scale. A moving coil is located inside a horseshoe magnet. Current flows through the coil from the circuit being tested. A reaction occurs between the electromagnetic field of the coil and the permanent magnetic field of the **horseshoe magnet**. This reaction causes the hand to move toward the right side of the scale. This moving-coil meter movement operates on the same principle as an electric motor. It can be used for single-function meters that measure only one quantity. It can also be used for **multifunction meters**, such as VOMs, that measure more than one quantity. The d'Arsonval meter movement can be used to measure voltage, current, or resistance. Resistors of proper value are connected to the meter movement for making these measurements. Working with analog meters is an excellent application of basic electrical problem solving.

3.2 Measurements

Measuring Resistance

Many important electrical tests may be made by measuring resistance. **Resistance** *is* opposition to the flow of current in an electrical circuit. The current that flows in a circuit depends on the amount of resistance in that circuit. Measurement of resistance in an electrical circuit is accomplished by using a meter, such as an ohmmeter or multimeter. Multimeters such as those shown in **Figure 3-9**. Multimeters are the most used meters for doing electrical work. A multimeter is used to measure resistance, voltage, or current. The type of measurement is changed by adjusting the range/function select switch to the desired measurement. This switch also is used to set the **maximum** quantity that can be measured.

In order to measure electrical quantities, either an analog meter or digital meter may be used. **Figure 3-9** compares the two types of meters. Note that the analog meter has a scale which is used to make a measurement. The digital meter provides a direct reading of the quantity measured, making it easier to interpret. Both types of meters are used by electrical technicians.

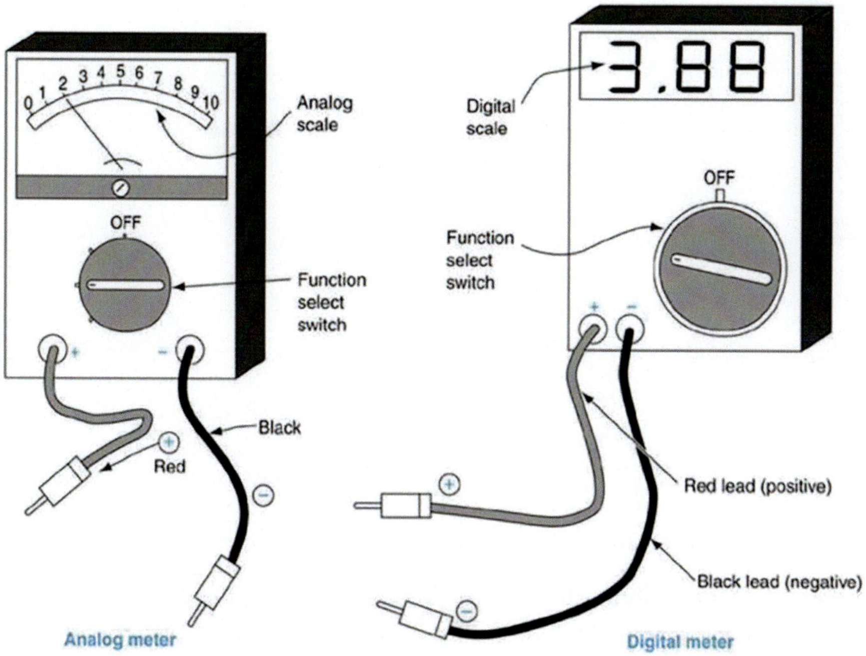

Figure 3-9 Multimeters – analog and digital.

Analog meters are discussed in this chapter to provide the basic rules of interpreting analog scales. The same rules of meter use apply for both analog and digital meters.

The same rules of meter use apply for analog and digital meters. Refer to **Figure 3-7** and note the controls of the VOM/multimeter similar to the one that will also be used to learn **analog scales** in this chapter. Even if digital meters are predominately used, analog scales should be studied.

Perhaps, the operation most simplified by the replacement of analog instrumentation with digital is the area of **resistance** measurement. When operating a digital multimeter to measure resistance, there are many fewer steps than with an analog meter. After the meter has been powered on, select the resistance function, and identify the proper jacks for the test lead placement. Often the resistance functions are not labeled with the word resistance but may use abbreviations such as the symbol omega (Ω). After removing power from the circuit and isolating the component to be measured, connect the leads of the digital meter across the component, in the same configuration as an analog meter. If the meter is not auto-ranging, the digital meter may indicate an out-of-range condition by displaying all

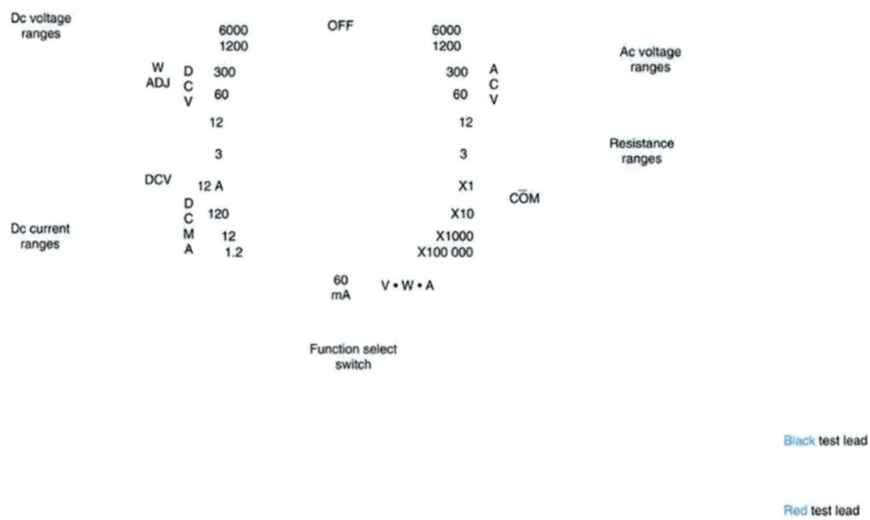

Figure 3-10 Ranges and functions of a VOM or multimeter.

dashes, a 1., or flashing the display. To correct this, change to the proper range. This change can be done without recalibration of the meter, unlike an analog meter, which must be "zeroed" before each use and whenever a range change is made. Digital ohmmeters do not require "zeroing" for typical use. After the meter is connected, the number on the LCD display is often the resistance measured, and no further calculation is necessary, unless it is to add a kilo or mega suffix. Should the user not be interested in resistance, but continuity, many digital meters offer a setting that "beeps" when continuity is found between the test leads.

Measuring Resistance, Analog Meters

Notice in the view of the ranges and scales of **Figure 3-10**. A rotary switch (see **Figure 3-7)** is in the center portion of the meter. Also notice that the lower right section contains ranges for measuring ohms or **resistance**. This type of meter is called a **multirange, multifunction meter**.

The ohms measurement ranges are divided into four portions: × 1, × 10, × 1000, and × 100,000. Most **multimeters** or VOMs are similar to the example shown. The meter may be adjusted to any of the four positions for measuring resistance. The **test leads** used with the VOM are ordinarily black and red. These colors are used to help identify which lead is the positive or negative side of the meter. This is important when measuring direct-current

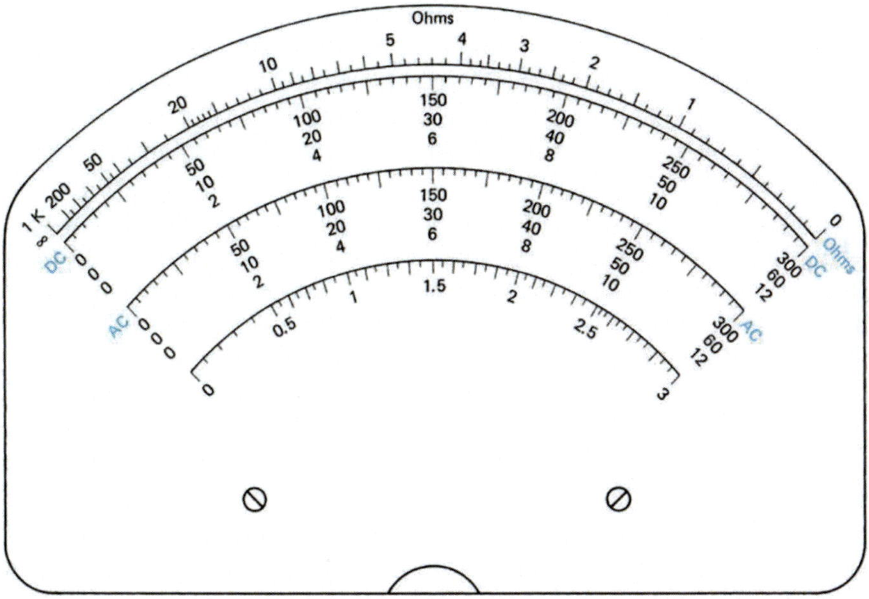

Figure 3-11 VOM (multimeter) scale.

(dc) values. Red indicates positive (+) polarity and black indicates negative (–) polarity.

Refer again to the diagram of the meter controls shown in **Figure 3-10**. The red test lead is put into the “jack” marked with “V-W-A” or volts-ohms-amps. The black test lead is put into the jack labeled “–COM” or negative common. The function selector switch should be placed on one of the **resistance ranges**. When the test leads are touched together or “shorted,” the meter needle moves from the left side of the meter to the right side. This test shows that the meter is operational.

The meter’s **scale** *is* also important. **Figure 3-10** shows the scale of a common type of VOM. Note that the top scale, from zero (0) to infinity (¥), is labeled “Ohms.” This scale is used for measuring ohms **only**. On most VOMs, the top scale is the resistance or ohms scale. To measure any resistance, first select the proper meter range. On the meter range shown in **Figure 3-10**, there are four ranges: ×1, ×10, ×1000, and ×100,000. These values are called **multipliers**. The ohmmeter must be properly “zeroed” before attempting to measure resistance accurately. To “zero” the ohmmeter properly, touch the two test leads together. This should cause the needle to move from infinity (¥) on the left to zero (0) on the right. Infinity represents a very high resistance. Zero represents a very low resistance. If the needle does not reach zero or goes past zero when the test leads are touched or shorted, the control marked "Ohms Adjust" is used. The needle is adjusted to zero when the test leads are touched together. The ohms adjust control is often indicated by "ADJ." The ohmmeter should always be zeroed prior to each resistance measurement and after changing ranges. If the meter is not zeroed, measurements will be incorrect.

A more accurate measurement of resistance occurs when the meter's needle stops somewhere between the middle of the ohms scale and zero. Choosing the proper range adjustments controls how far the needle moves. If the same range selected is × 1, this means that the number to which the needle points must be multiplied by 1. If the function select switch is adjusted to the ×100,000 range, it means that the number the needle points to is multiplied by 100,000. The meter needle should always move to near the center of the scale. Always zero the meter when changing ranges, and always multiply the number indicated on the scale by the multiplier of the range. Never measure the resistance of a component until it has been disconnected or the reading may be wrong. Voltage should **never** be applied to a component when measuring resistance.

Some examples with the meter range set on different multipliers are done as follows. The meter must be zeroed when a range is changed. The test leads are then placed across a resistance. Assume that the needle of the meter moves to point A on the scale of **Figure 3-12**. The resistance equals 7.5 × 1 = 7.5 W. Now change the meter range to ×1000. The reading at point B equals 5.5 × 1000 = 5500 W. At point C, the reading is 0.3 × 1000 = 300 W. The same procedure is used for the ×100,000 range. If the needle moves to 2.2 (point D) on the scale, the reading would be equal to 2.2 × 100,000 or 220,000 W. If the meter range is set on ×100,000 and the needle moves to 3.9 (point E) on the scale, the reading is 3.9 × 100,000 or 390,000 W.

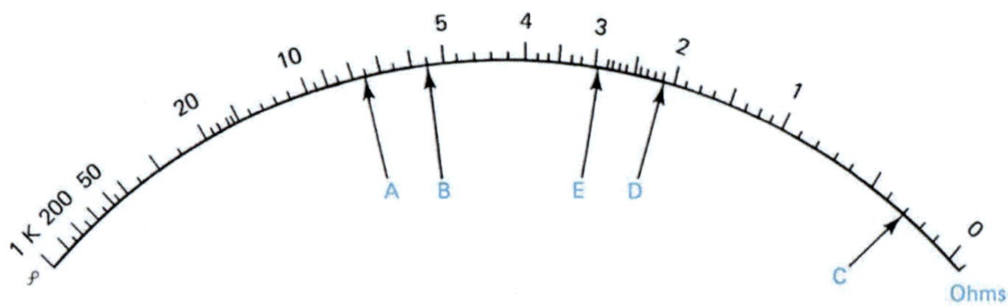

Figure 3-12 Examples of measuring resistance using the ohm's scale of a VOM (multimeter).

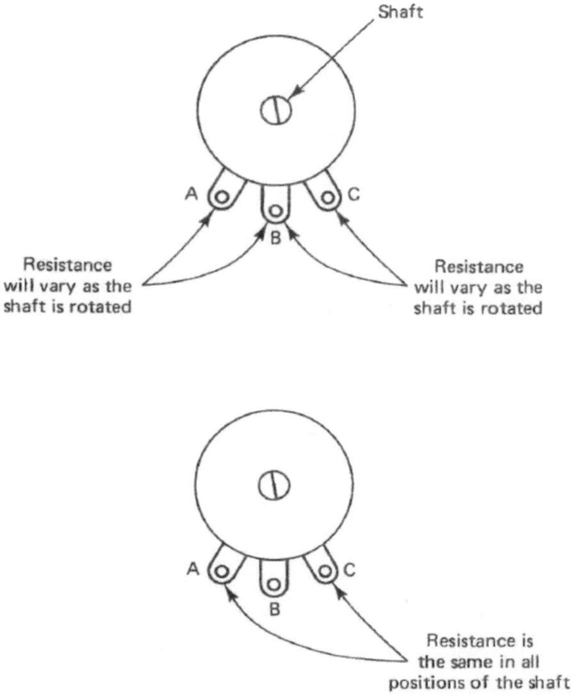

Figure 3-13 Measuring the resistance of a potentiometer.

Remember to zero the meter by touching the test leads together and using the W-ADJ control before making a resistance measurement. Each time the meter range is changed, the meter needle must be zeroed on the scale. If this procedure is not followed, the meter reading will not be accurate.

To learn to measure resistance, it is easy to use **color-coded resistors**. These resistors are small and easy to handle. Practice in the use of the meter to measure several values of resistors makes reading the meter much easier. A VOM may also be used to measure the resistance of a potentiometer, as shown in **Figure 3-13**. If the shaft of the pot is adjusted while the ohmmeter is connected to points A and C, no resistance change will take place. The resistance of the potentiometer is measured in this way. Connecting to points B and C or to points B and A allows changes in resistance as the shaft is turned. The potentiometer shaft may be adjusted both clockwise and counterclockwise. This adjustment affects the measured resistance across points B and C or B and A. The resistance varies from zero to maximum and from maximum back to zero as the shaft is adjusted.

Measuring Voltage

Voltage is applied to electrical equipment to cause it to operate. It is important to be able to measure voltage to check the operation of equipment. Many electrical problems develop due to either too much or too little voltage being applied to the equipment. A **voltmeter** is used to measure voltage in an electrical circuit. A voltmeter, or the **voltage function**, of a multimeter is used to measure voltage. When operating a digital multimeter to measure voltage, there are fewer steps than with an analog meter. After the meter has been turned on, select the proper voltage function and perhaps range for the voltage to be measured and identify the proper jacks for the test lead placement. Often the voltage functions are not labeled with the word voltage but may use abbreviations such as DCV or ACV. Other manufacturers use symbols to discern AC or DC voltage. Should the user not be certain what setting is the voltage function, the manual to the meter should be consulted.

Unlike an analog meter, when measuring DC voltage with a digital meter, proper **polarity** is ordinarily not a concern. If the polarity selected is in reverse, most digital voltmeters will display a minus or negative symbol on the left of the display to indicate reverse polarity. When measuring an unknown voltage, use the same rule as that with analog, and select the highest range first. After the meter is connected in parallel to the circuit, the number on the LCD display is often the voltage measured, and no further calculation is necessary, unlike an analog meter. Should the user inadvertently select a range that is too small, many meters change the LCD display to all dashes, –1, 1, or a flashing display to signify the range selected is too small. Often, this does not result in damage to the meter (unless the voltage measured exceeds the capacity of the meter), unlike an analog movement, which may sustain permanent damage in an out-of-range condition.

Alternating-current (AC) voltage is measured in the same way as DC voltage. The AC voltage scales and ranges on the VOM are used.

Measuring Voltage, Analog Meter

Refer to the controls of the VOM shown in **Figure 3-9(b)**. The voltage ranges shown are 3, 12, 60, 300, 1200, and 6000 V. Other VOMs have different ranges and scales. When the function select switch is adjusted to 3 V on the **DC volts range**, the meter measures up to 3 V. The same is true for the other ranges of DC voltage. The voltage value of each range is the maximum value of voltage that may be measured with the VOM set on that range. When making voltage measurements, adjust the function select switch to the highest range of DC

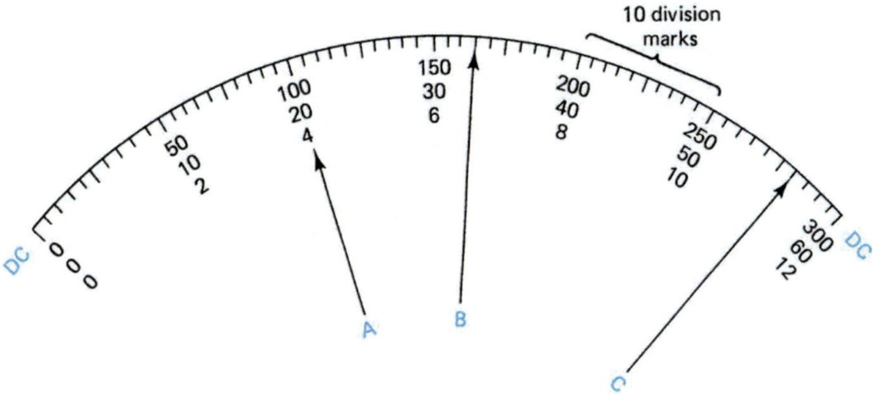

Figure 3-14 DC voltage scale of a VOM (multimeter).

voltage. Connect the red and black test leads to the meter by putting them into the proper jacks. The red test lead should be put into the jack labeled "V-W-A." The black test lead should be put into the jack labeled "–COM."

It is easy to become familiar with the part of the meter's scale that is used to measure DC voltage. Refer to the VOM scale of **Figure 3-10**. Note that the part of the scale below the ohms scale is the DC voltage scale. This scale is usually black. Note that there are three DC voltage scales: 0–12 V, 0–60 V, and 0–300 V. All DC voltages are measured using one of these scales. Note that each of the DC voltage ranges on the function select switch corresponds to a number on the right side of the meter scale **or** a number that can be easily multiplied or divided to equal the number on the function switch.

Note that when the 12-, 60-, or 300-V range is used, the scale is read directly. On these ranges, the number to which the needle points is the actual value of the voltage being measured. When the 3-, 1200-, or 6000-V range is used, the number to which the needle points must be multiplied or divided. If the meter's needle points to the number 50 while the meter is adjusted to the 60-V range, the measured voltage is 50 V. If the meter's needle points to the number 250 while the meter is adjusted to the 3-V range, the measured voltage is 2.5 V (250 ¸ 100 = 2.5). When the 1200-V range is used, the numbers on the 0- to 12-V scale are read and then multiplied by 100. Most VOMs have several scales. Some of these scales are read directly, whereas others require multiplication or division.

Before making any measurements, the proper **DC voltage range** is chosen. The value of the range being used is the maximum value of voltage that can be measured on that range. For example, when the range selected is 12 V, the maximum voltage the meter can measure is 12 V. Any voltage above

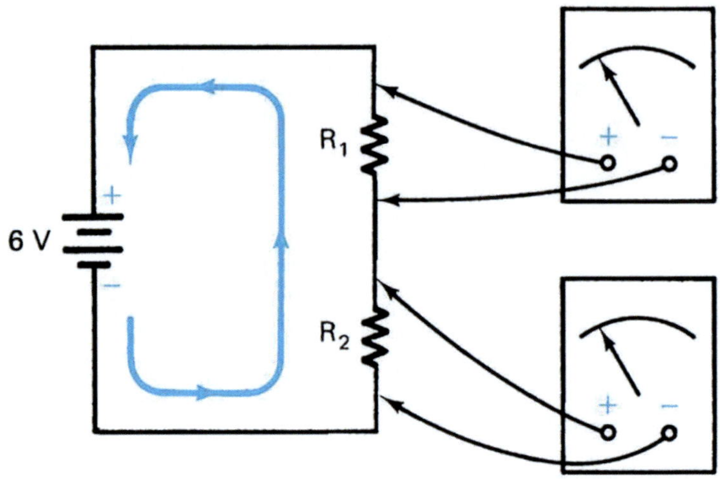

Figure 3-15 Measuring voltage drop in a DC circuit.

12 V could damage the meter. To measure a voltage that is unknown (no indication of its value), start by using the highest range on the meter. Then slowly adjust the range downward until a voltage reading is indicated on the right side of the meter scale.

Matching the meter polarity to the voltage polarity is important when measuring DC voltage. The meter needle moves backward, possibly damaging the meter if the polarities are not connected properly. Meter polarity is simple to determine. The **positive (+)** red test lead is connected to the positive side of the DC voltage being measured. The **negative (–)** black test lead is connected to the negative side of the DC voltage being measured. The meter is always connected across (in parallel with) the DC voltage being measured.

Some examples of **DC voltage measurements** with the meter set on the 3-V range are given below. If the test leads of the meter are placed across a voltage source and the meter's needle moves to point A on the scale of **Figure 3-14**, the DC voltage is equal to 100 divided by 100, or 1 V. The reading at point B is 165 divided by 100, or 1.65 V. At point C, the reading is 280 divided by 100, or 2.8 V. There is some difficulty in reading the voltage divisions on the scales. Look at the division marks from 200 to 250. The difference between 200 and 250 is 50 units (250 – 200 = 50). There are 10 division marks between 200 and 250. The voltage per division mark is 50 divided by 10, or 5 V per division. So, each division mark between 200 and 250 V is equal to 5 V. This procedure is like reading a ruler or other types of scales.

If the range switch is changed to the 12-V position, the voltage is read directly from the meter scale. For example, if the range is set on 12 V and the meter needle moves to point A on **Figure 3-14**, the voltage equals 4 V. The reading at point B equals 6.6 V. At point C, the reading is 11.2 V. The same procedure is used for all other ranges.

A certain amount of voltage is needed to cause electrical current to flow through a resistance in a circuit. This voltage is called **voltage drop**. Voltage drop is measured across any component through which current flows. The polarity of a voltage drop depends on the direction of current flow. Current flows from the negative polarity of a battery to the positive polarity. In **Figure 3-15**, the bottom of each resistor is negative. The negative test lead of the meter is connected to the bottom of the resistor. The positive test lead is connected to the top. The meters are connected as shown to measure each of the voltage drops in the circuit. If the meter polarity is reversed, the meter needle would move in the wrong direction.

Measuring Current

Current flows through a complete electrical circuit when voltage is applied. Many important tests are made by **measuring current** flow in electrical circuits. The current values in an electrical circuit depend on the amount of resistance in the circuit. Learning to use an ammeter to measure current in an electrical circuit is important. When operating a digital multimeter to measure current, there are fewer steps than with an analog-type meter. As with a voltage measurement, verify that the meter has been turned on, select the proper function and perhaps range for the current to be measured, and identify the proper jacks for the test lead placement.

Often, the current measuring function is not labeled with the word current but may use abbreviations such as DCA. Other manufacturers use symbols to discern AC or DC current. Should the user not be certain what function indicates current, she/he should refer to the meter's manual. As with any measurement, the range is the maximum amount that may be measured; therefore, unknown currents must be dealt with by starting with the highest range. Should the user inadvertently select a range that is too small, the digital meter will change the LCD display to signify this circumstance. Often, this does not result in damage to the meter (unless the measurement exceeds the capacity of the meter), but an **over-range** condition may damage an analog movement. Unlike an analog meter, when measuring DC current, proper polarity is not a concern. If the polarity selected is in reverse, most

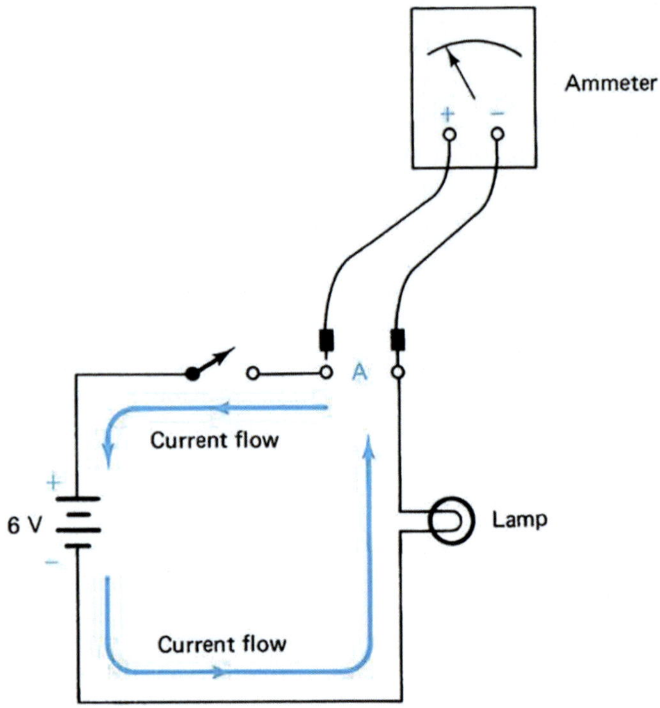

Figure 3-16 Meter connection *for* measuring direct current.

digital voltmeters will display a minus or negative symbol on the left of the display to indicate reverse polarity. After the meter is connected in series to the circuit as shown in **Figure 3-16**, the number on the LCD display is often the current measured, and no further calculation is necessary, except perhaps to add units such as milliamps.

Measuring Current, Analog Meters

Analog VOMs will also measure DC current. Refer to the ranges and functions of the VOM shown in **Figure 3-10**. The range/function select switch may be adjusted to any five ranges of direct current: 12 A, 120 mA, 12 mA, 1.2 mA, and 60 RA. For example, when the function select switch is placed in the 120-mA range, the meter is capable of measuring **up to** 120 mA of current. The value of the current set on the range is the maximum value that can be measured on that range. The function select switch should first be adjusted to the highest range of direct current. Current is measured by connecting

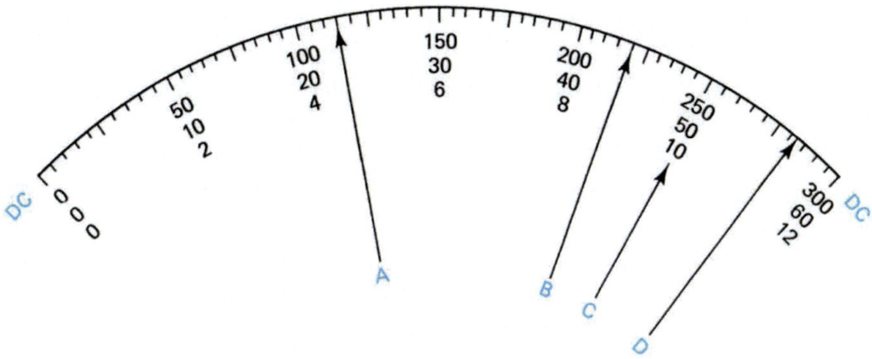

Figure 3-17 Direct-current scale of a VOM (multimeter).

the meter into a circuit as shown in **Figure 3-16**, which is referred to as connecting the meter in **series** with the circuit.

Current flows from a voltage source when some device that has resistance is connected to the source. When a lamp is connected to a battery, current flows from the battery through the lamp. In the circuit of **Figure 3-16**, electrons flow from the negative battery terminal, through

the lamp and back to the positive battery terminal. Electrons are so small that the human eye cannot see them, but their movement can be measured with an **ammeter**. As the voltage applied to a circuit increases, the current also increases. So, if 12 V is applied to the lamp in **Figure 3-16**, a larger current will flow through the lamp. If 24 V is applied to the same lamp, an even larger current will flow. As resistance gets smaller, current increases. Resistance is the opposition to current flow. When a circuit has more resistance, it has less current flow.

Refer to the **direct-current ranges** of the VOM shown in **Figure 3-10**. Note that the ranges begin with 12 A. The next ranges are for measuring 120 mA, 12 mA, 1.2 mA, and 60 μA. There are a total of five current ranges. The function select switch is adjusted to any of these five ranges for measuring direct current. When measuring current, always start with the meter set on its highest range. By practicing this procedure, it becomes a habit. This habit helps in using a meter properly. Always start on the highest range. Then move the range setting to a lower value if the meter needle only moves a small amount. The most accurate reading is when the meter needle is between the center of the scale and the right side.

The same scales on the VOM (see **Figure 3-17**) are often used for measuring direct-current or DC voltage. If the meter range is set on the 12-A range, the scale is read directly. The bottom DC scale, which has the number "12" on the right side, is used. Some examples are shown in **Figure 3-16** with the meter set on the 12-A range. At point A on the scale, the reading is 4.6 A. The reading at point B is 8.8 A.

The 60-μA range on the meter is for measuring very small currents. This range is also read directly on the meter scale. Note that the number "60" is the middle number on the right of the DC scale. When the meter is set on the 120-mA range, the meter will measure up to 120 mA of direct current. The readings on the scale are multiplied by 10 on this range setting. The readings at the points shown in **Figure 3-16** for the 120-mA range are:

- Point A = 46 (4.6 × 10) mA
- Point B = 88 (8.8 × 10) mA
- Point C = 100 (10 × 10) mA
- Point D = 113 (11.3 × 10) mA

Note that the reading at point D is between two of the scale divisions. Point D is halfway between the 11.2 and 11.4 divisions on the scale. So, the reading is 11.3 × 10, or 113 mA. The test lead **polarity** of the VOM is also important when measuring direct current. The VOM is connected to allow current to flow through the meter in the right direction. The negative test lead is connected nearest to the negative side of the voltage source. The meter is then connected **into** the circuit. To measure current, a wire is removed from the circuit to place the meter into the circuit. No voltage should be applied to the circuit when connecting the meter. The meter is placed in series with the circuit. Series circuits have one path for current flow.

The proper procedure for measuring current through point A in the circuit of **Figure 3-16** is:

- Turn off the circuit's voltage source by opening the switch.
- Set the meter to the highest current range (12 A).
- Remove the wire at point A.
- Connect the negative test lead of the meter to the terminal nearest to the negative side of the voltage source.
- Connect the positive lead to the end of the wire which was removed from point A.
- Turn on the switch to apply voltage to the circuit.
- Look at the meter needle to see how far it has moved up the scale.
- Adjust the meter range until the needle moves to between the center of the scale and the right side.

Always remember the following safety tips when measuring current. Turn off the voltage before connecting the meter, so as not to get an electrical shock. This is an important habit to develop. Always remember to turn off the voltage before connecting the meter. Set the meter to its highest current range. This ensures that the meter needle will not move too far to the right of the scale and possibly damage the meter.

A wire is disconnected from the circuit and the meter is put in series with the circuit. Always remember to disconnect a wire and reconnect the wire to one of the meter test leads. If a wire is not removed to put the meter into the circuit, the meter will not be connected properly.

Use the proper meter polarity. The negative test lead is connected so that it is nearest to the negative side of the voltage source. Similarly, the positive test lead is connected so that it is nearest to the positive side of the voltage source.

Summary

- Early measuring instruments were analog
- Analog instruments use a d'Arsonval movement
- Analog instruments are read on a scale
- Digital instruments typically have an LCD display
- When measuring resistance, digital meters require proper setting of function and, perhaps, range
- When measuring resistance, analog meters require zeroing, in addition to function and range
- Resistance measurements are made with the power removed
- When measuring resistance, the component must be isolated from the circuit
- When measuring voltage, digital meters require proper setting of function and range
- When measuring voltage, analog meters require proper setting of function and range, and a multiplier or divider may be needed when determining the actual value after reading the scale
- When measuring current, digital meters require proper setting of function and range
- When measuring current, analog meters require proper setting of function and range, and a multiplier or divider may be needed when determining the actual value after reading the scale

Self-examination/Answers

1. What are the primary types of electronic measurement instruments?
2. Which meter has less chance for human error?
3. Which meter can be damaged by hooking it up in reverse?
4. If a meter requires "zeroing," what meter are you using and what are you measuring?
5. When measuring ___ with an analog meter, polarity must be observed.
6. Digital meters typically have ___ displays.
7. When measuring an unknown quantity, meters should be set to the ___ setting.
8. Why is the ohms scale of an analog meter considered to be nonlinear?
9. Where on the ohms scale of an analog meter are the most accurate measurements found?
10. What is meant by the ×1000 range on the ohmmeter?

Answers

1. Digital and analog
2. Digital meter
3. Analog
4. Analog/Resistance
5. Voltage
7. Seven-segment LED or LCD
8. Highest
9. Unequal distance between units
10. Between zero and one on the scale
11. Multiply scale reading by 1000

Problems

Voltage Measurement Problems

Determine the voltage values using **Figure 3-14**.

DC Volts Range	*A*	*B*	*C*	*D*	*E*
2.5 V dc	(7) ___	(8) ___	(9) ___	(10) ___	(11) ___
10 V do	(12) ___	(13) ___	(14) ___	(15) ___	(16) ___

50 V dc	(17) ___	(18) ___	(19) ___	(20) ___	(21) ___
250 V dc	(22) ___	(23) ___	(24) ___	(25) ___	(26) ___
500 V dc	(27) ___	(28) ___	(29) ___	(30) ___	(31) ___
1000 V do	(32) ___	(33) ___	(34) ___	(35) ___	(36) ___

Current Measurement Problems

Determine the current values using **Figure 3-17**.

DC Amps Range	*A*	*B*	*C*	*D*	*E*
10A	(44) ___	(45) ___	(46) ___	(47) ___	(48) ___
2.5 A	(49) ___	(50) ___	(51) ___	(52) ___	(53) ___
500 mA	(54) ___	(55) ___	(56) ___	(57) ___	(58) ___
100 mA	(59) ___	(60) ___	(61) ___	(62) ___	(63) ___
50 mA	(64) ___	(65) ___	(66) ___	(67) ___	(68) ___
10 mA	(69) ___	(70) ___	(71) ___	(72) ___	(73) ___
2.5 mA	(74) ___	(75) ___	(76) ___	(77) ___	(78) ___
250 11A	(79) ___	(80) ___	(81) ___	(82) ___	(83) ___

Glossary

Ammeter
A meter used to measure current flow.

Continuity check
A test to see if a circuit is an open or closed path.

Multifunction meter
A meter that measures two or more electrical quantities, such as a volt-ohm-milliammeter (VOM), which measures voltage, resistance, and current. Such meters are also commonly called **multimeters**.

Multirange meter
A meter that has two or more ranges to measure an electrical quantity.

Ohmmeter
A meter used to measure resistance.

Polarity
The direction of an electrical potential (– or +) or a magnetic charge (north or south).

Potentiometer
A variable-resistance component used as a control device in electrical circuits.

Voltage drop
The electrical potential (voltage) that exists across two points of an electrical circuit.

Voltmeter
A meter used to measure voltage.

Volt-ohm-milliammeter (VOM)
A multifunction, multirange meter which is usually designed to measure voltage, current, and resistance (also called a **multimeter**.)

4

Ohm's Law and Series Electrical Circuits

To understand electrical fundamentals, it is necessary to know how to apply basic electrical theory. Electrical science is a somewhat mathematical discipline; however, the mathematics is easy to understand because it has practical applications. The basic theory used is called Ohm's law, which is important because it applies to all electrical circuits. The examples in this chapter are direct-current (DC) circuits.

Objectives

1. Define Ohm's law
2. Define a series circuit
3. Solve series circuit problems finding current, resistance, voltage, and power
4. Define voltage drop in a circuit
5. Solve circuit problems with resistors in series configurations

Chapter Outline

4.1 Ohm's Law
4.2 Characteristics of a Series Circuit
4.3 Applying Ohm's Law to Series Circuits
4.4 Applying the Power Formula to Series Circuits
4.5 Troubleshooting Series Circuits

4.1 OHM'S LAW

Ohm's law is the most basic and most used of all electrical theories. Ohm's law explains the relationship of voltage (the force that causes current to flow), current (the movement of electrons), and resistance (the opposition to current flow). Ohm's law is stated as follows: An increase in voltage increases current if resistance remains the same. Ohm's law stated another way is as follows: An increase in resistance causes a decrease in current if voltage remains the same. The electrical values used with Ohm's law are usually represented with capital letters. For example, voltage is represented with the letter *V*, current with the letter *I*, and resistance with the letter *R*. The mathematical relationship of the three electrical quantities is shown in the following formulas. These should be memorized. The Ohm's law circle in **Figure 4-1** is helpful to remember the formulas.

$V = I \times R$ – Voltage (*V*) equals Current (*I*) multiplied by Resistance (*R*)
$I = V/R$ – Current (*I*) equals Voltage (*V*) divided by Resistance (*R*)
$R = V/I$ – Resistance (*R*) equals Voltage (*V*) divided by Current (*I*)

Voltage (*V*) is measured in **volts**. **Current** (*I*) is measured in **amperes**. **Resistance** (*R*) is measured in **ohms**. If two electrical values are known, the third value can be calculated by using one of the formulas. Look at **Figure 4-2**.

Using the Ohm's law current formula, $I = V/R_1$ the calculated value of *I* in the circuit is

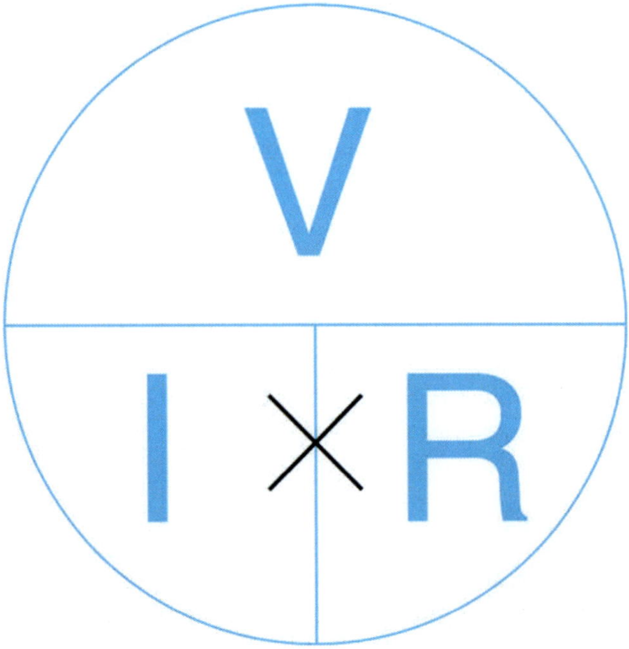

Figure 4-1 Ohm's law circle: *V* – voltage; *I* – current; *R* – resistance. To use the circle, cover the value you want to find and read the other values as they appear in the formula: $V = I \times R$; $I = V/R$; $R = V/I$.

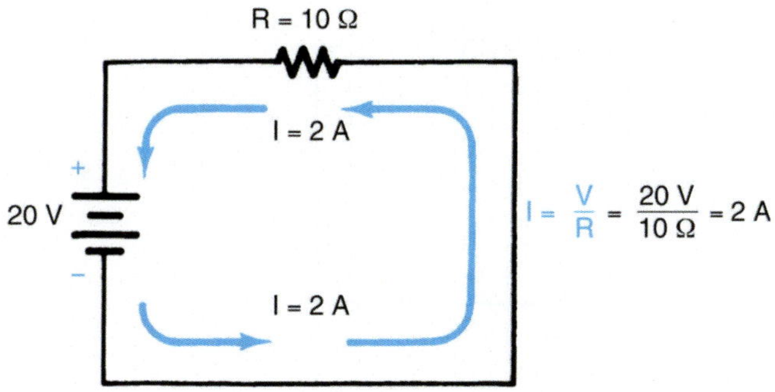

Figure 4-2 Ohm's law example.

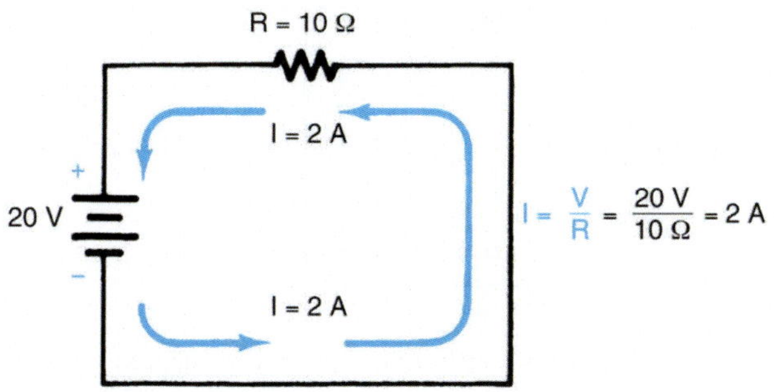

Figure 4-3 Ohm's law with voltage doubled.

$I = V/R$ = 10 V/10 W = 1 A.

If the voltage in the circuit is doubled as shown in **Figure 4-3**, the calculated current using the Ohm's law current formula ($I = V/R$) is

$I = V/R$ = 20 V/10 W = 2 A.

From this example, note that as voltage is increased, current also increases if resistance remains the same. Also, if voltage is doubled, current is doubled. If voltage is 10 times larger, the current becomes 10 times larger. This is called a **direct relationship**. When voltage increases, current also increases.

Now, look at **Figure 4-4(a)**. The calculated current flow in the circuit is

$I = V/R$ = 10 V/100 W = 0.1 A.

In **Figure 4-4(b)**, the 100-W resistor in the circuit is replaced with a 1000-W resistor. The calculated value of the current with 10 V applied to the 1000-W resistance is

$I = V/R$ = 10 V/1000 W = 0.01 A.

This is called an **inverse relationship**. As resistance increases, current decreases.

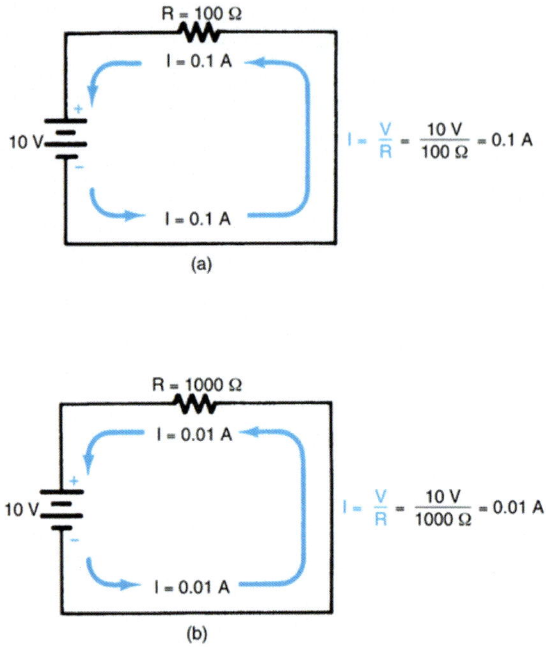

Figure 4-4 Effect of increasing resistance.

Ohm's law explains the relationship of voltage, current, and resistance in electrical units. The circle shown in **Figure 4-1** is used to help remember this relationship. An easy way to remember the **Ohm's law formulas** used to find voltage, current, and resistance values is to use this circle. To calculate the voltage in a circuit, cover the *V* on the circle. Note that *V* is equal to *I* times *R*. To find current, cover the *I* and note that *I* is equal to *V* over *R*. To find resistance, cover the *R* and note that *R* is equal to *V* over *I*. The circle is easy to remember. It helps in using Ohm's law to solve simple electrical problems.

Another example of using Ohm's law is shown in **Figure 4-5(a)**. To find the value of current that flows in this circuit, use the **Ohm's law circle** in **Figure 4-1**, cover the *I*, and find that *I* equals *V* over *R*. In this circuit, *I* equals 12 V divided by 2 W. So, the current is equal to 6 A.

$I = V/R$

$I = 12\ \text{V}/2\ \Omega$

$I = 6\ \text{A}.$

Another example of Ohm's law is shown in **Figure 4-5(b)**. The voltage of this circuit is equal to 10 V and the current is equal to 2 A. The resistance of the circuit is equal to 10 V divided by 2 A, or 5W.

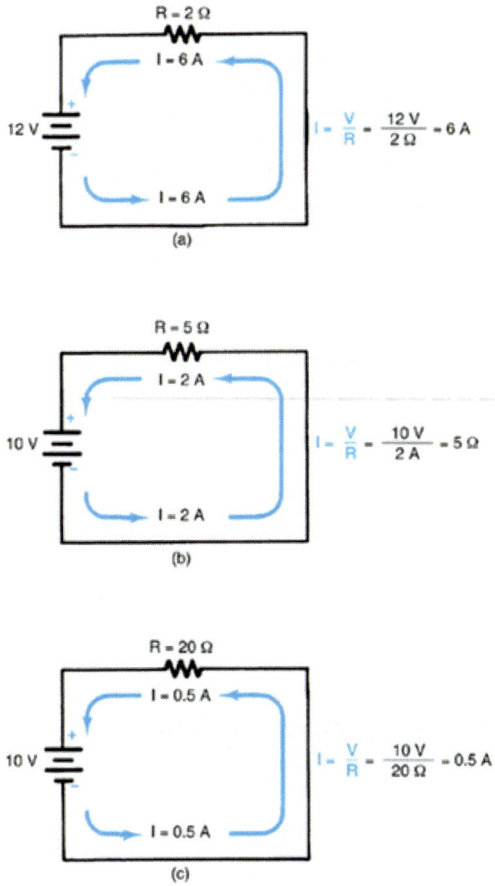

Figure 4-5 Ohm's law examples.

$R = V/I = 10\ \text{V}/2\ \text{A} = 5\ \Omega$.

Ohm's law also states that **if** the resistance of a circuit increases, the current will decrease if the voltage stays the same. An example of this is shown in **Figure 4-5(c)**. The resistance of the circuit is increased to 20 W. The current of the circuit is now equal to 10 V divided by 20 W,

or 0.5 A.

$I = V/R = 10\ \text{V}/20\ \Omega = 0.5\ \text{A}$.

The current in the previous circuit was 2 A. If the resistance of a circuit is increased four times, the current flow decreases to one-fourth its original value. Remember the inverse relationship of resistance and voltage.

In the previous example, current was calculated. Voltage and resistance of a circuit may also be calculated by using Ohm's law. In a circuit where

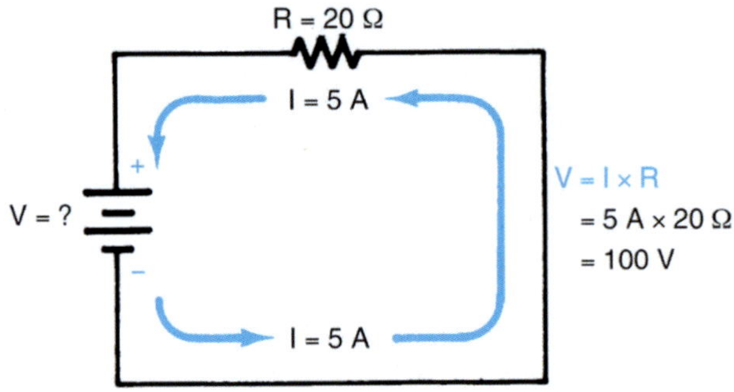

Figure 4-6 Using Ohm's law to find voltage.

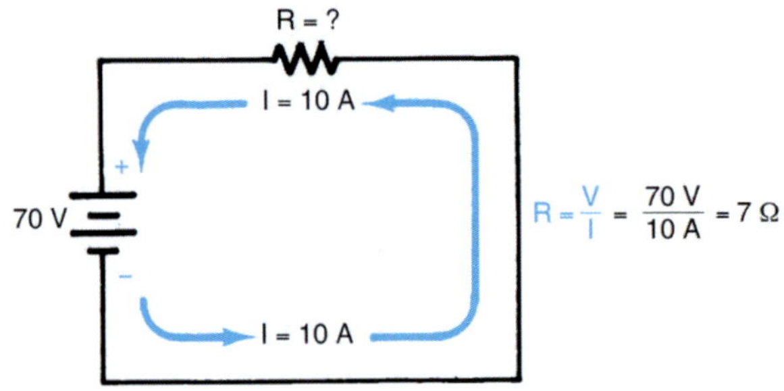

Figure 4-7 Using Ohm's law to find resistance.

current and resistance are known, Ohm's law can be used to find voltage. In the circuit shown in **Figure 4-6**, assume that resistance is equal to 20 W and the current is 5 A. Use the circle in **Figure 4-1** and cover the *V*. This shows that *V* equals *I* times *R*. So, the voltage required causing 5 A of current through a 20-W resistance is equal to 5 A times 20 W (100 V).

$V = I \times R = 5\text{ A} \times 20\ \Omega = 100\text{ V}.$

Ohm's law is also used to find the value of resistance in a circuit. Assume that a circuit has a known value of voltage and current. The value of resistance required to cause this value of current flow may be found. In the example of **Figure 4-7**, the voltage is equal to 70 V and the current equals 10 A. The resistance of the circuit is found by using the Ohm's law circle in **Figure 4-1**. Cover the *R* and find that *R* is equal to *V* over *I*. So, the resistance of this circuit is equal to 70 V divided by 10 A (7 W).

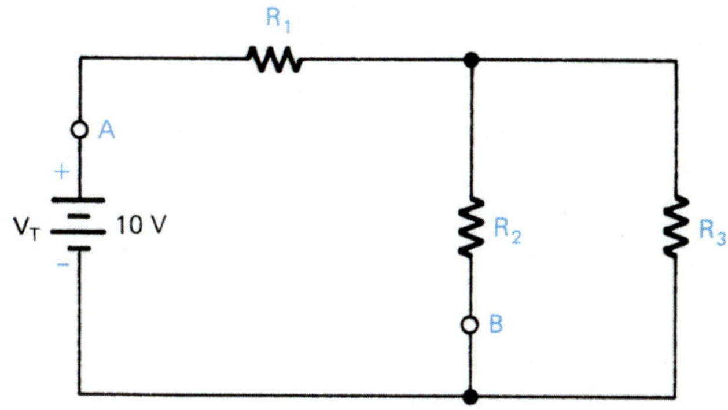

Figure 4-8 Ohm's law subscripts.

$R = V/I = 70\ \text{V}/10\ \text{A} = 7\ \Omega$

The following circuit examples use many **subscripts** (such as R_T, V_T, and I_I). It is common to use subscripts to identify electrical components in circuit diagrams. The circuit shown in **Figure 4-8** has three resistors and a battery. The resistors are labeled R_1, R_2, and R_3. The subscripts identify each of these three resistors. Subscripts also aid in making measurements. The voltage drop across resistor R_I is called voltage drop V_1. The term V "total" is represented by the subscript T, such as V_T. V_T is total voltage applied to a circuit. The current measurement I_2 is the current through resistor R_2 measured at point B. Total current (I_T) is measured at point A. The voltage drop across R_3 is called V_3. Subscripts are also valuable in troubleshooting and repair of equipment. It would be impossible to isolate problems in equipment without components that are easily identified.

4.2 Characteristics of Series Circuits

The three types of electrical circuits are **series circuits**, **parallel circuits**, and **combination series–parallel circuits**. The easiest type of circuit to understand is the series circuit. Series circuits are different from other types of electrical circuits. It is important to remember the characteristics of a series circuit. In series circuits, the resistors may appear "in a line" with each other on a schematic drawing, thereby representing one path. This path may consist of any number of resistors or electrical loads, so long as current must exit one component to enter the next, etc., with never seeing a divergence in the path.

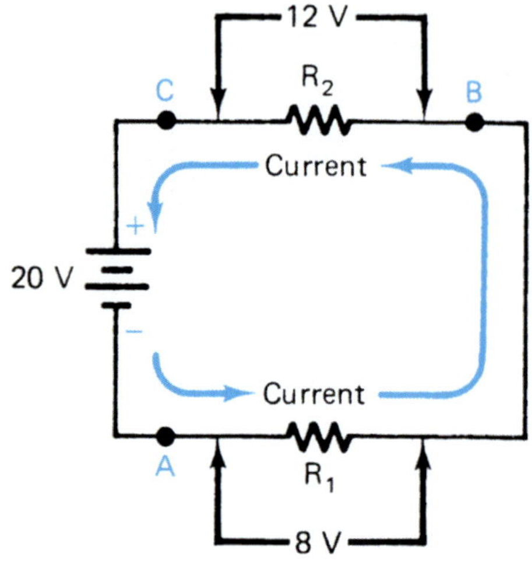

Figure 4-9 Series electrical circuit.

Current

Current or the amount of electrons flowing in a circuit is key to understanding the series circuit. The main characteristic of a series circuit is that it has only one path for current flow. Because there is only one current path, the current flow is the **same** value in any part of the circuit. Mathematically, this appears as

$I_T = I_1 = I_2 = I_3 = I_n$,

where I_T is the total current, I_1 is the value of current in the first component, I_2 the value of current in the second component, and so on until all the components (I_n) are included.

The voltages in the circuit depend on the resistance of the components in the circuit. When a series circuit is opened, there is no path for current to flow. Thus, the circuit would not operate.

In the circuit shown in **Figure 4-9**, current flows from the negative side of the voltage source, through resistor R_1, through resistor R_2, and then to the positive side of the voltage source. Since a series circuit has the same current everywhere in the circuit, an ammeter could be placed anywhere into the circuit to measure current. The value of current at point A is the same as the current at point B or point C. All of the current should measure exactly the same value.

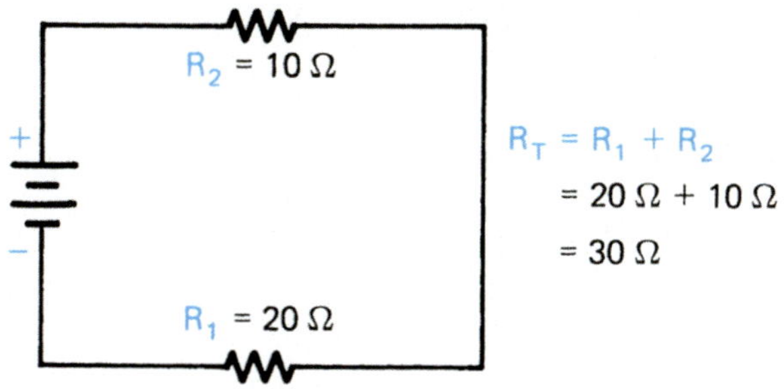

Figure 4-10 Finding total resistance in a series circuit.

Resistance

Resistors oppose current flow in a series circuit. A characteristic of **resistance** to current flow offered in a series circuit (R total or R_T) is found by adding the value of all the series resistors.

$R_T = R_1 + R_2 + R_3 + R_n$

where R_T is the total resistance, R_1 is the value of resistance in the first component, I_2 the second component, and so on until all the components (R_n) are included

In the circuit shown in **Figure 4-10**, the **total resistance** of the circuit is the sum of the two resistances. So, the total resistance is equal to 20 W plus 10 W, or 30 W.

$R_T = R_1 + R_2 = 20\ \Omega + 10\ \Omega = 30\ \Omega.$

Voltage

The **voltage** across any component in a circuit is called **voltage drop**. Voltage drops can be easily calculated using Ohm's law if the current and resistance are known. Simply multiply the value of the resistance by the value of the current and the result is the voltage drop. Voltage drops can also be measured by using a voltmeter.

In any series circuit, the sum of the **voltage drops** is equal to the voltage applied to the circuit. This can be shown mathematically as

$V_T = V_1 + V_2 + V_3 + V_4$,

where V_T is the total voltage, V_1 is the value of voltage drop in the first component, V_2 is the voltage drops of the second component, and so on until all the voltage drops (V_n) are included.

The circuit shown in **Figure 4-9** has voltage drops of 8 V plus 12 V, which is equal to 20 V.

$$V_T = V_1 + V_2$$
$$= 8\text{ V} + 12\text{ V}$$
$$= 20\text{ V}.$$

Power

The energy used, or **power**, can be calculated using the equation $P = I \times V$, where P stands for power in watts, I is current in amps, and V is voltage in volts. The relationship of power in a series circuit can be expressed mathematically as

$$P_T = P_1 + P_2 + P_3 + P_n,$$

where P_T is the total power used in watts, P_1 is the value of power used by the first component, P_2 is the power used by the second component, and so on until all components using power (P_n) are included.

Summary of Series Circuits Rules

There are several important characteristics of series circuits. Remember these basic rules for series circuits:

- The same current flows through each part of a series circuit.
- The total resistance of a series circuit is equal to the sum of the individual resistances.
- The voltage applied to a series circuit is equal to the sum of the individual voltage drops.
- The voltage drop across a resistor in a series circuit is directly proportional to the size of the resistor.
- If the circuit is broken at any point, no current will flow.

Ohm's law is used to explain how a series circuit operates. In the circuit of **Figure 4-11**, the total resistance is equal to 2 W plus 3W, or 5W.

$$R_T = R_1 + R_2$$
$$= 2\ \Omega + 3\ \Omega$$
$$= 5\ \Omega.$$

The applied voltage is 10 V. Current is equal to voltage divided by resistance, or $I = V/R$. In the circuit shown, current is equal to 10 V divided by 5 W, which is 2 A.

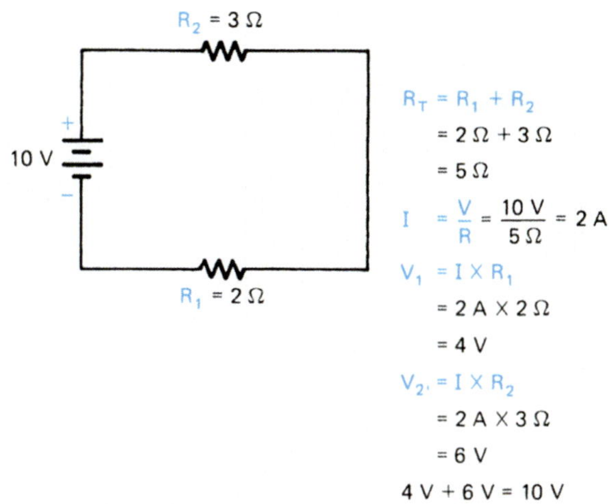

Figure 4-11 Using Ohm's law for a series circuit.

$$\begin{aligned} I &= V/R \\ &= 10\ V/5\ \Omega \\ &= 2\ A. \end{aligned}$$

If a current meter is connected into this circuit, the current measurement should be 2 A. **Voltage drops** across each of the resistors may also be found. Voltage is equal to current times resistance (*V* = *I* × *R*). The voltage drop across R_1 (V_1) is equal to the current through R_1 (2 A) times the value of R_1 (2 W). The voltage drop across R_1 equals 2 A × 2W, or 4 V.

$$\begin{aligned} V_1 &= I_1 \times R_1 \\ &= 2\ A \times 2\ \Omega \\ &= 4\ V. \end{aligned}$$

The voltage drop across R_2 (V_2) equals 2 A × 3W, or 6 V.

$$\begin{aligned} V_2 &= I_2 \times R_2 \\ &= 2\ A \times 3\ \Omega \\ &= 6\ V. \end{aligned}$$

The sum of these voltage drops is equal to the applied voltage. To check these values, add 4 V and 6 V, which is equal to 10 V.

$$\begin{aligned} V_T &= V_1 + V_2 \\ &= 4\ V + 6\ AV \\ &= 10\ V. \end{aligned}$$

If another resistance is added to a series circuit, as shown in **Figure 4-12**, resistance increases. Because there is more resistance, the current flow becomes smaller. The circuit now has R_3 (a 5-Ω resistor) added in series to R_1 and R_2. The **total resistance** is now 2 W + 3 W = 5 W, or 10 W, compared

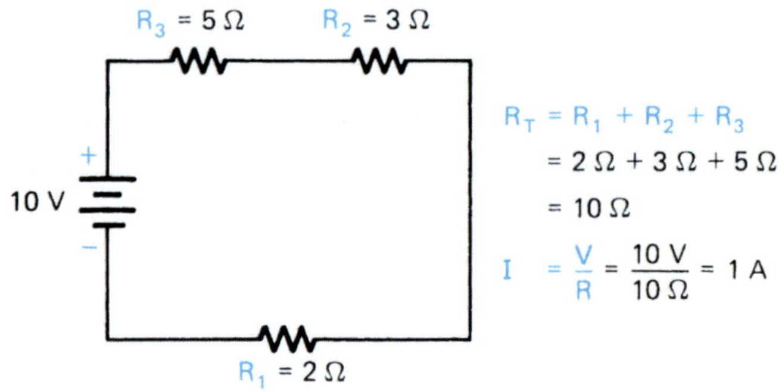

Figure 4-12 Effect of adding resistance to a series circuit.

with 5 W in the previous example. The current is now 1 A, compared with 2 A in the other circuit:

$I = V/R$ or I = 10 V/10 W = 1 A.

4.3 Applying Power Formulas to Series Circuits

To operate, electrical circuits depend on the relationship between energy, work, and power. A basic law of physics states that **energy** (the capacity to do work) can be neither created nor destroyed. It can, however, be converted from one form to another. The process of converting energy from one form to another is called **work**. The rate at which work is accomplished, or at which energy is converted, is called **power**. Power may be expressed in an energy unit called the **watt (W)** or in the power unit called **horsepower (hp)**. For electrical circuits, the **watt** is used exclusively. However, many electric motors are rated by horsepower and must be converted from horsepower to watts, which is accomplished by the following:

1 hp = 746 W.

Sample Problem: Work

Work is done when a force (F) is moved a distance (d), or

$W = F \times d,$

where

W = work in joules (J)

F = force in Newtons

d = distance the force moves in meters.

Given: An object with a mass of 22 kg is moved 55 m.

Find: The amount of work done when the object is moved.

Solution: The force of gravity acting on the object is equal to 9.8 (a constant that applies to objects on earth) multiplied by the mass of the object, or:

$$F = 9.8 \times 22 \text{ kg} = 215.6 \text{ Newtons}$$
$$W = F \times d$$
$$= 215.6 \times 55$$
$$\text{W} = 11{,}858 \text{ J.}$$

Sample Problem: Power

Power is the time rate of doing work, which is expressed as

$P = W/t$

where

P = power in watts

W = work done in joules

t = time taken to do the work in seconds.

Given: An electric motor is used to move an object along a conveyor line. The object has a mass of 150 kg and is moved 28 m in 8 seconds.

Find: The power developed by the motor in watts and horsepower units.

Solution:

Force (F) = 9.8 × mass

= 9.8 × 150 kg

F = 1470 Newton's

Work (W) = $F \times d$

= 1470 × 28 m

W = 41,160 J

Power (P) = W/t

= 41,160/8

P = 5145 W

Horsepower = P/746, because

1 hp = 746 W

hp = 5145/746 = 6.9 hp.

The energy–power–work relationship, given in units of watts for power and in joules for work, is as follows:

$P = W/t$ or power = energy/time

where: P = power (Watts)

W = work (energy converted) (Joules)

t = time (seconds).

As shown by this equation, 1 W of power is equal to 1 J of energy converted in 1 second of time. If the same amount of work is accomplished in less time, then more power is produced. If time is increased for a given amount of work, then the result is less power produced.

The **watt** is the basic unit of electrical power. Power (P) is equal to voltage (in volts) multiplied by current (in amperes). The formula is: $P = V \times I$.

The same equation for the **power law** can be found by looking at energy and charge. In electrical circuits, power is developed when current flows through a resistance. The source of energy in a circuit provides the energy to do the work of setting the electrons in motion and producing current flow. When electrons pass through a resistance, the moving electrons collide with the atoms in the resistance (sometimes called a load resistor), creating friction. In the process, energy is transferred from the electrons to the resistor. The resistor must give off, or **dissipate**, a corresponding amount of energy. Some energy is dissipated in the form of heat. The rate at which dissipation occurs determines the power developed for the circuit.

Voltage is equal to work energy (in joules) divided by the charge in coulombs. This equation is expressed as
$V = W/Q$ or voltage = work (energy)/charge.

Current is the charge that flows, divided by time. Current is expressed as

$I = Q/t$ or current = charge/time.

Rearranging each of these equations shows the following:

$W = VQ$ or work = voltage × charge

and

$t = Q/I$ or time = charge/current.

Power is calculated as

$P = W/t$.

Substituting from the rearranged equations shows the following:

$P = (VQ)\,(I/Q)$ or power = (work × charge) × (current/charge).

Canceling terms gives the following power formula:

$P = VI$ or power = voltage × current.

To determine an actual quantity of electrical energy, a factor that indicates **how long** a power value continued must be used. Such a unit of electrical energy is called a **watt-second**. It is the product of watts (W) and time (in seconds). The watt-second is a very small quantity of energy. It is more common to measure electrical energy in **kilowatt-hours (kWh)**. It is the kWh quantity of electrical energy that is used to determine the amount of electric utility bills. A kilowatt-hour is 1000 W in 1 h of time or 3,600,000 W per second.

In terms of voltage and current, **power (P)** is equal to voltage (in volts) multiplied by current (in amperes). The formula is: $P = V \times I$. This formula is an easy way to find electrical power. For example, a 120-V electrical outlet with 4 A of current flowing from it has a power value of

$P = V \times I$

or

120 V × 4 A = 480 W.

The unit of electrical power is the **watt**. In the example, 480 W of power is converted by the load portion of the circuit. Another way to find power is

$P = V^2/R$.

This formula is used when voltage and resistance are known, but current is not known. The formula $P = I^2 \times R$ is used when current and resistance are known.

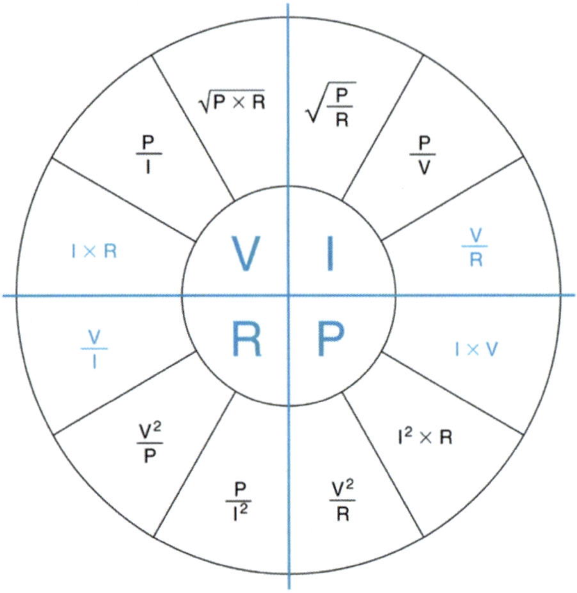

Figure 4-13 Formula circle to simplify calculating voltage, current, resistance, and power.

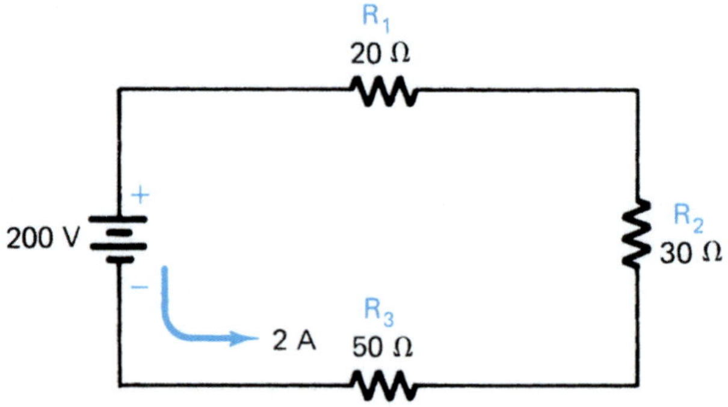

Figure 4-14 Power calculations.

Several formulas are summarized in **Figure 4-13**. The quantity in the center of the circle may be found by any of the three formulas along the outer part of the circle in the same part of the circle. This circle is handy to use for making electrical calculations for voltage, current, resistance, or power.

It is easy to find the amount of power converted by each of the resistors in a series circuit, such as the one shown in **Figure 4-14**. In the circuit shown, the amount of power converted by each of the resistors and the total power are found as follows:

1. Power converted by resistor R_1:
   $P_1 = I^2 \times R_1$
   $P_1 = 2^2 \times 20$ W
   $P_1 = 80$ W.

2. Power converted by resistor R_2:
   $P_2 = I^2 \times R_2$
   $P_2 = 2^2 \times 30$ W
   $P_2 = 120$ W.

3. Power converted by resistor R_3:
   $P_3 = I^2 \times R_3$.
   $P_3 = 2^2 \times 50$ W
   $P_3 = 200$ W.

4. Total power converted by the circuit:
   $P_T = P_1 + P_2 + P_3$
   $P_T = 80$ W + 120 W + 200 W
   $P_T = 400$ W

or

   $P_T = V_T \times I_T$
   $P_T = 200$ V × 2 A
   $P_T = 400$ W.

When working with electrical circuits, it is possible to check your results by using other formulas.

As an example, if an electrical heater operates on 120 V and has a resistance of 20 W, what is the cost to use the heater for 200 h at a cost of 5 cents per kWh?

1. $P = V_2/R = 120^2/20$ W = 14,400/20 W = 720 W = 0.72 kW.
2. There are 1000 W in a kilowatt (1000 W = 1 kW).
3. Multiply the kW that the heater has used by the hours of use:
   kW × 200 h = kilowatt-hours (kWh)
   0.72 × 200 h = 144 kWh.
4. Multiply the kWh by the cost:

   kWh × cost = 144 kWh × 0.05 = $7.20.

In some cases, it is desirable to determine the amount of power developed by a resistance when only its value and the amount of current flow are known. In this case, the quantity V must be eliminated from the equation and an equivalent value substituted. To determine power when the resistance and current are known, use the following power formula:

$P = VI.$

Because $V = IR$, we may substitute:

$P = (I \times R)\, I.$

Simplifying yields

$P = (I \times I)\, R.$

Because the value $I \times I$ is stated as I^2, this equation is written as

$P = I^2R$ or power = $(\text{current})^2$ (resistance).

4.4 Troubleshooting Series Circuits

Modern electronic systems are highly reliable but still subject to failure. An electronics technician must be capable of quickly and accurately **analyzing** a problem and making necessary repairs. It is necessary that a technician develop a logical, step-by-step approach to locate faults.

The first step in troubleshooting is to determine that a fault really exists. A technician must know the **purpose** of the electronic system and look for symptoms of trouble. Once the **symptom** is determined, possible **causes** must be considered and **tested** until the actual fault is located. The old saying that "if it is not broken, do not fix it" applies in troubleshooting. If no symptoms exist, testing should not be done.

Troubleshooting requires certain steps in the process. These steps consist of (1) determining from the operation of the equipment that a fault does exist; (2) checking the source voltage for its proper value; (3) determining the cause of zero, high, or reduced current; (4) making repairs to the system; and (5) checking the system for proper operation. Troubleshooting should involve an analytical approach. Through Ohm's law, the relationships among voltage, current, and resistance in a circuit are clear. The power equations make it possible to determine the amount of power developed in a circuit or in a component. Mathematical calculations make it possible to determine the theoretical circuit operation. Measurements in the circuit make it possible to determine if these values are true operational values.

In problems involving power conversion, two of the **senses** (sight and smell) are helpful. Overheated wiring or printed circuit boards often have a darkened appearance. This discoloration may be an indication of trouble.

This appearance should be carefully examined as a possible effect of a circuit problem. Resistors may overheat to the point where the surface looks charred. This makes it impossible to determine the color-coded value. A burned resistor may even break completely. A burned resistor may not be the actual trouble in a circuit but an effect of the trouble.

Normal **heating** in an electronic system does not produce charring or discoloration. A technician must learn the difference between normal heating and overheating. Some circuits operate at high temperatures. When troubleshooting electronic equipment, the sense of smell sometimes gives the first indication of trouble. Overheated insulation or resistors have a distinct odor which technicians will recognize. When this odor is perceived, it is a good idea to turn the equipment off as quickly as possible! If this is not done, serious damage to the equipment may occur. Components burning or smoking are usually the effect of a trouble, rather than the problem. Overheating occurs in a circuit when an overcurrent flows. An **overcurrent** is produced when the source voltage is too great, resistance is too great, or resistance is too small. The source voltage can be measured to determine its value. If it is correct, then something has happened to reduce the resistance. A possibility is a short-circuit path developed around a resistor. Once the trouble is determined, it must be corrected. Then the circuit must be checked.

Circuits operate through the control of electrical current. **Faults** occur that cause the current to be too low, too high, or zero. Once the fault is determined, possible causes must be analyzed. Ohm's law and the basic behavior of series circuits are used in the analysis.

The first step in troubleshooting circuits is checking the voltage source. Current flows as a result of the source. No circuit will operate without the proper source voltage. Also, be sure that the power switch is on. Batteries, when used as voltage sources, must be tested under full-load conditions.

Open Circuits

In a series circuit, zero current indicates that the circuit is open. An **open circuit** can be found using a voltmeter. A **voltage (IR) drop** is developed when current flows through a resistor. So, with no current flow, **all IR** drops are zero. The voltage across an open resistor will equal the source voltage. An open circuit may be caused by a broken wire or printed circuit board (PCB) strip. It may also be caused by a component that has overheated. This trouble is corrected by replacing the open component or repairing the break in the printed circuit board.

When a series circuit is opened, there is no longer a path for current flow. The circuit will not operate. In the circuit of **Figure 4-15**, if lamp 1 is burned out, its filament is **open**.

Because a series circuit has only one current path, that path is broken. No current flows in the circuit. Lamp 2 will not work either. If one light burns out, the others will go out also because the series current path is opened.

Short Circuits

If the current is too high in a series circuit, the resistance is too low. **Low resistance** can be caused by a shorted component or an alternate current path. A low-resistance alternate path may be caused by corrosion, solder bits, or wire clippings. A shorted resistor will have no voltage drop, while the other series components have voltage drops larger than normal. This problem can be corrected by replacing the shorted resistor or removing the alternate path. Solid-state components, such as transistors, which are discussed later, are more likely to short than resistors. Resistors sometimes burn open due to high currents.

Changed Resistor Values

If the current in a series circuit is too low, the resistance is too high. A possible cause is a resistor that has **increased** in value. This increase is usually caused by overheating. If a resistor increases in value, it has a larger voltage drop than normal. Others have lower voltage drops. This trouble is remedied by replacing the component that has changed value. The replacement should be exactly like the original. Never replace a resistor with one that has greater tolerance. The replacement should also have the same or higher wattage rating as the original.

Self-examination

1. The filament (coil) of a lamp (bulb) has burned out. If this lamp were part of many lamps in a series circuit, would all lamps go out?
2. Is a blown fuse considered a short circuit?
3. What voltage will a troubleshooter find at a good fuse?
4. What voltage will a troubleshooter find at a blown fuse?
5. Something has happened to a series circuit containing many resistors and a lamp. The troubleshooter notices the lamp is brighter than when

the circuit was operating normally. Is the troubleshooter suspect of a short or open component?

Summary

- Some simple series electrical circuit examples have been discussed in this chapter. They become easy to understand after practice with each type of circuit. It is important to understand the characteristics of series, parallel, and combination circuits.
- Current has specific rules it follows in a series circuit.
- Voltage has specific rules it follows in a series circuit.
- Resistance has specific rules it follows in a series circuit.
- Power has specific rules it follows in a series circuit.
- When troubleshooting, shorted and open components behave in a particular manner in series circuits.
- A series circuit has only one path for current.
- Because there is only one path for current, all current in a series circuit is equal.
- All the voltages in a series circuit add to the total voltage.
- All the resistances in a series circuit add to the total resistance.
- All the powers in a series circuit add to the total power.

Formulas

$I_T = I_1 = I_2 = I_3 = I_n$
$R_T = R_1 + R_2 + R_3 + R_n$
$V_T = V_1 + V_2 + V_3 + V_n$
$P_T = P_1 + P_2 + P_3 + P_n$

Self-examination/Answers

1. State Ohm's law for voltage.
2. State Ohm's law for current.
3. State Ohm's law for resistance.
4. What does the abbreviation R_T stand for?
5. What does the abbreviation V_T stand for?
6. State the way in which a series circuit is identified in relationship to current.
7. State the mathematical relationship of current in all devices found in a series circuit.

8. State the mathematical relationship of voltages across all devices in a series circuit.
9. State the mathematical relationship of resistance in a series circuit.
10. Work can be defined as _____.
11. Energy can be defined as _____.
12. Electrical power is measured in units of _____.
13. One horsepower is equal to ____ W.
14. The filament (coil) of a lamp (bulb) has burned out. If this lamp were part of many lamps in a series circuit, would all lamps go out?
15. Is a blown fuse considered a short circuit?
16. What voltage will a troubleshooter find at a good fuse?
17. What voltage will a troubleshooter find at a blown fuse?
18. Something has happened to a series circuit containing many resistors and a lamp. The troubleshooter notices the lamp is brighter than when the circuit was operating normally. Is the troubleshooter suspect of a short or open component?
19. What is the relationship of voltage and current in a circuit?
20. What is the relationship of resistance and current in a circuit?

Answers

1. $V = I \times R$, $I = V/R$, $R = V/I$
2. $I = V/R$
3. $R = V/I$
4. The total resistance
5. The total voltage
6. One path for current flow
7. Current stays the same or is equal everywhere in the circuit
8. Voltage drops at all components add to the source (or total) voltage
9. Resistances of all components add to the total resistance
10. Process of converting energy from one form to another
11. The ability to do work
12. Watts
13. 746
14. Yes, all lamps would go out
15. No, a blown fuse is an open circuit
16. Nearly 0 V
17. The source (total) voltage

18. Likely a resistor has some short
19. Voltage increases cause current increases (direct proportion)
20. Resistance increases cause current decreases (inverse proportion)

Glossary

Current
The movement of electrical charge; the flow of electrons through an electrical circuit.

Difference in potential
The voltage across two points of a circuit.

Equivalent resistance
A resistance value that would be the same value in a circuit as two or more parallel resistances of a circuit.

Ohm's law
The law that explains the relationship of voltage, current, and resistance in electrical circuits.

Power (P)
The rate of doing work in electrical circuits, found by using the equation $P = I \times V$.

Reciprocal
1 divided by a value, such as 1/R.

Resistance (R)
Opposition to the flow of current in an electrical circuit; its unit of measurement is the ohm (W).

Series circuit
A circuit that has one path for current flow.

Total current
The current that flows from the voltage source of a circuit.

Total resistance
The total opposition to current flow of a circuit, which may be found by removing the voltage source and connecting an ohmmeter across the points where the source was connected.

Total voltage
The voltage supplied by a source.

Voltage (V)
The electrical force or pressure that causes current to flow in a circuit.

Voltage drop
The voltage across two points of a circuit, found by using the equation $V = I \times R$.

Watt (W)
The unit of measurement of electrical power; the amount of power converted when 1 A of current flows under a pressure of 1 V.

5

Parallel Circuits

In the previous chapter, it was noted that there are at least three types of electrical circuits: series, parallel, and circuits that combine characteristics of both parallel and series. The previous chapter identified series circuits by the number of paths for current. **Parallel circuits** may have many paths for current to flow but have only one voltage. In the previous chapter, all of the rules and formula regarding how current, voltage, resistance, and power behaved in a series circuit. This chapter will identify how the same characteristics (current, voltage, resistance, and power) function in a parallel circuit.

Objectives

Upon the completion of this chapter, you should be able to:

1. Define a parallel circuit
2. Solve parallel circuit problems finding current
4. Solve parallel circuit problems finding resistanceSolve parallel circuit problems finding voltage
8. Solve parallel circuit problems finding powerDefine voltage drop in a circuitSolve parallel circuit problems

Chapter Outline

5.1 Characteristics of a parallel circuit
5.2 Applying Ohm's law to parallel circuits
5.3 Applying the power formula to parallel circuits
5.4 Parallel circuit measurements
5.5 Troubleshooting parallel circuits

5.1 Characteristics of a Parallel Circuit

Parallel circuits are different from series circuits in several ways. A parallel circuit has two or more paths for current to flow from the voltage source. In **Figure 5-1**, path 1 is from the negative side of the voltage source, through R_1, and back to the positive side of the voltage source. Path 2 is from the negative side of the voltage source, through R_2, and back to the positive side of the voltage source. Also, path 3 is from the negative side of the power supply, through R_3, and back to the voltage source.

Voltage in a Parallel Circuit

In a parallel circuit, the **voltage is the same** across every component of the circuit. In **Figure 5-1**, the voltage across points A and B is equal to 10 V. This is the value of the voltage applied to the circuit. By following point A to point C, it can be seen that these two points are connected together. Point B and point D are also connected. So, the voltage from points A to B will be the same as the voltage from points C to D. From these findings, we can determine the rule for voltage in a parallel circuit. That formula is $V_T = V_1 = V_2 = V_3 = V_n$

where V_T is the voltage total, and the voltage across each component t in the parallel circuit is represented by V_1, V_2, and V_3 for as many components present in the circuit (V_n).

Current in a Parallel Circuit

Another characteristic of a parallel circuit is that the sum of the currents through each path equals the total current that flows from the voltage source.

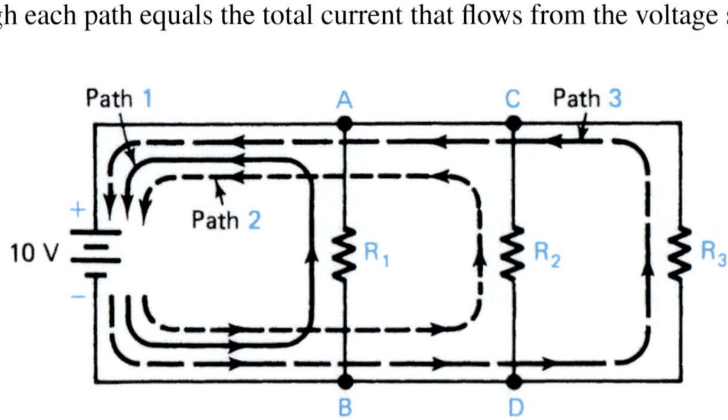

Figure 5-1 Parallel electrical circuit.

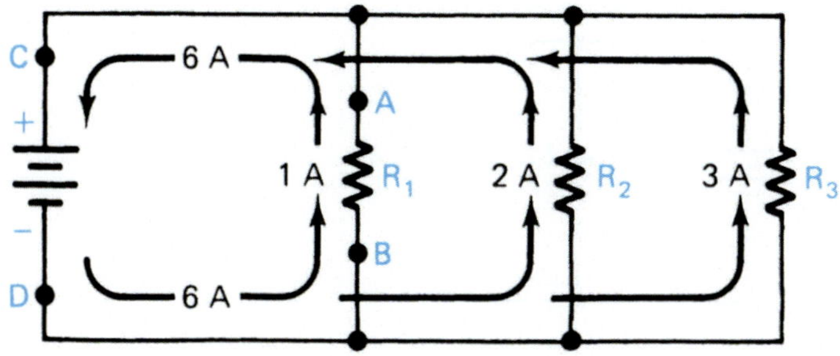

Figure 5-2 Current flow in a parallel circuit.

In **Figure 5-2**, the currents through the paths are 1, 2, and 3 A. One ampere of current could be measured through R_1 at point A or point B. Two amperes of current could be measured through R_2 and 3 A through R_3. The total current is equal to 6 A. This value of total current could be measured at point C or point D in the circuit. Remember that the current is the same in every part of a series circuit, and current **divides** through each branch in a parallel circuit. More resistance causes less current to flow through a parallel **branch**. A branch is a parallel path through a circuit.

Based on this example, the formula for current flow in a parallel circuit is$I_T = I_1 + I_2 + I_3 + I_n$

where I_T is the current total, and the current through each component t in the parallel circuit is represented by I_1, I_2, and I_3 for as many components present in the circuit (I_n).

Resistance in a Parallel Circuit

The **total resistance** of a parallel circuit is more difficult to calculate than for a series circuit. The formula used is

$$1/R_T = 1/R_1 + 1/R_2 + 1/R_3 + \cdots.$$

This is called an inverse or **reciprocal formula**. Refer to the example of **Figure 5-3**. When trying to find the total resistance of a parallel circuit, first write the formula. Note that 1 is divided by each value. Next, divide each resistance value by 1 and write the values. Then add these values to get a value of 1 divided by total resistance. Do not forget to divide the value obtained into 1 to find the total resistance of the circuit. Also, always remember that the total resistance in a parallel circuit is **less** than any individual resistance in the circuit. In the example shown, 1.33 W is less than 2 or 4 W.

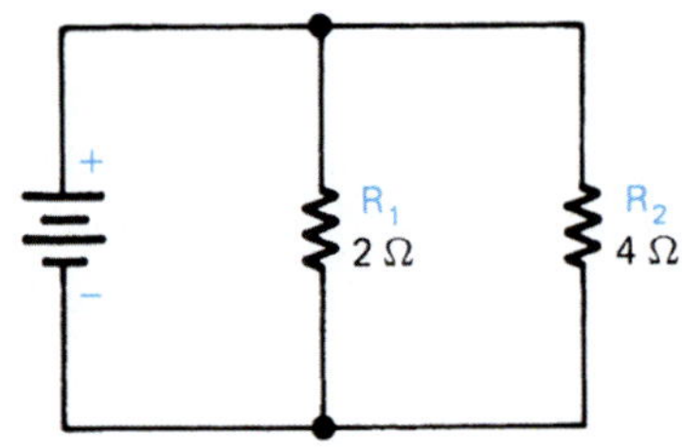

$$\frac{1}{R_T} = \frac{1}{R_1} + \frac{1}{R_2} + \ldots$$
$$= \frac{1}{2} + \frac{1}{4}$$
$$= 0.5 + 0.25$$
$$= 0.75$$
$$R_T = \frac{1}{0.75} = 1.33\ \Omega$$

or

when there are only *two* resistances:

$$R_T = \frac{R_1 \times R_2}{R_1 + R_2} = \frac{2\ \Omega \times 4\ \Omega}{2\ \Omega + 4\ \Omega} = \frac{8}{6} = 1.33\ \Omega$$

Figure 5-3 Finding total resistance of a parallel circuit.

If there are **only two paths** in a parallel circuit, it is easier to find the total resistance. This formula can be used:

$$R_T = R_1 \times R_2/R_1 + R_2.$$

Note that these two resistances are multiplied and then added. The product of the two is put at the top of the formula and the sum is put at the bottom. When using this formula, it is not necessary to divide the resistance values into 1. Remember that this formula may be used only when there are two resistances in a parallel circuit. If there are more than two resistances, the reciprocal formula must be used.

Another simple method of finding **total resistance** in a parallel circuit is when **all the resistance values are the same**. An example of a parallel circuit with all resistances being the same is a circuit of lights connected in parallel. Each lamp has the same resistance. When all resistances are equal, to find total resistance, divide the resistance value of each resistor by the number of paths (refer to **Figure 5-4**). If five 20-W resistors are connected in parallel, the total resistance is equal to 20 divided by 5, or 4 W.

When one of the components of a parallel path is **opened**, the rest of the circuit will continue to operate. Remember that in a series circuit, when one component is opened, no current will flow in the circuit. Since the same voltage is applied to all parts of a parallel circuit, the circuit will operate unless the path from the voltage source is broken (refer to **Figure 5-5**). If lamp 3 has a filament burned out, this causes an open circuit. However, lamps 1 and 2 will continue to operate. Some types of Christmas tree lights are connected in parallel. If one of these lights burns out, the rest of them will still burn.

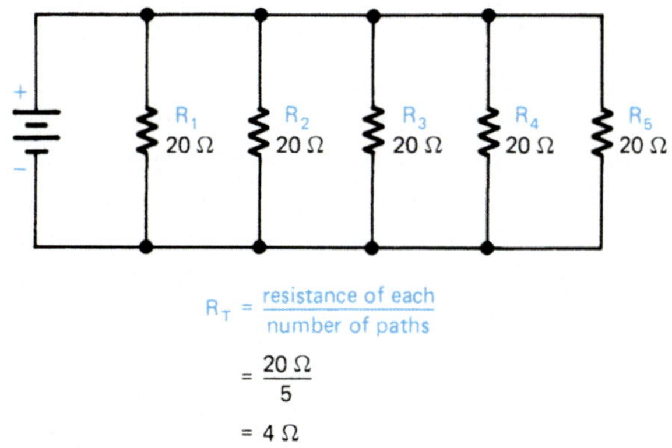

Figure 5-4 Finding total resistance when all resistances are the same.

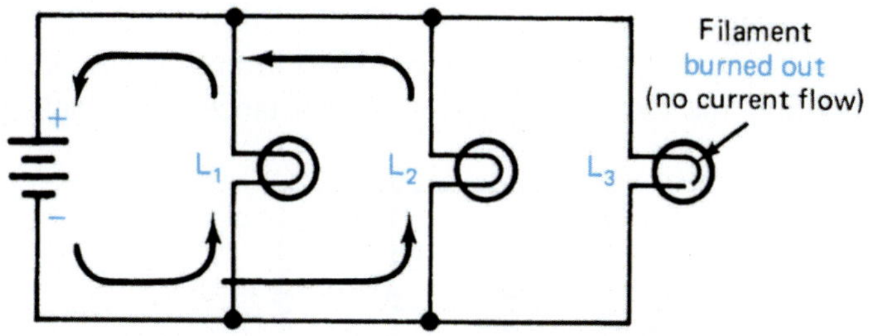

Figure 5-5 Three lamps connected in parallel.

Power in Parallel Circuits

Power in parallel circuits behaves exactly the same as in series circuits, and it adds.

$P_T = P_1 + P_2 + P_3 + P_n$.

Please refer to Section 5.1.4 for the application of this formula.

Section 5.1: Review Questions

5.1.1 State the voltage formula for parallel circuits
5.1.2 State the current for parallel circuits
5.1.3 State the resistance formula for parallel circuits
5.1.4 State the power formula for parallel circuits

5.2 Applying Ohm's Law to Parallel Circuits

Because of the multiple paths for current to follow, particular attention must be paid to each path and component to properly apply Ohm's law. A sample problem with a parallel circuit is shown in **Figure 5-6**. This circuit has a source voltage of 10 V and resistors of 5, 10, and 20 W. First, find the total resistance of the circuit. The **total resistance** of this circuit is 2.85 W. The total opposition to current flow which this circuit has is 2.85 W.

Next, find the **current** that flows from the voltage source. In this circuit, the total current is found by dividing the voltage by the total resistance. So, 10 V divided by 2.85 W = 3.5 A of current. This means that 3.5 A of current flows from the voltage source and divides into the three paths of the circuit.

Now, find the **current through each path** of the circuit. The current from the voltage source divides into each path of the circuit. A lower resistance in a path causes more current to flow through the path. Current is equal to voltage divided by resistance. The current through each path is found by dividing the voltage (10 V) by the resistance of the path. Note that

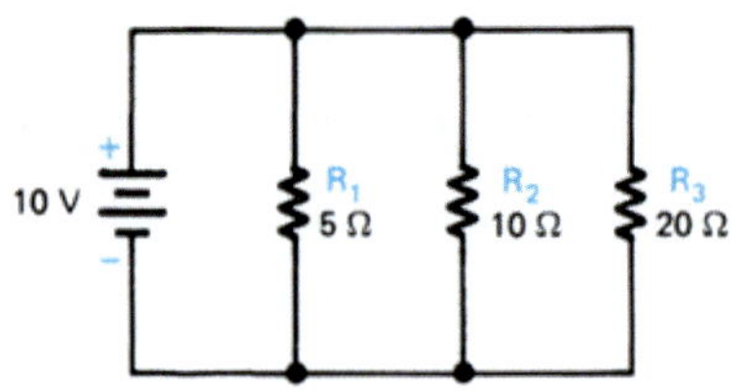

$$\frac{1}{R_T} = \frac{1}{R_1} + \frac{1}{R_2} + \frac{1}{R_3}$$
$$= \frac{1}{5} + \frac{1}{10} + \frac{1}{20}$$
$$= 0.2 + 0.1 + 0.05$$
$$= 0.35$$
$$R_T = \frac{1}{0.35} = 2.85\ \Omega$$
$$I_T = \frac{V}{R_T} = \frac{10\ \text{V}}{2.85\ \Omega} = 3.5\ \text{A}$$
$$I_1 = \frac{V}{R_1} = \frac{10\ \text{V}}{5\ \Omega} = 2\ \text{A}$$
$$I_2 = \frac{V}{R_2} = \frac{10\ \text{V}}{10\ \Omega} = 1\ \text{A}$$
$$I_3 = \frac{V}{R_3} = \frac{10\ \text{V}}{20\ \Omega} = 0.5\ \text{A}$$
$$I_T = I_1 + I_2 + I_3$$
$$= 2\ \text{A} + 1\ \text{A} + 0.5\ \text{A}$$
$$= 3.5\ \text{A}$$

Figure 5-6 Sample parallel circuit problem.

the same voltage is across each path. The current values are 2, 1, and 0.5 A. These are added to equal the total current. The total current of 3.5 A is equal to 2 A × 1 A × 0.5 A.

5.3 Applying the Power Formula to Parallel Circuits

Power in parallel circuits is found in the same way as for series circuits; it adds to total power. In the circuit of **Figure 5-6**, let us change the value of R_3 to 10 Ω and calculate the power converted by each of the resistors. The total power of the parallel circuit are found as follows:

1. Power converted by resistor R_1:
   $P_1 = V_2/R_1 = 30^2/5 = 900/5 = 180$ W.
2. Power converted by resistor R_2:
   $P_2 = V_2/R_2 = 30^2/10 = 900/10 = 90$ W.
3. Power converted by resistor R_3:
   $P_3 = V_2/R_3 = 30^2/10 = 900/10 = 90$ W.
4. Total power converted by the circuit:
   $P_T = P_1 + P_2 + P_3 = 180$ W + 90 W + 45 W = 315 W.

5.4 Parallel Circuit Measurements

For example, please look at the circuit displayed in **Figure 5-8**.

There are two possible paths for current to flow. To measure current through path 1 in the parallel circuit of **Figure 5-7**, use the following procedure:

1. Make sure that no voltage is applied to the circuit by opening the switch.
2. Remove wires 1 and 2 from point A.
3. Set the meter to the highest current range.
4. Connect wires 1 and 2 to the positive test lead of the meter.
5. Connect the negative test lead of the meter to point A.
6. Turn on the switch to apply voltage to the circuit.
7. Adjust the meter, if necessary, to a lower range to get an accurate reading.
8. Read the current value on the scale of the meter.

To measure the resistance of a parallel circuit, first **remove** the voltage source. Prepare the meter to measure resistance. Be sure to zero the meter. Connect the meter across the points where the circuit **was** connected to the

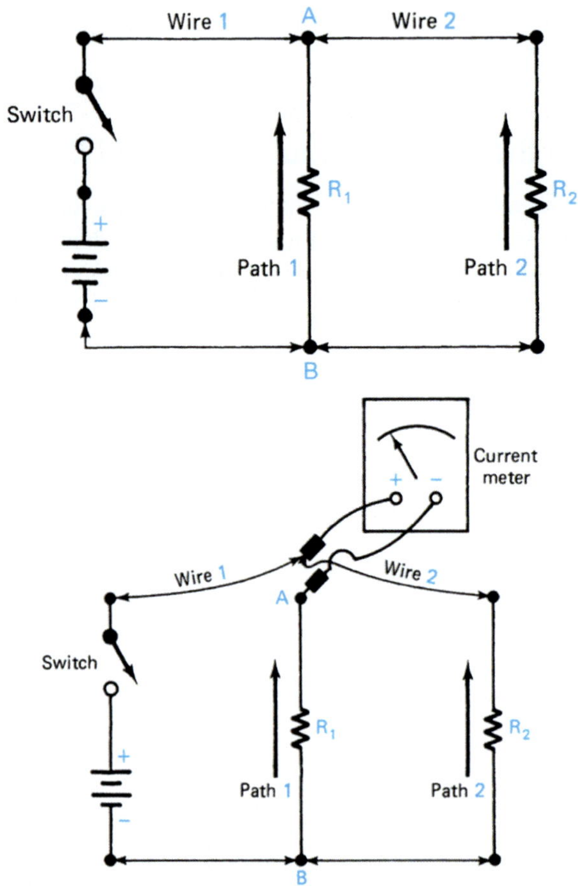

Figure 5-7 Making measurements in a parallel circuit: (a) original circuit; (b) circuit set up to measure current through path 1.

voltage source [points A and B in **Figure 5-7(a)**]. Adjust the meter range, if necessary, to get an accurate resistance reading. Be sure to zero the meter each time a change of ranges is made. Once the proper range is selected, accurately read the resistance on the meter scale.

Parallel circuits have more than one current path connected to the same voltage source. A parallel-circuit example is shown in **Figure 5-8**. Start at the voltage source (V_T) and look at the circuit. Two complete current paths are formed. One path is from the voltage source, through resistor R_1, and back to the voltage source. The other path is from the voltage source, through resistor R_2, and back to the voltage source. Remember that the **source voltage** divides

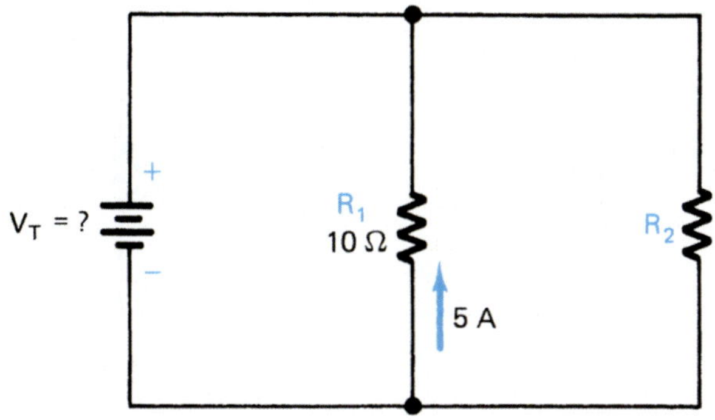

Figure 5-8 Finding power values in a parallel circuit.

across the resistors in a series circuit. In a parallel circuit, the same voltage is across all the resistors.

Assume that the current through resistor R_1 in **Figure 5-8** is 5 A and the value of the resistor is 10 W. The voltage across resistors R_1 and R_2 and the source voltage (V_T) may be found. Using Ohm's law, the voltage across resistor R_1 may be found:

$$V_T = I_1 \times R_1 = 5\text{A} \times 10\text{ W} = 50\text{ V}.$$

Because this is a parallel circuit, the voltages are the same. The source voltage in this circuit is equal to 50 V. Ohm's law shows that the current in a circuit is **inversely proportional** to the resistance of the circuit. This means that when resistance increases, current decreases. The division of current in parallel-circuit paths is based on this fact. Remember that in a series circuit, the current is the same through each part. The total current in a parallel circuit divides through the paths based on the value of resistance in each path.

The flow of **current** in a parallel circuit is shown in **Figure 5-9**. **Figure 5-9(a)** shows a **series** circuit. The total current passes through one resistor (R_1). The amount of current is

$$I_1 = V_T/R_1 = 10\text{ V}/10\text{ W} = 1.0\text{ A}.$$

Figure 5-9(b) shows the same circuit with another resistor (R_2) connected in parallel across the voltage source. The current through R_2 is the same as that through R_1 because their resistances are equal:

$$I_2 = V_T/R_2 = 10\text{ V}/10\text{ W} = 1.0\text{ A}.$$

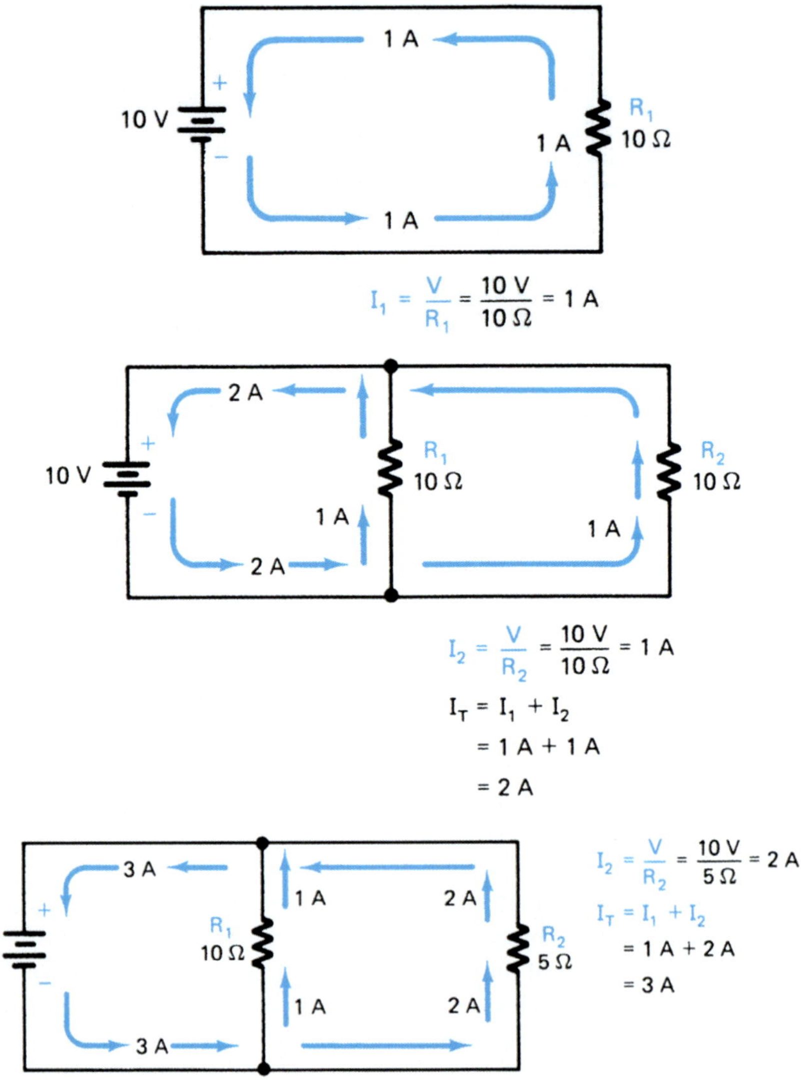

Figure 5-9 Current flow in a parallel circuit: (a) one path; (b) two paths; (c) R_2 changed to 5 Ω.

Because 1 A of current flows through each of the two resistors, a total current of 2 A flows from the source. The division of currents through the resistors is shown here. Each current path in the parallel circuit is called a **branch**. Each branch carries part of the total current which flows from the source. Changing the value of any resistor in a parallel circuit has no effect on the current in the other branches. However, a change in resistance does affect the total current. If R_2 of **Figure 5-9(b)** is changed to 5 W [**Figure 5-9(c)**], the total current would increase:

$I_1 = V_T/R_1 = 10$ V/10 W = 1.0 A
$I_2 = V_T/R_2 = 10$ V/5 W = 2.0 A
$I_T = I_1 + I_2 = 1.0$ A + 2.0 A = 3.0 A.

The **total resistance** of a parallel circuit is **not** equal to the sum of the resistors as in series circuits. When more branches are added to a parallel circuit, the total current (I_T) increases. In a parallel circuit, the total resistance is **less than any of the branch resistances**. As more parallel resistances are added, the total resistance of the circuit decreases.

There are several ways to find the **total resistance** of parallel circuits. The method used depends on the type of circuit. Some common methods for finding total parallel resistance are shown here.

1. **Equal resistors**: When two or more equal-value resistors are connected in parallel, their total resistance (R_T) is

   $$R_T = \frac{\text{value of one resistance}}{\text{number of paths}}$$

   If four 20-W resistors are connected in parallel, their total resistance is 20 divided by 4 = 5 W; also, two 50-W resistors in parallel – R_T = 50 divided by 2 = 25 W.
2. **Product over the sum**: Another shortcut for finding total resistance of two parallel resistors is called the product-over-sum method. This method is
   $R_T = R_1 \times R_2/R_1 + R_2$.
   For example, to find the total resistance (R_T) of a 10-W and a 20-W resistor connected in parallel:
   $R_T = R_1 \times R_2/R_1 + R_2 = 10 \times 20/10 + 20 = 200/30 = 6.67$ W.
   Note that the total resistance of a 10-W and a 20-W resistor is *less than the smallest* individual resistor. The product-over-sum method can be used only for *two* resistances in parallel.
3. **Reciprocal method**: Most circuits have more than two resistors of unequal value. The reciprocal method must then be used to find total resistance. The reciprocal method is as follows:
   $1/R_T = 1/R_1 + 1/R_2 + 1/R_3 + \cdots$ (continued for the number of resistances in the circuit). For example, if 1-, 2-, 3-, and 4-Ω resistors are connected in parallel:

$$
\begin{aligned}
1/R_T &= 1/1 + 1/2 + 1/3 + 1/4 \\
&= 1 + 0.5 + 0.33 + 0.25 \\
&= 2.08. \\
R_T &= 1/2.08 \\
&= 0.48\ \text{W}.
\end{aligned}
$$

5.5 Troubleshooting Parallel Circuits

The basic principles of **troubleshooting** for series circuits can also be applied to parallel circuits. A parallel-circuit failure also results in a current that is too high, too low, or zero. The first step in troubleshooting is to check the voltage source. Check for blown fuses, a tripped circuit breaker, or a disconnected power cord. No circuit will operate without the proper source voltage. Zero current is the result of an open circuit. The only other cause of zero current flow in a parallel circuit is if all loads are open.

High current is caused by resistance that is too low. The resistance of a parallel circuit decreases as paths for current are **added**. If any component in a parallel circuit "**shorts**," then a short is across **all** components. The current would then become extremely high. The circuit protection device (circuit breaker or fuse) would then open to protect the source from damage. Shorted resistors are uncommon. A cause of high current could be a resistor that has decreased in value. In this case, one parallel branch conducts much greater current than normal. This increases the total current. It is necessary to measure the branch currents to determine which resistor has changed value. The faulty resistor can then be replaced to correct the trouble.

Low current is likely to be caused by a resistor that has opened. The full source voltage is measured across an open resistor. This measurement is useless in a parallel circuit as the source voltage is across all resistors. It is necessary to make current measurements to locate the fault.

Note: If measuring the resistance of a branch, it is necessary to disconnect one end of the resistor. Otherwise, the parallel combination of other branches gives a false reading on an ohmmeter.

A parallel circuit with four 40-Ω resistors is being tested. Current in the first resistor is significantly higher than the other three. What is likely the cause of the problem?

Summary

- There are several important characteristics of parallel circuits. Remember these basic rules for parallel circuits:

- There are two or more paths for current flow.
- Voltage is the same across each component of the circuit.
- The sum of the currents through each path is equal to the total current that flows from the source.
- Total resistance is found by using the following formula: $1/R_T = 1/R_1 + 1/R_2 + 1/R_3 + \cdots$.
- If one of the parallel paths is broken, current will continue to flow in all the other paths.

Formulas

$$V_T = V_1 = V_2 = V_3 = V_n$$
$$1/R_T = 1/R_1 + 1/R_2 + 1/R_3 + \cdots I_T = I_1 + I_2 + I_3 + I_n$$
$$P_T = P_1 + P_2 + P_3 + P_n$$

Self-examination/Answers

1. State the voltage formula for parallel circuits.
2. State the current for parallel circuits.
3. State the resistance formula for parallel circuits.
4. State the power formula for parallel circuits.
5. T/F. A large resistor will have a large voltage drop in a parallel circuit. Why?
7. T/F. If a parallel circuit has three paths, the path with the smallest resistance will have the greatest current. Why?
9. T/F. When making a voltage measurement on a parallel circuit, all voltages should sum to the source voltage? Why or why not?
10. T/F. A one-million (MΩ) resistor is in parallel with a one-thousand (KΩ) and a one (1 Ω) resistor, the total resistance of the circuit is larger than 1 million ohms. Why or why not?
12. T/F. Current in parallel circuits is measured in series. Why or why not?
13. What is a parallel circuit?
14. What are the voltage, current, and resistance characteristics of parallel circuits?
15. Explain the three ways used to find total resistance of parallel circuits.
16. How is total voltage of a parallel circuit measured?
17. How is total current of a parallel circuit measured?
18. How is total power of a parallel circuit measured?

Answers

1. Voltages are equal
2. Current adds
3. Reciprocal formula
4. Power adds
5. False, voltages are equal in a parallel circuit
6. True, current takes path of least resistance
7. False, voltages are equal in a parallel circuit
8. True, current takes path of least resistance
9. False. Current is always measured in series, no matter what kind of circuit is being tested
10. A circuit with multiple branches connected across one source
11. Voltage is the same, current adds, and resistance needs special formula
12. Reciprocal formula, product over sum, and division *b* number of resistors (assuming all resistors are of equal size)
13. Voltage is measured in parallel across the component being tested
14. Current is measured in series with the component being tested
15. Resistance is measured with the component isolated from the parallel circuit

Summary

- Parallel circuits are commonly used
- Parallel circuits have two or more paths
- Parallel circuits have the same voltage across each path
- Parallel circuits do not have to have equal current in each path
- Parallel circuits with equal resistors have the same current in each path
- Calculating resistance in parallel circuits requires a special formula

Glossary

Branch
A path of a parallel circuit.

Branch current
The current through a parallel branch.

Branch resistance
The total resistance of a parallel branch

Branch voltage
The voltage across a parallel branch

Circuit
A path through which electrical current flows.

Inverse
The value of 1 divided by some quantity, such as $1/R_T$ for finding parallel resistance.

Parallel circuit
A circuit that has two or more current paths.

Reciprocal.
See inverse.

6

Series–Parallel Circuits and Applications

Electrical circuits are basic to all electrical systems. To understand electricity and electronics, it is necessary to know how to apply basic electrical theory. Electricity and electronics is a somewhat mathematical discipline. The mathematics is easy to understand because it has practical applications. The basic theory used is called Ohm's law, which is important because it applies to the basic theory of electrical circuits. All examples in this chapter are direct-current (DC) circuits. Alternating-current (AC) circuits are more complex. AC circuits are studied in subsequent chapters of this book.

Objectives

Upon completion of this chapter, you will be able to:

1. Solve basic electrical problems using a calculator
2. Define Ohm's law and the power equation
3. Solve problems finding current, voltage, and resistance
4. Calculate power using the proper power formulas
5. Define voltage drop in a circuit
6. Solve circuit problems with resistors in different configurations
7. Define maximum power transfer in electrical circuits
8. Design a voltage-divider circuit by calculating proper values of resistance
9. Solve electrical circuit problems using Kirchhoff's voltage law, superposition, equivalent circuit methods, and bridge-circuit simplification

Chapter Outline

6.5 Specialized Circuit Applications
6.6 Problem-solving Methods

6.1 Combination Electrical Circuits

Combination electrical circuits consist of both series and parallel parts. They are sometimes called **series–parallel circuits**. Almost all electrical equipment has combination circuits rather than only series circuits or only parallel circuits. However, it is important to understand series *and* parallel circuits in order to work with combination circuits.

A **combination circuit** has both a series part and a parallel part; so remember that **a series circuit** has only **one path** for current flow from the voltage source and a **parallel circuit** has **two or more paths**. In the circuit shown in **Figure 6-1**, R_1 is in series with the voltage source and R_2 and R_3 are in parallel. There are many different types of combination circuits. Some have only one series component and many parallel components. Others have many series components and only a few parallel components.

In the circuit shown in **Figure 6-2**, R_1 and R_2 are in **series** with the voltage source. R_3–R_4 and R_5–R_6 are in **parallel**. Note the two paths in each parallel part of the circuit. The total current of the circuit flows through each series part of the circuit. In this circuit, the current is the same through resistors R_1 and R_2.

Another combination circuit is shown in **Figure 6-3**. R_1, R_2, and R_3 are in **series** with the voltage source. The **total current** *(I_T)* flows through R_1, R_2, and R_3. At point A, the current divides through the **parallel paths** of R_4, R_5, and R_6. The currents I_4, I_5, and I_6 flow through the parallel paths.

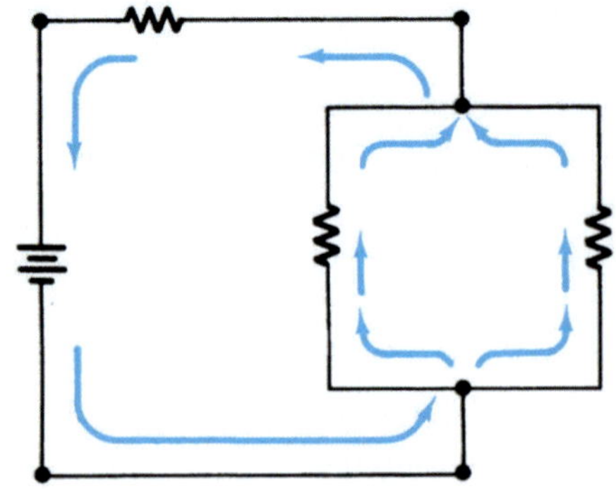

Figure 6-1 Simple combination circuit.

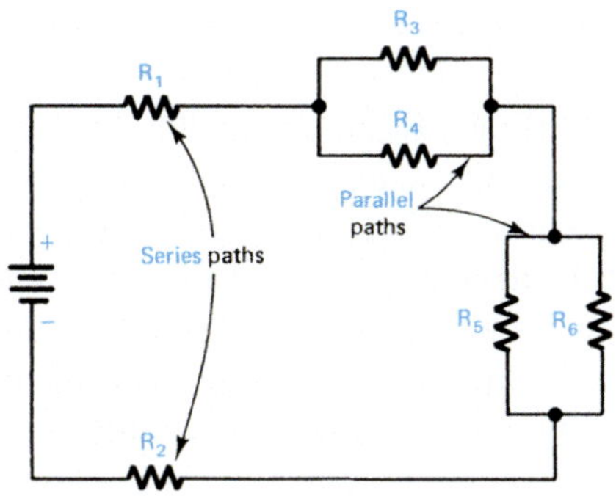

Figure 6-2 Combination circuit.

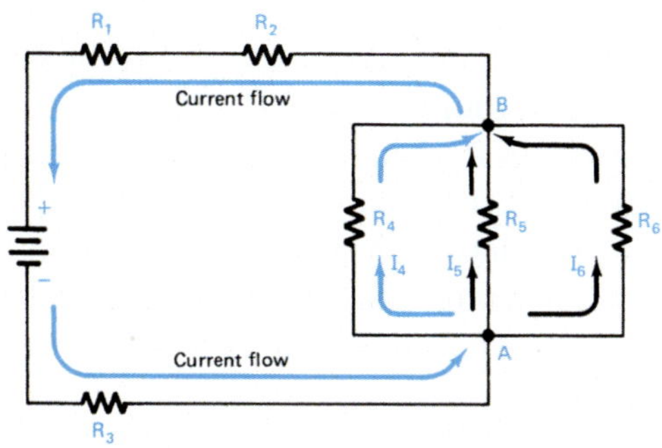

Figure 6-3 Current paths in a combination circuit.

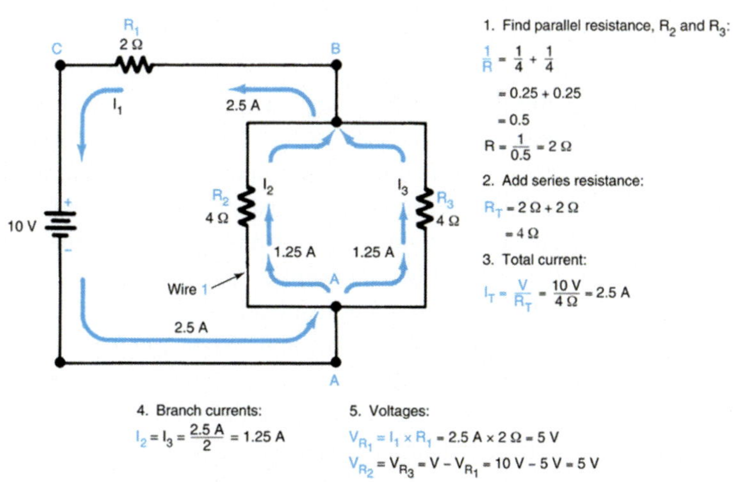

Figure 6-4 Combination circuit example.

To find the **total resistance** of a combination circuit, the series resistance is added to the parallel resistance. In the circuit shown in **Figure 6-4**, R_1 is the only series part of the circuit. R_2 and R_3 are in parallel. First, the parallel resistance of R_2 and R_3 is found. Then the series resistance and the parallel resistance are added to find the total resistance. Total resistance of this circuit is 4 Ω.

To find the **total current** in the circuit, the same method that was discussed before is used. Current is equal to voltage divided by resistance. The current that flows from the voltage source in this circuit is 2.5 A. The total current also flows through each series part of the circuit. So, 2.5 A flows through R_1. In this circuit, it is easy to find the current flow through R_2 and R_3. Because R_2 and R_3 are equal resistance values, the same current flows through each of them. There is 2.5 A of current flowing to point A of the circuit. This 2.5 A divides into two paths through R_2 and R_3. To find the current through R_2 and R_3, divide the current coming into point A (2.5 A) by the number of paths (2). The current through R_2 **and** R_3 is 1.25 A.

To find **voltage across R$_1$**, multiply the current through R_1 by the value of R_1. The voltage across R_1 is equal to 2.5 A times 2 W, or 5 V. This is the voltage from point C to point B. The voltage across R_2 or R_1 is also across points B and A. The voltage across points B and A is found by subtracting the voltage across R_1 from the applied voltage. The voltage across R_2 and R_3 is equal to 10 – 5 V, or 5 V.

6.2 Combination Circuit Measurements

To measure the **total resistance** of the circuit of **Figure 6-4**, first remove the voltage source from the circuit. Prepare the meter to measure resistance. Be sure to zero the meter. Connect the meter across points A and C. These points are where the voltage source **was** connected into the circuit. Adjust the meter range if necessary to get an accurate resistance reading. Once the proper range is selected, read the measured resistance on the meter scale.

To measure the **current** through R_1, first make sure that no voltage is applied to the circuit. *R*emove the wire at point C. Set the meter to the highest current range. Connect the negative meter test lead to point C and the positive test lead to the positive power source terminal. Now apply voltage to the circuit. Adjust the meter, if necessary, to a lower range to get an accurate current reading. Then read the current value on the scale of the meter. The current through R_1 should equal 2.5 A. This value is the same as the total current (I_T) of the circuit because R_1 is a series resistance.

To measure the **current through R_2** in this combination circuit, use the same procedure as for any parallel path. Use the following procedure:

- Make sure that no voltage is applied to the circuit.
- Remove wire 1 from point A.
- Set the meter to its highest current range.
- Connect wire 1 to the positive test lead of the meter.
- Connect the negative test lead of the meter to point A.
- Apply voltage to the circuit.
- Adjust the meter, if necessary, to a lower current range to get an accurate reading.
- Read the current value on the scale of the meter. The current through R_2 should equal
- 1.25 A.

To measure the **voltage across R_1**, connect the negative meter lead to point B and the positive lead to point C. The voltage should equal 5 V. The voltage across R_2 equals the voltage across R_3 since they are in parallel. The negative meter lead is connected to point A and the positive lead to point B. The voltage across R_2 and R_3 should equal 5 V.

6.3 Kirchhoff's Laws

Ohm's law shows the relationship of voltage, current, and resistance in electrical circuits. **Kirchhoff's voltage law (KVL)** is also important in solving electrical problems. A German scientist named Gustav Kirchhoff is given credit for discovering this effect. He found that the sum of voltage drops around any closed-circuit loop must equal the voltage applied to that loop. **Figure 6-5(a)** shows a simple series circuit to illustrate this law. This law holds true for any series-circuit loop.

Kirchhoff's current law (KCL*)* is also important in solving electrical problems, especially for parallel circuits. The current law states that at any junction of electrical conductors in a circuit, the total amount of current entering the junction must equal the amount of current leaving the junction. **Figure 6-5(b)** shows some examples of Kirchhoff's current law. The use of KVL and KCL as circuit problem-solving techniques is included at the end of this chapter.

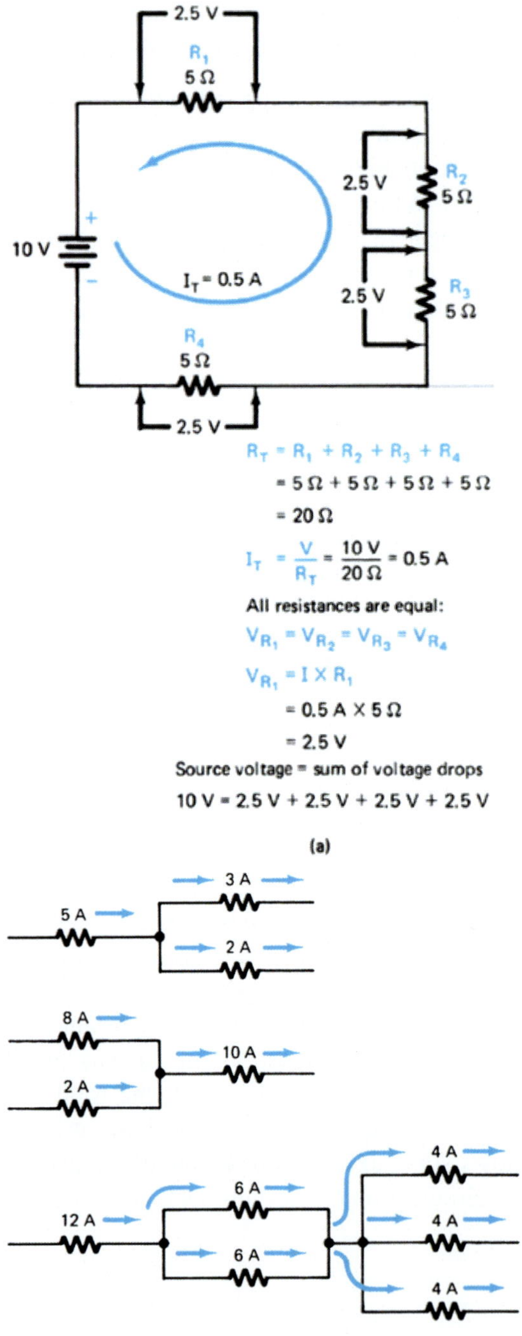

Figure 6-5 Kirchhoff's laws: (a) voltage law example; (b) current law examples.

6.5 Examples of Combination Circuits

Most circuits have both series and parallel parts and are called **combination circuits**. The solution of problems for this type of circuit is done by combining series- and parallel-circuit rules.

Look at the circuit in **Figure 6-6**.

The value that should first be calculated is the **resistance** of R_2 and R_3 in parallel. When this quantity is found, it can be added to the value of the series resistor (R_1) to find the total resistance of the circuit:

$R_T = R_1 + R_2 \, \text{l l} \, R_3$

("l l" means R_2 is in parallel with R_3)

$= 30\ \Omega + 10\ \Omega \times 20\ \Omega/10\ \Omega + 20\ \Omega = 30\ \Omega + 6.67\ \Omega = 36.67\ \Omega.$

When the **total resistance (R_T)** is found, the **total current (I_T)** may be found:

$I_T = V_T/R_T = 40\ \text{V}/36.67\ \text{W} = 1.09\ \text{A}.$

Note that the **total current** flows through resistor R_1 because it is in series with the voltage source. The **voltage drop** across resistor R_1 is

$V_1 = I_T \times R_1 = 1.09\ \text{A} \times 30\ \text{W} = 32.7\ \text{V}.$

The applied voltage is 40 V and 32.7 V are dropped across resistor R_1. The remaining voltage is dropped across the two parallel resistors (R_2 and R_3): 40 V - 32.7 V = 7.3 V across R_2 and R_3. The currents through R_2 and R_3 are

$I_2 = V_2/R_2 = 7.3\ \text{V}/10\ \Omega = 0.73\ \text{A}$

$I_3 = V_3 / R_3 = 7.3\ \text{V}/20\ \Omega = 0.365\ \text{A}.$

Another type of **combination circuit** is shown in **Figure 6-7**. Resistors R_2 and R_3 are in series. When they are combined by adding their values, the

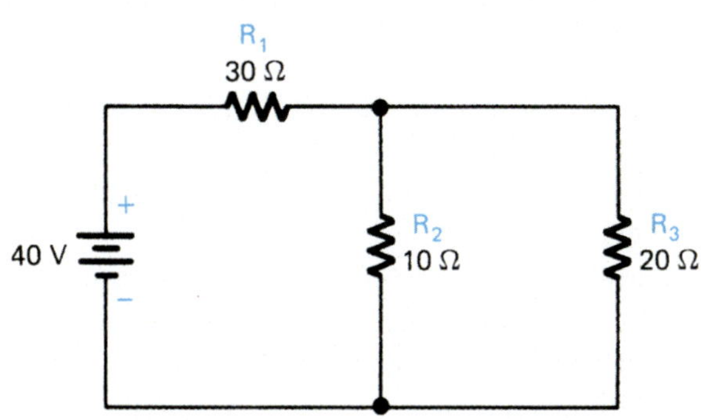

Figure 6-6 Combination circuit example.

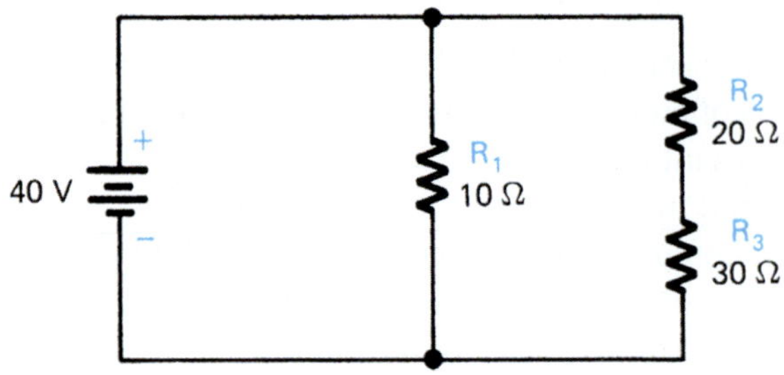

Figure 6-7 Combination circuit example.

circuit becomes a two-branch parallel circuit. Total resistance (R_T) is found by using the product-over-sum method:

$R_T = 10\ \Omega \times 50\ \Omega/10\ \Omega + 50\ \Omega = 500/60 = 8.33\ \Omega$.Total current is

$I_T = V_T/R_T = 40\ \text{V}/8.33\ \Omega = 4.8\ \text{A}$.

The voltage across R_1 is 40 V because it is in parallel with the voltage source. Voltage drops across R_2 and R_3 are

$I_2 = V_T/R$ of branch $= 40\ \text{V}/50\ \Omega = 0.8\ \text{A}$

$V_2 = I \times R_2$

$= 0.8\ \text{A} \times 20\ \Omega = 16\ \text{V}$

$V_3 = I \times R_3$

$= 0.8\ \text{A} \times 30\ \Omega = 24\ \text{V}$.

Note that the sum of V_2 and V_3 (16 V + 24 V) is equal to the **source voltage**.

Combination circuit problems may be solved by using the following step-by-step procedure:

- Combine series and parallel parts to find the total resistance of the circuit.
- Find the total current that flows through the circuit.
- Find the voltage across each part of the circuit.
- Find the current through each resistance of the circuit.
- Steps 3 and 4 must often be done in combination with each other, rather than doing one and then the other.
- Another helpful technique is to redraw the circuit.

6.5 Specialized Circuit Applications

Maximum Power Transfer

An important consideration in electrical circuits is called **maximum power transfer**. Maximum power is transferred from a voltage source to a load when the load resistance (R_L) is equal to the **internal resistance** of the source (R_s). The source resistance limits the amount of power that can be applied to a load. For example, as a flashlight battery gets older, its internal resistance

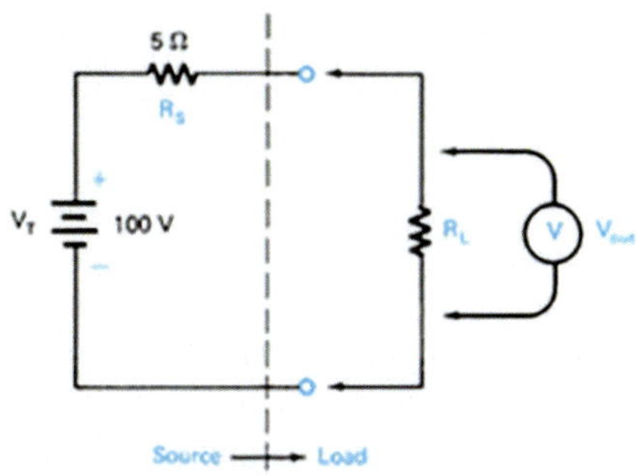

R_L	I_L	V_{out}	Power Output (W)
0	$\frac{100\ V}{5\ \Omega} = 20\ A$	20 A × 0 Ω = 0 V	20 A × 0 V = 0 W
2.5 Ω	$\frac{100\ V}{7.5\ \Omega} = 13.3\ A$	13.3 A × 2.5 Ω = 33.3 V	13.3 A × 33.3 V = 444 W
5 Ω	$\frac{100\ V}{10\ \Omega} = 10\ A$	10 A × 5 Ω = 50 V	10 A × 50 V = 500 W
7.5 Ω	$\frac{100\ V}{12.5\ \Omega} = 8\ A$	8 A × 12.5 Ω = 60 V	8 A × 60 V = 480 W
10 Ω	$\frac{100\ V}{15\ \Omega} = 6.7\ A$	6.7 A × 10 Ω = 67 V	6.7 A × 67 V = 444 V

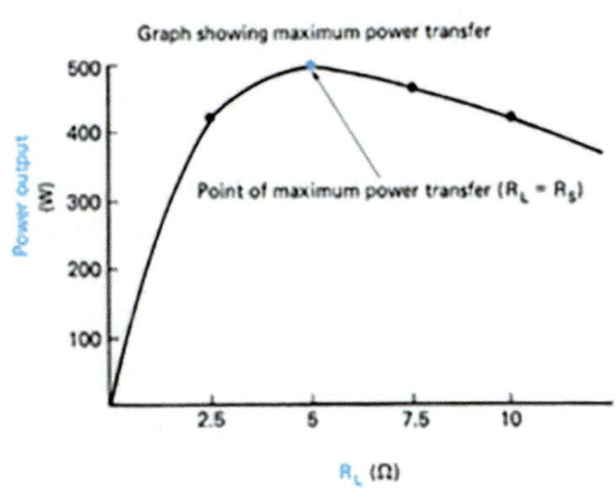

Figure 6-8 Problem that shows maximum power transfer.

increases. This increase in the internal resistance causes the battery to supply less power to the lamp load. Thus, the light output of the flashlight is reduced.

Figure 6-8 shows an example that illustrates **maximum power transfer**. The source is a 100-V battery with an internal resistance of 5 W. The values of I_L, V_{out}, and power output (P_{out}) are calculated as follows:

$$I_L = V_T / R_S + R_L \quad V_{out} = I_L \times R_L \quad P_{out} = I_L \times V_{out}.$$

Note the graph in **Figure 6-38** showing that maximum power is transferred from the source to the load when $R_L = R_s$. This is an important circuit design consideration for power sources, amplifier circuits, microphones, or practically any type of electronic circuit.

Voltage-divider Circuits

The simple series circuit of **Figure 6-9(a)** is a **voltage divider**. Voltage division takes place due to voltage drops across the three resistors. Because each of the three resistors has the same value (1 kW), the voltage drop across each is 3 V. Thus, a single voltage source is used to derive three separate voltages.

Another method used to accomplish voltage division is the **tapped resistor**. This method relies on the use of a resistor which is wire wound and has a tap onto which a wire is attached. The wire is attached so that a certain amount of the total resistance of the device appears from the tap to the outer terminals. For example, if the tap is in the center of a 100-W wire-wound resistor, the resistance from the tap to either outer terminal is 50 W. Tapped resistors often have two or more taps to obtain several combinations of fixed-value resistance. **Figure 6-39(b)** shows a tapped

resistor used as a voltage divider. In the example, the voltage outputs are each 3 V, derived from a 6-V source.

A common method of **voltage division** is shown in **Figure 6-9(c)**. Potentiometers are used as voltage dividers in volume control circuits of radios and televisions. They may be used to vary voltage from zero to the value of the source voltage. In the example, the voltage output may be varied from 0 to 1.5 V. It is also possible to use a voltage-divider network and a potentiometer to obtain many variable voltage combinations, as discussed in the following section.

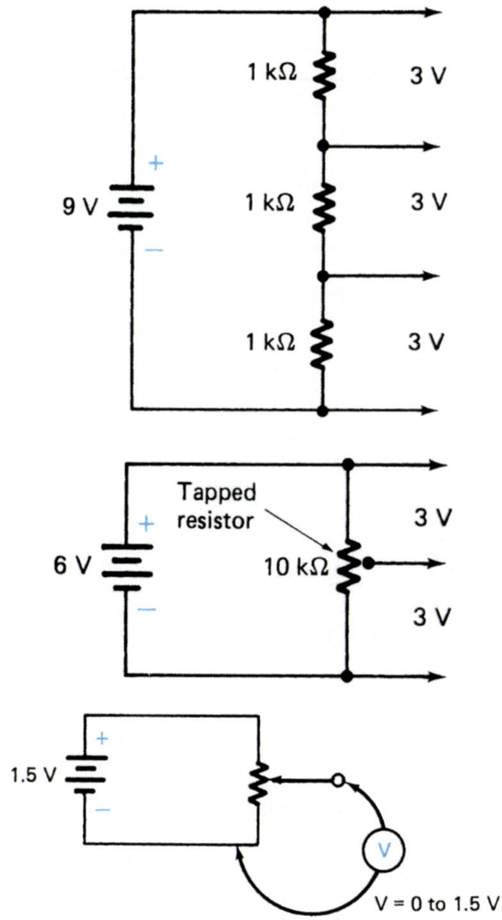

Figure 6-9 Voltage-divider circuits: (a) series dc circuit used as a voltage divider; (b) tapped resistor used as a voltage divider; (c) potentiometer used as a voltage divider.

Voltage-divider Design

The design of a **voltage-divider circuit** is a good application of basic electrical theory. Refer to the circuit of **Figure 6-10**. Resistors R_1, R_2, and R_3 form a voltage divider to provide the proper voltage to three known loads. The loads could be transistors of a 9-V portable radio, for example. The operating voltages and currents of the load are constant. The values of R_1, R_2, and R_3 are calculated to supply proper voltage to each of the loads. The value

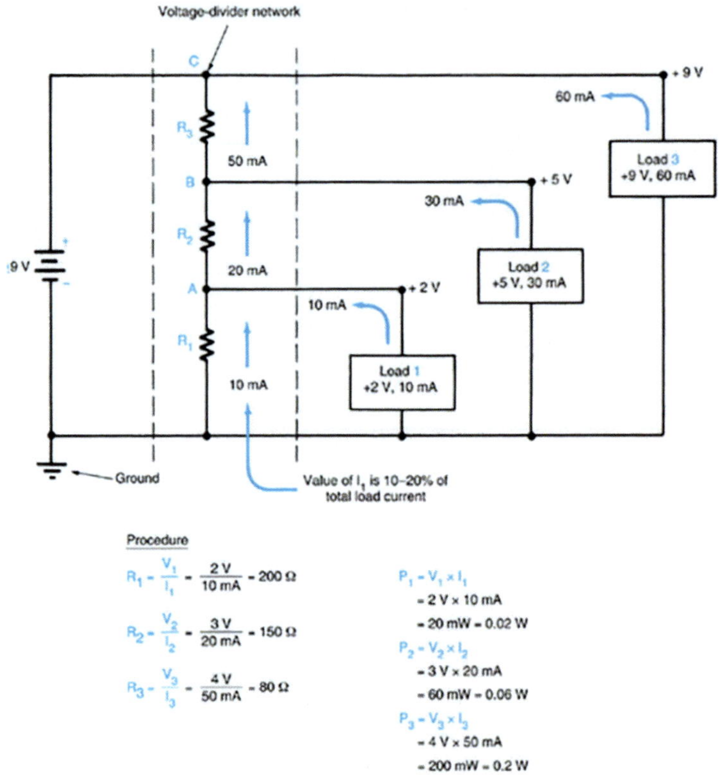

Figure 6-10 Voltage-divider design.

of current through R_1 is selected as 10 mA. This value is ordinarily 10%–20% of the total current flow to the loads (10 + 30 + 60 mA = 100 mA and 10% of 100 mA = 10 mA).

To calculate the values of R_1, R_2, and R_3, the **voltage** across each resistor and the current through each resistor must be known. Start with R_1 at the bottom of the circuit. The current through R_1 is given (10 mA). The voltage across R_1 is 2 V since the ground is a zero-voltage reference, and 2 V must be supplied to load 1. The value of R_1, as shown in the procedure of **Figure 6-10**, must be 200 W (2 V ¸10 mA).

Resistor R_2 has a voltage of 3 V across it. Point A has a potential of +2 V and point B has a potential of +5 V for load 2. The **difference in potential** or **voltage drop** is, therefore, 5 - 2 V = 3 V. The current through R_2 is 20 mA. A current of 10 mA flows up through R_1 and 10 mA flows to point A from load 1. These two currents (10 + 10 mA = 20 mA) combine and flow through R_2. The value of R_2 must be 150W (3 V ¸ 20 mA). Resistor R_3 has a voltage of 4

V across it (9 - 5 V = 4 V). The current through R_3 is 50 mA because 20 mA flows upward through R_2 and 30 mA flows from load 2 to point B (20 + 30 mA = 50 mA). The value of R_3 must be 80 W (4 V ¸ 50 mA).

With the calculated values of R_1, R_2, and R_3 used as a **voltage-divider network**, the proper values of 0.02, 0.06, and 0.2 W are calculated in **Figure 6-10**. Often, a *safety factor* is used to ensure that power values are large enough. A safety factor is a multiplier used with the minimum power values. For example, if a safety factor of 2 is used, the minimum power values for the circuit would become $P_1 = 0.02\ W \times 2 = 0.04\ W$, $P_2 = 0.06\ W \times 2 = 0.12\ W$, and $P_3 = 0.2\ W \times 2 = 0.4\ W$.

Voltage-division Equation

The voltage-division equation, often called the **voltage-divider rule**, is convenient to use with voltage-divider circuits when the current delivered from the voltage divider is negligible. The voltage-divider equation and a sample problem are shown in **Figure 6-11**. This equation applies to series circuits. The voltage (V_x) across any resistor in a series circuit is equal to the ratio of that resistance (R_X) to total resistance (R_T) multiplied by the source voltage (V_T).

Negative Voltage Derivation

Voltage-divider circuits are often used as power sources for other types of electronic circuits. In circuit design and analysis, reference is often made to **negative** voltage. The concept of a negative voltage is made clear in **Figure 6-12**. Voltage is ordinarily measured with respect to a ground reference point. The circuit ground is shown at point A. Point E in **Figure 6-12** is connected to the negative side of the power source. Point A, where the ground reference is connected, has a higher potential than point E. Therefore, the voltage across points A and E is -50 V.

Voltage Division with a Potentiometer

A typical circuit design problem using a **potentiometer** is shown in **Figure 6-13**. A given value of 10 k1 is used as the potentiometer. The desired variable voltage from the potentiometer center terminal to ground is 5–10 V. The values of R_1 and R_3 are calculated to derive the desired variable voltage from the potentiometer.

$$V_x = \frac{R_x}{R_T} \times V_T$$

V_x is voltage across a resistance

V_T is total voltage applied

R_x is the resistance where V_x is measured

R_T is the total resistance of the voltage-divider network

(a)

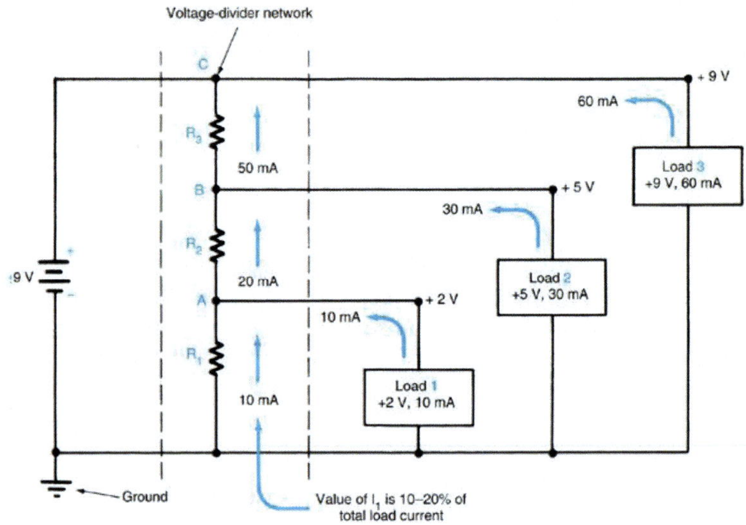

Procedure

$R_1 = \frac{V_1}{I_1} = \frac{2\ V}{10\ mA} = 200\ \Omega$

$R_2 = \frac{V_2}{I_2} = \frac{3\ V}{20\ mA} = 150\ \Omega$

$R_3 = \frac{V_3}{I_3} = \frac{4\ V}{50\ mA} = 80\ \Omega$

$P_1 = V_1 \times I_1$
$= 2\ V \times 10\ mA$
$= 20\ mW = 0.02\ W$

$P_2 = V_2 \times I_2$
$= 3\ V \times 20\ mA$
$= 60\ mW = 0.06\ W$

$P_3 = V_3 \times I_3$
$= 4\ V \times 50\ mA$
$= 200\ mW = 0.2\ W$

Figure 6-11 Voltage-divider design.

The current flow in a voltage-divider network is established by the value of R_1 (10 kW) and the range of voltage variation (5–10 V = 5 V variation). The current flow calculation in the circuit of **Figure 6-13** is shown in the procedure. Since $I = V/R$, the current through R_2 and the other parts of this series circuit is 0.5 mA. Once the current is found, values of R_1 and R_3 may be found as shown in the procedure. **Figure 6-13** shows an easy method of

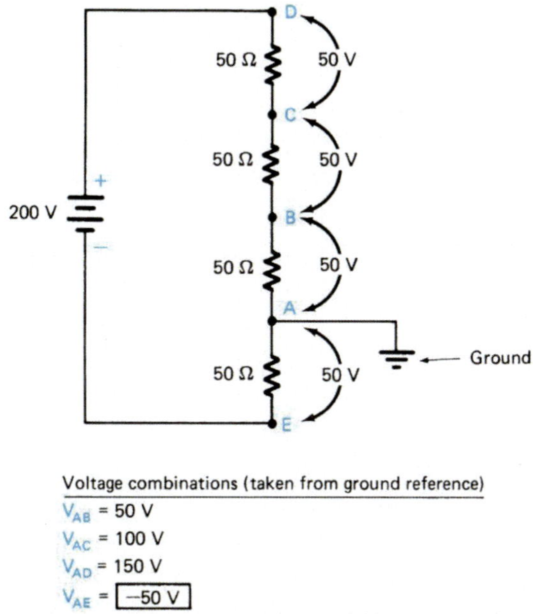

Figure 6-12 Negative voltage derived from a voltage divider.

determining voltage drops. A network of resistances in series can be thought of as a scale. In the example, the voltage at point A is +5 V and the voltage at point B is + 10 V. The difference in potential is 5 V (10 - 5 V = 5 V). This is similar to reading a scale.

6.6 Problem-solving Methods

Several different techniques may be applied to solve electrical circuit problems. Some of these methods include the following:

- **Kirchhoff's voltage law (KVL) –** an algebraic procedure which may be used to find current flow in electrical circuits for single voltage sources and for circuits which have two or more voltage sources.
- **Superposition** – a non-algebraic procedure which may be used to find current flow in electrical circuits for single voltage sources and for circuits which have two or more voltage sources.
- **Equivalent circuits** – simplified circuits, which include **Thevenin** and **Norton** equivalent circuit applications, may be used to solve circuit problems.
- **Bridge-circuit simplification** – a procedure which may be used to make problem solving with bridge circuits easier to accomplish.

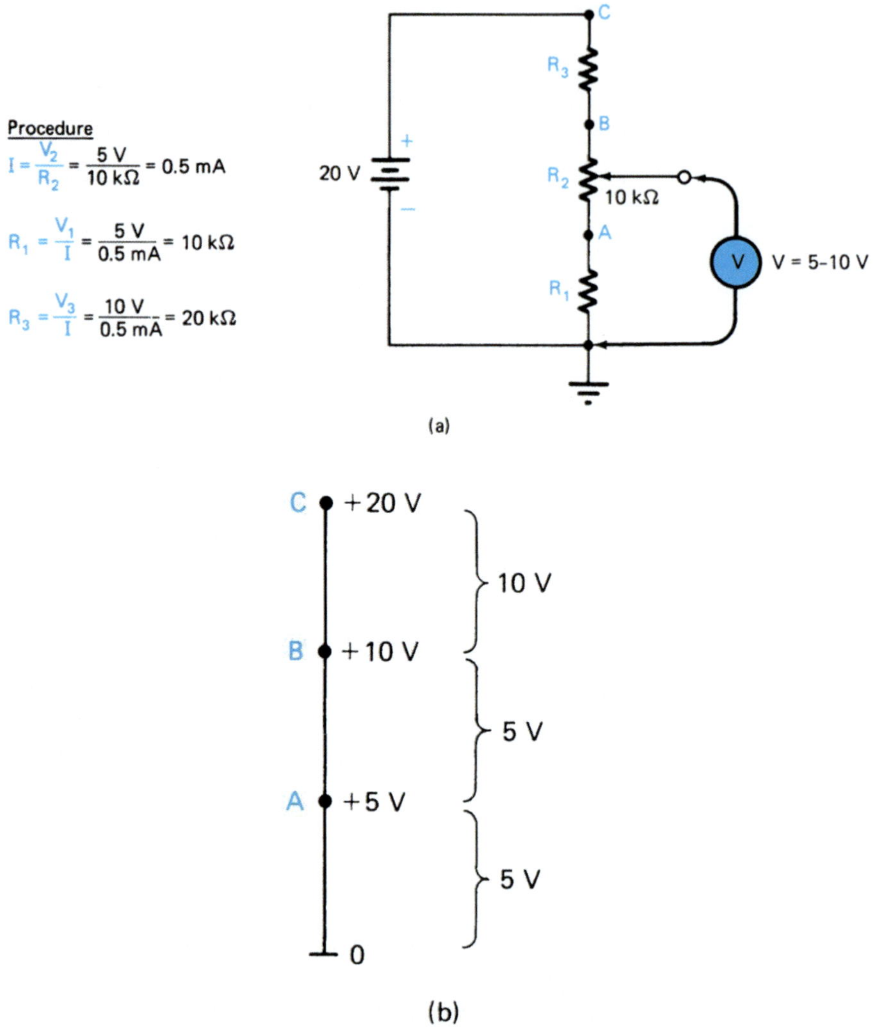

Figure 6-13 Voltage-divider design problem: (a) circuit; (b) voltage values.

Kirchhoff's Voltage Law Methodology

Kirchhoff's voltage law is illustrated in two different ways in **Figure 6-14**. The voltage law may be stated in two ways: (1) the sum of voltage drops in a closed-looped circuit is equal to the source voltage; and (2) the algebraic sum of the voltage sources and voltage drops in a closed-loop path is equal to zero. The first method deals with the voltage drops in a closed-loop, or series, path. The sum of the voltage drops across the components is equal to the

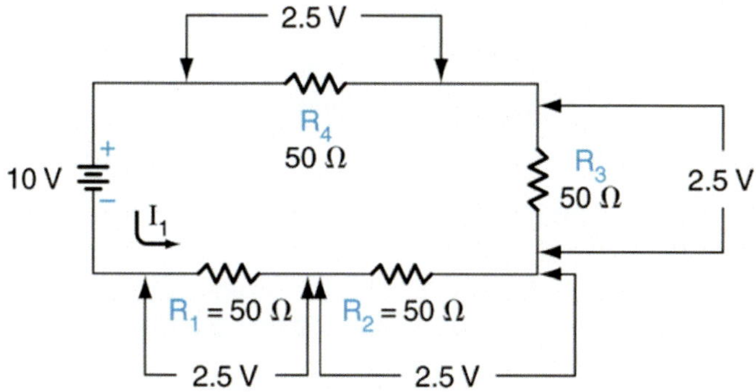

(a) Voltage drop procedure:

$$10\ V = 2.5\ V + 2.5\ V + 2.5\ V + 2.5\ V$$

(b) Algebraic procedure:

$$10\ V - 50\ I_1 - 50\ I_1 - 50\ I_1 - 50\ I_1 = 0$$

$$10\ V - 200\ I_1 = 0$$

$$I_1 = \frac{10\ V}{200} = 0.05\ A$$

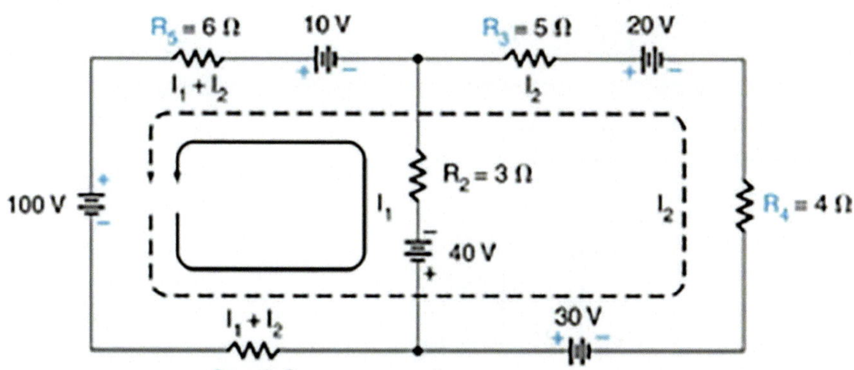

(a) Circuit

(b) Procedure

1. Assign a direction of current flow (– to +) through each path. Start with the largest voltage source.

2. Develop an equation for loop 1 (solid line):

$$100\ V - 2(I_1 + I_2) + 40\ V - 3\ I_1 - 10\ V - 6(I_1 + I_2) = 0$$

$$100\ V - 2\ I_1 + 2\ I_2 + 40\ V - 3\ I_1 - 10\ V - 6\ I_1 + 6\ I_2 = 0$$ (Multiply $I_1 + I_2$ terms)

Equation for loop 1 ⟶ $130\ V - 11\ I_1 - 8\ I_2 = 0$ (Combine like terms)

3. Develop an equation for loop 2 (dashed line):

$$100\text{ V} - 2(I_1 + I_2) + 30\text{ V} - 4\,I_2 - 20\text{ V} - 5\,I_2 - 10\text{ V} - 6(I_1 + I_2) = 0$$
$$100\text{ V} - 2\,I_1 - 2\,I_2 + 30\text{ V} - 4\,I_2 - 20\text{ V} - 5\,I_2 - 10\text{ V} - 6\,I_1 - 6\,I_2 = 0$$

(Multiply $I_1 + I_2$ terms)

Equation for loop 2 → $\boxed{100\text{ V} - 8\,I_1 - 17\,I_2 = 0}$ (Combine like terms)

4. Place the two equations together and eliminate either I_1 or I_2 to solve for one unknown term:

$$(\times 8)\ 130\text{ V} - 11\,I_1 - 8\,I_2 = 0$$
$$(\times 11)\ 100\text{ V} - 8\,I_1 - 17\,I_2 = 0$$

Multiply top equation by 8 and bottom equation by 11 to eliminate I_1

5. Rewrite the equations after they have been multiplied:

$$1040\text{ V} - 88\,I_1 - 64\,I_2 = 0$$
$$(-)\ 1100\text{ V} \overset{(+)}{-} 88\,I_1 \overset{(+)}{-} 187\,I_2 = 0$$
$$-\ 60\text{ V} \qquad +123\,I_2 = 0$$

(These cancel)

6. Subtract bottom equation from top equation

7. Solve for I_2:

$$-60\text{ V} + 123\,I_2 = 0$$
$$+123\,I_2 = 60\text{ V}$$
$$I_2 = \frac{60\text{ V}}{123} = 0.488\text{ A}$$

8. Substitute value of I_2 in one of the original equations to solve for I_1:

$$100\text{ V} - 8\,I_1 - 17\,I_2 = 0$$
$$100\text{ V} - 8\,I_1 - 17(0.488) = 0$$
$$100\text{ V} - 8\,I_1 - 8.3\text{ V} = 0$$
$$91.7\text{ V} - 8\,I_1 = 0$$
$$-8\,I_1 = -91.7\text{ V}$$
$$I_1 = \frac{-91.7\text{ V}}{8}$$
$$I_1 = 11.46\text{ A}$$

9. Add I_1 and I_2 to get I_T:

$$I_1 + I_2 = I_T$$
$$11.46\text{ A} + 0.488\text{ A} = 11.948\text{ A}$$

Figure 6-14 Kirchhoff's voltage law: (a) voltage drop procedure; (b) algebraic procedure.

source voltage. In the example of **Figure 6-14**, voltage drops are written as $R \times I$, such as $50 \times I_1 = 50I_1$. The current in a loop is given an algebraic value (I_1). Remember that any voltage drop is equal to $I \times \text{R}$.

The algebraic procedure of **Kirchhoff's voltage law** involves setting up a simple equation for a circuit loop. Values of current flow in circuits

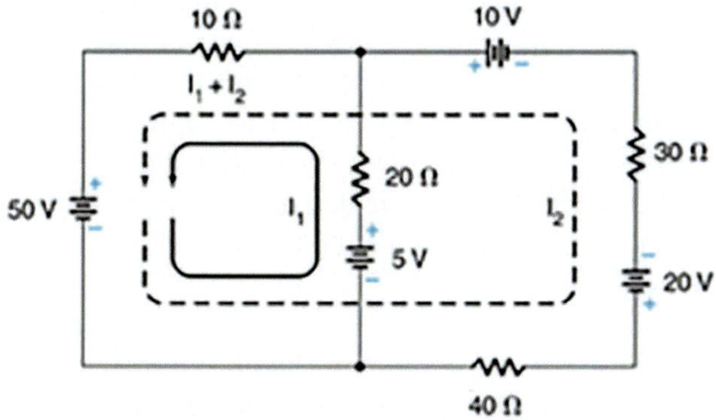

1. Equation for loop 1:

$$50\text{ V} - 5\text{ V} - 20\, I_1 - 10(I_1 + I_2) = 0$$
$$45\text{ V} - 20\, I_1 - 10\, I_1 - 10\, I_2 = 0$$
$$\boxed{45\text{ V} - 30\, I_1 - 10\, I_2 = 0}$$

2. Equation for loop 2:

$$50\text{ V} - 40\, I_2 + 20\text{ V} - 30\, I_2 - 10\text{ V} - 10(I_1 + I_2) = 0$$
$$60\text{ V} - 40\, I_2 - 30\, I_2 - 10\, I_1 - 10\, I_2 = 0$$
$$\boxed{60\text{ V} - 80\, I_2 - 10\, I_1 = 0}$$

3. Combine equations and cancel one unknown term:

$$45\text{ V} - 30\, I_1 - 10\, I_2 = 0$$
$$(\times 3)\ 60\text{ V} - 10\, I_1 - 80\, I_2 = 0 \leftarrow \text{Multiply this equation by 3}$$

$$45\text{ V} - 30\, I_1 - 10\, I_2 = 0$$
$$(-)\ 180\text{ V} \ominus 30\, I_1 \ominus 240\, I_2 = 0 \leftarrow \text{Subtract bottom equation from top equation}$$
$$-135\text{ V} \qquad +230\, I_2 = 0$$
$$230\, I_2 = 135\text{ V}$$
$$I_2 = \frac{135\text{ V}}{230} = 0.587\text{ A}$$

4. Substitute value of I_2 in one equation to solve for I_1:

$$60\text{ V} - 10\, I_1 - 80\, I_2 = 0$$
$$60\text{ V} - 10\, I_1 - 80(0.587) = 0$$
$$60\text{ V} - 10\, I_1 - 47\text{ V} = 0$$
$$13\text{ V} - 10\, I_1 = 0$$
$$-10\, I_1 = -13\text{ V}$$
$$I_1 = \frac{-13\text{ V}}{10} = 1.3\text{ A}$$

5. Add $I_1 + I_2$ to get I_T:

$$I_1 + I_2 = I_T$$
$$1.3\text{ A} + 0.587\text{ A} = 1.887\text{ A}$$

Figure 6-15 Kirchhoff's voltage law example.

may be found by using this procedure. In a circuit that has only one voltage source, it is easier to use Ohm's law for series circuits to find current flow. The advantage of the algebraic method for problem solving is that currents in multiple-source circuits may be easily calculated. Ohm's law cannot be used to find the current flow through each of the paths shown in the circuit of **Figure 6-15(a)**. The algebraic procedure derived from Kirchhoff's voltage law allows the calculation of current in a circuit with more than one voltage source [see **Figure 6-15(b)**].

The method used in **Figure 6-15** may be used for **multiple-voltage-source** problems that have two current loops. The first step in this method is to assign directions of current flow (from – to +) in the circuit. When there are several sources, start with the largest voltage source. To avoid confusion, the current paths should be marked so that they appear different. The example uses a solid line for path 1 (I_1) and a dashed line for path 2 (I_2). When both paths pass through a resistance, the current is called $I_1 + I_2$.

An equation is developed for each of the **circuit loops**, based on Kirchhoff's voltage law. Each current is followed from the largest voltage source in the direction of the current arrow. Voltage sources must be given the proper sign when setting up an equation. When the current direction is to + through the source, a negative (-) sign is used in the equation. A positive (+) sign is used when the direction of the current arrow is from + to - through the source. The equation for each loop is developed using simple algebraic procedures, as shown. Practice in using this method makes it a convenient way to calculate current flow in a circuit with two current loops and two or more voltage sources. Another example is shown in **Figure 6-16**.

Superposition Method

An alternate method for finding current flow in circuits with two or more voltage sources is called the **superposition method**. This non-algebraic method involves some rather lengthy, but simple, calculations. **Multiple-voltage-source** circuits may be broken down into as many individual circuits as there are voltage sources. For instance, a circuit with two voltage sources is reduced to two individual circuits. Each voltage source is considered separately, with other voltage sources **short-circuited** for making current calculations. In this way, the contribution of each voltage source to the current flow in the circuit may be determined.

For a **two-source circuit**, one of the individual circuits is **superimposed** onto the other using the following procedure.

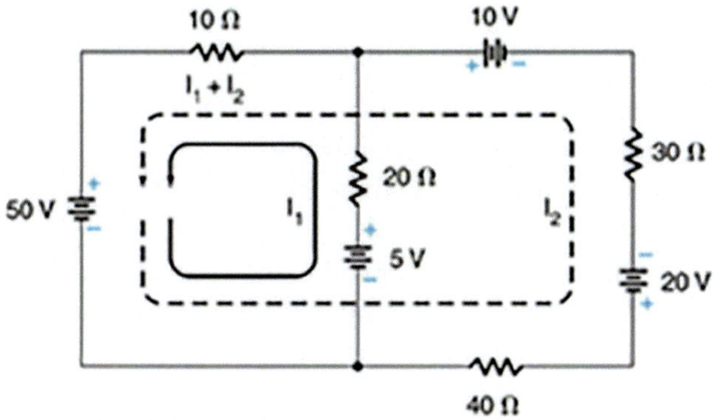

1. Equation for loop 1:

$$50\,V - 5\,V - 20\,I_1 - 10(I_1 + I_2) = 0$$
$$45\,V - 20\,I_1 - 10\,I_1 - 10\,I_2 = 0$$
$$\boxed{45\,V - 30\,I_1 - 10\,I_2 = 0}$$

2. Equation for loop 2:

$$50\,V - 40\,I_2 + 20\,V - 30\,I_2 - 10\,V - 10(I_1 + I_2) = 0$$
$$60\,V - 40\,I_2 - 30\,I_2 - 10\,I_1 - 10\,I_2 = 0$$
$$\boxed{60\,V - 80\,I_2 - 10\,I_1 = 0}$$

3. Combine equations and cancel one unknown term:

$$45\,V - 30\,I_1 - 10\,I_2 = 0$$
$$(\times 3)\ 60\,V - 10\,I_1 - 80\,I_2 = 0 \leftarrow \text{Multiply this equation by 3}$$

$$45\,V - 30\,I_1 - 10\,I_2 = 0$$
$$(-)\ 180\,V \overset{(+)}{-} 30\,I_1 \overset{(+)}{-} 240\,I_2 = 0 \leftarrow \text{Subtract bottom equation from top equation}$$
$$-135\,V \qquad +230\,I_2 = 0$$
$$230\,I_2 = 135\,V$$
$$I_2 = \frac{135\,V}{230} = 0.587\,A$$

4. Substitute value of I_2 in one equation to solve for I_1:

$$60\,V - 10\,I_1 - 80\,I_2 = 0$$
$$60\,V - 10\,I_1 - 80(0.587) = 0$$
$$60\,V - 10\,I_1 - 47\,V = 0$$
$$13\,V - 10\,I_1 = 0$$
$$-10\,I_1 = -13\,V$$
$$I_1 = \frac{-13\,V}{10} = 1.3\,A$$

5. Add $I_1 + I_2$ to get I_T:

$$I_1 + I_2 = I_T$$
$$1.3\,A + 0.587\,A = 1.887\,A$$

Figure 6-16 Kirchhoff's voltage law example problem.

Figure 6-17 shows a circuit with two voltage sources. The procedure for finding current flow through each component in the circuit is as follows:

- Short-circuit one power source and use basic Ohm's law procedure to find current flow through each component.
- Record the amount of current and the direction of flow through each component on this circuit.
- Short-circuit the other power source and use basic **Ohm's law** procedures to find current flow through each component.
- Record the amount of current and the direction of flow through each component on this circuit.
- Find the current flow through each component by looking at the direction of flow through each individual circuit. If the directions through components of both circuits are the same, **add** the values. If the directions of current flow are opposite, **subtract** the values.
- Record the amount and direction of current flow on the original circuit. The current flows in the direction of the **largest** flow in an individual circuit.

The **superposition method** can also be used for circuits with more than two sources. A four-source circuit, for example, would require four individual circuits superimposed to find resultant current values. Current flows in the same directions would be added and those in opposite directions would be subtracted. Direction of current flow is in the direction of the largest sum of currents through the path.

Equivalent Circuit Methodology

The previous sections have dealt with relatively simple circuit applications. Simplification of more complex circuits may be accomplished by **applying equivalent circuit methods**. Several equivalent circuit methods, sometimes called complex circuit theorems or **network theorems**, may be utilized to simplify complex circuits. This section deals primarily with the Thevenin and Norton equivalent circuit applications for solving complex electrical circuit problems.

Thevenin Equivalent Circuit Method

The **Thevenin equivalent circuit method** is used to simplify electrical circuits. A French engineer, M. L. Thevenin, developed this method, which

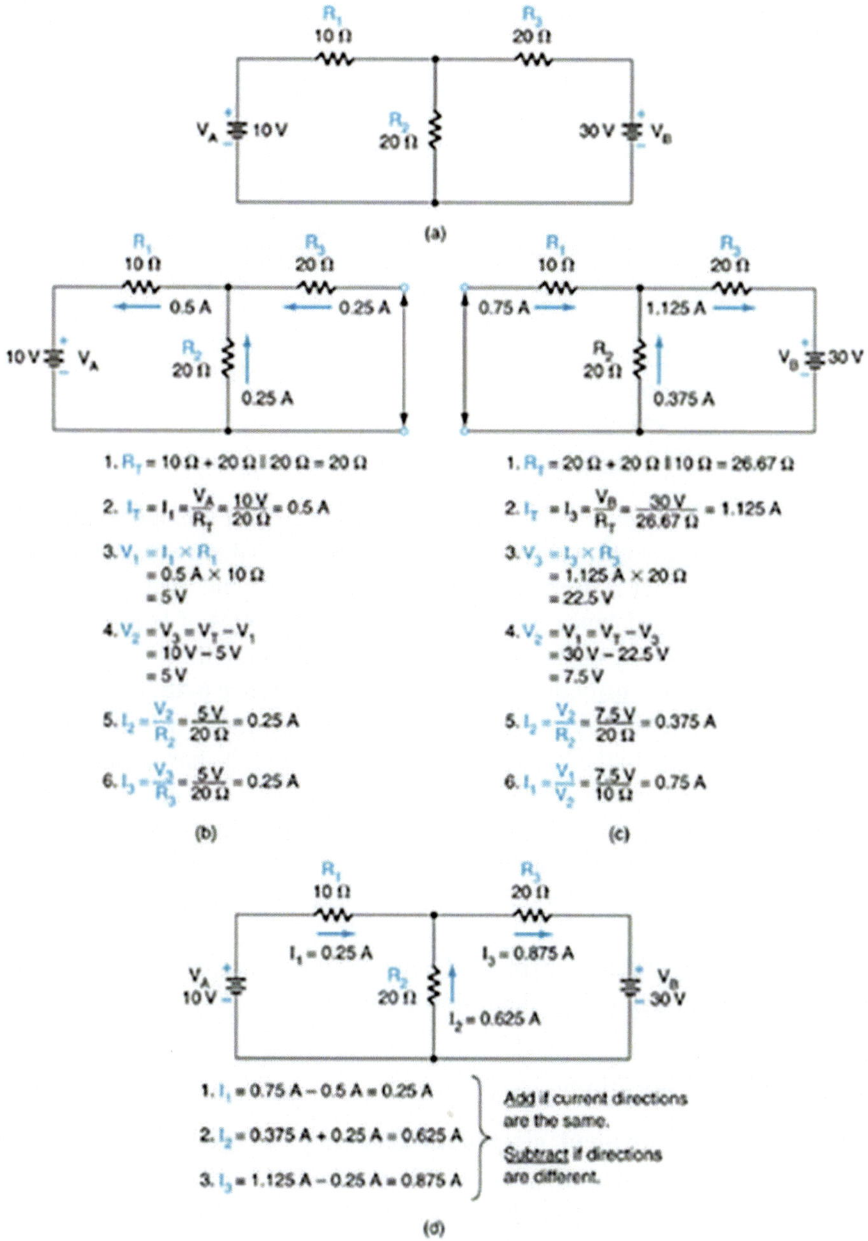

Figure 6-17 The superposition method: (a) original circuit; (b) circuit with 30-V source shorted; (c) circuit with 10-V source shorted; (d) original circuit with currents recorded.

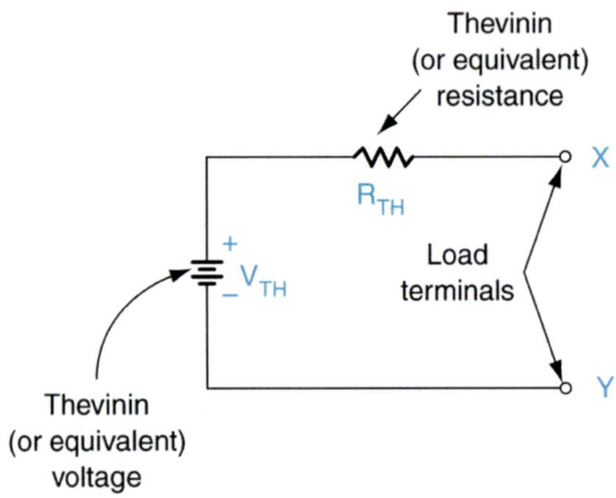

Figure 6-18 Thevenin equivalent circuit.

allows a complex circuit to be reduced to one equivalent voltage source and series resistance for purposes of calculation or lab experimentation. It is a practical method that is used to calculate load currents and load voltages for any value of load resistance. Working with varying values of load resistance is greatly simplified by using the Thevenin equivalent circuit method.

The **Thevenin equivalent circuit** is shown in **Figure 6-18**. It is called an **equivalent circuit** because it is equivalent to a more complex circuit (as seen by a load connected to the circuit). Remember that circuits have a **source** and a **load**. A complex circuit is reduced to one with a single voltage source (V_{TH}) and a series resistance (R_{TH}). These values are called the equivalent voltage and equivalent resistance. The load is connected to the load terminals of the circuit, which are labeled as points X and Y. The procedure for simplifying circuits using the

Thevenin method is explained next. Some examples are shown in **Figures 6-19** through **6-2** on the following pages.

Single-source Problem

The **Thevenin equivalent circuit method** may be used for simplifying circuits which have one voltage source. **Figure 6-19** shows a circuit with one voltage source and the calculations used to obtain an equivalent circuit.

The procedure for finding V_{TH} and R_{TH} is as follows:

- Find V_{TH}.
- Remove the load from the circuit, leaving terminals X and Y open.

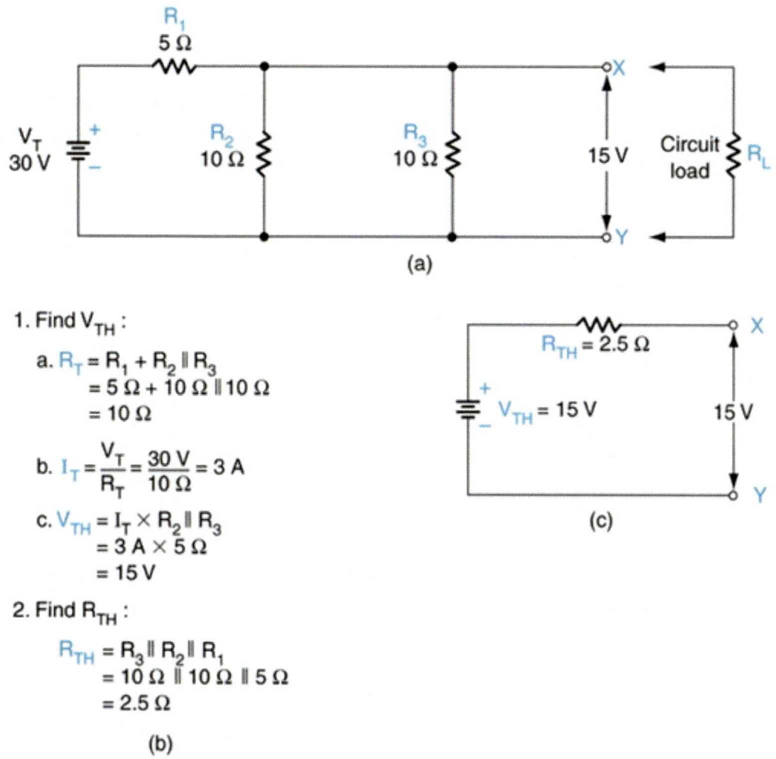

Figure 6-19 Using the Thevenin method for a one-source circuit: (a) original circuit; (b) problem-solving procedure; (c) Thevenin equivalent circuit.

- Use basic Ohm's law procedures to find the voltage across the load (X and Y) terminals.
- The voltage across the load is the equivalent voltage (V_{TH}).
- When resistances are in series with the load, they are disregarded when finding V_{TH}. V_{TH} is an **open-circuit voltage**; therefore, maximum voltage in the current loop containing the load appears across the load terminals (X and Y).
- Find R_{TH}.
- Replace the source with a short circuit.
- Remove the load from the circuit, leaving terminals X and Y open.
- "Look into" the circuit **from the load terminals** to determine the circuit configuration as seen by a load connected to the load terminals. Some examples of determining circuit configurations as seen from the load are shown in **Figure 6-20**.

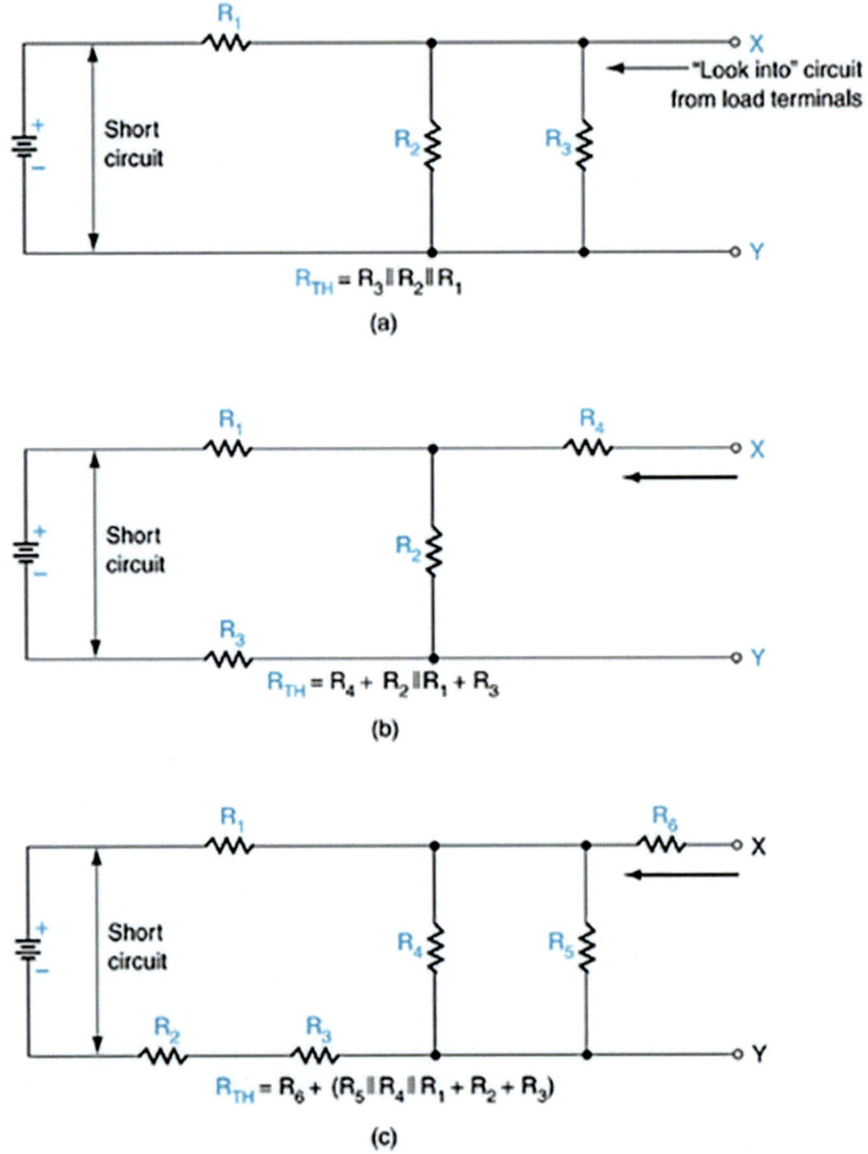

Figure 6-20 Determining circuit configuration for finding R_{TH}.

After the equivalent circuit has been developed, it is simple to calculate values of load current (I_L) and voltage output across to load resistance (V_{out}). **Figure 6-21** shows the calculations of several values of I_L and V_{out} with a Thevenin equivalent circuit.

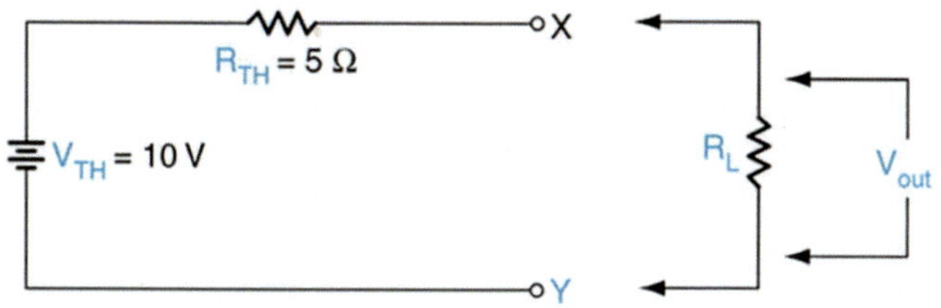

R_L	$I_L = \frac{V_{TH}}{R_{TH} + R_L}$	$V_{out} = I_L \times R_L$
5 Ω	$\frac{10\ V}{10\ \Omega} = 1\ A$	1 A × 5 Ω = 5 V
10 Ω	$\frac{10\ V}{15\ \Omega} = 0.667\ A$	0.667 A × 10 Ω = 6.67 V
15 Ω	$\frac{10\ V}{20\ \Omega} = 0.5\ A$	0.5 A × 15 Ω = 7.5 V
20 Ω	$\frac{10\ V}{25\ \Omega} = 0.4\ A$	0.4 A × 20 Ω = 8 V

Figure 6-21 Calculating load current and voltage output.

Two-source Problem

The **Thevenin equivalent circuit** is easy to apply to a circuit that has two voltage sources. Consider the circuit of **Figure 6-22**, which has 10-V and 2-V sources. Note that the load terminals (X and Y) are in the center of the diagram. To find the R_{TH}, look from the load terminals into the circuit. R_1 and R_2 are in parallel, as seen from the load terminals.

The **Thevenin equivalent voltage (V_{TH})** is found by looking at the difference in potential across the circuit resistances. The same procedure as single-source circuits is used; however, the potential at point X must be found. The potential at point X in the example is the V_{TH}. The difference in potential across points A and B is 8 V (10 - 2 V). The 8-V value is then used to find the current that would flow through R_1 and R_2. Once the current is calculated, the voltage across either resistance may be found. The voltage across R_1 may be **subtracted** from the potential at point A to find V_{TH}. Also, the voltage across R_2 may be **added** to the potential at point B to determine V_{TH}.

Note that if the polarity of V_2 is reversed, the difference in potential across R_1 and R_2 would become 12 V[10 V - (-2 V)]. This would cause the value of V_{TH} to change also.

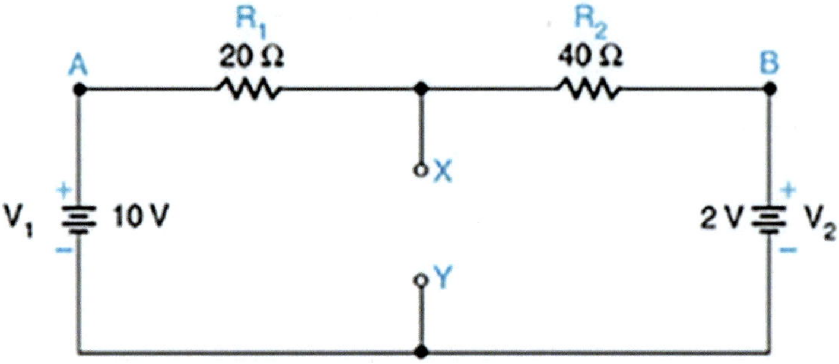

1. Find R_{TH}:

$$R_{TH} = R_1 \parallel R_2 = 20\ \Omega \parallel 40\ \Omega = 13.33\ \Omega$$

2. Find V_{TH}:

Potential at point A = +10 V

Potential at point B = +2 V

Difference in potential = 10 V − 2 V = 8 V

$$\text{Current flow} = \frac{8\ \text{V}}{60\ \Omega} = 0.133\ \text{A}$$

$$V_1 = I \times R_1 = 0.133\ \text{A} \times 20\ \Omega = 2.67\ \text{V}$$

$$V_{TH} = 10\ \text{V} - V_1 = 10\text{V} - 2.67\ \text{V} = 7.33\ \text{V}$$

or

$$V_2 = I \times R_2 = 0.133\ \text{A} \times 40\ \Omega = 5.33\ \text{V}$$

$$V_{TH} = 2\ \text{V} + V_2 = 2\ \text{V} + 5.33\ \text{V} = 7.33\ \text{V}$$

(a)

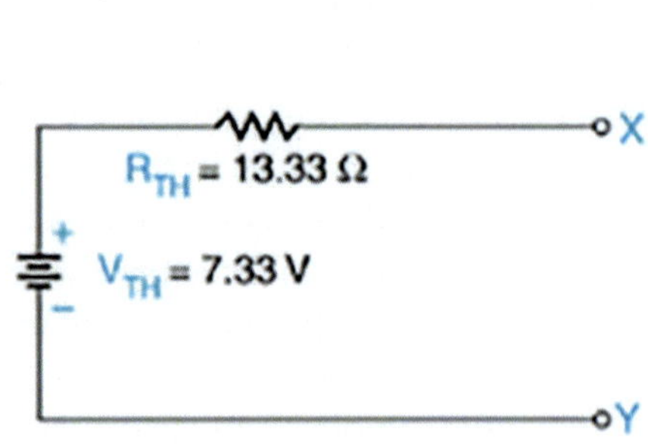

(b)

Figure 6-22 Two-source Thevenin equivalent circuit: (a) problem-solving procedures; (b) Thevenin equivalent circuit.

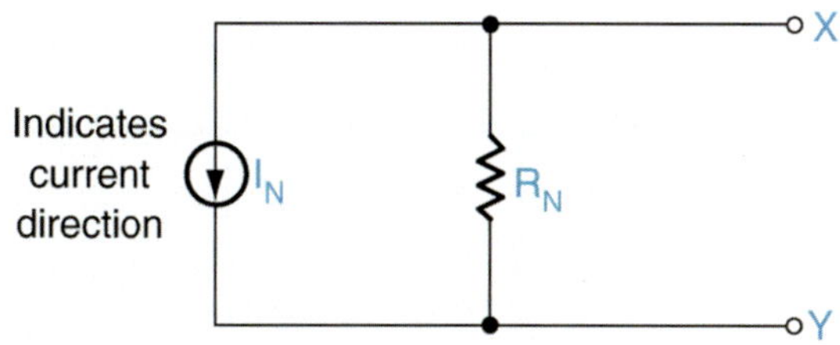

Figure 6-23 Norton equivalent circuit.

Remember that Thevenin equivalent circuits are used to reduce more complex circuits into a circuit that has one equivalent voltage source and one series resistance. They are very helpful in simplifying the procedure for calculating **load current** and **voltage output** of circuits that have several values of **load resistance**.

Norton Equivalent Circuit Method

Another method of simplifying circuits is the **Norton equivalent circuit method**, which is shown in **Figure 6-23**. The Norton current (I_N) is the maximum current that will flow from the source. It is calculated when $R_L = 0$ W. The Norton resistance (R_N) is calculated in the same way as Thevenin resistance (R_{TH}).

The **Norton method** allows the reduction of a circuit to a constant **current source (I_N)** and an **equivalent resistance (R_N)** in parallel. **Figure 6-24(a)** shows a sample procedure for applying the Norton method.

The procedure for developing a Norton equivalent circuit is as follows:

- Find I_N **by** short-circuiting the load (X–Y) terminals.
- Calculate the current that will flow from the source when the load resistance is equal to 0 W (short circuit). This is the **Norton current (I_N)**.
- Label the direction of current flow from the source with an arrow on the equivalent circuit diagram.
- Find R_N by using the same procedure as outlined for finding R_{TH}.
- Label the value of R_N on the equivalent circuit diagram.

The **load current (I_L)** that will flow from a circuit may easily be calculated by applying the Norton equivalent circuit. The formula used to calculate load current values from the equivalent values of R_N, and I_N is

$$I_L = I_N \times R_N/R_N + R_L.$$

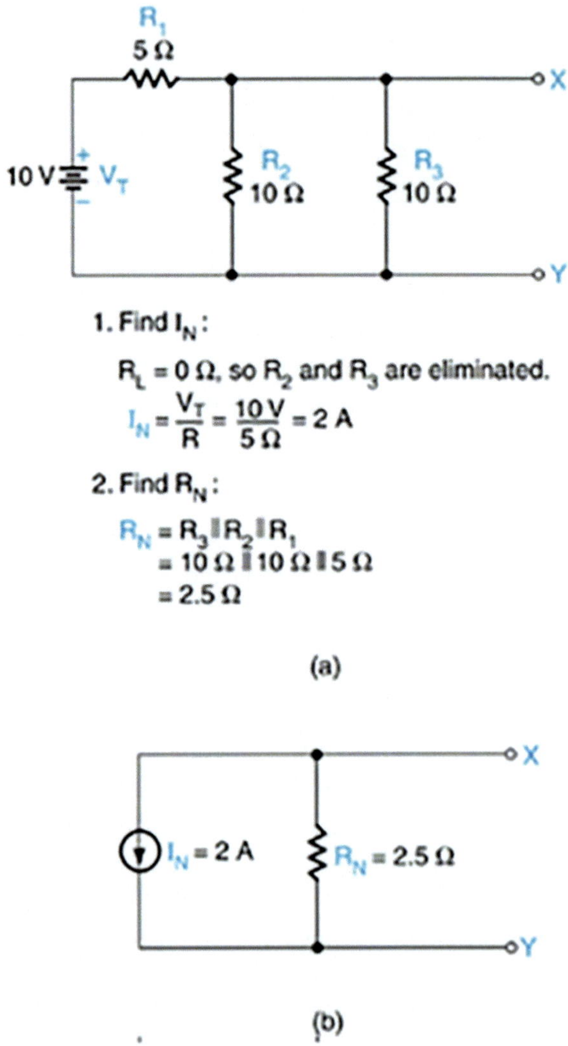

Figure 6-24 Norton equivalent circuit procedure: (a) problem-solving procedure; (b) Norton equivalent circuit.

Bridge-circuit Simplification

A **bridge circuit** is shown in **Figure 6-25**. Bridge circuits are used for several applications, particularly in electrical measurement. A bridge circuit may be designed to measure electrical component values by comparing an **unknown**

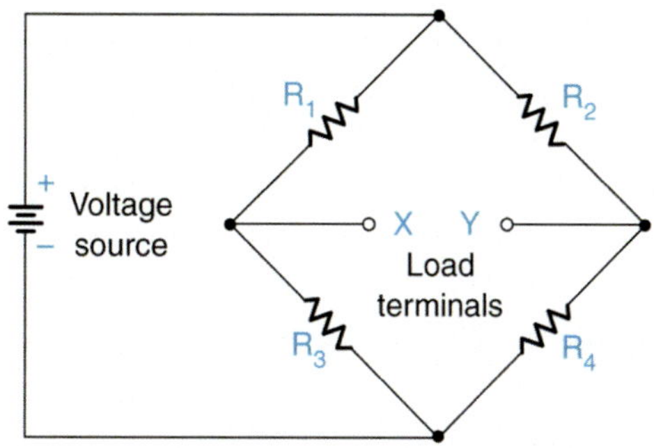

Figure 6-25 Bridge circuit.

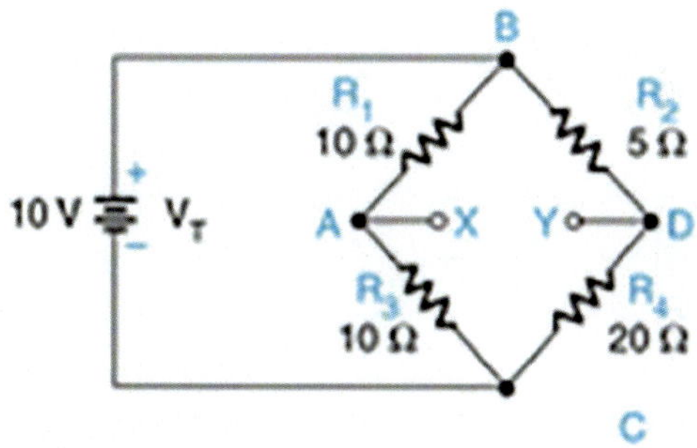

Problem-solving procedure

1. Find R_{TH}:

a. Resistor configuration looking from load terminals with voltage source circuited is:

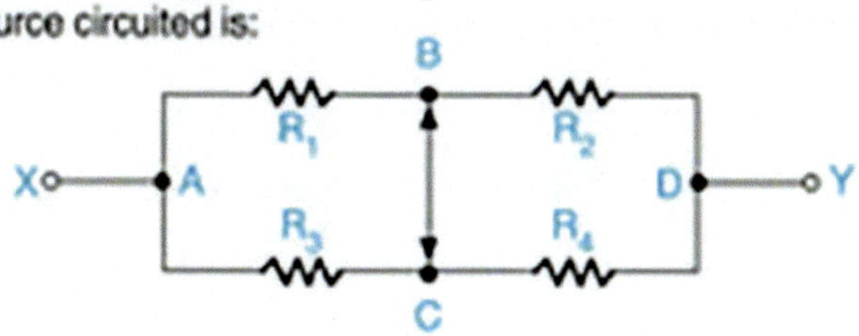

b. This configuration is the same as:

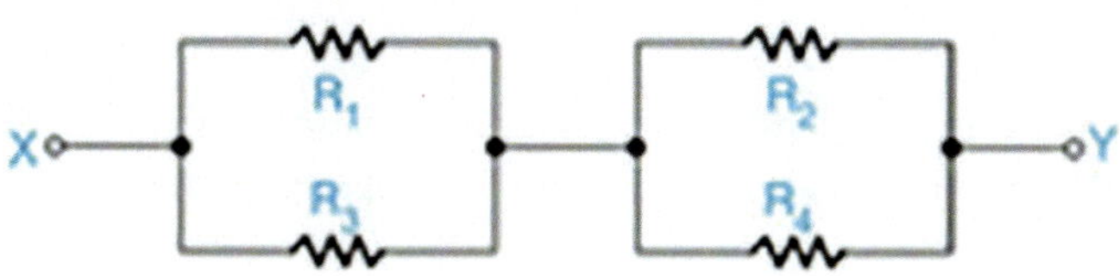

c. Therefore, $R_{TH} = R_1 \parallel R_3 + R_2 \parallel R_4$

$$= 10\ \Omega \parallel 10\ \Omega + 5\ \Omega \parallel 20\ \Omega$$

$$= 5\ \Omega + 4\ \Omega = 9\ \Omega$$

2. Find V_{TH}:

a. Disregard $R_2 - R_4$ path to find $I_1 = I_3$:

$$I_1 = I_3 = \frac{V_T}{R_1 + R_3} = \frac{10\ V}{20\ \Omega} = 0.5\ A$$

b. Disregard $R_1 - R_3$ path to find $I_2 = I_4$:

$$I_2 = I_4 = \frac{V_T}{R_2 + R_4} = \frac{10\ V}{25\ \Omega} = 0.4\ A$$

c. Calculate V_3:

$$V_3 = I_3 \times R_3 = 0.5\ A \times 10\ \Omega = 5\ V$$

d. Calculate V_4:

$$V_4 = I_4 \times R_4 = 0.4\ A \times 20\ \Omega = 8\ V$$

e. Subtract V_3 from V_4 to find V_{TH}:

$$V_{TH} = V_4 - V_3 = 8\ V - 5\ V = 3\ V$$

3. Complete the Thevinin equivalent circuit diagram:

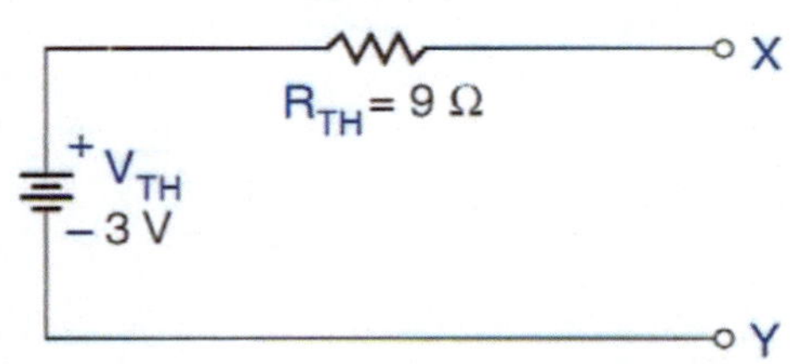

Figure 6-26 Simplification of a bridge circuit.

value with a known or standard value. Other applications of bridge circuits include rectification circuits, which convert alternating current into direct current.

Bridge circuits are difficult to analyze using Ohm's law techniques. The easiest method to use in developing a **Thevenin equivalent circuit is** to simplify the analysis of this type of circuit.

Use the following procedure to calculate the value of **Thevenin equivalent voltage (V_{TH})** and **equivalent resistance (R_{TH})** for the bridge circuit shown in **Figure 6-26**.

- Find R_{TH} by removing the load resistance from the circuit.
- Look **into** the circuit from the load (X and Y) terminals to determine R_{TH}. The power supply terminals should be replaced by a **short circuit**.
- The circuit arrangement of the four resistors is shown in **Figure 6-26(b)**.
- Calculate the $\mathbf{R_{TH}}$ of this arrangement and label its value on the equivalent circuit diagram.
- Find V_{TH} by disregarding R_2 and R_4 and calculate the current that would flow through R_1 and R_3 if R_2 and R_4 were disconnected from the circuit.
- Disregard R_1 and R_3 and calculate the current that would flow through R_2 and R_4 if R_1 and R_3 were disconnected from the circuit.

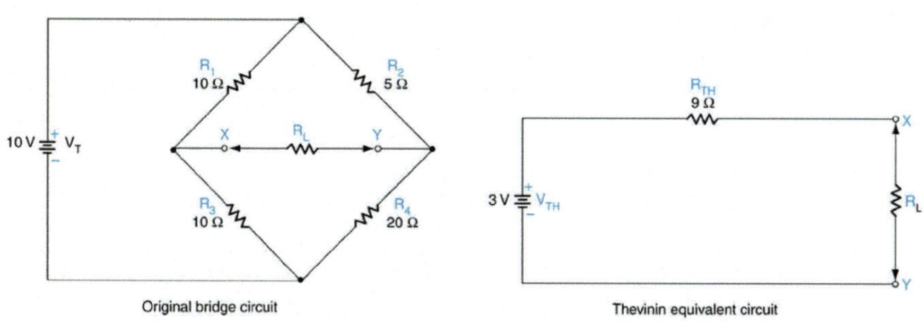

Load current and output voltage calculations:

R_L	$I_L = \frac{V_{TH}}{R_{TH} + R_L}$	$V_{out} = I_L \times R_L$
3 Ω	$\frac{3\,V}{12\,\Omega} = 0.25\,A$	0.25 A × 3 Ω = 0.75 V
6 Ω	$\frac{3\,V}{15\,\Omega} = 0.2\,A$	0.2 A × 6 Ω = 1.2 V
9 Ω	$\frac{3\,V}{18\,\Omega} = 0.16\,A$	0.16 A × 9 Ω = 1.44 V
12 Ω	$\frac{3\,V}{21\,\Omega} = 0.143\,A$	0.143 A × 12 Ω = 1.71 V

Figure 6-27 Calculating load current and voltage output of a bridge circuit.

- Calculate the voltage drop across R_3 ($V_3 \times I_3$) with the current determined in the first step.
- Calculate the voltage drop across R_4 due to the current determined in the second step
- ($V_4 = I_4 = R_4$).
- Subtract V_3 from V_4. This is the difference in potential (**voltage drop**) across points X
- and Y. This value is the $\mathbf{V}_{TH}$ of the circuit, which should be labeled on the equivalent circuit diagram.

The use of a **Thevenin equivalent circuit** greatly simplifies the calculation of **load current** and **voltage output** of bridge circuits. Note in **Figure 6-27** that addition of a load resistance to a bridge circuit produces a complex circuit configuration. Simplification of a bridge circuit using the Thevenin method provides an easy way to analyze bridge circuits.

Summary

- A series circuit has one path for current flow.
- A parallel circuit has two or more paths for current flow.
- A combination (series–parallel) circuit has both series and parallel paths.
- Meters are placed in series to measure current.
- To measure current, place the meter across a resistor, observing polarity.
- Problem-solving methods include Kirchhoff's laws (KVL and KCL), superposition, Thevenin, and Norton equivalent methods and bridge-circuit simplification.
- A voltage-divider circuit supplies multiple voltage outputs to load circuits.
- A negative voltage is delivered as negative with respect to ground.
- Currents are added in parallel branches to find total current.
- The sum of individual currents equals total current in combination circuits.
- In series circuits, the individual sums equal source voltage.

Self-examination/Answers

1. What is a series circuit?
2. What is a parallel circuit?

3. What is a combination circuit?
4. How is total resistance of a series circuit measured?
5. How is total resistance of a parallel circuit measured?
6. How is total current of a series circuit measured?
7. How is total current of a parallel circuit measured?
8. How is voltage drop measured for a series circuit?
9. How is voltage drop measured for a parallel circuit?
10. What are Kirchhoff's laws?What is a voltage-divider circuit?
12. What is meant by a negative voltage?
13. How does the total current of a combination circuit compare with the current through its series components?
15. How does the sum of the currents in the parallel paths of a combination circuit compare with the total current of the circuit?
17. How does the sum of the voltage across the series components and the parallel paths of a combination circuit compare with the source voltage?

Answers

1. One path for current
2. Two or more paths for current
3. Has both series and parallel parts
4. Meter in series with voltage source
5. Remove voltage terminals; place meter across terminals
6. Remove one voltage terminal; Place meter in series with terminal
7. Same as # 6
8. Place meter across resistor, observing polarity
9. Same as # 8
10. KVL and KCL
11. Supplies multiple voltage outputs to load circuits
12. Voltage delivered negative with respect to ground
13. Currents are added
14. Sum of individual currents equals total current
15. The individual sums equal source voltage

Problems

Combination Circuit Problems

Solve each of the following combination circuit problems.

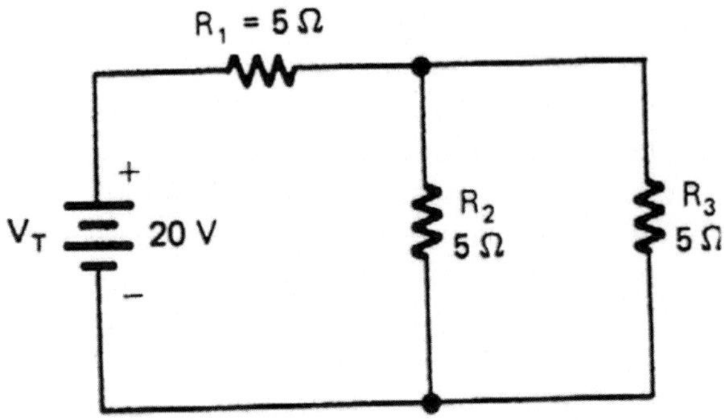

Figure 6-28 Combination circuit problem.

Find each of the following for **Figure 6-28**:

Total resistance (R_T) = W.
Total current (I_T) = A.
Voltage across R_1 (V_1) = V.
Total power (P_T) = __________ W.
Current through R_2 (I_2) = ______ A.
Voltage across R_2 (V_2) = V.

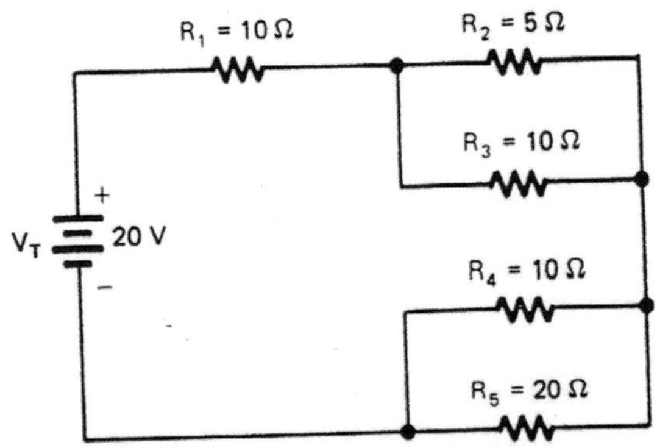

Figure 6-29 Combination circuit problem.

Find each of the following for **Figure 6-29**:

Total resistance (R_T) = W.
Total current (I_T) = A.
Total power (P_T) = W.
Voltage across R_1 (V_1) = ________ V.
Current through R_2 (I_2) = A.
Current through R_4 (I_4) = A.
Voltage across R_5 (V_5) = V.

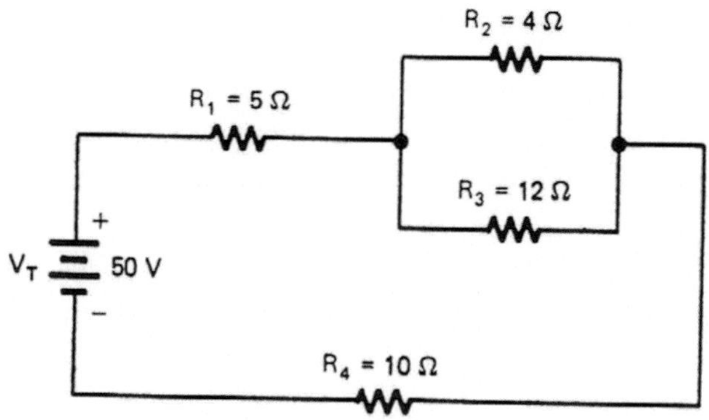

Figure 6-30 Combination circuit problem.

Find each of the following for **Figure 6-30**:

Total resistance (R_T) = W.
Total current (I_T) = A.
Voltage across resistor R_1 (V_1) = V.
Voltage across resistor R_4 (V_4) = V.
Current through resistor R_2 (I_2) = A.
Current through resistor R_3 (I_3) = A.

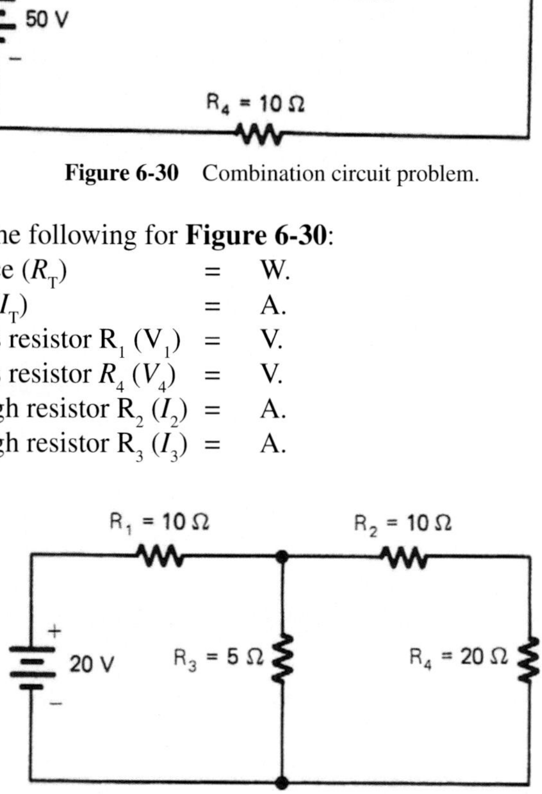

Figure 6-31 Combination circuit problem.

Find each of the following for **Figure 6-31**:

Total resistance (R_T) = W.
Total current (I_T) = A.
Voltage across resistor R_1 (V_1) = V.
Voltage across resistor R_2 (V_2) = V.
Voltage across resistor R_3 (V_3) = V.
Current through resistor R_3 (I_3) = A.
Current through resistor R_2 (I_2) = A.
Total power (P_T) = W.
Power converted by R_3 (P_3) = W.

Voltage-divider Circuit Problem

Two transistors are to be supplied the following DC voltages and currents: 5 V at 10 mA and 3 V at 5 mA. Use the circuit in **Figure 6-32** to determine the values of R_1, R_2, and R_3 required for this voltage-divider circuit.

R_1 = ______ □□□□. R_2 = ______ □□□□. R_3 = □.

Determine the minimum power rating of each resistor in the voltage divider of **Figure 6-32**:

P_{R1} = ______ W; P_{R2} = W; P_{R3} = W.

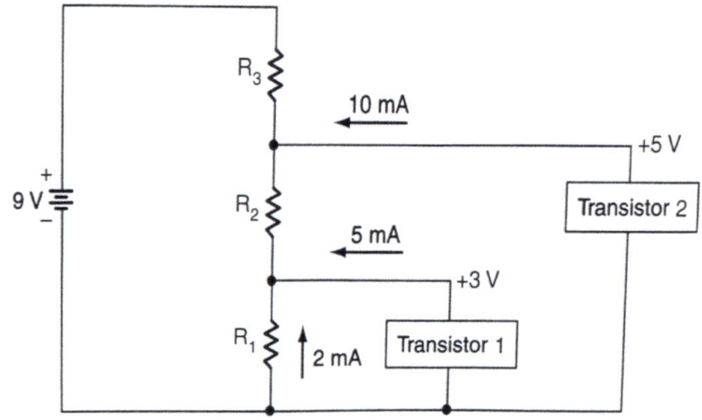

Figure 6-32 Voltage-divider problem.

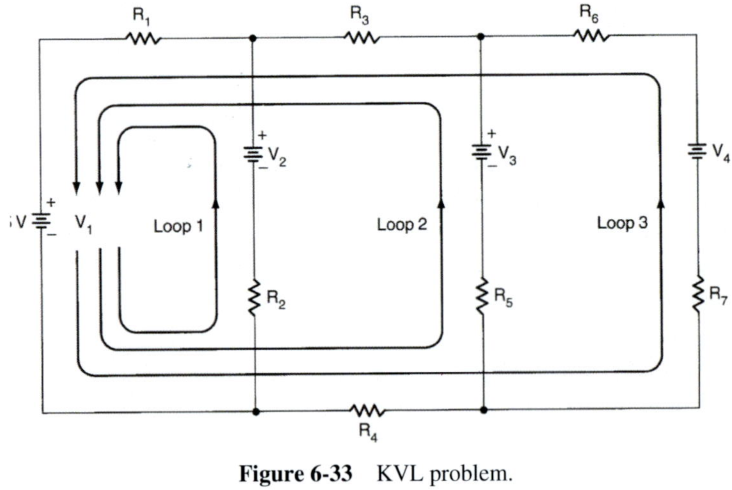

Figure 6-33 KVL problem.

Kirchhoff's Voltage Law (KVL) Problem

Refer to **Figure 6-79**. Set up equations for loops 1, 2, and 3 of the circuit.

Kirchhoff's Current Law (KCL) Problem

Solve each of the following problems by applying **Kirchhoff's current law** to the circuit shown in **Figure 6-34**.

I_1 in (a) = A. I_2 in (b) = A. I_3 in (c) = A. I_4 in (d) = _______ A.

Maximum Power Transfer Problem

Solve the following **maximum power transfer** problems using **Figure 6-35**.

Find the values of load current (I_L), voltage output (V_{out}), and power output (P_{out}) for the circuit of **Figure 6-35**, using load resistance values of 0, 1, 2, 3, 4, 5, 6, 7, and 8 W.

Draw a power transfer curve using the values obtained for **Figure 6-35**. Plot power output (watts) on the vertical axis and load resistance (R_L) on the horizontal axis.

Find the values of V_{out} and P_{out} for the circuit of **Figure 6-35** if the value of input voltage is changed to 20 V, and R_S is changed to 3 W. Use the same R_L values as the preceding problem.

On a sheet of **graph paper**, plot the relationship of R_L (horizontal axis) and power (vertical axis) for the values given here:

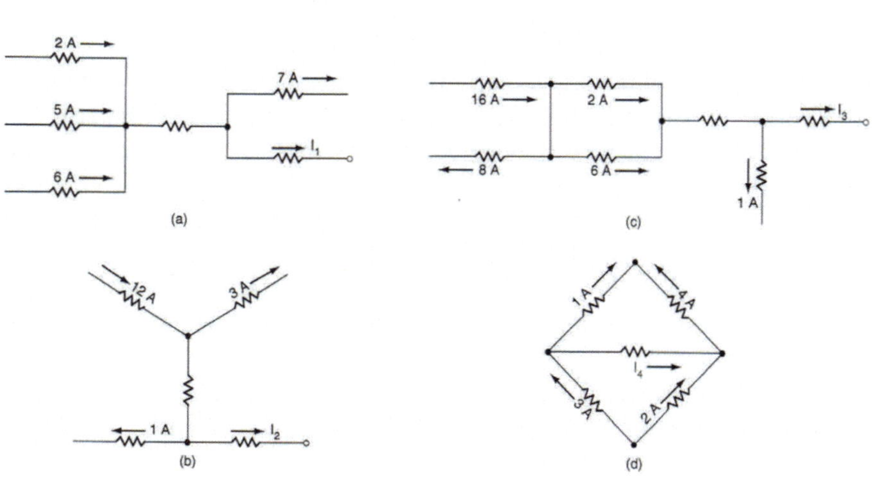

Figure 6-34 KCL problem.

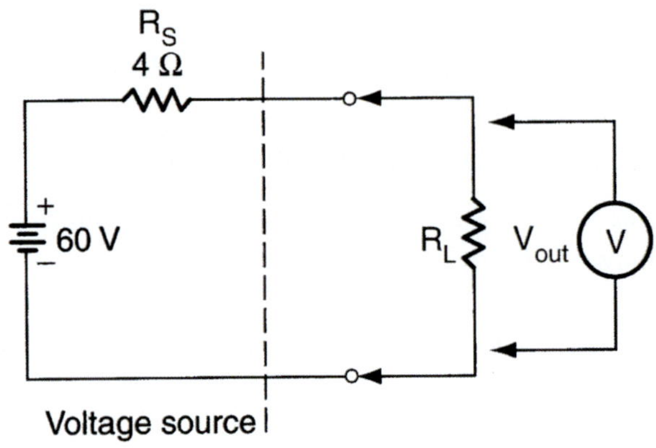

Figure 6-35 Maximum power transfer problem.

R_L	P_{out}
0 Ω	0 W
250 Ω	4500 W
500 Ω	5000 W
750 Ω	4800 W
1000 Ω	4400 W

Superposition Problems

Solve each of the following problems by applying the superposition method.

Find the current flow through R_1, R_2, and R_3 in the circuit of **Figure 6-36**.

Find the current flow through R_1, R_2, R_3, R_4, and R_5 in the circuit of **Figure 6-37**.

Thevenin Equivalent Circuit Problems

Solve each of the following problems by applying **Thevenin's equivalent circuit method**.

Find the Thevenin voltage (V_{TH}) and Thevenin resistance (R_{TH}) for the circuit of **Figure 6-85**. Sketch the **Thevenin equivalent circuit**.

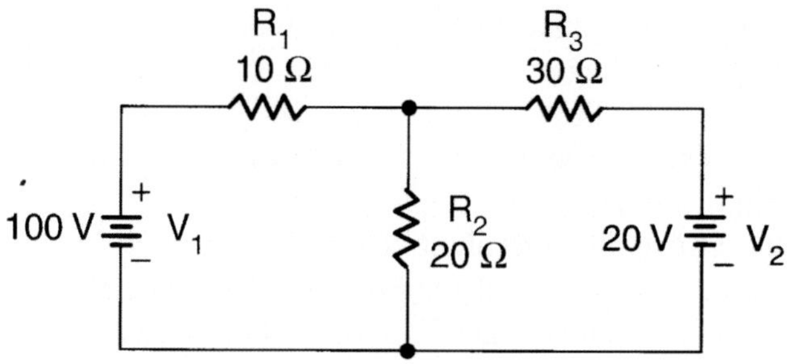

Figure 6-36 Superposition problem.

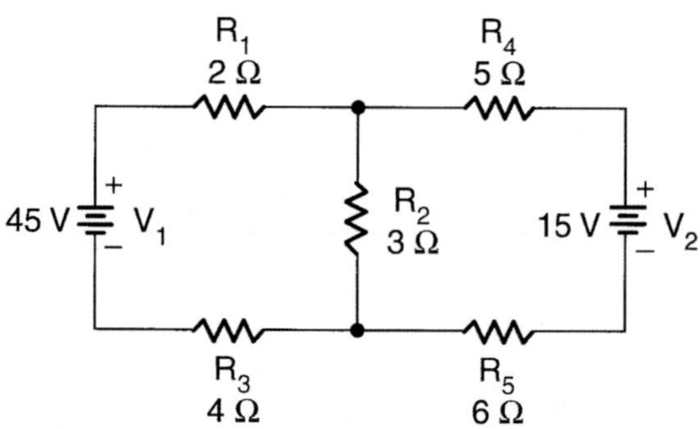

Figure 6-37 Superposition problem – two-source.

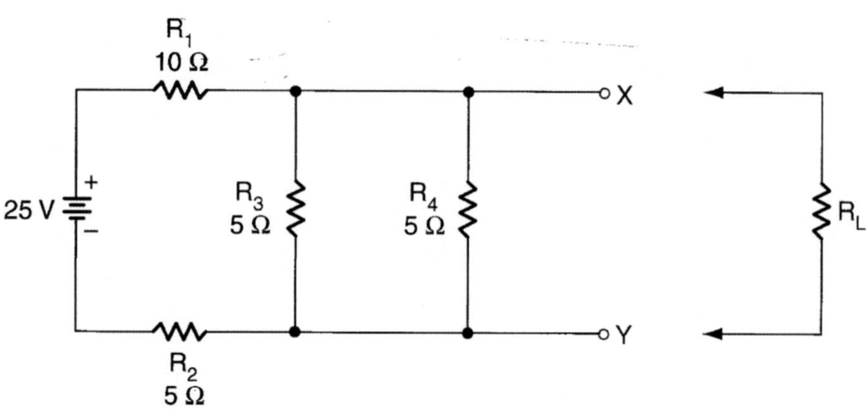

Figure 6-38 Thevenin equivalent circuit problem.

Refer to the values obtained for **Figure 6-38**. Calculate the values of load current (I_L) and voltage output (V_{out}) for load resistance values of: (a) 2 W, (b) 3 W, (c) 4 W, and (d) 5 W.

Find the **Thevenin voltage (V_{TH})** and **Thevenin resistance (R_{TH})** for the circuit of **Figure 6-86.**

Sketch the **Thevenin equivalent circuit**.

Refer to the values obtained for **Figure 6-39**. Calculate the values of load current (I_L) and output voltage (V_{out}) for load resistance values of: (a) 20 I, (b) 30 W, and (c) 50 f.

Find the values of V_{TH} and R_{TH} for the **two-source** circuit of **Figure 6-40.**

Calculate I_L and V_{out} for R_L values of: (a) 10 W, (b) 20 W, and (c) 30 W (for **Figure 6-40**).

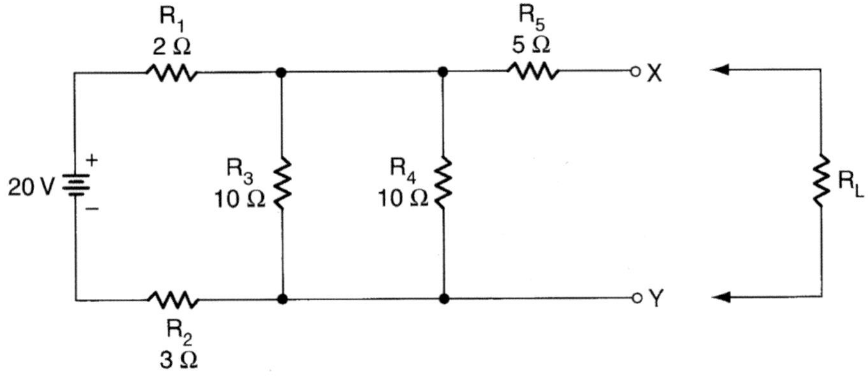

Figure 6-39 Thevenin equivalent circuit problem.

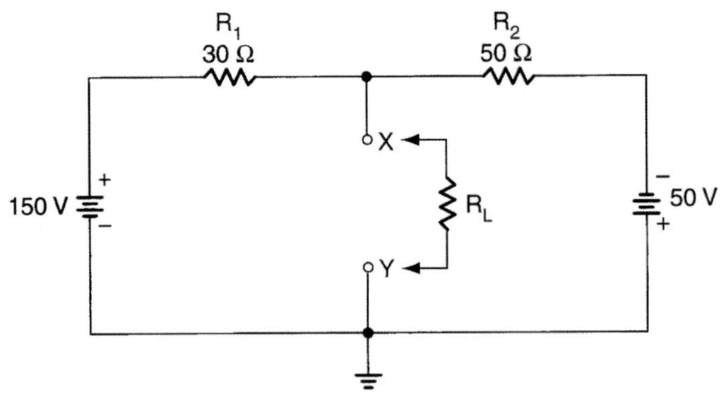

Figure 6-40 Thevenin equivalent circuit problem.

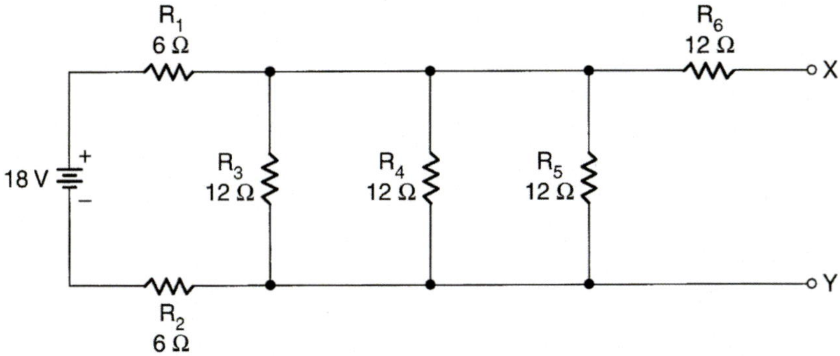

Figure 6-41 Norton equivalent circuit problem.

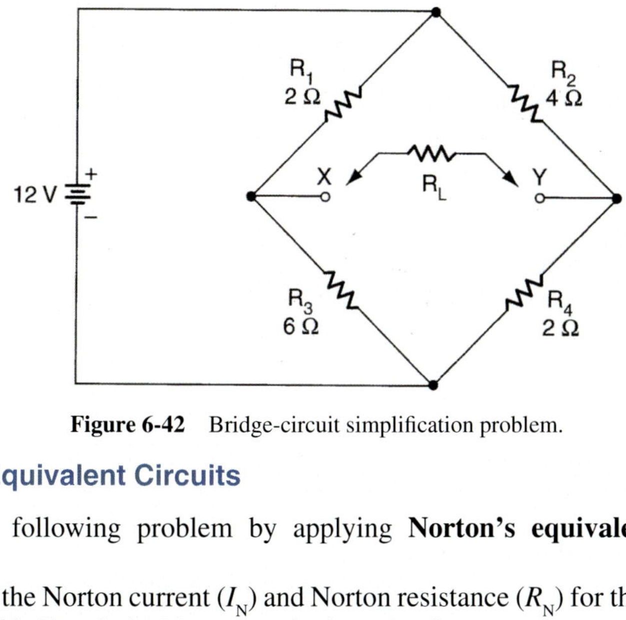

Figure 6-42 Bridge-circuit simplification problem.

Norton Equivalent Circuits

Solve the following problem by applying **Norton's equivalent circuit method**.

Find the Norton current (I_N) and Norton resistance (R_N) for the circuit of **Figure 6-41**. Sketch the Norton equivalent circuit.

Bridge-circuit Simplification

Solve the following problems by applying **bridge-circuit simplification**.

Find the values of V_{TH} and R_{TH} for the bridge circuit of **Figure 6-42**.

Refer to the values obtained for **Figure 6-42**. Calculate the values of load current (I_L) and output voltage for R_L values of: (a) 3 W, (b) 5 W, and (c) 8 W.

Glossary

Branch
A path of a parallel circuit.

Branch current
The current through a parallel branch.

Branch resistance
The total resistance of a parallel branch.

Branch voltage
The voltage across a parallel branch.

Circuit
A path through which electrical current flows.

Combination circuit
A circuit that has one portion connected in series with the voltage source and another part connected in parallel.

Complex circuit
See combination circuit.

Current
The movement of electrical charge; the flow of electrons through an electrical circuit.

Difference in potential
The voltage across two points of a circuit.

Directly proportional
When one quantity increases or decreases, causing another quantity to do the same.

Equivalent resistance
A resistance value that would be the same value in a circuit as two or more parallel resistances of a circuit.

Inverse
The value of 1 divided by some quantity, such as $1/R_T$ for finding parallel resistance.

Inversely proportional
When one quantity increases or decreases, causing another quantity to do the opposite.

Kirchhoff's current law
The sum of the currents flowing into any point or junction of conductors of a circuit is equal to the sum of the currents flowing away from that point.

Kirchhoff's voltage law
In any loop of a circuit, the sum of the voltage drops is equal to the voltage supplied to that loop; or taken with proper signs (- or +), the algebraic sum of the voltage sources and voltage drops in a circuit is equal to zero.

Ohm's law
The law that explains the relationship of voltage, current, and resistance in electrical circuits.

Parallel circuit
A circuit that has two or more current paths.

Power (P)
The rate of doing work in electrical circuits, found by using the equation $P = I \times V$.

Reciprocal
See Inverse.

Resistance (R)
Opposition to the flow of current in an electrical circuit; its unit of measurement is
the ohm (W).

Series circuit
A circuit that has one path for current flow.

Total current
The current that flows from the voltage source of a circuit.

Total resistance
The total opposition to current flow of a circuit, which may be found by removing the voltage source and connecting an ohmmeter across the points where the source was connected.

Total voltage
The voltage supplied by a source.

Voltage
The electrical force or pressure that causes current to flow in a circuit.

Voltage drop
The voltage across two points of a circuit, found by using the equation $V = I \times R$.

Watt (W)
The unit of measurement of electrical power; the amount of power converted when 1 A of current flows under a pressure of 1 V.

7

Magnetism and Electromagnetism

Magnetism has been a topic of study for many years. Some metals in their natural state attract small pieces of iron. This property is called magnetism. Materials that have this ability are called natural magnets. The first magnets used were called lodestones. Now, artificial magnets are made in many different strengths, sizes, and shapes. Magnetism is important because it is used in electric motors, generators, transformers, relays, and many other electrical devices. The earth itself has a magnetic field like a large magnet.

Electromagnetism is magnetism which is brought about due to electrical current flow. There are many electrical machines which operate because of electromagnetism. This chapter deals with magnetism, electromagnetism, and some important applications.

Chapter Outline

7.1 Permanent Magnets
7.2 Magnetic Field Development
7.3 Electromagnetism
7.4 Magnetic Theory
7.5 Magnetic Devices
7.6 Magnetic Terminology

Objectives

1. Define terms relative to magnetism
2. Explain the operation of various magnetic devices
3. State Faraday's law for electromagnetic induction
4. List three factors that affect the strength of electromagnets
5. Apply the left-hand rule for determining polarity
6. Describe the construction of a relay and solenoid

7. Define the terms *residual magnetism*, *permeability*, *retentivity*, *magnetic saturation*, and *magnetizing force*
8. Describe the domain theory of magnetism

7.1 Permanent Magnets

Magnets are made of iron, cobalt, or nickel materials, usually in an alloy combination. An alloy is a mixture of these materials. Each end of the magnet is called a pole. If a magnet were broken, each part would become a magnet. Each magnet would have two poles. Magnetic poles are always in pairs. When a magnet is suspended in air so that it can turn freely, one pole will point to the north pole of the earth. The earth is like a large permanent magnet. This is why compasses can be used to determine direction. The north pole of a magnet will attract the south pole of another magnet. A north pole repels another north pole, and a south pole repels another south pole. The two laws of magnetism are: (1) like poles repel, and (2) unlike poles attract.

The magnetic field patterns when two permanent magnets are placed end to end are shown in **Figure 7-1**. When the magnets are farther apart, a smaller force of attraction or repulsion exists. This type of permanent magnet is called a bar magnet.

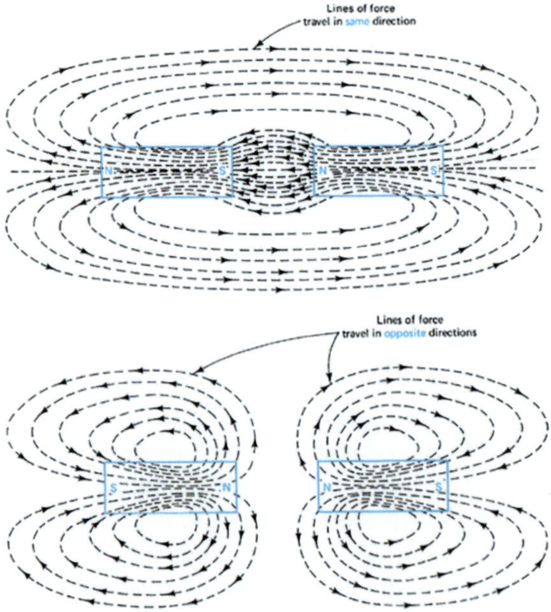

Figure 7-1 Magnetic field patterns when magnets are placed end to end.

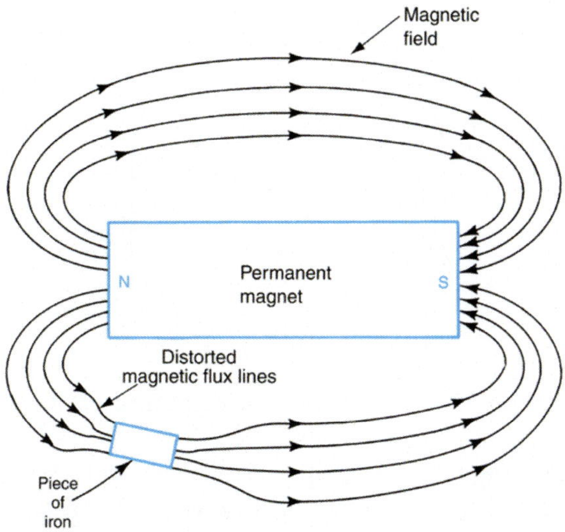

Figure 7-2 Magnetic flux lines distorted while passing through a piece of iron.

Some materials retain magnetism longer than others. Hard steel holds its magnetism much longer than soft steel. A **magnetic field** is set up around any magnetic material. The field is made up of lines of force or magnetic flux. These magnetic flux lines are invisible. They never cross one another, but they always form individual closed loops around a magnetic material. They have a definite direction from the north pole to the south pole along the outside of a magnet. When **magnetic flux lines** are close together, the magnetic field is strong. When magnetic flux lines are farther apart, the field is weaker. The magnetic field is strongest near the poles.

Lines of force pass through all materials. It is easy for lines of force to pass through iron and steel. Magnetic flux passes through pieces of iron as shown in **Figure 7-2**.

When magnetic flux passes through a piece of iron, the iron acts like a magnet. Magnetic poles are formed due to the influence of the flux lines. These are called induced poles. The induced poles and the magnet's poles attract and repel each other. Magnets attract pieces of soft iron in this way. It is possible to temporarily magnetize pieces of metal by using a **bar magnet**. If a magnet is passed over the top of a piece of iron several times in the same direction, the soft iron becomes magnetized. It will stay magnetized for a short time.

When a **compass** is brought near the **north pole** of a magnet, the north-seeking pole of the compass is attracted to it. The polarities of the magnet may

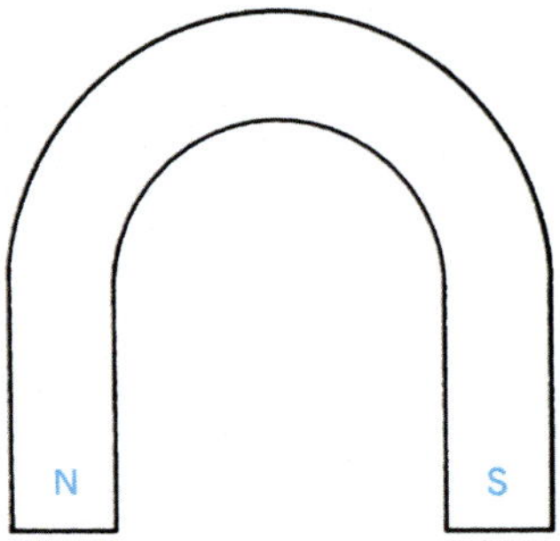

Figure 7-3 Horseshoe magnet.

be determined by observing a compass brought near each pole. Compasses detect the presence of magnetic fields.

Horseshoe magnets are similar to bar magnets. They are bent in the shape of a horseshoe, as shown in **Figure 7-3**. This shape gives more magnetic field strength than a similar bar magnet because the magnetic poles are closer. The magnetic field strength is more concentrated into one area. Many electrical devices use horseshoe magnets.

A magnetic material can lose some of its magnetism if it is jarred or heated. People must be careful when handling equipment that contains **permanent magnets**. A magnet also becomes weakened by loss of magnetic flux. Magnets should always be stored with a keeper, which is a soft-iron piece used to join magnetic poles. The keeper provides the magnetic flux with an easy path between poles. The magnet will retain its greatest strength for a longer period of time if keepers are used. Bar magnets should always be stored in pairs with a north pole and a south pole placed together. A complete path for magnetic flux is made in this way.

7.2 Magnetic Field Development

Magnetic field around a conductor

Current-carrying conductors produce a **magnetic field**. It is possible to show the presence of a magnetic field around a current-carrying conductor. A compass is used to show that the magnetic flux lines are circular in shape. The conductor is in the center of the circular shape. The direction of the current flow and the magnetic flux lines can be shown by using the **left-hand rule** of magnetic flux. A conductor is held in the left hand as shown in **Figure 7-4(a)**. The thumb points in the direction of current flow from negative to positive. The fingers will then encircle the conductor in the direction of the **magnetic**

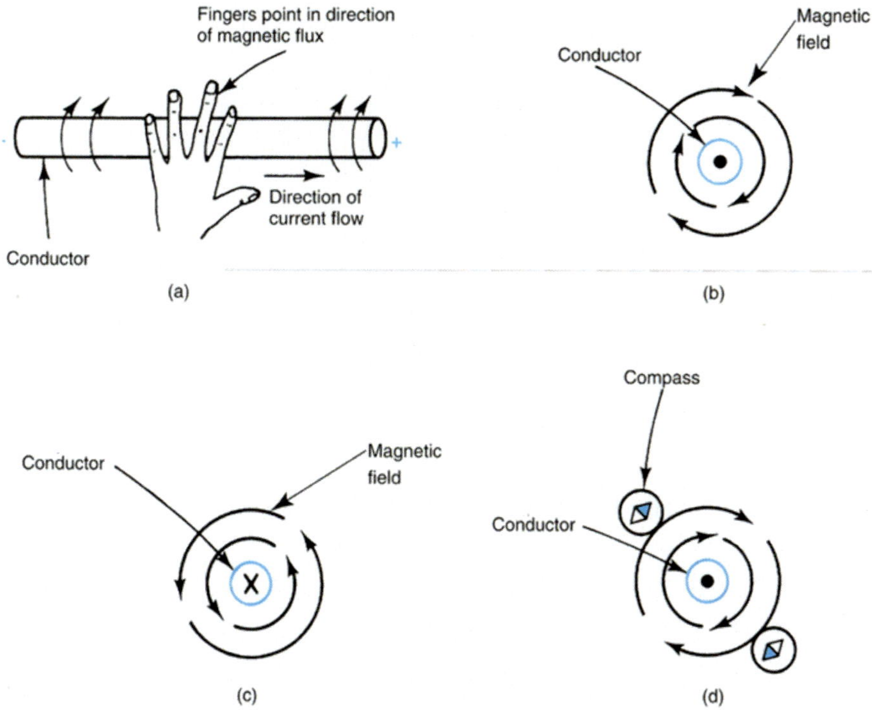

Figure 7-4 Magnetic fields: (a) left-hand rule of magnetic flux; (b) cross section of a conductor with current flow toward the observer; (c) cross section of a conductor with current flow away from the observer; (d) compass aligns tangent to the circular magnetic lines of force.

flux lines. A circular magnetic field is produced around a conductor. The field is stronger near the conductor and becomes weaker farther away from the conductor. A cross-sectional end view of a conductor with current flowing toward the observer is shown in **Figure 7-4(b)**. Current flow toward the observer is shown by a circle with a dot in the center. Note that the direction of the magnetic flux lines is clockwise. This can be verified by using the left-hand rule.

When the direction of current flow through a conductor is reversed, the direction of the magnetic lines of force is also reversed. The cross-sectional end view of a conductor in **Figure 7-4(c)** shows a current flow in a direction away from the observer. Note that the direction of the **magnetic lines of force** is now counterclockwise. The presence of magnetic lines of force around a current-carrying conductor can be observed by using a compass. When a compass is moved around the outside of a conductor, the needle will align

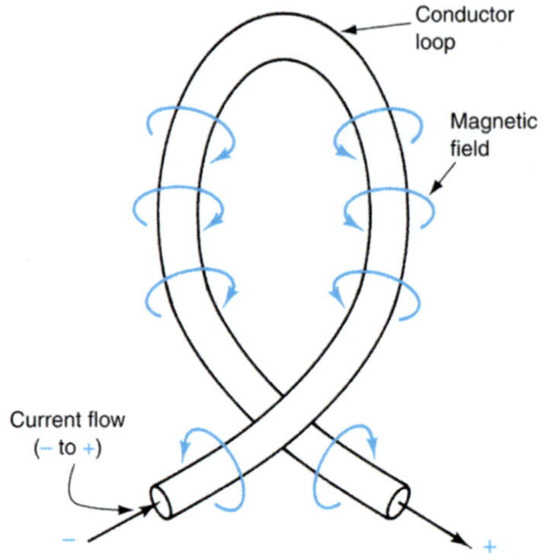

Figure 7-5 Magnetic field around a loop of wire.

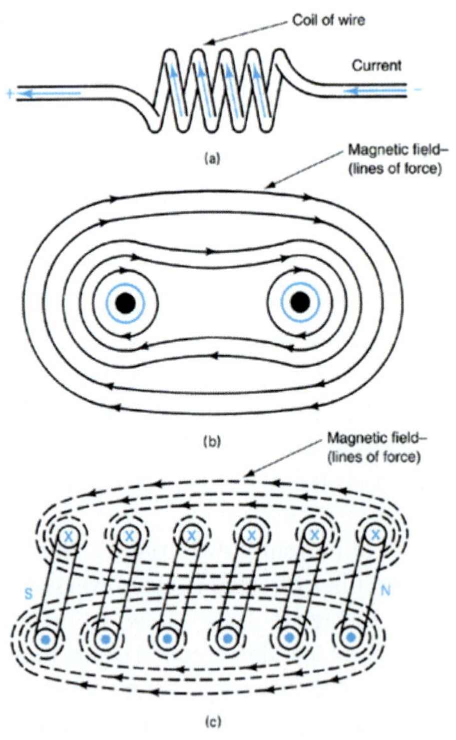

Figure 7-6 Magnetic field around a coil: (a) coil of wire showing current flow; (b) lines of force around two loops that are parallel; (c) cross section of a coil showing lines of force.

itself tangent to the lines of force as shown in **Figure 7-4(d)**. The needle will not point toward the conductor. When current flows in the opposite direction, the compass **polarities** reverse. The compass needle will align itself tangent to the conductor.

Magnetic Field around a Coil

The magnetic field around one loop of wire is shown in **Figure 7-5**. Magnetic flux lines extend around the conductor as shown. Inside the loop, the magnetic flux is in one direction.

When many loops are joined together to form a coil, the magnetic flux lines surround the coil as shown in **Figure 7-6**. The field around a coil is much stronger than the field of one loop of wire. The field around the coil is the same shape as the field around a bar magnet. A coil that has an iron or steel core inside it is called an **electromagnet**. A core increases the magnetic flux density of a coil.

7.3 Electromagnetism

Electromagnets are produced when current flows through a coil of wire as shown in **Figure 7-7**. The **north pole** of a coil of wire is the end where the lines of force exit. The **south pole** is the end where the lines of force enter the coil. This is like the field of a bar magnet. To find the north pole of a coil, use the left-hand rule for polarity, as shown in **Figure 7-8**. Grasp the coil with the left hand. Point the fingers in the direction of current flow through the coil. The thumb points to the north polarity of the coil. When the polarity of the voltage source is reversed, the magnetic poles of the coil will also reverse. The poles of an electromagnet can be checked with a compass. The compass is placed near a pole of the electromagnet. If the north-seeking pole of the compass points to the coil, that side is the north side.

Electromagnets have several turns of wire wound around a soft-iron core. An electrical power source is then connected to the ends of the turns of wire. When current flows through the wire, magnetic polarities are produced at the ends of the soft-iron core. The three basic parts of an electromagnet are (1) an iron core, (2) wire windings, and (3) an electrical power source. Electromagnetism is made possible by electrical current flow which produces a magnetic field. When electrical current flows through the coil, the properties of magnetic materials are developed.

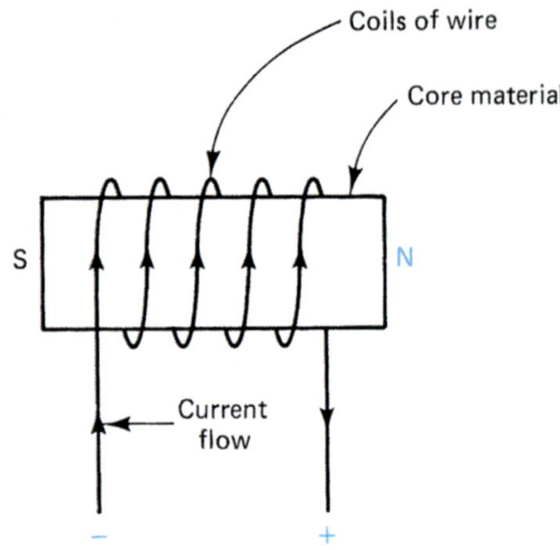

(a)

(b)

Figure 7-7 Electromagnets: (a) pictorial; (b) an electromagnet in operation [(b) courtesy of O.S. Walker Co.].

The **magnetic strength** of an electromagnet depends on three factors: (1) the amount of current passing through the coil, (2) the number of turns of wire, and (3) the type of core material. The number of magnetic lines of force is increased by increasing the current, by increasing the number of turns of wire, or by using a more desirable type of core material. The magnetic strength of electromagnets is determined by the **ampere-turns** of each coil.

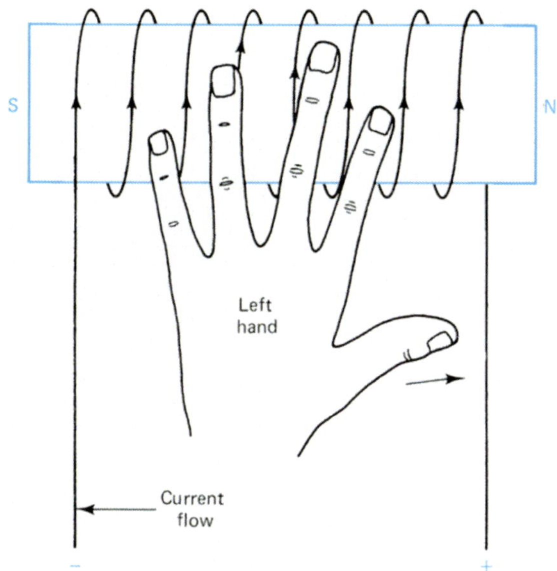

Figure 7-8 Left-hand rule for finding the polarities of an electromagnet.

The number of ampere-turns is equal to the current in amperes multiplied by the number of turns of wire ($I \times N$). For example, 200 ampere-turns is produced by 2 A of current through a 100-turn coil. One ampere of current through a 200-turn coil would produce the same magnetic field strength. **Figure 7-9** shows how the magnetic field strength of an electromagnet changes with the number of ampere-turns.

The **magnetic field strength** of an electromagnet also depends on the type of core material. Cores are usually made of soft iron or steel. These materials will transfer a magnetic field better than air or other nonmagnetic materials. Iron cores increase the **flux density** of an electromagnet. **Figure 7-10** shows that an iron core causes the magnetic flux to be denser.

An electromagnet loses its field strength when the current stops flowing. However, an electromagnet's core retains a small amount of magnetic strength after current stops flowing. This is called **residual magnetism** or leftover magnetism. It can be reduced by using soft-iron cores or increased by using hard-steel core material. Residual magnetism is very important in the operation of some types of electrical generators.

In many ways, electromagnetism is similar to magnetism produced by **natural magnets** such as bar magnets; however, the main advantage of electromagnetism is that it is easily controlled. It is easy to increase the strength of an electromagnet by increasing the current flow through the coil,

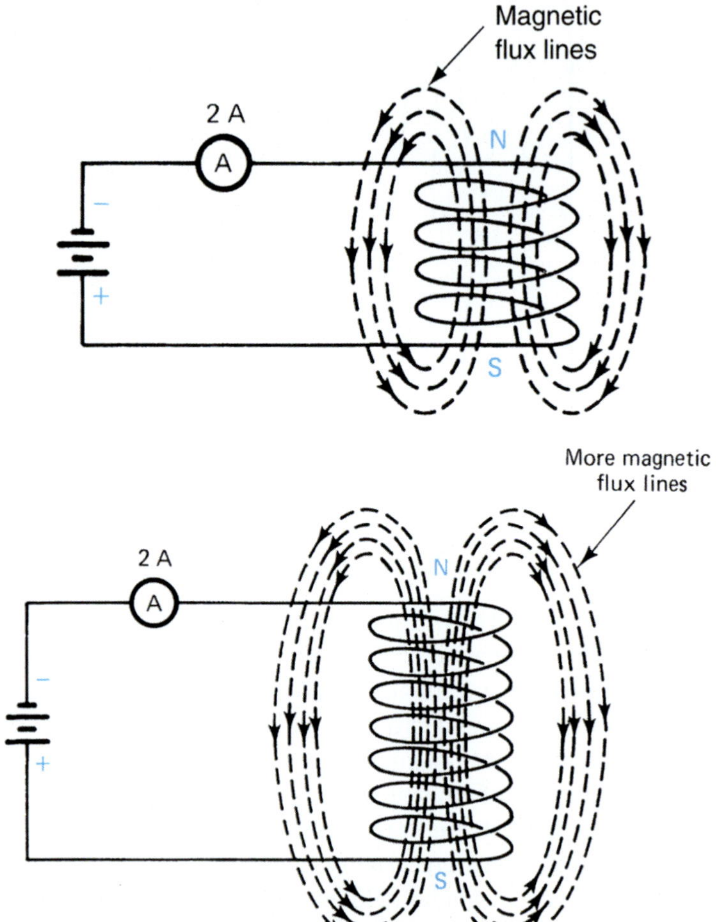

Figure 7-9 Effect of ampere-turns on magnetic field strength: (a) five turns and two amperes = 10 ampere-turns; (b) eight turns and two amperes = 16 ampere-turns.

which is done by increasing the voltage applied to the coil. The second way to increase the strength of an electromagnet is to have more turns of wire around the core. A greater number of turns produces more magnetic lines of force around the electromagnet. The strength of an electromagnet is also affected by the type of core material used. Different alloys of iron are used to make the cores of electromagnets. Some materials aid in the development of **magnetic lines of force** to a greater extent. Other types of core materials offer greater resistance to the development of magnetic flux around an electromagnet.

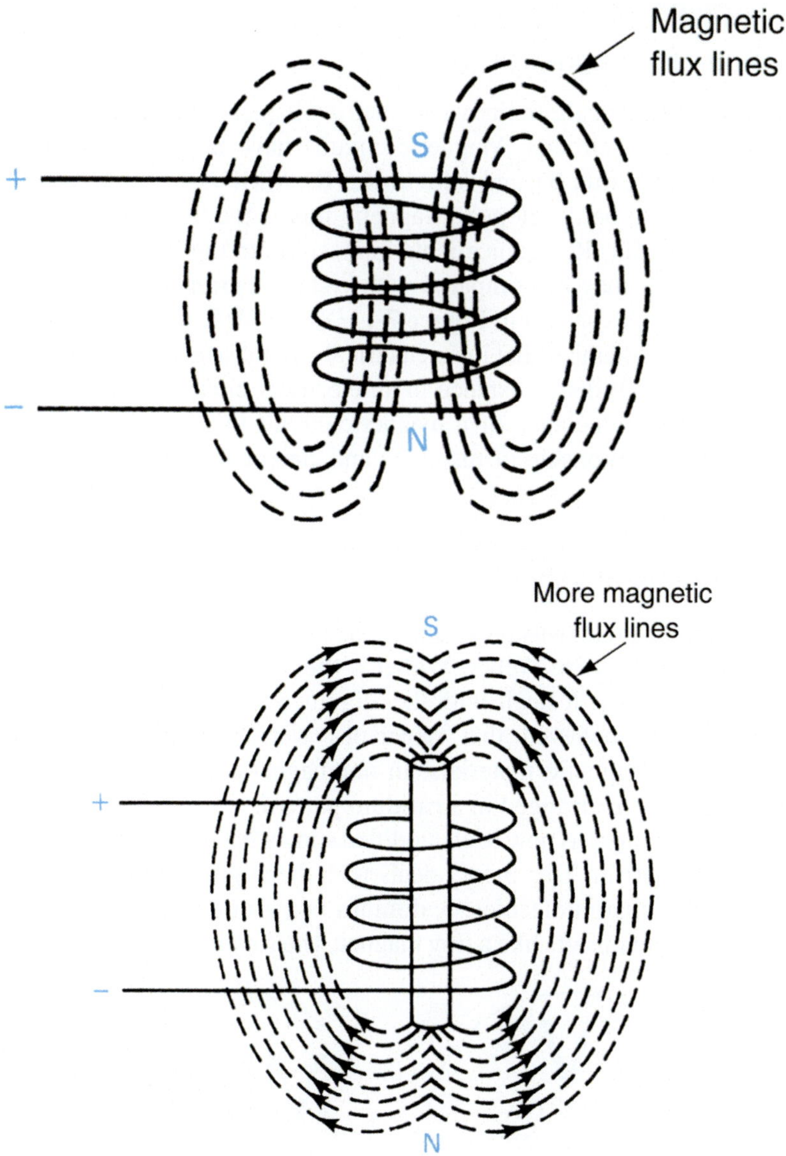

Figure 7-10 Effect of an iron core on magnetic strength: (a) coil without a core; (b) coil with a core.

7.4 Magnetic Theory

Ohm's Law for Magnetic Circuits

Ohm's law for electrical circuits was studied in Chapter 3. A similar relationship exists in magnetic circuits. Magnetic circuits have **magnetomotive force (MMF)**, **magnetic flux (f)**, and **reluctance (R)**. MMF is the force that causes a magnetic flux to be developed. Magnetic flux is the lines of force around a magnetic material. Reluctance is the opposition to the flow of a magnetic flux. These terms may be compared with voltage, current, and resistance in electrical circuits, as shown in **Figure 7-11**. When MMF increases, magnetic flux increases. Remember that in an electrical circuit, when voltage increases, current increases. When resistance in an electrical circuit increases, current decreases. When reluctance of a magnetic circuit increases, magnetic flux decreases. The relationship of magnetic and electrical terms in **Figure 7-11** is important to learn.

Domain Theory of Magnetism

A theory of magnetism was presented in the nineteenth century by a German scientist named Wilhelm Weber. **Weber's theory** of magnetism was called the molecular theory. It dealt with the alignment of molecules in magnetic materials. Weber believed that molecules were aligned in an **orderly** arrangement in magnetic materials. In nonmagnetic materials, he believed that molecules were arranged in a **random** pattern.

Weber's theory has now been modified somewhat to become the **domain theory** of magnetism. This theory deals with the alignment of "domains" in materials rather than molecules. A **domain** is a group of atoms (about 10^{15} atoms). Each domain acts like a tiny magnet. The rotation of electrons around

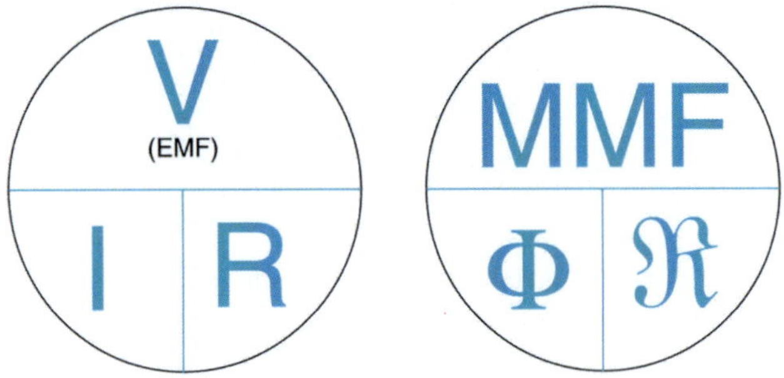

Figure 7-11 Relationship of magnetic and electrical terms.

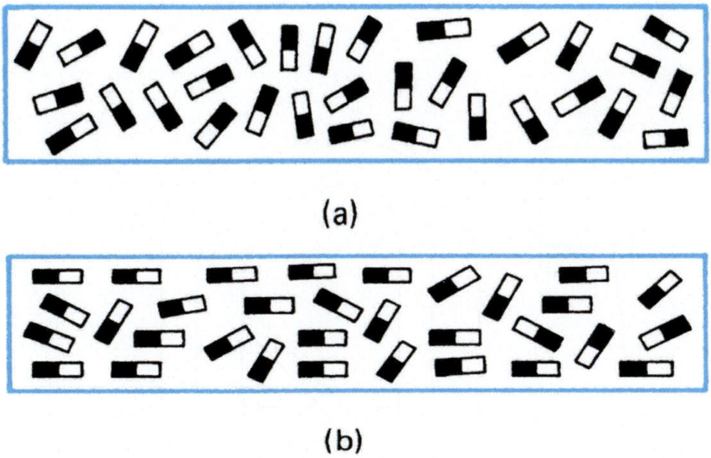

Figure 7-12 Domain theory of magnetism: (a) unmagnetized; (b) slightly magnetized; (c) fully magnetized saturation.

the nucleus of these atoms is important. Electrons have a negative charge. As they orbit around the nucleus of atoms, their electrical charge moves. This moving electrical field produces a magnetic field. The polarity of the magnetic field is determined by the direction of electron rotation.

The domains of magnetic materials are atoms grouped together. Their electrons are believed to spin in the same directions. This produces a magnetic field due to electrical charge movement. **Figure 7-12** shows the arrangement of domains in magnetic, nonmagnetic, and partially magnetized materials. In nonmagnetic materials, half of the electrons spin in one direction and half in the other direction. Their charges cancel each other out. There is no magnetic field produced because the charges cancel. Electron rotation in magnetic materials is in the same direction. This causes the domains to act like tiny magnets that align to produce a magnetic field.

Electrical Production

A scientist named Michael Faraday discovered in the early 1830s that electricity is produced from magnetism. He found that if a magnet is placed inside a coil of wire, electrical current is produced when the magnet is **moved**. Faraday found that electrical current is caused by magnetism and motion. **Faraday's law** is stated as follows: when a coil of wire moves across the lines of force of a magnetic field, electrons flow through the wire in one direction. When the coil of wire moves across the magnetic lines of force in the opposite direction, electrons flow through the wire in the opposite direction.

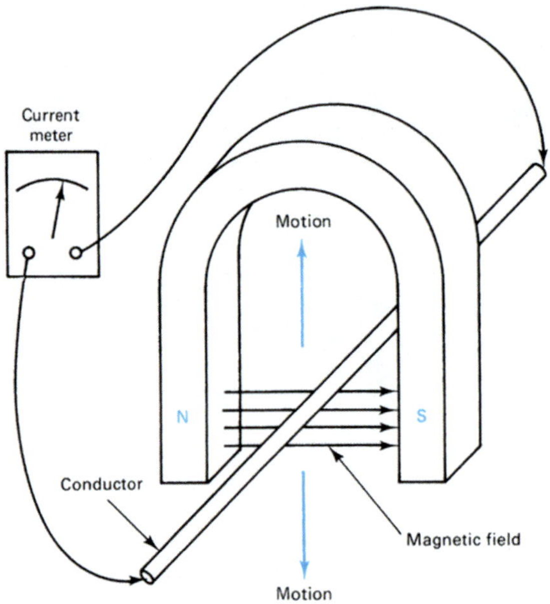

Figure 7-13 Faraday's law: electrical current is produced when there is relative motion between a conductor and a magnetic field.

This law is the principle of electrical power generation produced by magnetism. **Figure 7-13** shows Faraday's law as it relates to electrical power generation. Current flows in a conductor placed inside a magnetic field only when there is motion between the conductor and the magnetic field. If a conductor is stopped while moving across the magnetic lines of force, current stops flowing. The operation of electrical generators depends on conductors moving across a magnetic field. This principle is called **electromagnetic induction** and is discussed in more detail in a subsequent chapter.

7.5 Magnetic Devices

Many types of electrical devices operate due to the effects of magnetism or electromagnetism. Among these devices are relays, solenoids, and magnetic motor contactors.

Relays

Relays are electrical devices that rely on magnetism to operate. They control other equipment such as motors, lights, or heating elements. Relays are important devices. They are used to start the operation of other equipment. They use a small amount of electrical current to control a larger current, such as the current through a motor.

The basic construction and symbols of a relay are shown in **Figure 7-14**. A relay has an electromagnetic coil with electrical power applied to its two external leads. When the power is turned on, the electromagnet is energized. The electromagnet part of the relay controls a set of contacts. The contacts are called **normally open (NO)** or **normally closed (NC)**, depending on their conditions when the electromagnet is not energized. There is also **common contact**.

(a)

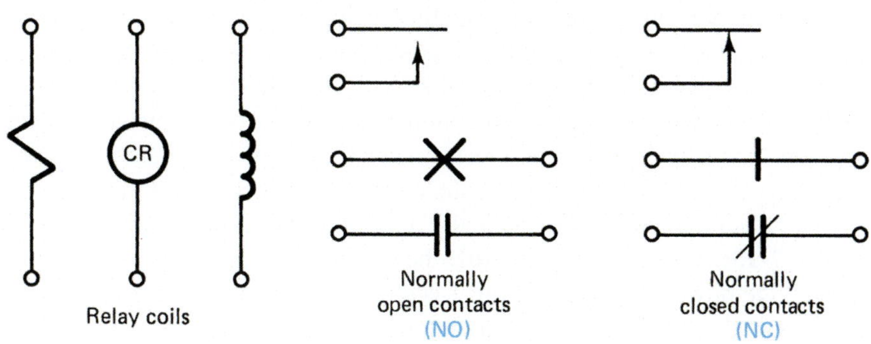

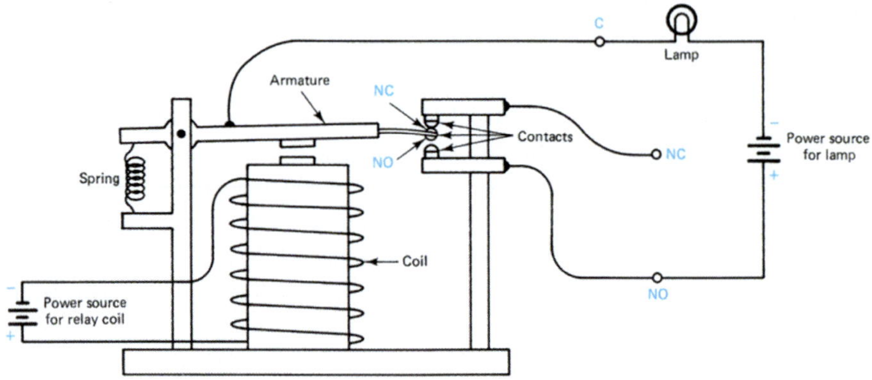

Figure 7-14 Construction and symbols of a relay: (a) pictorial; (b) symbols; (c) illustration [(a) Courtesy of Airpax Corp., Cambridge Division].

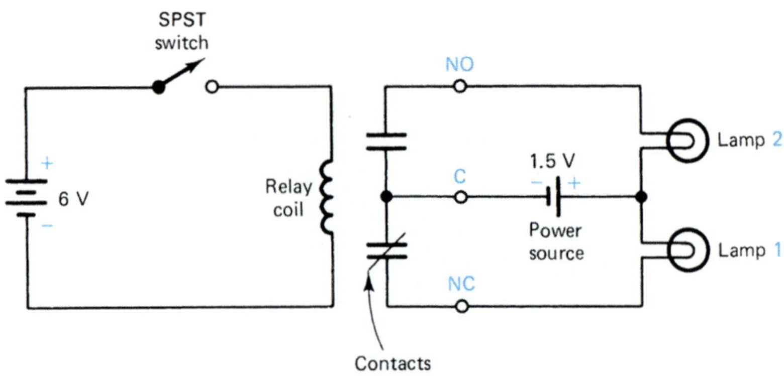

Figure 7-15 Schematic diagram of a relay circuit.

If a lamp and its power source are connected in series with the common and normally open contact as shown in **Figure 7-14(b)**, the lamp will be off when the relay is not energized. Note that the lamp or any load connected to the relay contacts requires a separate power source. If the relay is energized by applying power to it, the common contact is attracted to the normally open contact by magnetic energy. The common contact is built onto a movable armature that moves when the electromagnet is energized. When the relay is energized, the light connected to the normally open contact will turn on.

In a similar way, the normally closed contacts are used (refer to **Figure 7-15**). When the relay is off, the circuit from the common terminal through the power source and lamp 1 is closed. This causes lamp 1, which is in series with the normally closed contacts, to be turned off. Also note that the power source for the lamps is in the common line because it is common to both of the other contacts. When the relay is turned on, the lamps will change states. Lamp 2, connected to the normally open contact, will be turned on, while lamp 1, connected to the normally closed contact, will be turned off. Note, in **Figure 7-14(b)**, how the common contact moves from the NC contact so that it touches the NO contact. This shows the basic operation of a relay with normally open and normally closed contacts. Such a relay is common. It is called a **double-pole single-throw (DPST)** relay. There are many other types used.

The coil resistance of a relay is determined by the size of the wire used to wind the coil and the number of windings. A coil with only a few turns of large-diameter wire has a low resistance. This causes a high current flow through the relay coil. If a relay coil has many turns of small-diameter wire, it has a high resistance. Remember that small wire has high resistance and large wire offers lower resistance to current flow. Coil resistance is usually marked on a relay. It may also be measured with a meter.

There are some important current ratings for relays. Two of these ratings are called **pickup current** and **dropout current**. These ratings are usually specified on the relay. They may also be found by using a variable power source and a current meter to measure the actual values. When the voltage applied to the relay coil is increased to a point where the relay turns on, the current indicated on the meter is the pickup current. Pickup current is the minimum current required to energize the relay. If the applied voltage is decreased until the relay deenergizes, the meter indicates dropout current. Dropout current is the minimum current that will keep the relay energized. Frequently, pickup voltage and dropout voltage are specified. The **contact current** rating is also important. A large current usually flows through the relay contacts and the load connected to the relay. This rating is the maximum current that can safely flow through the contact circuit.

Solenoids

Solenoids are similar to relays as they also use electromagnetic coils. They are sometimes called **actuators**. Instead of having contacts, a solenoid has a plunger that moves when the coil is energized. The back of the plunger is attached to a spring. It causes the plunger to return to its original position when the solenoid deenergizes. The movement of the plunger of the solenoid is used to activate some type of load connected to it. A solenoid is shown in **Figure 7-16**. When the solenoid is energized, it moves the plunger in the

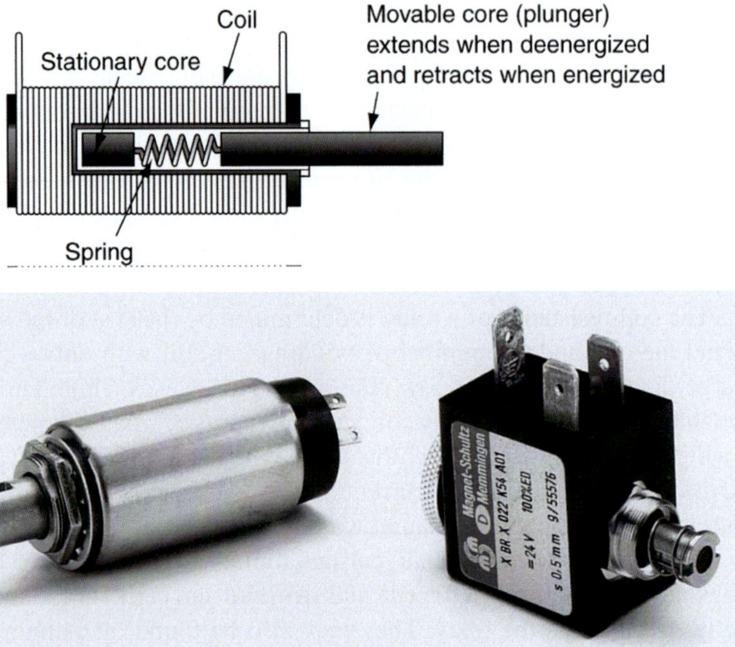

Figure 7-16 Solenoid devices [(b) courtesy of Magnet-Schultz of America, Inc.].

center. A solenoid could be used to open a control valve to allow liquid from a tank to flow into a container. When the solenoid is deenergized, the control valve would close and stop the flow of liquid.

Magnetic Motor Contactors

An important type of relay is an electric **motor contactor**. A magnetic motor contactor is shown in **Figure 7-17**. The motor contactor is a control element that starts and stops motors. It operates due to electromagnetic relay action. A "start" pushbutton switch is pressed to close a contactor. This completes a low-current path through the contactor coil. The contactor coil produces a magnetic field that causes a set of contact points to close. The movement of a part called an **armature** completes an electrical path between the power line and a motor. When this action takes place, the motor will start.

Releasing the **start button** does not deenergize the contactor coil. A path to the coil's voltage source is completed through the "stop" pushbutton switch. The motor continues to run as long as electrical power is applied. Stopping

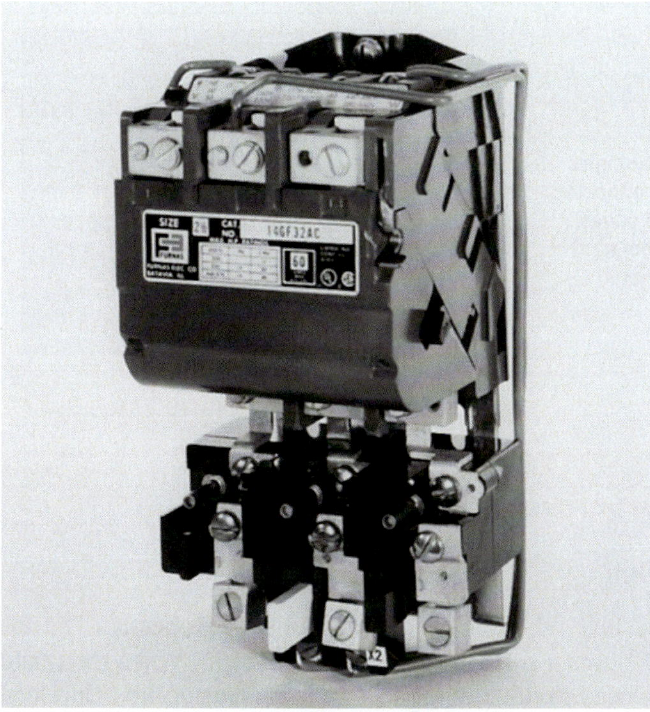

Figure 7-17 Magnetic motor contactor (courtesy of Furnas Electric Co.).

a contactor-controlled motor is achieved by pushing the **stop button**. This opens the contactor coil's voltage source. The coil will deenergize and cause the armature to move. The contact points then break contact. This removes the power from the motor. The path becomes open, and the motor will stop. Motor contactors are designed to latch in place. This holds it in place once the coil is energized. This character is important for motor control.

Magnetic Circuit Breaker

The **magnetic circuit breaker** shown in **Figure 7-18** is an application of an electromagnet. Circuit breakers are used as safety devices to protect electrical circuits from excess current flow. If the current rating of the magnetic circuit breaker is exceeded, magnetic attraction will pull the movable arm of the device to open the circuit. Also study the diagram of **Figure 7-17** to observe the action that occurs. Increases in current cause an increase in magnetic flux, which attracts the iron arm and opens the contacts.

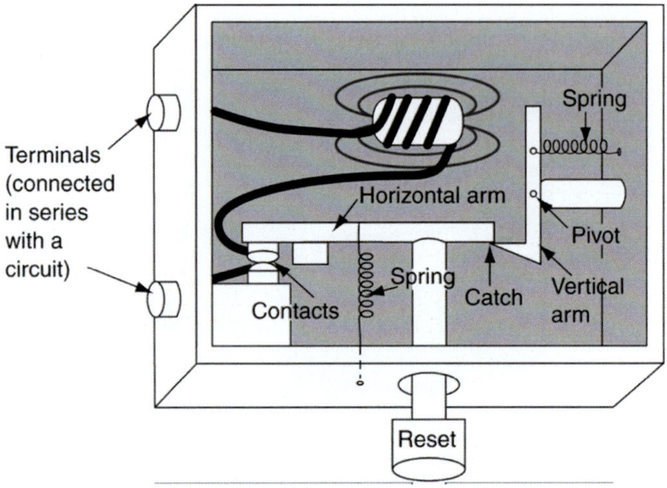

Figure 7-18 Magnetic circuit breaker.

Electric Bell

An electric bell relies on electromagnetism to operate. The example of **Figure 7-19** uses a soft-iron material for its core. When the **pushbutton** is pressed, a closed circuit through the electromagnet occurs. The electromagnet is energized, and a magnetic field is produced which attracts the soft-iron striker. The striker makes momentary contact with the gong and is then pulled back by the spring. The electromagnetic circuit is then closed again, causing the striker to make contact with the gong. This action continues to cause the bell to ring as long as the pushbutton is pressed.

Reed Switches/Relays

Reed switches and **reed relays** are also types of electromagnetic devices. These devices consist of two magnetic strips mounted in a glass or plastic enclosure, as shown in **Figure 7-20**. Note that the reed switch needs an external magnetic field to cause it to operate. They are often used on the doors or windows of buildings to indicate an open condition. A common application is in home security systems.

Reed relays have contacts which are normally open. The contact closes when the relay is energized. This device has its own coil which, when energized, causes the contacts to close. When either the reed switch or reed relay is energized, opposite polarities are induced in the overlapping reed blades, causing attraction to occur.

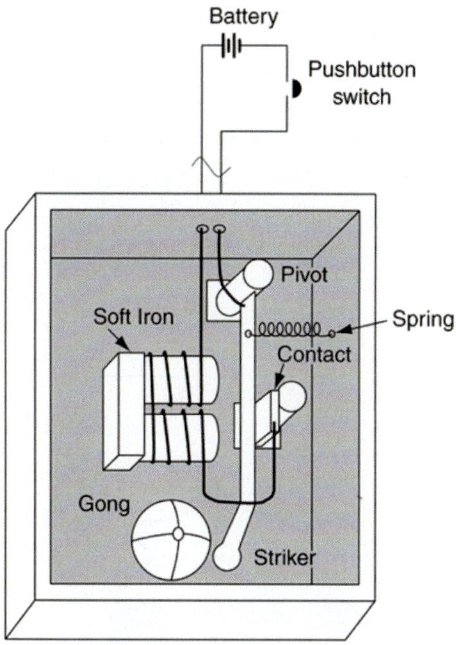

Figure 7-19 Electric bell.

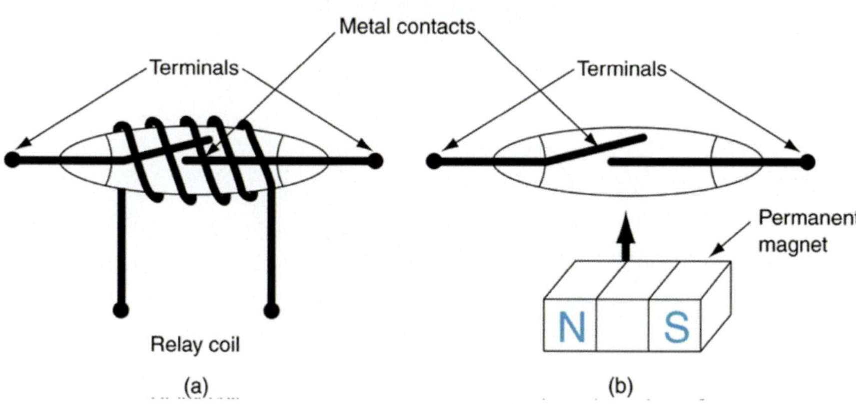

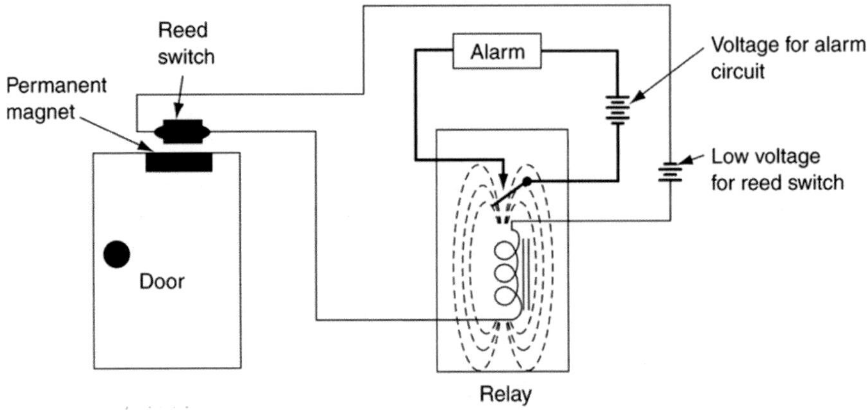

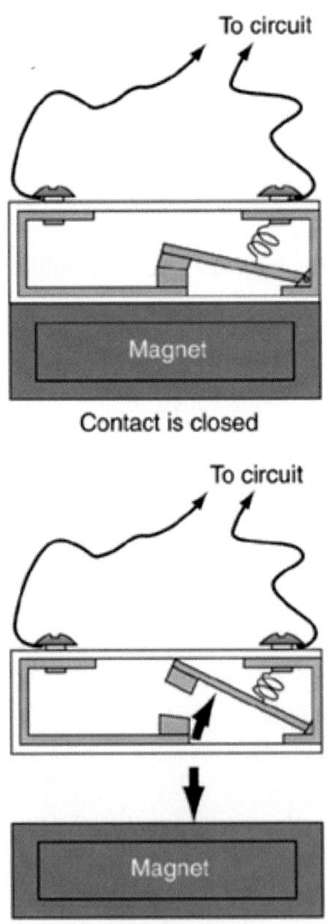

Figure 7-20 Reed devices: (a) reed relay; (b) reed switch; (c) reed switch as part of a home security system; (d) open and closed reed switch contacts.

Analog Meter Movement

The **analog meter movement** or **d'Arsonval movement**, which is used by analog meters to measure current, voltage, or resistance, is a type of electromagnetic device. The use of this device to measure electrical quantities is discussed in Chapter 3. The movement shown in **Figure 7-21** has a pointer which moves when the electromagnetic coil is energized by current flow. The movement of the pointer is in proportion to the strength of the electromagnetic field around the moving coil. Thus, this device is called a "moving coil" meter movement and can be used to measure electrical values. The basic principle of this movement is that increases in current cause increased electromagnetic field strength.

Magnetic Recording

Magnetic recording, which has been used for many years, is another example that relies on electromagnetic fields. **Figure 7-22** shows an electromagnetic recording head which records sound information on a magnetic tape by magnetizing the tape as it passes. This process is used in cassette tape

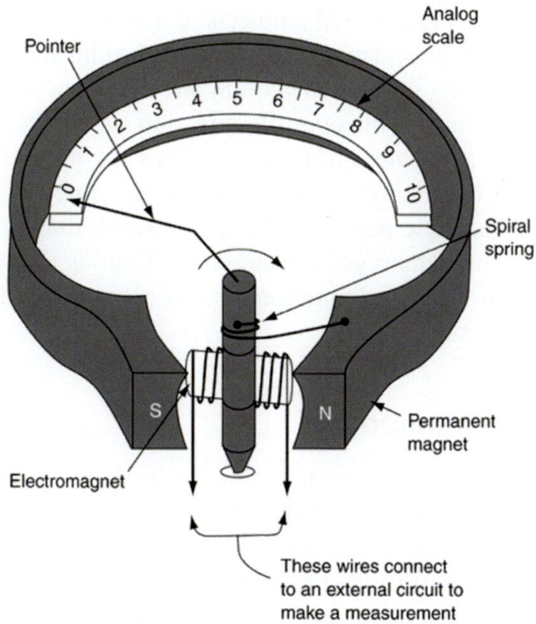

Figure 7-21 Moving coil meter movement.

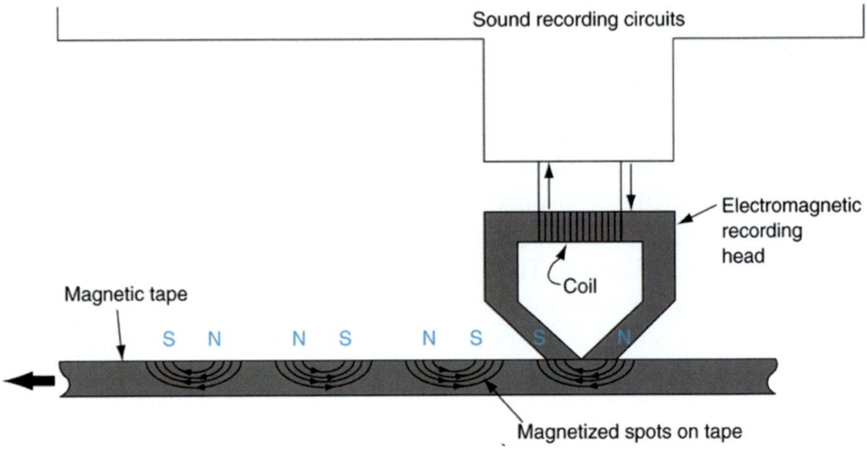

Figure 7-22 Electromagnetic tape recording principle.

recorders. Reversing the direction of current flow in the recording coil causes magnetic polarities to reverse.

Electromagnetic Speakers

Another common type of electromagnetic device is the **speaker**. The cone of the speaker shown in **Figure 7-23** connects to a hollow cylinder with a coil wound around it. A permanent magnet is positioned within the electromagnetic coil. When current flows through the coil in one direction, the coil will move to the right. Current flow in the opposite direction causes the coil to move in the other direction. Movement of the coil causes the flexible diaphragm to move in a direction based on current flow direction. More coil current causes a stronger electromagnetic field, which will cause the diaphragm to move a greater distance. The diaphragm vibrates in and out as the intensity of the sound input changes. The air vibrations that occur due to this action cause sound waves to be produced.

7.6 Magnetic Terminology

Several basic terms are important for understanding electromagnetic principles. One **magnetic line of force** is called a **maxwell**. The amount of **magnetic flux (f)** is measured in a unit called the **weber**. A weber is equal to 10^8 (100,000,000) lines of force. **Flux density** is equal to the number of lines of force per square meter and is measured in the unit **tesla**.

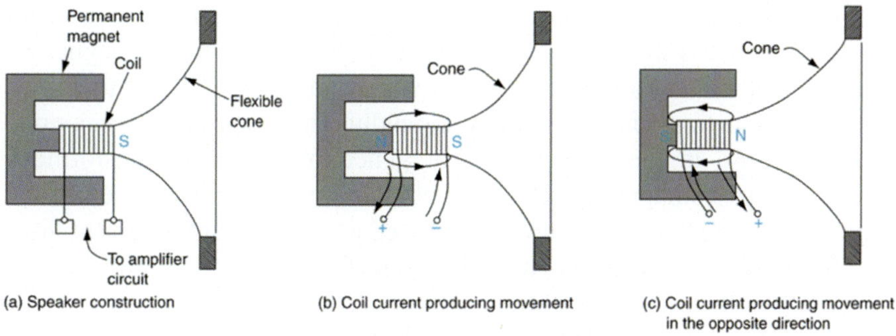

Figure 7-23 Electromagnetic speaker.

magnetic flux (f) = number of lines of force in weber

and

flux density (B) = magnetic flux (f)/area (A).

Magnetomotive force (MMF) is the magnetic effect that causes a magnetic field to be produced. MMF or ampere-turns is equal to the current through a coil multiplied by the number of turns of wire in the coil.

MMF = $I \times N$ (ampere-turns)

where MMF = magnetomotive force, in ampere-turns

I = current, in amperes

N = number of turns in the coil.

The length of a coil is also a factor that affects the field strength. The term **magnetizing force (H)** is used to express the magnetic field strength and is calculated as follows:

H = MMF/l

where H = magnetizing force in ampere-turns/meter (m)

MMF = magnetomotive force

l = length of coil, in meters (m).

Reluctance (Â) is the opposition to the development of a magnetic field in an electromagnet.

reluctance (Â) = MMF (magnetomotive force)/f (magnetic flux) (measured in ampere-turns/weber).

The relationship of MMF, f, and $\hat{A}$ in magnetic circuits to V, I, and R in electrical circuits shown in **Figure 7-11** should be reviewed.

Residual magnetism is an important effect in the operation of some types of electrical equipment. Residual magnetism is the ability of an electromagnet to hold a small magnetic field after electrical current is turned off. A small magnetic field remains around an electromagnet after it is demagnetized. This magnetic field is very weak.

Permeability (m) is the ability of a magnetic material to transfer magnetic flux. It is the ability of a material to magnetize and demagnetize. Soft-iron material has a **high** permeability – it transfers magnetic flux easily. Soft iron magnetizes and demagnetizes rapidly. This makes soft iron a good material to use in the construction of generators, motors, transformers, and other electromagnetic devices. Permeability is similar to electrical conductance, which is a measure of how well a material allows current flow.

Magnetic permeability (m) = 1/*Â* (reluctance).

Electrical conductance (G) = 1/R (resistance).

A related term is **relative permeability (m_r)**, which is a comparison of the permeability of a material to the permeability of air (1.0). Suppose that a material has a relative permeability of 1000. This means that the material will have 1000 times more magnetic flux than an equal amount of air. The relative permeability of materials is shown in **Table 7-1**.

Another magnetic term is **retentivity**. The retentivity of a material is its ability to retain a magnetic field after a magnetizing force is removed. Some materials will retain a magnetic flux for a long time. Other materials lose their magnetic flux almost immediately after the magnetizing force is removed.

Magnetic saturation is important in the operation of electrical equipment that has electromagnets, especially generators. Saturation is best explained by the curve shown in **Figure 7-24**, which is called a magnetization or **B-H curve**. The curve shows the relationship between a **magnetizing force (H)** and flux density (*B*). Note that as a magnetizing force increases, flux density also increases. **Flux density** is the amount of lines of flux per unit area of a material. An increase in flux density occurs

Table 7-1 Permeabilities (μ) and relative per Permeabilities (μ_r) of verious meterials

Material	*Relative Permeability* (μ_r)	*Permeability* (μ)
Air	1	1.26×10^{-6}
Nickel	50	6.28×10^{-5}
Cobalt	60	7.56×10^{-5}
Cast iron	90	1.1×10^{-4}
Machine steel	450	5.65×10^{-4}
Transformer iron	5,500	6.9×10^{-3}
Silicon iron	7,000	8.8×10^{-3}

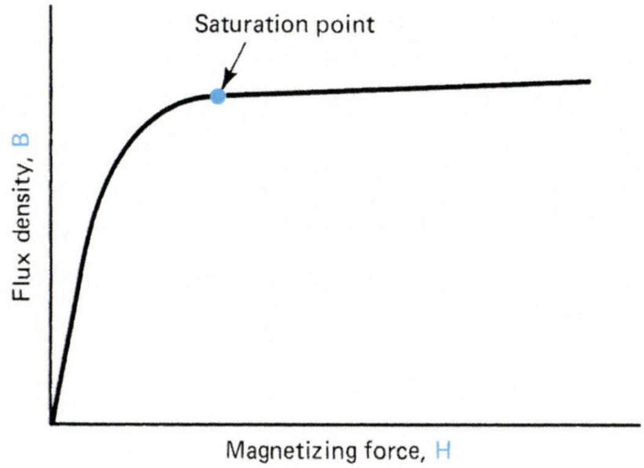

Figure 7-24 Magnetization or *B-H* curve.

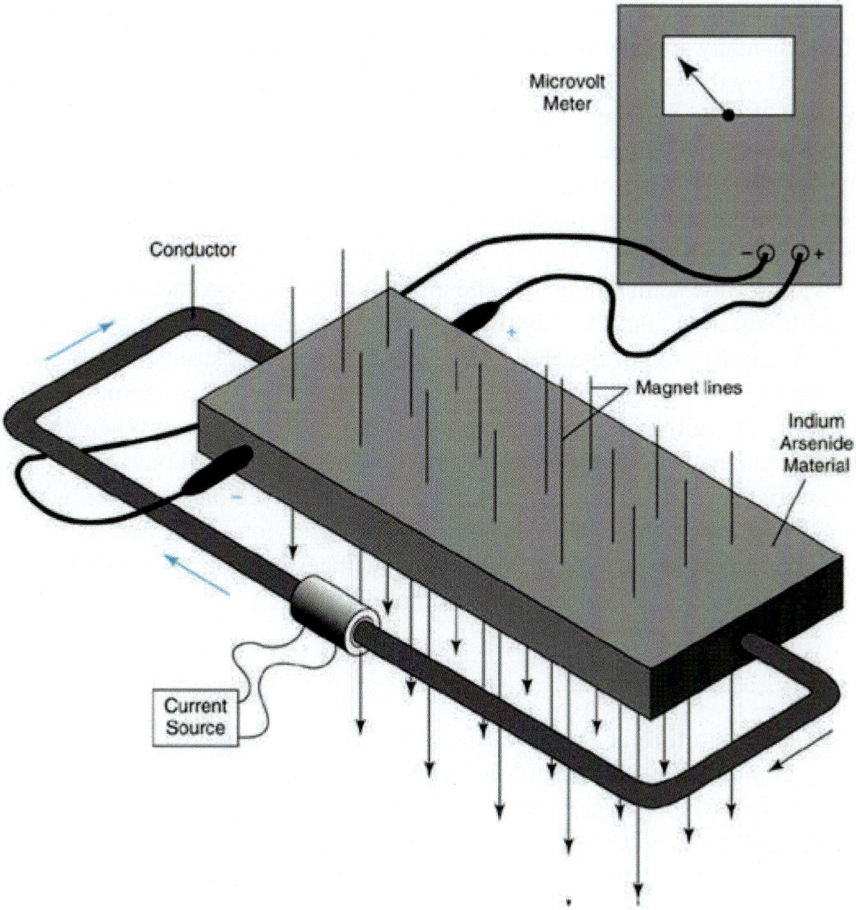

Figure 7-25 Illustration showing the Hall effect.

until magnetic saturation is reached. At the saturation point, the maximum alignment of domains within the material has taken place. Beyond saturation, an electromagnet is not capable of more magnetic field strength. The *B-H* curve is a straight line beyond the saturation point. Note the shape of the *B-H* curve. The magnetization or setting up of the magnetic field (*B*) lags the magnetizing force (*H*) due to friction of the molecules of a material. This time lag between magnetizing force (*H*) and the development of magnetic flux (*B*) is called **hysteresis**.

Hall Effect

Magnetism can be used other than in generators to produce voltage. When a magnetic field is placed at right angles to a current-carrying conductor, a small voltage is produced across the conductor known as the **Hall effect**. This effect has many applications in switching and measurement circuits. Voltages produced by this process are small (in the microvolt range) when conductors are used. When a semiconductor material such as indium arsenide is used, the voltage produced can be as much as 100 mV.

The Hall effect is illustrated in **Figure 7-25**. A block of indium arsenide has a small current flowing through it. If a magnetic field is placed perpendicular to the direction of current flow, a voltage is produced across the width of the semiconductor. The amount of voltage produced is directly proportional to the flux density of the magnetic field.

An application of the Hall effect is as a **sensor** in instruments designed to measure the strength of magnetic fields. One such device is known as a **gauss meter**. Within its measuring probe is a Hall effect device. When placed within a magnetic field, it produces a voltage directly proportional to the flux density of the field. This voltage is sensed and displayed on the meter, which is calibrated in gauss. There are many other uses for Hall effect devices, including position sensors for machines, switches for computer keyboards, and electronic ignition systems of automobiles.

Magnetic Levitation

Magnetic levitation is an interesting phenomenon. "Maglev" has been used for developing trains that do not ride on rails, but "levitate" above the rails on a magnetic cushion. Experimental models of this type of train have reached speeds of 300 miles per hour with a smooth ride. The forward motion of the train is produced by the attraction of fields with opposite magnetic polarities and the repulsion of fields with like magnetic polarities. For this type of application, the magnetic fields must be quite strong.

Rare Earth Magnets

Rare earth permanent magnets may be used for several applications. They may be used to produce a uniform, variable magnetic field with about one-tenth the size of iron-core electromagnets. They also do not require special power supplies. Magnetic fields play an important role in numerous industrial processes. For many industrial applications, rare earth permanent magnets offer advantages of smaller size, lower cost, and ease of operation. A permanent or "hard" magnet is one that remains constant in terms of its magnetization and direction. Rare earth permanent magnets, unlike steel magnets, can be formed into any shape without demagnetizing. Rare earth magnets are typically made of a material called neodymium; however, other materials have been tested to ensure proper magnetic characteristics. Applications include brushless direct current (DC) motors and actuators, cordless tools, and computer drive systems.

Summary Alnico is an alloy of aluminum, nickel, iron, and cobalt used to make permanent magnets.

- Ampere-turns of a coil are the amperes of current times the number of turns of wire.
- The movable part of a relay is called an armature, and the core is an iron or steel material around which coils are wound.
- When the atoms of a metal core material are aligned in the same pattern so that no more magnetic lines of force can be developed, the core is said to be saturated.
- The domain theory of magnetism assumes groups of atoms produced by movement of electrons align themselves in groups called "domains" in magnetic materials.
- A coil of wire wound on an iron core so that as current flows through the coil causing it to become magnetized forms an electromagnet.
- Invisible lines of force that extend around a magnetic material are called magnetic flux.
- Magnetic flux density is measured in Gauss.
- Force that produces it.
- The basic laws of magnetism are that (1) like magnetic poles repel and (2) unlike magnetic poles attract.

- A magnet is a metallic material, usually iron, nickel, or cobalt, which has magnetic properties.
- Magnetic lines of force that extend from a north pole and enter a south pole to form a closed loop around the outside of a magnetic material is called a magnetic field.
- Permeability is the ability of a material to conduct magnetic lines of force as compared with air.
- A relay is an electromagnetically operated switch.
- The opposition of a material to the flow of magnetic flux is called reluctance.
- The magnetism that remains around a material after the magnetizing force has been removed is called residual magnetism.
- A solenoid is an electromagnetic coil with a metal core that moves when current passes through the coil.

Self-examination/Answers

1. What were the first magnets called?
2. What is electromagnetism?
3. What three materials are used in the construction of permanent magnets?
4. What are the two laws of magnetism?
5. Why should magnets be stored in a "keeper"?
6. What are the three basic parts of an electromagnet?
7. What are three ways to increase the strength of an electromagnet?
8. What is residual magnetism?
9. What are NO and NC contacts of relays?
10. What are pickup current and dropout current ratings of relays?

Answers

1. Lodestone
2. Magnetic field developed by current flow through conductors
3. Nickel, iron, cobalt
4. Weber's theory, domain theory
5. Retains the strength of the magnetic field
6. Metal core, turns of wire, power source
7. Increase number of turns of wire, higher source voltage, change core material
8. Retained magnetism after energizing a coil

9. Normally open and normally closed
10. Minimum current required to energize a relay; minimum current required to allow a relay to deactivate

Glossary

Alnico
An alloy of aluminum, nickel, iron, and cobalt used to make permanent magnets.

Ampere-turn
The unit of measurement of magnetomotive force (MMF); amperes of current times the number of turns of wire.

Armature
The movable part of a relay.

Coefficient of coupling (k)
A decimal value that indicates the amount of magnetic coupling between coils.

Core
Iron or steel materials of internal sections of electromagnets around which coils are wound.

Core saturation
When the atoms of a metal core material are aligned in the same pattern so that no more magnetic lines of force can be developed.

Coupling
The amount of mutual inductance between coils.

Domain theory
A theory of magnetism that assumes groups of atoms produced by movement of electrons align themselves in groups called "domains" in magnetic materials.

Electromagnet
A coil of wire wound on an iron core so that as current flows through the coil, it becomes magnetized.

Flux (f)
Invisible lines of force that extend around a magnetic material.

Flux density
The number of lines of force per unit area of a magnetic material or circuit.

Gauss
A unit of measurement of magnetic flux density.

Gilbert
A unit of measurement of magnetomotive force (MMF).

Hysteresis
The property of a magnetic material that causes actual magnetizing action to lag behind the force that produces it.

Laws of magnetism
(1) Like magnetic poles repel; (2) unlike magnetic poles attract.

Lines of force
Same as magnetic flux; *see* **Flux**.

Lodestone
The name used in early times for natural magnets.

Magnet
A metallic material, usually iron, nickel, or cobalt, which has magnetic properties.

Magnetic circuit
A complete path for magnetic lines of force from a north to a south pole.

Magnetic field
Magnetic lines of force that extend from a north pole and enter a south pole to form a closed loop around the outside of a magnetic material.

Magnetic flux
See **Flux.**

Magnetic materials
Metallic materials such as iron, nickel, and cobalt which exhibit magnetic properties.

Magnetic poles
Areas of concentrated lines of force on a magnet which produce north and south polarities.

Magnetic saturation
A condition that exists in a magnetic material when an increase in magnetizing force does not produce an increase in magnetic flux density in the material.

Magnetomotive force (MMF)
A force that produces magnetic flux around a magnetic device.

Magnetostriction
The effect that produces a change in shape of certain materials when they are placed in a magnetic field.

Natural magnet
Metallic materials that have magnetic properties in their natural state.

Permanent magnet
Bars or other shapes of materials that retain their magnetic properties.

Permeability (m)
The ability of a material to conduct magnetic lines of force as compared with air.

Polarities
See **Magnetic poles.**

Relay
An electromagnetically operated switch.

Reluctance
The opposition of a material to the flow of magnetic flux.

Residual magnetism
The magnetism that remains around a material after the magnetizing force has been removed.

Retentivity
The ability of a material to retain magnetism after a magnetizing force has been removed.

Solenoid
An electromagnetic coil with a metal core that moves when current passes through the coil.

8

Sources of DC Electrical Energy

Electrical energy sources **convert** some other form of energy into electrical energy. Batteries and electrical generators are two major sources of electrical energy. Batteries convert **chemical energy** into **electrical energy**. The types of electrical generators include single-phase AC generators, three-phase AC generators, and DC generators. Electrical generators rely on the principle of electromagnetic induction to convert mechanical energy into electrical energy. Batteries, generators, and other sources of electrical energy are discussed in this chapter. Some sources produce direct current (DC) energy, whereas others produce alternating current (AC) energy. We will concentrate on DC (direct current) sources in this chapter.

Chapter Outline

Objectives

Upon completion of this chapter, you will be able to:

1. Describe basic types of batteries
2. Properly connect batteries in series, parallel, and combination configurations
3. Explain the purposes of different configurations of battery connections
4. Describe light, heat, pressure, and mechanical sources of electrical energy
5. State Faraday's law for electromagnetic induction

6. Explain the factors that affect the generation of voltage
7. Describe direct current generators

8.1 Chemical Sources

Conversion of chemical energy into electrical energy occurs through **chemical cells**. When two or more cells are connected in series or parallel (or a combination of both), they form a battery. A cell is made of two different metals immersed in a liquid or paste called an **electrolyte**. Chemical cells are

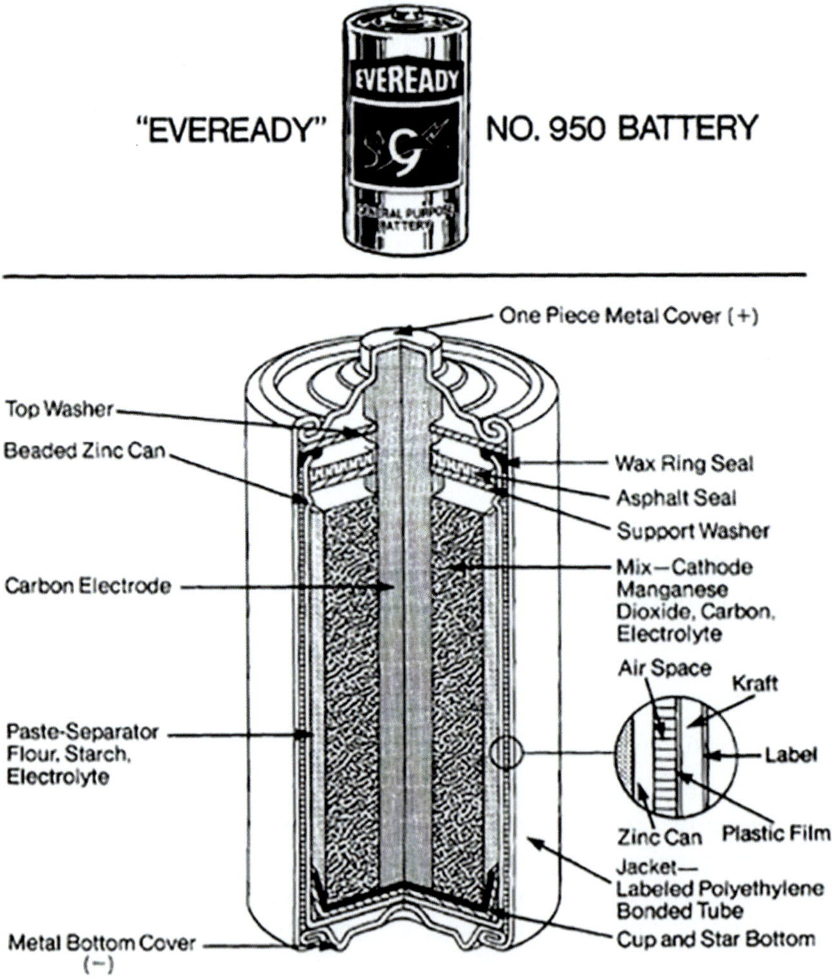

Figure 8-1 Carbon-zinc cell (courtesy of Union Carbide Corp.).

either primary or secondary cells. **Primary cells** are usable only for a certain time. **Secondary cells** are renewed after being used to produce electrical energy once again, which is known as **charging**. Both primary and secondary cells have many uses.

Primary Cells

The operation of a **primary cell** involves the placing of two unlike materials called **electrodes** into the solution, or **electrolyte**. When the materials of the cell are brought together, their molecular structures change. During this chemical change, atoms may either gain additional electrons or leave behind some electrons. These atoms then have either a positive or a negative electrical charge. They are called **ions**. Ionization of atoms allows a chemical solution of a cell to produce an electrical current.

A **load** device such as a lamp may be connected to a cell. Electrons flow from one of the cells' electrodes to the other through the electrolyte material. This creates an electrical current flow through the load. Current leaves the cell through its negative electrode. It passes through the load device and then goes back to the cell through its positive electrode. A complete circuit exists between the cell (source) and the lamp (load).

The **voltage output** of a primary cell depends on the electrode materials used and the type of electrolyte. The familiar carbon-zinc cell shown in **Figure 8-1** produces approximately 1.5 V. The negative electrode of this cell is the zinc container. The positive electrode is a carbon rod. A paste material acts as the electrolyte. It is placed between the two electrodes. This type of cell is called *a* **dry cell**.

Many types of **primary cells** are used today. The **carbon-zinc cell** is a cell that is low in cost and available in many sizes. Applications are mainly for portable equipment and instruments. For uses that require higher voltage or current than one cell can deliver, several cells are combined in series, parallel, or series–parallel connections. **Carbon-zinc batteries** are available in many voltage ratings.

An **alkaline (zinc-manganese dioxide) cell** is similar to a carbon-zinc cell and has a voltage per cell of 1.5 V. They supply higher-current electrical loads. They have much longer lives than carbon-zinc cells of the same types.

Another type of **primary cell** is the lithium type. **Lithium batteries** have an extremely long life and provide a leakproof, high-energy source for a wide range of applications. Lithium batteries typically operate at approximately 1.9-V output. Other types of lithium batteries are developed to make them compatible with 1.5-V applications.

Secondary Cells

Chemical cells that may be reactivated by charging are called **secondary cells** or storage cells. Common types are the lead-acid, nickel-cadmium, and nickel-metal-hydride (NiMH) cells.

Lead-acid cells: The lead-acid cell of **Figure 8-2** is a secondary cell. The **electrodes** of lead-acid cells are made of lead and lead peroxide. The positive plate is lead peroxide (PbO_2). The negative plate is lead (Pb). The **electrolyte** is sulfuric acid (H_2SO_4). When the lead-acid cell supplies current to a load, the chemical process is written as

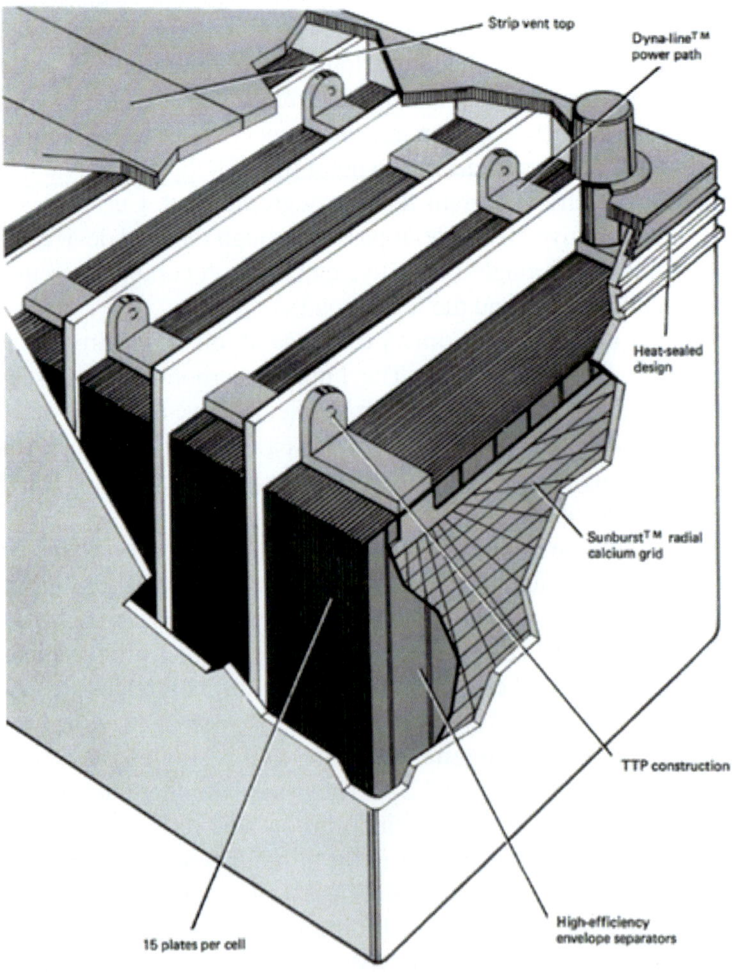

Figure 8-2 Lead-acid cells of a battery (courtesy of Exide Corp.).

$Pb + PbO_2 + 2H_2SO_4 - 2PbSO_4 + 2H_2O$.

The **sulfuric acid** ionizes to produce four positive hydrogen ions (H^+) and two negative sulfate (SO_4) ions. A negative charge is developed on the lead plate when an SO^-_4 ion combines with the lead plate to form lead sulfate ($PbSO_4$). The positive hydrogen ions (H^+) combine with electrons of the lead peroxide plate. They become neutral hydrogen atoms. The H^+ ions also combine with the oxygen (O) of the lead peroxide plate to become water (H_2O). The lead peroxide plate then has a positive charge. A lead-acid cell has a voltage between electrodes of about 2.1 V when fully charged.

Cells **discharge** when supplying current for a long time. They are no longer able to develop an output voltage when discharged. Cells may be charged by causing direct current to flow through the cell in the opposite direction. The chemical process of charging is written as

$2PbSO_4 + 2H_2O – Pb + PbO_2 + 2H_2SO_4$

or

2 parts lead sulfate + 2 parts water yields lead + lead peroxide + 2 parts sulfuric acid.

The original condition of the chemicals is reached by charging. The chemical reaction is reversible.

The amount of charge of a **lead-acid cell** is measured by a **specific gravity** test. A **hydrometer** is used to test the electrolyte solution. The specific gravity of a liquid is an index of how heavy a liquid is compared with water. Pure sulfuric acid has a **specific gravity** of 1.840. The dilute sulfuric acid of a fully charged lead-acid cell varies from 1.275 to 1.300. During the discharge of the cell, water is formed, which reduces the specific gravity of the electrolyte. A specific gravity of between 1.120 and 1.150 indicates a fully charged cell, as measured with a **hygrometer**.

The capacity of a battery made of lead-acid cells is given by an **ampere-hour rating**. A 50-ampere-hour battery is rated to deliver 50 A for 1 h, 25 A for 2 h, or 12.5 A for 4 h. The ampere-hour rating is an approximate value. It depends on the rate of discharge and the operating temperature of the battery.

Nickel-Cadmium Cells

Another type of secondary cell is a **nickel-cadmium (NiCd) cell**. These cells are available in many sizes. They are often used in portable equipment. The **positive plate** of this cell is nickel hydroxide. The **negative plate** is cadmium hydroxide. The **electrolyte** is made of potassium hydroxide. These cells have a long life. A fully charged nickel-cadmium cell has a voltage of approximately 1.25 V.

Nickel-Metal-Hydroxide Cells

Nickel-metal-hydride (NiMH) batteries are another type of rechargeable cell. Compared with nickel-cadmium types, NiMH batteries charge faster. In a nickel-metal-hydride battery, the positive electrode is made of nickel and the negative electrode is made of hydrogen-storing metal alloys. This type of battery will typically permit 500 charge–discharge cycles and recharge in 1.5 h. They operate at 1.2 V, which is approximately the same as nickel-cadmium batteries.

Secondary cells have many uses. Storage batteries are used in some buildings to provide emergency power when a power failure occurs. Standby systems are needed, especially for lighting when power is off. Automobiles use storage batteries for their everyday operation. Many types of instruments and portable equipment use batteries for power. Some instruments use **rechargeable** secondary cells and others use primary cells. The use of batteries for portable equipment is becoming increasingly important due to the increased use of electronic equipment such as personal computers (PCs), facsimiles (FAXs), headphone stereos, portable compact disc (CD) players, calculators, security systems, cordless and cellular telephones, power tools, and many other applications.

Other Batteries

The demand for smaller and lighter cellular phones, computers, and cordless tools is causing batteries to be made smaller and lighter. Nickel-cadmium (NiCd) is still used as is nickel-metal-hydride (NiMH), lithium-ion (Li-ion), and lithium-polymer (Li-polymer). Portable electronic equipment has used **NiCd batteries**, as has audio and visual communications equipment, motorized devices, and hobby equipment; however, they require frequent recharging. Advanced cellular phones and portable computers now require longer run times than previous portable equipment. NiMH has replaced NiCd in certain applications. NiMH allows decreased size at a reasonable cost without major environmental concerns. Manufacturers are trying to develop batteries in even smaller sizes and longer run times. Li-polymer could provide for even smaller and lighter batteries and other products in the future. Research is currently ongoing to produce efficient batteries for electric vehicles (EVs).

8.2 Battery Connections

Each electrical circuit requires a voltage source. One source for DC circuits is a cell or **battery**. The arrangement of the cells in a circuit depends on the load requirements of voltage and current. If the voltage must be high, cells are connected in series.

Series Connection

The voltage of a single primary cell, or **dry cell**, is 1.5 V. When the voltage required by a load is higher than 1.5 V, it is necessary to use more than one

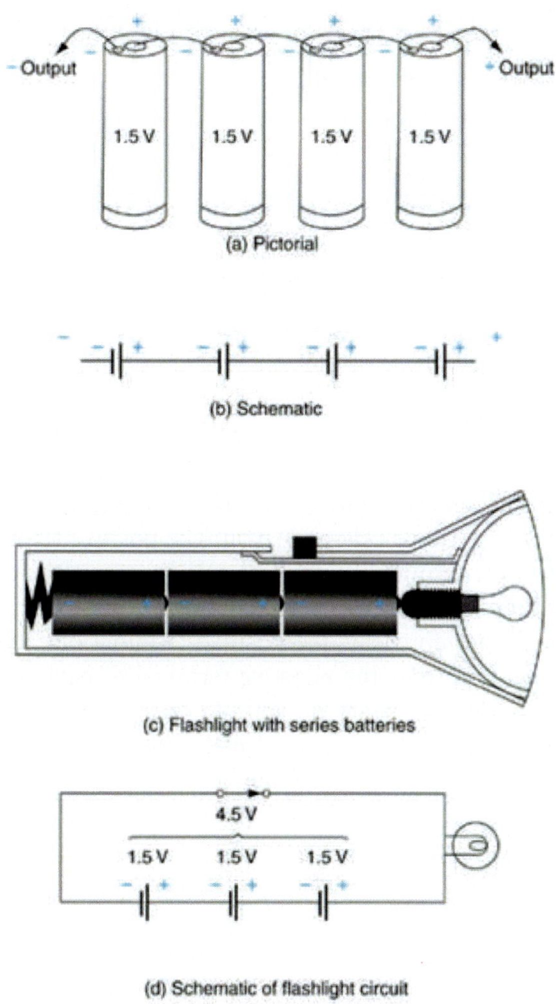

Figure 8-3 Series voltage connection: (a) pictorial; (b) schematic; (c) flashlight with series batteries; (d) schematic of flashlight circuit.

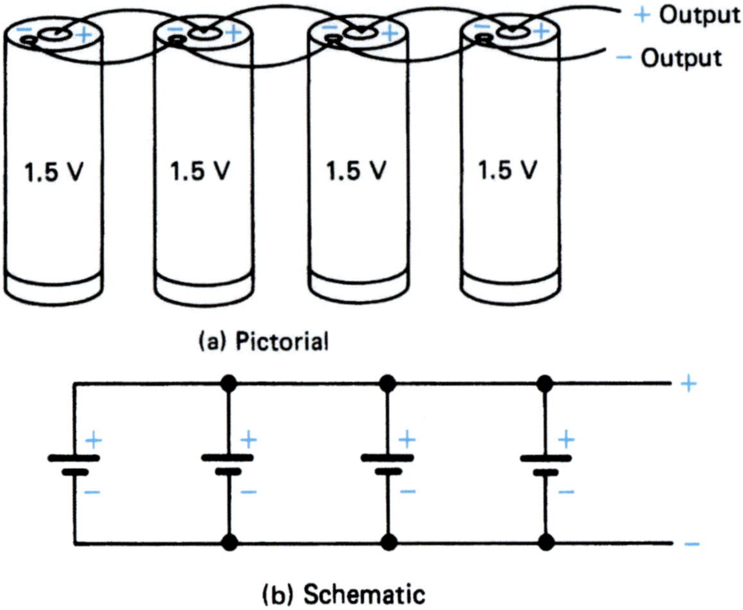

Figure 8-4 Parallel voltage connection: (a) pictorial; (b) schematic.

cell, and the cells must be connected in **series**, as shown in **Figure 8-3**. The **negative terminal** of the first cell is connected to the **positive terminal** of the second cell. The negative terminal of the second cell is connected to the positive terminal of the third cell, and so on. The positive terminal of the first cell and the negative terminal of the last cell become the output terminals for the circuit. **Figure 5-6(b)** is a schematic drawing of four cells in series. The long vertical line represents the positive terminal of each cell, and the short vertical line represents the negative terminal of each cell. When cells are connected in series, the same amount of current flows through each cell. The total voltage of the cells connected in series is equal to the sum of the voltages of the individual cells.

Parallel Connection

If the current requirement of a circuit is high, cells are connected in **parallel**. Cells are also rated by current. The current rating of cells is based on a cell's capacity or ability to furnish a certain amount of current for a length of time. The lifetime of cells can be increased by connecting more cells in parallel. **Figure 8-4** shows four 1.5-V dry cells connected in parallel. All the positive terminals are connected together. Likewise, all the negative terminals are connected together. **Figure 8-7(b)** shows a schematic drawing of the four cells in parallel. When cells are connected in parallel, the total current capacity is equal to the sum of the currents of the individual cells. Also, when cells are connected in parallel, the voltage applied to the circuit is the same as the voltage of one cell. Only voltage sources of the **same** voltage rating can be connected in parallel.

Combination (Series–Parallel) Connection

If both the voltage and current requirements of an electrical circuit are higher than the rated voltage and current of a single cell, it is necessary to use three or more cells in a series–parallel or combination circuit. For example, two 1.5-V dry cells could be connected in series, negative to positive; then these two are connected in parallel, negative to negative, and positive to positive. This would provide a 3-V output. When the cells are connected in series–parallel, the voltage applied to the circuit is equal to the **sum** of the cells connected in series. Also, when the cells are connected in series–parallel, the total current capacity is equal to the **sum** of the current ratings of the cells connected in parallel.

8.3 Light Sources

Light is a form of energy that is easily converted to electrical energy. The device used to convert light energy into electrical energy is called a **photovoltaic cell** or a **solar cell**. Solar cells are used in space programs. They collect the rays of the sun and convert them into electrical energy. Solar cells are used to power circuits that control space satellites, lunar modules, and other spacecraft.

A **solar cell**, shown in **Figure 8-9**, is usually made of two layers of material. The electrical characteristics of these materials are altered by the addition of other elements called **impurities**. When a solar cell is exposed to light, the two materials interact, producing an excess of electrons on one layer. A negative charge is, thus, developed. The other layer then has a deficiency of electrons or a positive charge. This imbalance in the electrons causes a difference of potential (voltage) between the two layers. The difference in potential depends on the amount of light falling on the cell. The voltage is used to cause current to flow through a load connected to the cell. Thus, light acts as a source of electrical energy.

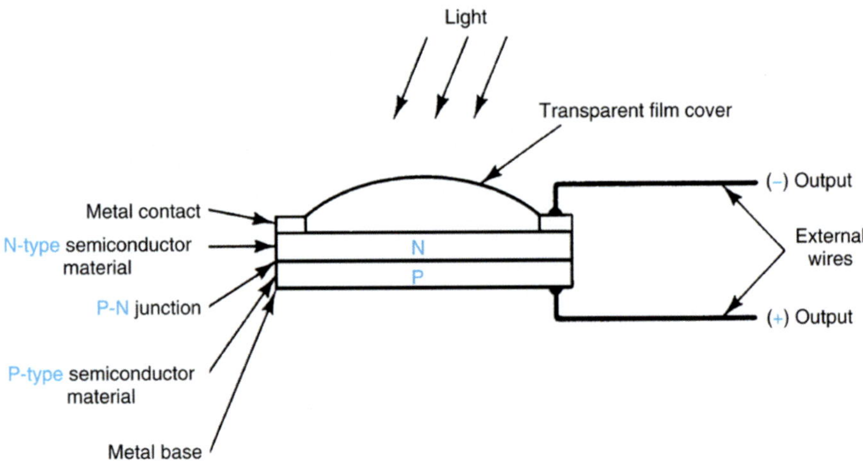

Figure 8-5 Solar cell.

8.4 Heat Sources

Electrical energy is converted into heat energy when food is cooked in an oven and homes are heated. Similarly, heat can be converted into electrical energy. Thomas J. Seebeck, a German scientist, discovered in the early 1800s that heat could be converted into electrical energy. He found that when the ends of two different types of metals are connected together and heated, a small DC voltage is created at the open ends (refer to **Figure 8-6**). The amount of DC voltage depends on the amount of heat being applied and the kind of metal used at the connected ends of the two pieces of metal. Converting heat into electrical energy by this method is known as the **Seebeck effect**. This effect is used to measure temperature and for thermostats.

When metals are heated, their electrons tend to move away from the areas being heated, which causes electrons to be more concentrated in a cool area than in a heated area. When two different types of metals are connected together and heated at their junction, electrons in both metals tend to move away from the heat. Because the two metals are different, there are more electrons at the cool end of one metal than at the other. This causes the metal with the most electrons to have a negative (-) charge. Compared with the other metal, the one with the least electrons is positive (+). The difference in charge between the two cooler ends of the metals develops a voltage. A small voltage, usually in millivolts, is produced. Devices used to convert heat energy into electrical energy are called **thermocouples**.

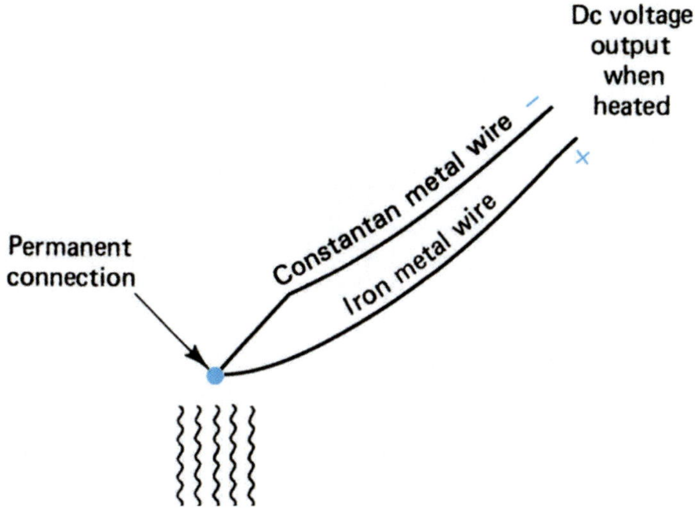

Figure 8-6 Operating principle of a thermocouple.

8.5 Pressure Sources

Electrical energy is produced by **mechanical energy** in electrical generators. Mechanical energy is used to rotate prime movers that drive electrical generators. Mechanical energy in the form of pressure is also used as a source of electrical energy. The change of mechanical pressure into electrical energy is called the **piezoelectric effect**. Certain crystal materials may be compressed as pressure is applied to the surfaces. A voltage is created between their top and bottom surfaces. The amount of voltage is determined by the amount of pressure – the greater the pressure, the greater the voltage; the less the pressure, the less the voltage will be for any **piezoelectric crystal**. These pressure-sensitive crystals have been used as phonograph cartridges to change the pressure applied by the grooves of the record into a voltage. They are also used as pressure sensors to sense and measure pressure in security and industrial systems.

When these crystalline materials are subjected to a mechanical pressure, electrical energy is developed across the material. Crystals such as **quartz** and **Rochelle salt** have this characteristic. An application of the piezoelectric principle is the cartridge and needle assembly used for vinyl recording sound systems. The cartridge contains a crystalline material. The crystal vibrates according to the size of the grooves of a phonograph record. The needle is attached to the cartridge to connect the record grooves to the crystal. The crystalline material produces a voltage due to the mechanical vibrations or

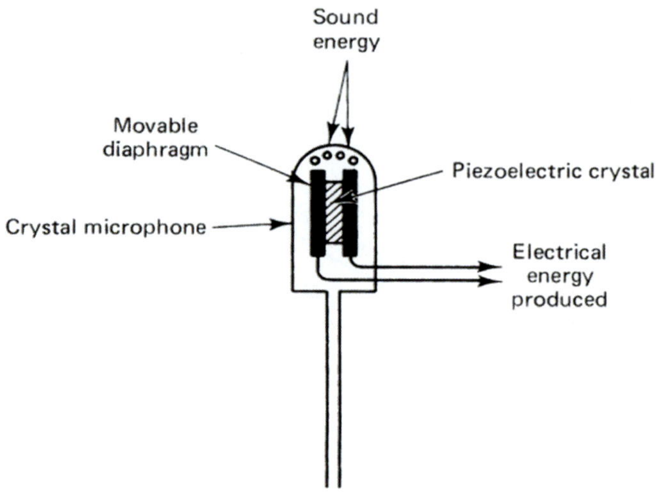

Figure 8-7 Piezoelectric principle of a microphone.

pressure changes. These small voltage changes are then amplified by a sound system.

It is also possible to convert pressure in the form of **sound** into electrical energy with piezoelectric crystals. This is done with **crystal microphones** (see **Figure 8-7**) as sound waves are used to cause vibration of a piezoelectric crystal. A voltage is developed across the crystal and amplified by the sound system. Higher sound pressure causes more voltage output to be produced.

8.6 Electromagnetic Sources

Alternating current is used in greater quantities than **direct current**; however, many important operations depend on direct current (DC) power. Industries use direct current power for many operations. Electroplating and variable-speed motor drives are two examples of direct current use. Direct current energy is used to start automobiles, in **electric vehicles** (EVs) and for many types of portable equipment used in homes. Three-phase or single-phase AC power is easily converted to direct current. Direct current is available from primary and secondary chemical cells. Direct current generators are also used to supply DC power for specialized applications.

Direct current generators are used to convert mechanical energy into direct current electrical energy. The parts of a simple direct current generator are shown in **Figure 8-8**. The principle of operation of direct current generators is similar to alternating current generators. **Electromagnetic**

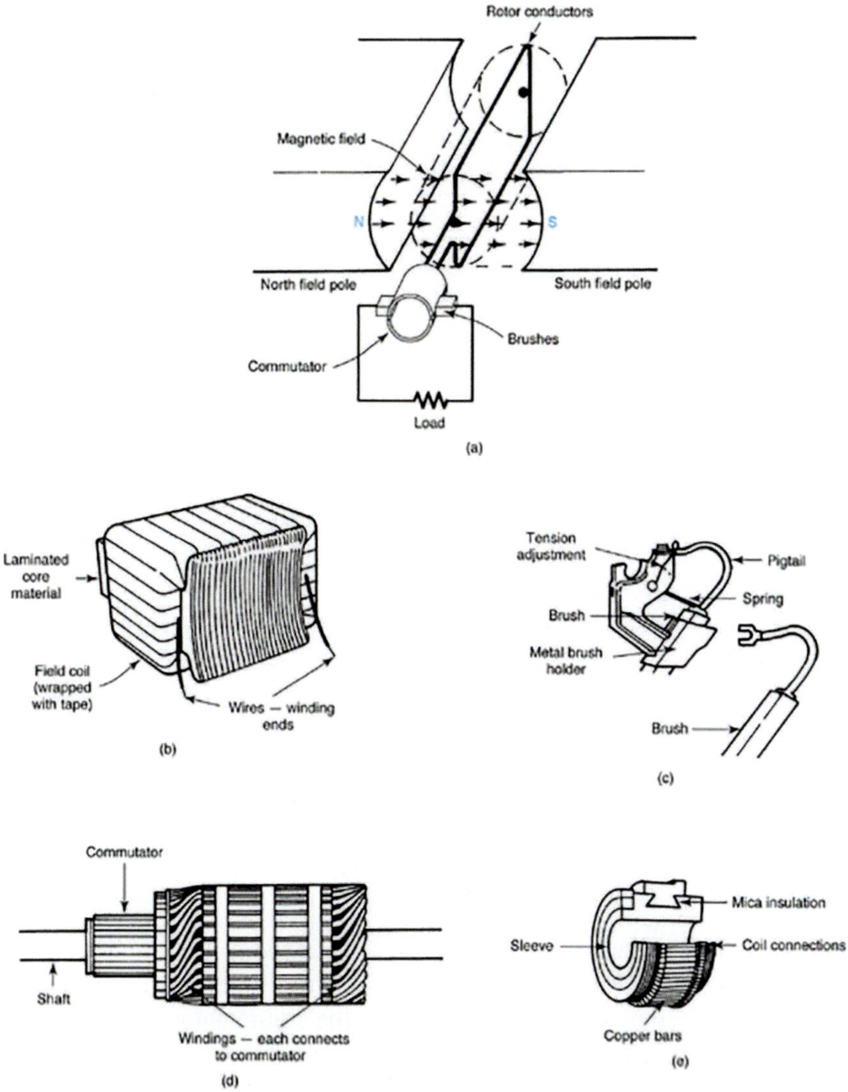

Figure 8-8 Parts of a direct current generator: (a) basic parts; (b) electromagnetic pole pieces; (c) brush assembly; (d) rotor; (e) commutator detail.

induction causes a voltage to be generated. **Armature** coils rotate through a **magnetic field**. North and south polarities of **permanent magnets** or **electromagnets** are used as the field. As the coils rotate, electromagnetic induction causes a current to be induced into them. The current induced in the conductors is an alternating current. This AC is converted to a form of

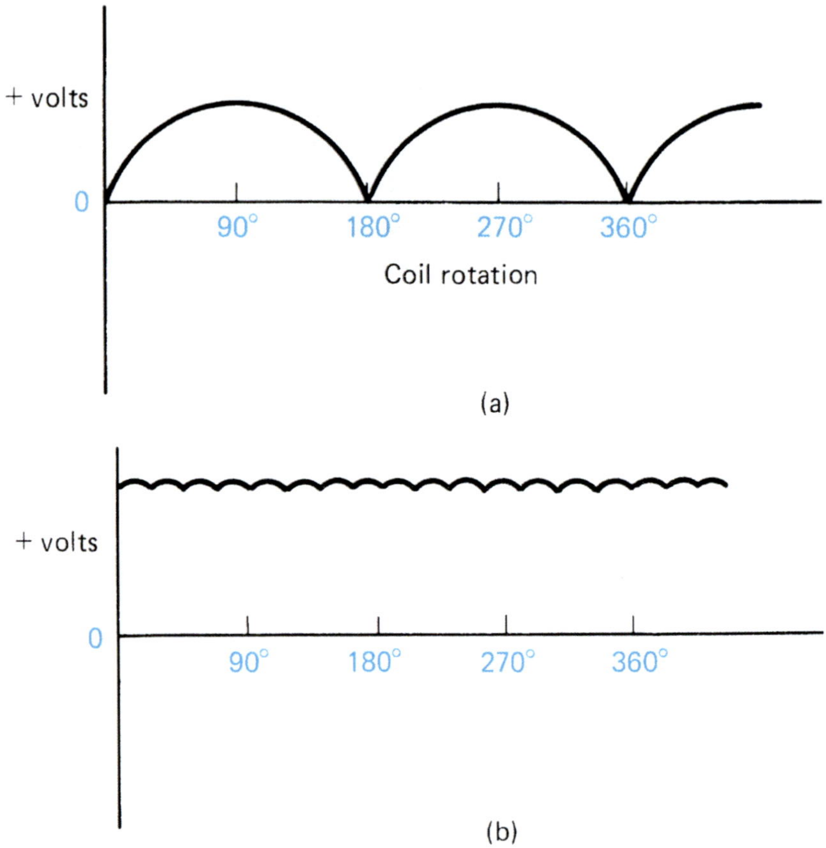

Figure 8-9 Current flow from the coils of a DC generator: (a) single coil – pulsating direct current; (b) many coils – more pure direct current.

direct current. The conversion from AC to DC is done by a commutator. The **commutator** has segments that are insulated from one another. The ends of the armature conductors are connected to the commutator segments. The purpose of a commutator is to reverse the armature coil connection to the external load at the same time that the alternating current induced into the armature coils is reversed. This causes direct current to be applied to the load.

The current that flows from one coil would appear as shown in **Figure 8-9(a)** and is called **pulsating direct current**. By using many turns of wire in the armature, the voltage output is a smoother direct current. This type of output is shown in **Figure 8-9(b)**. The voltage developed depends on (1) the strength of the magnetic field, (2) the number of coils in the armature, and (3) the speed of rotation. By increasing any of these factors, voltage output is increased. This is the same as AC generators.

Sample Problem: Voltage Output of a DC Generator

Voltage output of a DC generator can be expressed as

$V_0 = Z \times n \times F/60$

where

V_0 = voltage developed across the generator brushes in volts

Z = total number of armature conductors

n = speed of rotation in r/min

F = magnetic flux per pole in weber.

Given: A four-pole DC generator rotates at 1200 r/min. The armature has 36 slots, and each coil has four turns of wire. The magnetic flux per pole is 0.05 weber.

Find: The voltage output of the generator.

Solution: Since each turn has two conductors, and 36 slots are used in the armature core, Z = 36 coils × 2 coils per turn × 4 turns of wire per coil = 288 conductors.

288 × 1200 × 0.05/60

= 288 V.

Direct current generators are made in several types. One type is a **permanent magnet DC generator**. On this type of generator, permanent magnets are used to develop the magnetic field. Electromagnets are ordinarily used to develop the magnetic field of DC generators. A source of direct current must be applied to the electromagnetic coils. The most common method of developing a magnetic field is for part of the generator DC output to be used to supply the field. The three major classifications of direct current generators are (1) **permanent magnet generators**, (2) **separately excited generators**, and (3) **self-excited generators**. The self-excited types are classified according to the method used to connect the armature windings to the field windings. This is done in one of the following ways: (1) **series**, (2) **parallel (shunt)**, or (3) **compound**. A shunt-wound DC generator is shown in **Figure 8-10**.

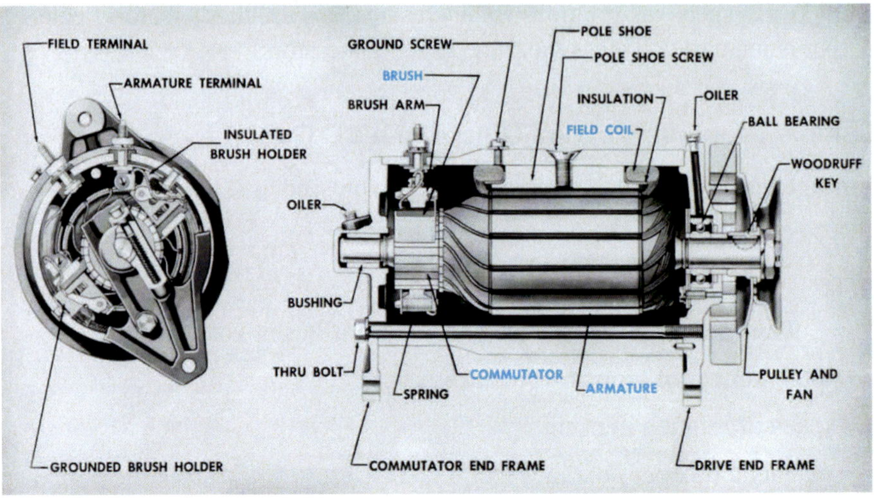

Figure 8-10 Cutaway of a shunt-wound DC generator (courtesy of Delco-Remy).

8.7 Direct Current (DC) Generators

Permanent Magnet DC Generators

A diagram of a **permanent magnet DC generator** is shown in **Figure 8-24**. The **rotor** conductors are connected to a **commutator** and **brushes**. The magnetic field is developed by permanent magnets made of Alnico or some other alloy. **Alnico** is an alloy of **al**uminum, **ni**ckel, **i**ron, and **co**balt. Several permanent magnets can be used together to create a stronger magnetic field.

The **armature** of a permanent magnet DC generator has many turns of insulated wires. The armature rotates inside the permanent magnetic field. An induced voltage is then developed and applied to a load device. Applications for this type of DC generator usually require low amounts of power. A permanent magnet DC generator is sometimes called a "**magneto**."

Separately Excited DC Generators

When large amounts of direct current electrical energy are needed, generators with electromagnetic fields are used. Strong magnetic fields are produced by electromagnets. It is also possible to control the strength of the field, which is done by varying the current through the field windings. The output of a generator is easily controlled in this way.

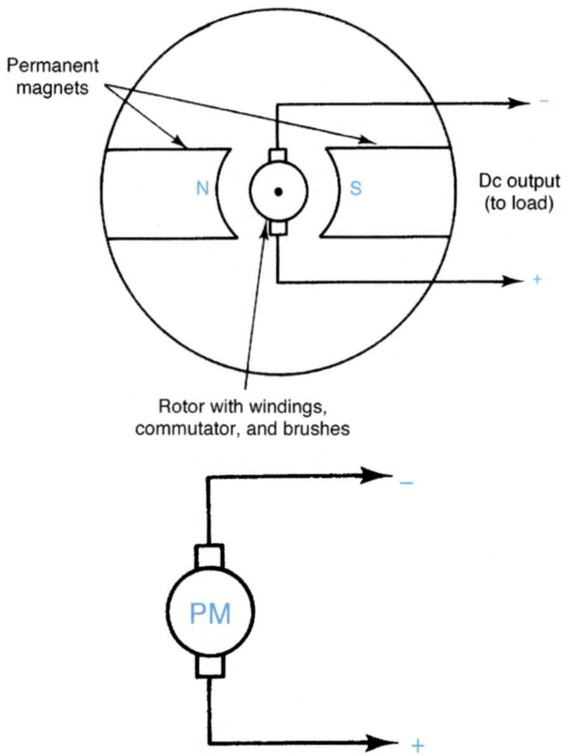

Figure 8-11 Permanent magnet DC generator: (a) pictorial; (b) schematic.

Direct current that is applied to the coils to develop an electromagnetic field is called **exciting current**. When DC exciting current comes from a source separate from the generator, it is called a **separately excited DC generator**. This type of generator is shown in **Figure 8-12**.

Storage batteries are sometimes used to supply DC exciting current to separately excited DC generators. The field circuit is not connected to the armature circuit. The separately excited DC generator has a constant output voltage. Changes in load affect the armature current, but they do not change the strength of the magnetic field. The output voltage of a separately excited DC generator is varied by adjusting the current flow through the field coils. A large **rheostat** in series with the field coils can be used to control current flow and adjust output voltage. Separately excited DC generators are used when precise voltage control is needed. Certain industrial processes require this precision. The cost of separately excited DC generators is usually high. Another disadvantage is that a separate direct current electrical energy source is needed.

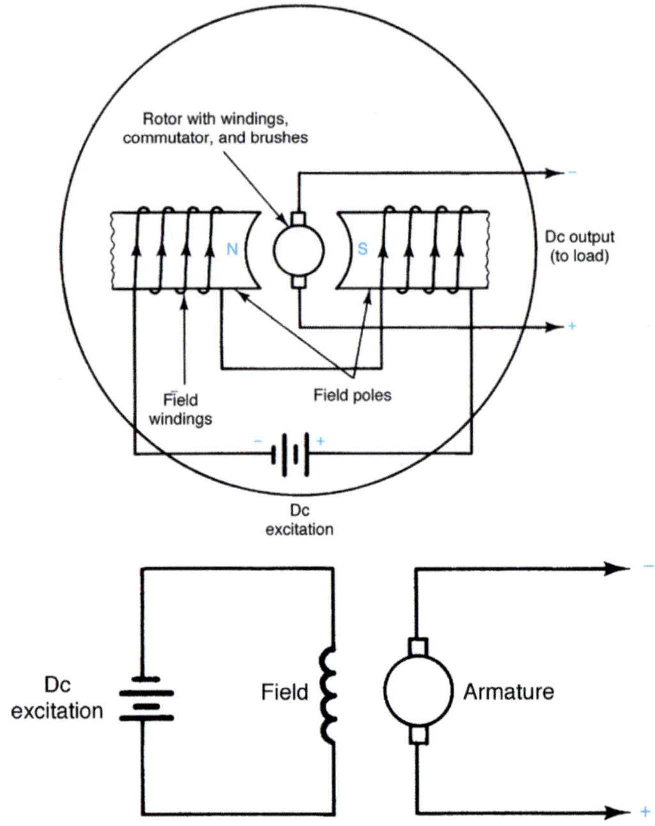

Figure 8-12 Separately excited DC generator: (a) pictorial; (b) schematic.

Self-excited, Series-wound DC Generators

DC generators produce direct current; so it is possible to take part of a generator's output to use as exciting current for the field coils. Generators that use part of their own output to supply DC exciting current are called **self-excited DC generators**. The method of connecting the armature windings and field windings together determines the type of generator. The armature and field windings may be connected in series, parallel (shunt), or series–parallel (compound). These are the three types of self-excited DC generators which could be designed.

Series-wound DC generators have armature windings connected in series with the field windings and the load as shown in **Figure 8-13**. In the series-wound DC generator, the total current flows through the load and through the field coils and armature.

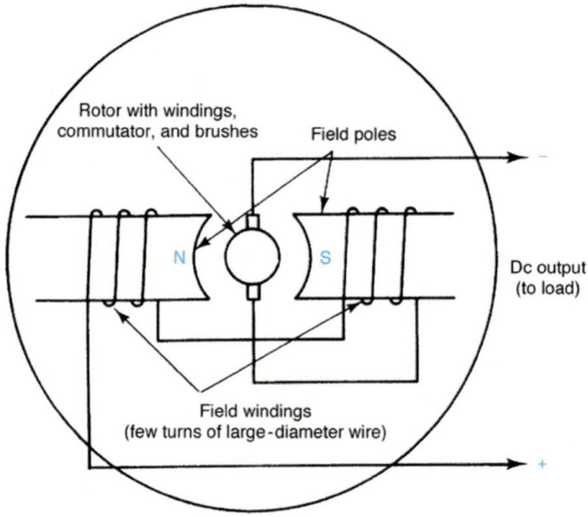

Figure 8-13 Series-wound DC generator pictorial.

The **field coils** are wound with low-resistance wire which has a few turns of large-diameter wire. An **electromagnetic field** is produced by the current flow through the coils. Remember that current flow is the same in all parts of a series circuit. If the load is disconnected, no current would flow through the generator. The field coils retain a small amount of magnetism after they are deenergized, which is called **residual magnetism**. Due to residual magnetism, current begins to flow as soon as the generator operates again. As the current increases, the magnetic flux of the field also increases. The output voltage rises as current flow increases. An output graph of a series-wound DC generator is shown in **Figure 8-14**. The peak of the curve shows magnetic saturation of the field coils. At this point, the "domains" of the coils have maximum alignment. Beyond this point, an increase in load current causes a decrease in output voltage due to energy losses which occur. The output of a series-wound DC generator varies with changes in load current. Self-excited, series-wound generators have only a few applications.

Self-excited, Shunt-wound DC Generators

When the field coils, armature circuit, and load are connected in **parallel**, a shunt-wound DC generator is formed. **Figure 8-15** shows a **shunt-wound DC generator**. The **armature current** developed by the generator (I_A) has two paths. One path is through the load (I_L) and the other is through the field coils (I_F). The shunt-wound DC generator is designed so that the field current

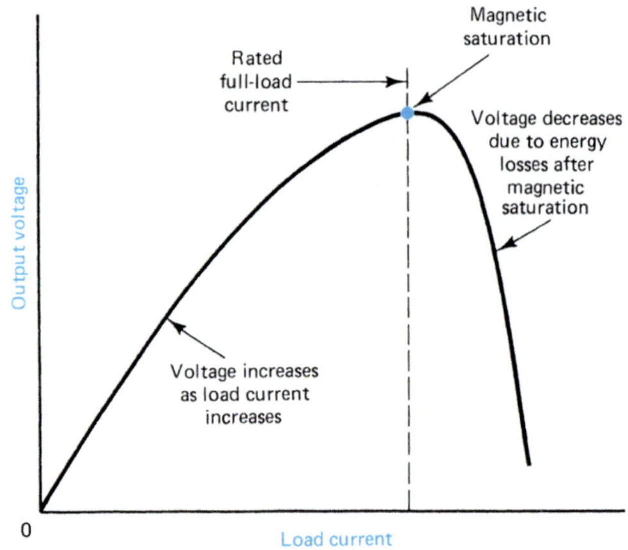

Figure 8-14 Output graph of a series-wound DC generator.

is not more than 10% of the total armature current (I_A). This is so that most of the generated armature current (I_A) will flow to the load.

A strong **electromagnetic field** must be produced. Also, the field current must be low. The field coils are wound with many turns of wire. They rely very little on the amount of field current to produce a strong magnetic field. The small-diameter wires limit the field current to a low value due to their high resistance. When **no load** is connected to a shunt-wound DC generator, a voltage is still generated. The voltage supplies energy to the field coils. Residual magnetism in the field coils is important for shunt-wound DC generators also. When a shunt generator is turned on, current flows in the armature and field circuit due to residual magnetism.

As current **increases**, the output voltage increases until magnetic saturation occurs.

When a load is connected to a DC shunt generator, the armature current (I_A) increases. The current increases the voltage ($I \times R$) drop of the armature. This causes a slightly smaller output voltage. Increases in load current cause slight decreases in output voltage. With load currents less than the rated value, the voltage is nearly constant. Large load currents cause the output voltage to drop sharply due to energy losses. Self-excited, shunt-wound DC generators are used when a fairly constant output voltage is needed.

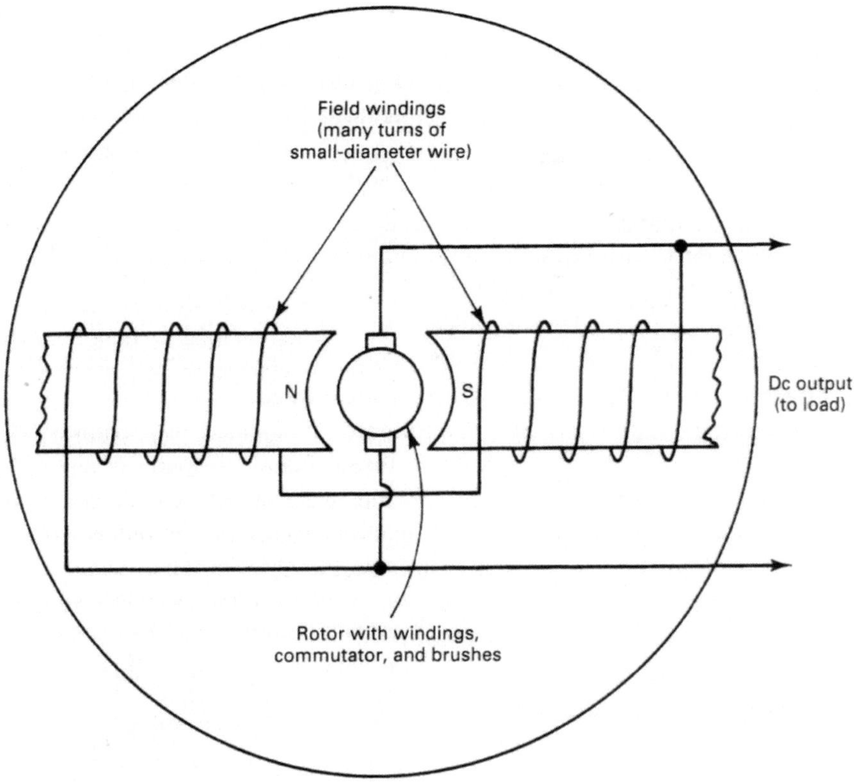

(a)

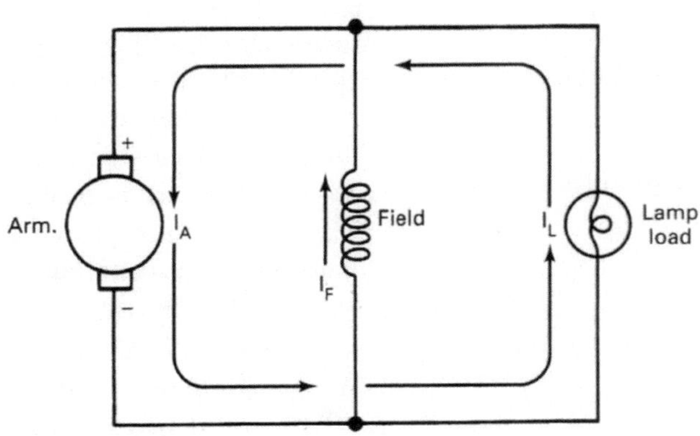

(b)

Figure 8-15 Shunt-wound DC generator: (a) pictorial; (b) schematic.

Self-excited, Compound-wound DC Generators

Compound-wound DC generators have two sets of field windings. One set is made of low-resistance coils connected in series with the armature circuit. The other set is made of high-resistance coils connected in parallel with the armature circuit. A compound-wound DC generator is shown in **Figure 8-16**.

The output voltage of a series-wound DC generator **increases** with increases in load current. The output voltage of a shunt-wound DC generator **decreases** slightly with increases in load current. A **compound-wound DC generator** has both series and shunt windings. Its output voltage is almost constant regardless of load current. The series field windings set up a magnetic field to counteract the voltage reduction caused by the voltage ($I \times R$) drop of the armature circuit. This produces a constant voltage.

A constant output voltage is produced by a **flat-compounded** DC generator. The no-load voltage is equal to the rated full-load voltage of a flat-compounded generator. No-load voltage is the output when there is no load connected to the generator. Full-load voltage is the output when the rated value of load is connected to the circuit. A compound-wound DC generator with full-load voltage greater than no-load voltage is called an **over compounded** generator. A generator with full-load voltage less than no-load voltage is called an **under compounded** generator. Output graphs for the three types of compound generators are shown in **Figure 8-17**.

Compound-wound DC generators can be made so that the series and shunt fields either aid or oppose each other. If the polarities of the coils on one side are the same, the magnetic fields aid each other. This type is called a **cumulative** compound DC generator. A **differential** compound DC motor has opposite polarities on both sides. The **cumulative** generator is mostly used. Compound-wound DC generators are used for applications that require constant voltage output.

DC Generator Operating Characteristics

DC generators have a characteristic known as **armature reaction**. Current flow through the armature windings produces a circular magnetic field. These fields react with the main field as shown in **Figure 8-18**. A magnetic field is produced which tends to distort the main magnetic field. As load current increases, armature current also increases. The increase in armature current causes more armature reaction to occur. Armature reaction causes sparking between the brushes and commutator. The theoretical switching point of the current to the load occurs at the generator's **neutral plane**. This is a plane or

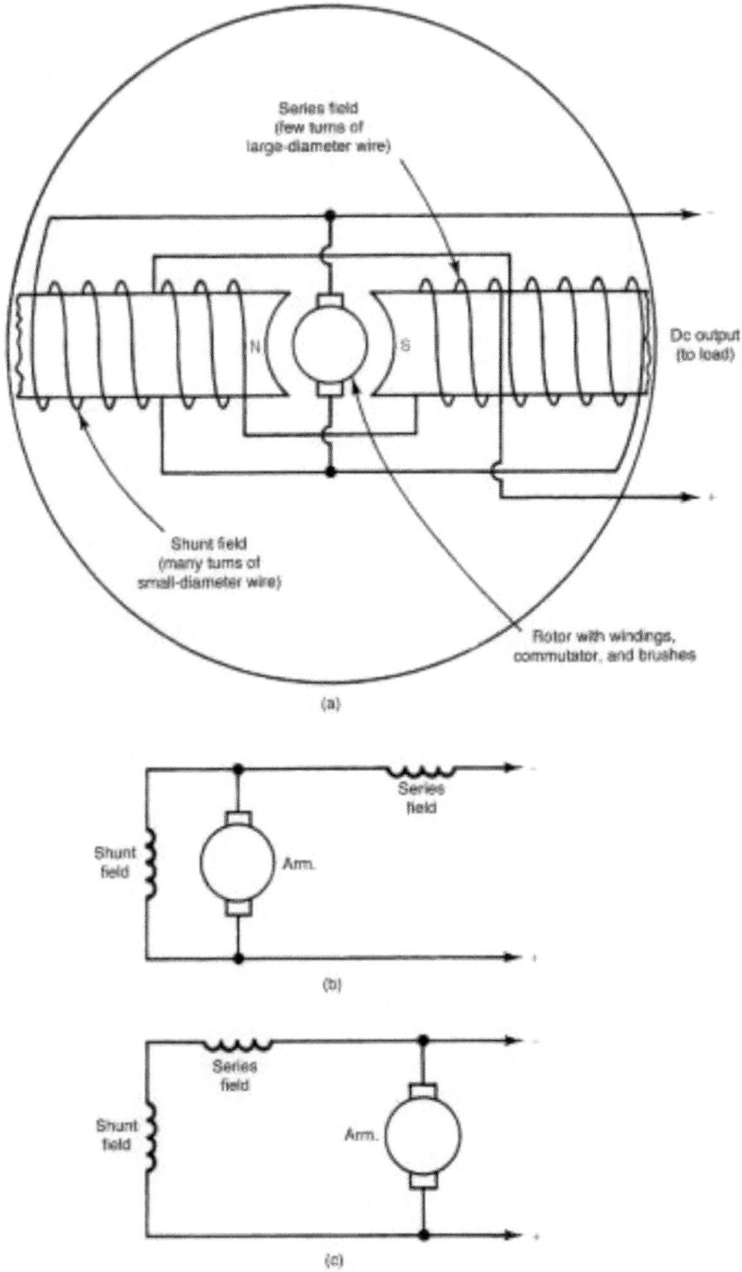

Figure 8-16 Compound-wound DC generator: (a) pictorial; (b) short shunt generator – shunt field is connected across the armature only; (c) long shunt generator – shunt field is connected across the armature and series field.

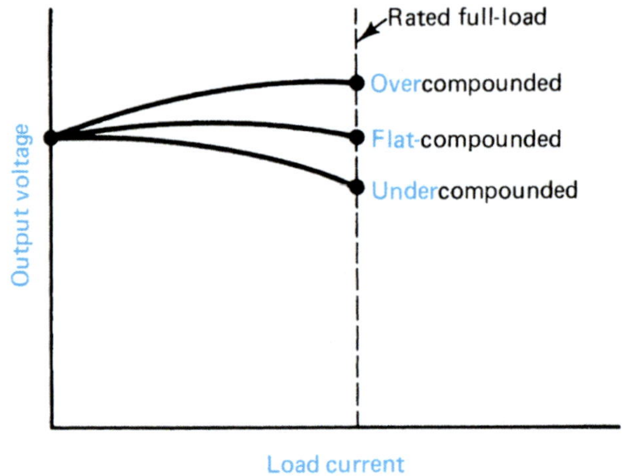

Figure 8-17 Output graph for three types of compound-wound DC generators.

position where no voltage is induced into the armature conductor connected to the commutator. Also, no distortion of the main magnetic field takes place. The switching position is perpendicular to the main magnetic field. Armature reaction causes the main magnetic field to become distorted. The new switching position occurs when a small voltage is induced into an armature coil. This new switching position is called the **running neutral plane**.

Armature reaction is reduced by using windings called **interpoles** between the main field coils. These coils are connected in series with the armature circuit. An increase in armature current causes a stronger magnetic field around the interpoles. Its field counteracts the distortion of the main field caused by armature reaction.

Generator **power output** is usually rated in kilowatts. This rating is the electrical power generating capacity of a generator. Ratings are specified by the manufacturer on the nameplate of a generator. Other ratings include output voltage, speed, and temperature limits. Generators are made in many sizes. As the load of a generator is increased, the voltage drop due to increased current flowing through the armature resistance increases. The output voltage then decreases. The amount of voltage change depends on the type of generator. The amount of change in output voltage from no-load value to rated full-load value is called **voltage regulation**. Voltage regulation is found by using the following formula:

$\% \text{VR} = V_{NL} - V_{FL}/V_{FL} \times 100$

where

% VR is the voltage regulation

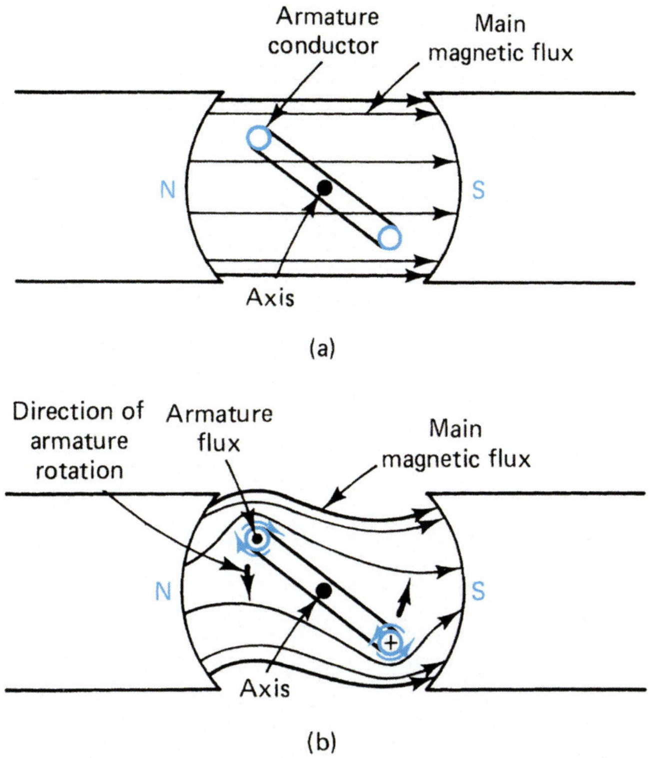

Figure 8-18 Effect of armature reaction in DC generators: (a) main magnetic field with no current flow through the armature windings; (b) distortion of the main magnetic field with current flow through the armature windings.

V_{NL} is the voltage with no load connected

V_{FL} is the rated full-load voltage of the generator.

The **efficiency** of a generator is the ratio of its power output in watts and its power input in horsepower. The efficiency of a generator is found by using this formula:

$\%$ efficiency $= P_{out}/P_{in} \times 100$

where P_{in} is the power input in horsepower and P_{out} is the power output in watts. Horsepower must be converted to watts. Since 1 hp = 746 W, multiply the horsepower by 746.

Sample Problem:

Given: A DC generator that has a no-load output voltage of 122.5 V and a rated full-load voltage of 120.0 V.

Find: The voltage regulation of the alternator.

Solution:

$$\%VR = V_{NL} - V_{FL} \times 100/V_{FL} \times 100$$
$$= 122.5 - 120/120 \times 100$$
$$= 0.02 = 2\%.$$

Sample Problem:

Given: A DC generator that has a power output of 22 MW and a power input of 35,000 horsepower.

Find: The efficiency of the alternator.

Solution:

$$\%\ Eff = P_{out}/P_{in} \times 100$$
$$= 22{,}000{,}000\ W/35{,}000\ hp \times 746 \times 100$$
$$\%\ Eff = 84\%.$$

To convert horsepower to watts, remember that 1 hp = 746 watts. The efficiency of a generator usually ranges from 70% to 85%.

Summary

- Alternating current (AC) is the current produced when electrons move first in one direction and then in the opposite direction.
- An alternator is a rotating machine that generates AC voltage.
- An armature is the part of a generator into which current is induced.
- A battery is an electrical energy source consisting of two or more cells connected together.
- A dry cell is a nonliquid cell that produces DC voltage by chemical action.
- Electromagnetic coils that develop the magnetic fields of electrical machines are referred to as field coils and the laminated metal core material field poles.
- A generator is a rotating electrical machine that converts mechanical energy into electrical energy.
- A prime mover supplies the mechanical energy to rotate an electrical generator.

- A brush is a sliding contact made of carbon and graphite, connected between the commutator and the power source of a DC motor or load of a DC generator.
- The process of changing AC induced into a generator rotor, referred to as induced current, in a DC generator converted to pulsating DC and applied to a load circuit is called commutation.
- A commutator is an assembly of copper segments that provide a method of connecting rotating coils to the brushes of a DC generator.
- A DC generator core is laminated soft iron or steel used to reduce heat loss due to eddy currents in the internal construction of the magnetic circuit.
- The rotating-armature method is used when a generator has DC voltage applied to produce a field to the stationary part (stator) of the machine and voltage is induced into the rotating part (rotor)
- The rotating-field method used when a generator has DC voltage applied to produce a field to the rotor of the machine and voltage is induced into the stator coils.
- A neutral plane is the theoretical switching position of the commutator and brushes of a DC generator or motor which occurs when no current flows through the armature conductors and the main magnetic field has least distortion and the running neutral plane is the actual switching position of the commutator and brushes of a DC generator or motor that shifts the theoretical neutral plane due to armature reaction.
- Regulation is the measure of the amount of voltage change that occurs in the output of a generator due to changes in load.
- The left-hand rule is used to determine the **direction of the magnetic field** around a single conductor or to determine the **polarity of an electromagnetic coil**.
- Lenz's law states that the induced voltage in any circuit is always in such a direction that it will oppose the force that produces it.
- The flow of electrons in one direction from negative (–) to positive (+) is called direct current (DC).
- The ratio of output power to input power of a generator is called efficiency (efficiency = power output (watts)/power input (watts).
- The positive (anode) or negative (cathode) part of a cell is called an electrode.
- The solution used in a cell that produces ions is called an electrolyte.
- An ion is an atom that has lost or gained electrons, making it a negative or positive charge.

- A lead-acid cell is a secondary cell that has positive and negative plates made of lead peroxide and lead and a liquid electrolyte of sulfuric acid mixed with water.
- A photovoltaic cell that produces DC voltage when light shines onto its surface.
- Piezoelectric effect is the property of certain crystal materials to produce a voltage when pressure is applied to them.
- A primary cell cannot be recharged, and a secondary cell or storage cell can be charged with a charger.
- A thermocouple is a device that has two pieces of metal joined together so that when its junction is heated, a voltage is produced.
- A voltaic cell produces voltage due to two metal electrodes that are suspended in an electrolyte.

Self-examination/Answers

1. List five sources of electrical energy.
2. What are the two types of electrical current produced by energy sources called?
3. What are two classifications of chemical cells, and how are they different?
4. What three factors determine the amount of voltage induced into a conductor inside a magnetic field?
5. What are the differences between alternator slip rings and a commutator used for DC electrical generators?
6. What are the two methods used to produce voltage output in generators?
7. What are the parts of a DC generator?
8. What are three types of DC generators?
9. What is generator voltage regulation?
10. What is generator efficiency?

Answers

1. Chemical, light, heat, pressure, mechanical
2. DC and AC
3. Primary (not rechargeable) and secondary (rechargeable)
4. Magnetic field strength, number of conductors, relative speed of rotation

5. Slip rings are one circular section and split rings are segmented
6. Rotating armature and rotating field
7. Armature, field, commutator, prime mover
8. Permanent magnet, self-excited, separately excited
9. Ratio of full-load and no-load voltage output
10. Ratio of power output to power input

Glossary

Direct current
The flow of electrons in one direction from negative (–) to positive (+).

Dry cell
A nonliquid cell that produces DC voltage by chemical action.

Eddy current
Induced current in the metal parts of electrical machines which causes heat losses.

Efficiency
The ratio of output power to input power.
efficiency = power output (watts)/power input (watts).

Electrode
A specific part of a unit such as the cathode of a semiconductor cell.

Electrolyte
The solution used in a cell which produces ions.

Field coils
Electromagnetic coils that develop the magnetic fields of electrical machines.

Field pole
Laminated metal that serves as the core material for field coils.

Frequency
The number of AC cycles per second, measured in hertz (Hz).

Generator
A rotating electrical machine that converts mechanical energy into electrical energy.

Hydrometer
An instrument used to measure the specific gravity or "charge" of the electrolyte of a storage battery.

Induced current
The current that flows through a conductor due to magnetic transfer of energy.

Induced voltage
The potential that causes induced current to flow through a conductor which passes through a magnetic field.

Ion
An atom that has lost or gained electrons, making it a negative or positive charge.

Laminations
Thin sheets of soft iron or steel used in the construction of electrical machines to reduce heat losses due to eddy currents.

Lead-acid cell
A secondary cell that has positive and negative plates made of lead peroxide and lead and has a liquid electrolyte of sulfuric acid mixed with water.

Left-hand rule
(1) To determine the **direction of the magnetic field** around a single conductor, point the thumb of the left hand in the direction of current flow (- to +) and the fingers will extend around the conductor in the direction of the magnetic field; (2) to determine the **polarity of an electromagnetic coil**, extend the fingers of the left hand around the coil in the direction of current and the thumb will point to the north polarity; (3) to determine the **direction of induced current flow** in a generator conductor, hold the thumb, forefinger, and middle finger of the left hand at right angles to one another, point the thumb in the direction of motion of the conductor, the forefinger in the direction of the magnetic field (N to S), and the middle finger will point in the direction of induced current.

Lenz's law
The induced voltage in any circuit is always in such a direction that it will oppose the force that produces it.

Neutral plane
The theoretical switching position of the commutator and brushes of a DC generator or motor which occurs when no current flows through the armature conductors and the main magnetic field has least distortion.

Photovoltaic cell
A cell that produces DC voltage when light shines onto its surface.

Piezoelectric effect
The property of certain crystal materials to produce a voltage when pressure is applied to them.

Primary cell
A cell that cannot be recharged.

Prime mover
A system that supplies the mechanical energy to rotate an electrical generator.

Pulsating DC
A voltage or current value that rises and falls with current flow always in the same direction.

Regulation
A measure of the amount of voltage change that occurs in the output of a generator due to changes in load.

Rotating-armature method
The method used when a generator has DC voltage applied to produce a field to the stationary part (stator) of the machine and voltage is induced into the rotating part (rotor).

Rotating-field method
The method used when a generator has DC voltage applied to produce a field to the rotor of the machine and voltage is induced into the stator coils.

Rotor
The rotating part of an electrical generator or motor.

Running neutral plane
The actual switching position of the commutator and brushes of a DC generator or motor which shifts the theoretical neutral plane due to armature reaction.

Secondary cell
A cell that can be recharged by applying DC voltage from a battery charger.

Sine wave
The waveform of AC voltage.

Single-phase AC generator
A generator that produces single-phase AC voltage in the form of a sine wave.

Slip rings
Copper rings mounted on the end of a rotor shaft and connected to the brushes and rotor windings.

Specific gravity
The weight of a liquid as compared with the weight of water, which has a value of 1.0.

Split-ring commutator
See **Commutator**.

Stator
The stationary part of an electrical generator or motor.

Storage battery
See **Secondary cell**.

Thermocouple
A device that has two pieces of metal joined together so that when its junction is heated, a voltage is produced.

Three-phase AC generator
A generator that produces three AC sine-wave voltages that are separated in phase.

Voltaic cell
A cell that produces voltage due to two metal electrodes that are suspended in an electrolyte.
Power lines extend from the remaining beginnings or ends.

Part II

AC (Alternating Current)

9

AC (Alternating Current) Electrical Fundamentals

Much of the electrical energy used today is called **alternating current (AC)**. Most of the electrical equipment and appliances used in homes operate from the alternating current energy delivered by power lines. Alternating current electricity has many applications in homes, industries, and commercial buildings. Electrical power plants in our country produce alternating current or AC electricity. Most power plants have huge steam turbines that rotate AC generators. These generators produce three-phase AC that is distributed by long distance power transmission lines to the places where the electrical power is used. Industries and large commercial buildings use **three-phase AC**. Homes use single-phase AC power. Alternating current is the most common form of electrical energy used.

Objectives

1. Compare DC and AC waveforms
2. Determine the characteristics of AC waveforms – time period, frequency, instantaneous value, peak value, average value, effective value, and phase
3. Determine the phase difference between multiple waveforms
4. Use vector (phasor) diagrams for representing AC waveforms
5. Analyze the operation of resistive AC circuits

Chapter Outline

9.1 Sinusoidal AC Waveforms
9.2 Non-sinusoidal Waveforms
9.3 Single- and Three-phase AC
9.4 Resistive AC Circuits
9.5 Measuring AC Voltages

9.1 Sinusoidal AC Waveforms

DC and AC voltage **waveforms** are shown in **Figure 9-1.** The waveform of a DC voltage [**Figure 9-1(a)**] is a straight line or unidirectional voltage. The direction of electron current flow is from negative to positive through a DC circuit.

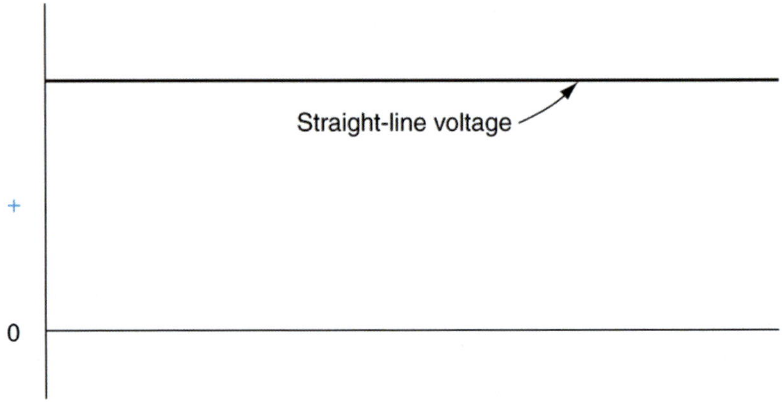

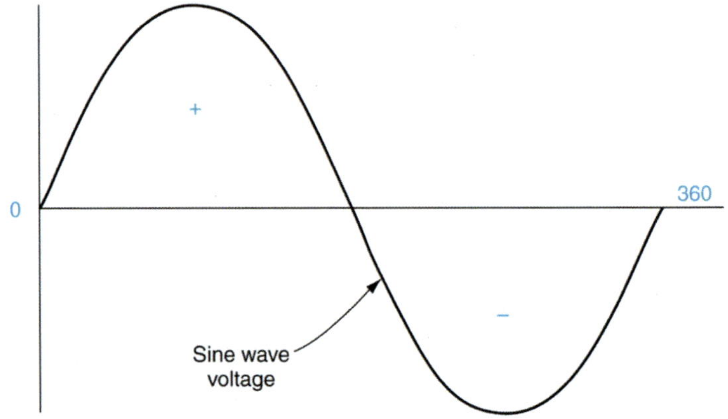

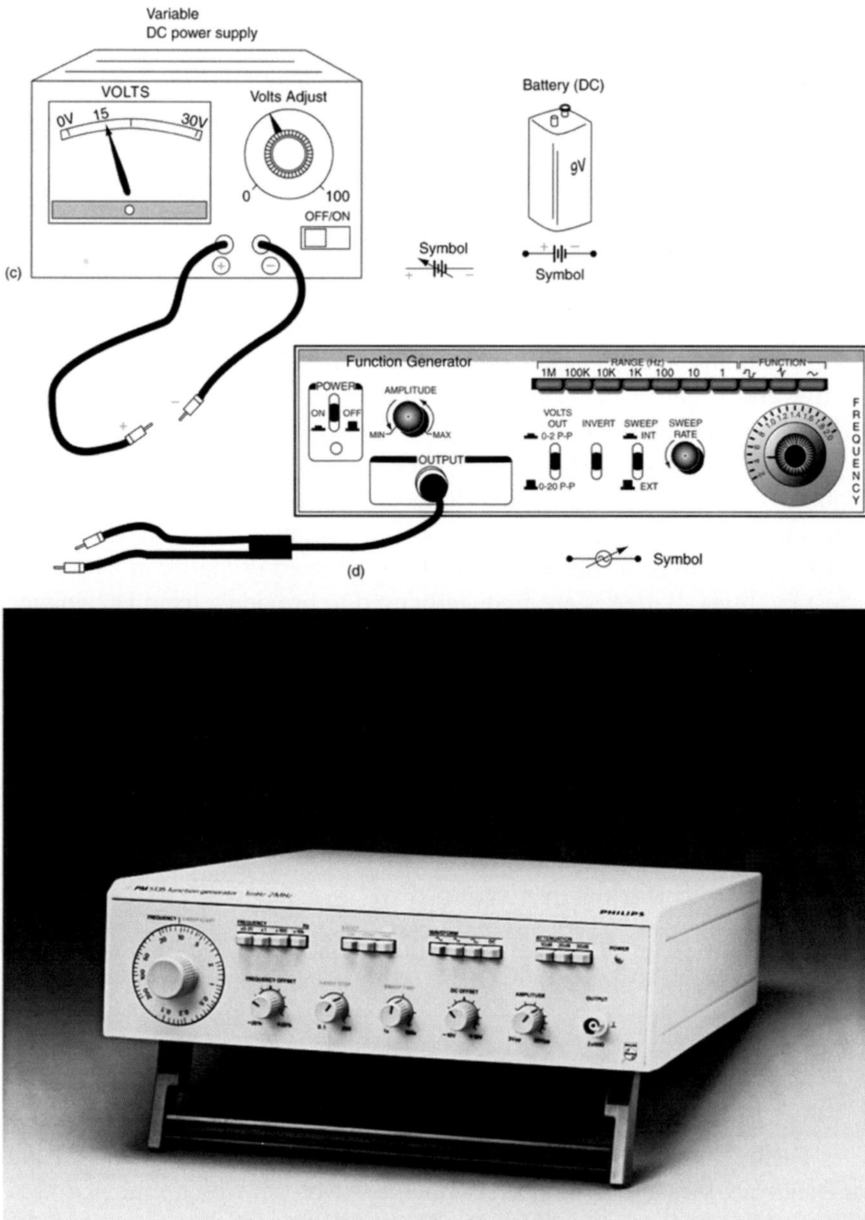

Figure 9-1 Comparison of (a) DC waveform, (b) AC waveform, (c) DC variable power supply and battery-sources of DC, (d) function generator – a source of AC, (e) function generator – a common type of laboratory equipment (courtesy of Fluke Corp.).

When an AC source is connected to some type of load, current direction changes several times in a given unit of time. Remember that direct current (DC) flows in one direction only. A waveform of one **cycle** of alternating current is compared with a DC waveform in **Figure 9-1(a)**. This waveform is called an **AC sine wave**. The time required to complete one complete cycle is called its **time period (T)**. When the AC generator shaft rotates one complete revolution, or 360°, one AC sine wave is produced. Note that the AC sine wave has a **positive peak** at 90° and then decreases to zero at 180°. It then increases to a **peak negative** voltage at 270° and then decreases to zero at 360°. The cycle then repeats itself. Current flows in one direction during the positive part and in the opposite direction during the negative half-cycle. The positive and negative half-cycles are referred to as alternations. Thus, one cycle has two alternations.

DC voltage waveforms, as shown in **Figure 9-1(a)**, can be developed by a source of DC voltage such as a **battery** or variable DC power supply. This is shown in **Figure 9-1(a)**. A variable DC power supply can be used to adjust the DC voltage to some desired value. They are used primarily in laboratory or test facilities. Batteries are frequently used to provide a fixed DC voltage for portable equipment.

Alternating current (AC) is produced by generators at electrical power plants. This AC voltage is in the form of sine waves shown in **Figure 9-1(b)**. In a laboratory, variable AC is often provided by a function generator. This is shown in **Figure 9-2(b)**. The function generator produces variable voltage output at a wide range of frequencies.

Frequency and Period of a Waveform

The **frequency** of an AC waveform is the reciprocal of the time period (*T*) and is expressed as follows:

$$f = \frac{1}{T}.$$

The frequency is measured in hertz (Hz). **Hertz** is the international unit for frequency measurement. For example, a waveform with a time period of 2 seconds has a frequency of ½ or 0.5 Hz.

Alternatively, if the **time period** is known, the **frequency (f)** of the AC waveform can be determined using the following formula:

$$T = \frac{1}{f}.$$

AC generators at power plants in the United States operate at a frequency of 60 cycles per second, or 60 Hz. If 60 AC sine waves are produced every second, a speed of 60 revolutions per second is needed. This produces a frequency of 60 cycles per second. The time period of the waveform can be determined:

$$T = \frac{1}{60} = 0.016667\ \text{s} = 16.667\ \text{ms}.$$

Figure 9-2 shows five cycles of alternating current. If the time required for an AC generator to produce five cycles were 1 s, the frequency of the AC would be 5 cycles per second.

AC Amplitude Voltage Values

AC voltage is measured with either an analog or a digital voltmeter (multimeter). The polarity of the meter leads is not important because AC changes direction. Remember that polarity is important when measuring DC because direct current flows only in one direction. Many analog multimeters do not measure AC current. They have ranges for AC voltage only.

Instantaneous Voltage

The value of a waveform at any given instant of time is called its **instantaneous value**, or V_i. Over a complete time period or cycle the instantaneous value of the waveform changes from 0 to a positive peak (V_p) value and then returns

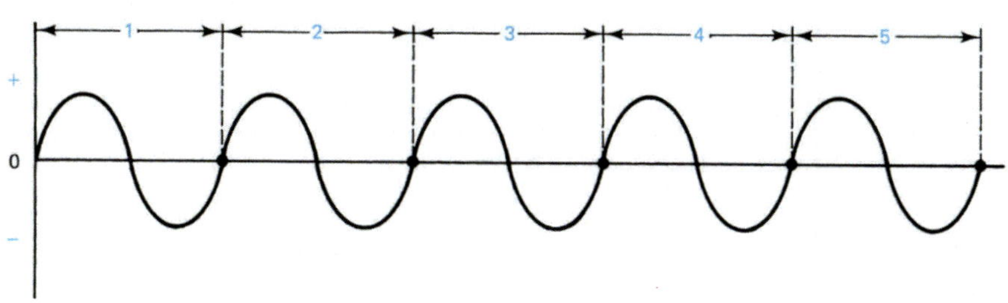

Figure 9-2 Five cycles of alternating current.

to 0 during the first **AC alternation**. The second alternation reverses polarity, changes from 0 to the negative peak (V_{-p}), and then returns to 0. This cycle repeats.

Peak Voltage

Figure 9-3 shows some of the voltage values commonly associated with alternating current. Among these are peak positive, peak negative, and peak-to-peak AC values. Peak positive is the maximum positive voltage reached during a cycle of AC. Peak negative is the maximum negative voltage reached. Peak-to-peak is the voltage value from peak positive to peak negative. These values are important to know when working with communications circuitry.

Average Voltage

The average value (V_{avg}) of an AC sinusoidal waveform over a complete cycle is the mathematical average of all instantaneous voltages that occur at each period of time throughout an alternation. This value evaluates to zero because the positive part of the waveform in the first alternation is equal and opposite in value to the waveform in the second alternation. This can be observed in **Figure 9-1(b)** where the positive (+) and negative (–) portions of the waveforms are indicated. Since the average value of a sinusoidal AC waveform over a complete cycle is zero, the average value is determined over a half-cycle instead. The average value is equal to 0.636 times the peak value

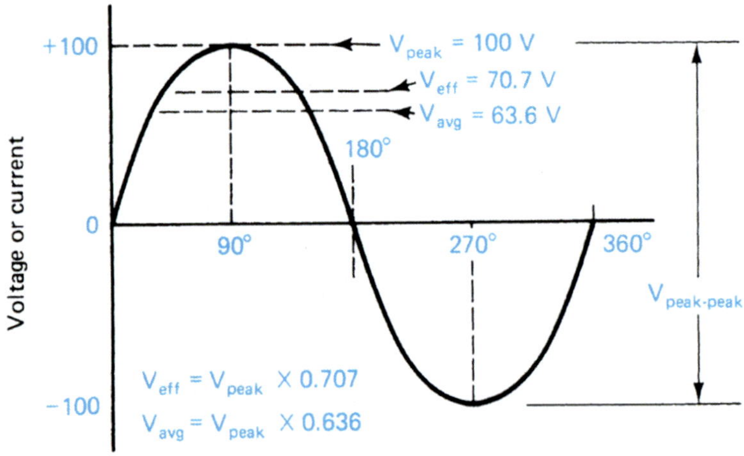

Figure 9-3 Voltage values of an AC waveform.

for one-half cycle (alternation) AC voltage. To determine the **average AC value** over a half-cycle, when the peak value is known, the following formula is used:

Average value = 0.636 × Peak value.

For an AC voltage waveform with a peak (V_p) of 100 V, as shown in **Figure 9-3**, the peak value is multiplied by the 0.636 conversion factor, and the average voltage is found to be about 63.6 V.

Effective or RMS Voltage

A more practical way of measuring the amplitude of an AC waveform uses the amount of energy converted by a source regardless of whether it is DC or AC. A circuit for doing so is shown in **Figure 9-4**. When the switch is in position 1, as shown in **Figure 9-4(a)**, only the 10-V DC source is connected to the lamp and it converts a certain amount of the energy for lighting the lamp.

When the switch is moved to position 2, as shown in **Figure 9-4(b)**, the lamp is connected to an AC source. The AC source that converts the same amount of energy in lighting the lamp as the DC source is said to have the same effective voltage value as the equivalent DC source. This effective value is called the RMS or root mean square value. The effective value of AC is defined as the AC voltage that will do the same amount of work as a DC voltage of the same value. Thus, in **Figure 9-4(b)**, when the switch is placed in position 2, 10-V AC effective value is applied to the lamp. The lamp should produce the same amount of brightness with 10-V AC effective value as with 10-V DC applied when the switch is in position 1. When AC voltage is measured with a meter, the resulting reading is the RMS or effective value.

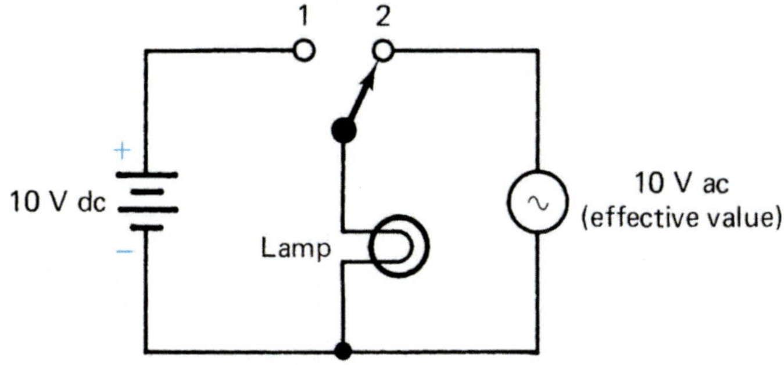

Figure 9-4 Comparison of effective AC voltage and DC voltage.

The most important AC value is the **effective or measured AC value**. This value is less than the peak positive value. RMS refers to the mathematical method used to determine effective voltage. **RMS voltage** and **effective voltage** are the same. To determine the RMS or effective AC value when the peak value is known, the following formula is used:

RMS or Effective value = 0.707 × Peak value.

For an AC voltage waveform with a peak (V_p) of 100 V, as shown in **Figure 9-5**, the peak value is multiplied by the 0.707 conversion factor, and the effective or **RMS voltage** is found to be about 70.7 V. To determine the peak AC value, when the measured RMS or effective value is known, the following formula is used:

Peak value = 1.41 × Effective value.

A common AC effective voltage value is 120 V. For an AC waveform with an RMS value of 120 V, the peak value can be found by multiplying the RMS value by 1.41. The **peak voltage** is found to be about 170 V. The voltage rating of electronic devices must be greater than the peak AC voltage applied to them. If 120-V AC is the measured voltage applied to a device, the peak voltage is about 170 V. So, the device must be rated over 170 V rather than 120 V.

Phase Shift

The **phase difference** between two waveforms refers to difference in the angular position between corresponding points on both waveforms. Two or more sinusoidal AC waveforms may have the same time period (T) and, hence, the same frequency but could still differ in their starting positions. The location on the time-axis, which a waveform first crosses which increasing in amplitude, may be referred to as its first **zero-crossing position**. After achieving the positive peak value, the amplitude of the waveform decreases and crosses the time-axis. This marks its second zero-crossing position. If two or more waveforms reach their zero and maximum values simultaneously, they are "in phase." Thus, multiple waveforms having the same time period, and which also have the same zero crossing positions, are said to be in phase. **Figure 9-5(a)** shows two voltage waveforms, V_1 and V_2, which are in phase or in step with each other.

When the starting zero-crossing location of waveforms that have the same time period does not coincide, the waveforms are said to be **out-of-phase**. In such cases, it is conventional to choose one of the waveforms as the reference and determine the phase or angular difference in the starting position. Phase difference is given in degrees. Usually, the waveform that

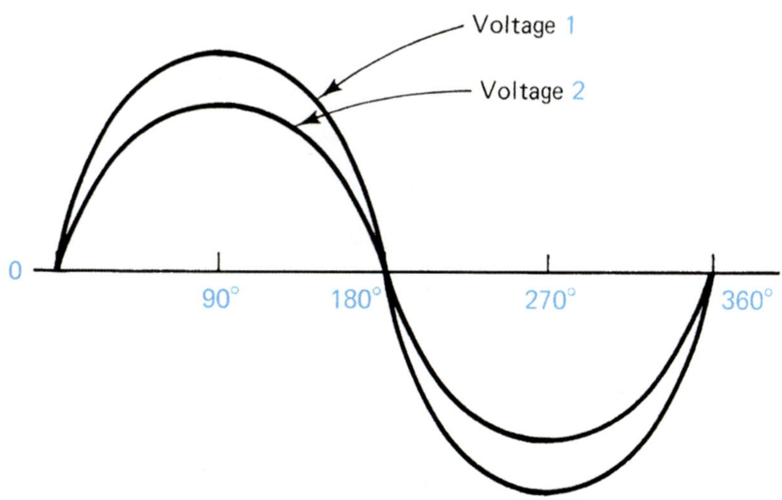

Figure 9-5 Two AC voltages that are in phase.

first crosses the time-axis while going positive or negative is chosen as the reference. This waveform is said to **lead** the other waveform by the angular difference in the starting positions. **Figure 9-6(b)** shows the voltage waveform V_1 leading V_2 by 90°. Alternatively, V_2 **lags** behind V_1 by 90°. It can be noted that waveforms that have an earlier first zero-crossing location will also have an earlier second zero-crossing location. The phase difference between waveforms with the same time period can, thus, be determined by examining either of the zero-crossing locations.

Vector (Phasor) Diagrams

Vectors are straight lines that have specific direction and length (magnitude). A **sinusoidal waveform** can be developed by rotating a vector. This rotating vector is commonly referred to as a **phasor diagram**. They are used to represent voltage and current values in AC circuits. The understanding of the vector diagram is helpful when working with AC circuits. Rather than using waveforms to show phase relationships, it is easier to use vector diagrams.

A horizontal line is drawn when beginning a vector diagram. An arrow head is often used to indicate the **direction** of the vector. The other end serves as **the reference point** and is commonly designated as "O" or origin. The vector $\overrightarrow{OA}$ or v_{OA}, for example, indicates a voltage vector directed from point "O" to "S." This vector is assumed to rotate at the same frequency as the waveform it represents.

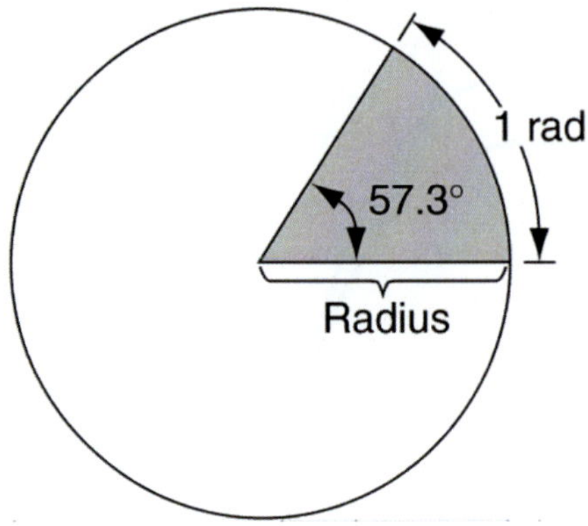
1 rad
57.3°
Radius

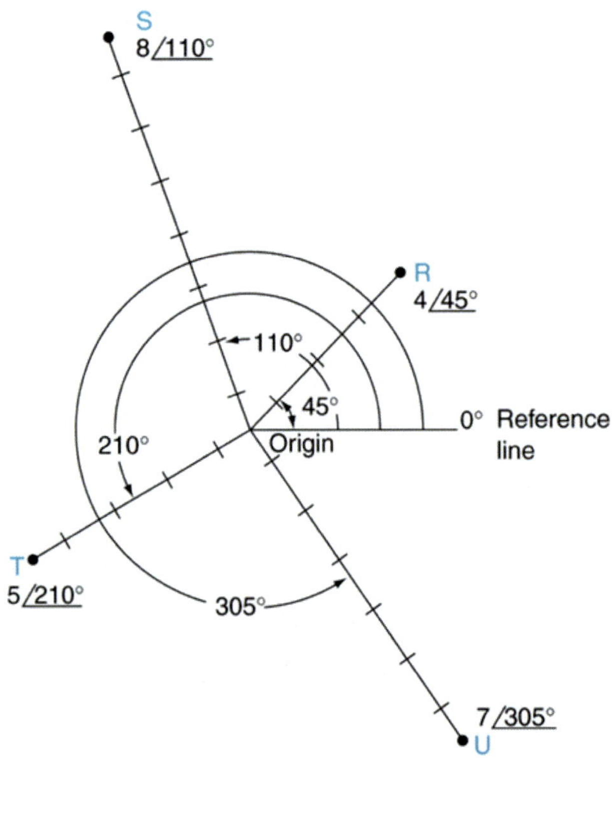
S
8/110°
R
4/45°
110°
45°
0° Reference line
210°
Origin
T
5/210°
305°
7/305°
U

Radian/Degree Conversion

Radians can be converted to degrees using the equation:

$$\text{rad} = \left(\frac{\pi \text{ rad}}{180^\circ}\right) \times \text{degrees}$$

Similarly, degrees can be converted to radians with the equation:

$$\text{degrees} = \left(\frac{180^\circ}{\pi \text{ rad}}\right) \times \text{rad}$$

Degrees (°)	*Radians (rad)*
0	0
45	$\pi/4$
90	$\pi/2$
135	$3\pi/4$
180	π
225	$5\pi/4$
270	$3\pi/2$
315	$7\pi/4$
360	2π

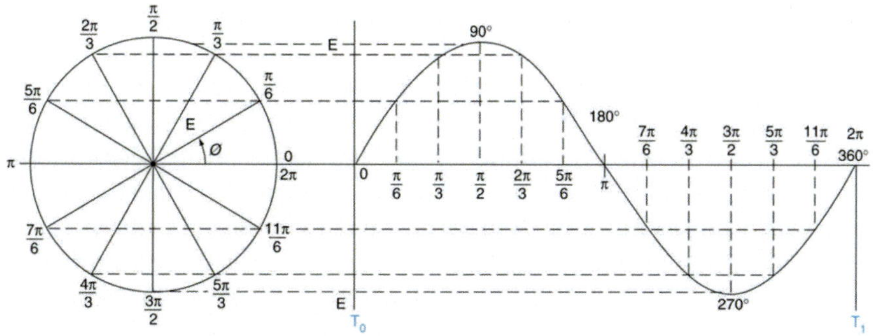

Figure 9-6 Radian values: (a) relation of radians to degrees and polar coordinate diagram; (b) radian/degree conversions formulas and table of radian/degree conversions; (c) sine-wave development.

For example, if a voltage vector $\overrightarrow{OA}$ is rotated about the center point "O" at a given frequency (f) or corresponding time period (T). Since one complete rotation corresponds to 360° and also to 2π radians, it is possible to convert between degrees and radian measures. One **radian** corresponds to 57.3° as shown in **Figure 9-6(a)**. The formulas for converting degrees to radians and vice versa are also given in **Figure 9-6(b)**.

To convert degrees into radians:

$$\text{radians} = \left(\frac{\pi}{180°}\right) \times \text{degrees.}$$

$$\text{For example, } 30° \text{ is } \left(\frac{\pi}{180°}\right) \times 30° = \frac{\pi}{6} \text{ radians.}$$

To convert radians into degrees:

$$\text{degree}s = \left(\frac{180°}{\pi}\right) \times \text{radians.}$$

$$\text{For example, } \frac{\pi}{6} \text{ radians is } \left(\frac{180°}{\pi}\right) \times \frac{\pi}{6} = 30°.$$

Commonly used degree and radian measures are given in **Figure 9-6(b)**. **Figure 9-6(c)** illustrates how a rotating vector can be used to sketch a corresponding sinusoidal waveform which begins at 0° and ends at 360°. This is done by plotting the height of the vector as observed on the corresponding angular measure while the vector is rotating about the origin. The rotation is always assumed to occur in the **counter-clockwise** or CCW direction. When the rotation begins at 0°, the height of the vector is 0. As the rotation continues and reached π/6 **radians** or 30°, the height of the vector on the corresponding 30° location voltage waveform is plotted. Note that the peak of the voltage occurs at π/2 or 90°, and the height of the voltage waveform is also the complete length of the vector. In a similar way, the rest of the waveform is completed. As can be seen, it is convenient to represent a sinusoidal waveform by a rotating vector. Only the start position of the vector needs to be shown.

Differences in phase between two or more waveforms can be easily represented using **phasor diagrams** (discussed in a subsequent chapter). For example, two voltage waveforms with the same time period (T) and phase, but different amplitudes can be represented as vectors with different lengths, having the same start point and direction.

9.2 Non-sinusoidal Waveforms

Up to now, the only AC waveform that has been discussed is the **sine wave**. Except for DC, the sine wave is the simplest of all waveforms. In the study of electronic fundamentals, you will encounter many waveforms that do not have the simple structure of the sine wave. These are known as **non-sinusoidal** or **complex waves**. A non-sinusoidal wave does not follow the sine curve in amplitude variations. Its form is not necessarily symmetrical, and it may be composed of more than one frequency.

One common non-sinusoidal wave is the **square wave**. **Figure 9-7** shows the graph of a square wave. At point A of **Figure 9-7**, the voltage has risen from zero to a positive 100-V level. The voltage remains at the positive level through the period from A to B and then drops to zero at point C. The voltage remains zero for a time identical to the period that it was positive. At point D, the wave period is complete, and the cycle begins again. The **period** of the square wave is the time from point A to point D. Its frequency, known as its fundamental frequency, is the reciprocal of the period:

$$f = \frac{1}{T}.$$

Thus, frequency equals 1 divided by the time period. Like the sine wave, a square wave has a peak and a peak-to-peak value.

Harmonics

A **square wave** commonly has two voltage levels: a minimum and a maximum with instantaneous changes in amplitude between the two levels. A square wave can be created using a combination of AC waveforms. Such a waveform will contain a **fundamental frequency** and odd **harmonics**. A harmonic is a sine wave whose frequency is a (whole number) multiple of the fundamental frequency. For example, for a frequency of 5 kHz, the **first harmonic** is 5 × 1 kHz, which is the fundamental frequency. The **second harmonic** is 2 x 5 kHz, or 10 kHz. The **third harmonic** is 15 kHz. The **fourth harmonic** is 20 kHz, and so on. The even multiples are known as even harmonics, and the odd multiples are known as odd harmonics. Thus, the harmonic content of a square wave is the first, third, fifth, seventh, ninth, and so on. A square wave can be formed by adding the odd harmonics to a fundamental frequency. The harmonics will not have the same amplitude as the fundamental. The third harmonic's amplitude is one-third of the fundamental, the fifth harmonic's amplitude is one-fifth of the fundamental, and so on.

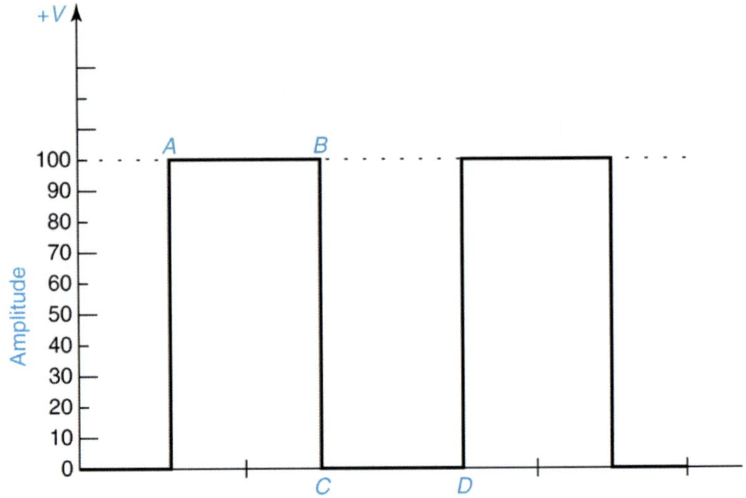

Figure 9-7 A square-wave voltage.

The **square wave** is shown in **Figure 9-8(a)**. In **Figure 9-8(b)**, a **sine wave** with the same frequency as the **fundamental frequency** of the square wave is labeled (A). Its **third harmonic** is labeled (B). The algebraic sum of the fundamental and the third harmonic is labeled (C). Note that the sides begin to show a sharper rise and the top is beginning to flatten. In **Figure 9-8(c)**, the **fifth harmonic** is added algebraically to waveform (C), forming waveform (*D*). This waveform is the fundamental plus the third and fifth harmonics. In **Figure 9-8(d)**, the **seventh harmonic** is added algebraically to waveform (*D*) to form waveform (*E*), which is the fundamental plus the third, fifth, and seventh harmonics. Note that the sides of waveform (*D*) are becoming vertical, and the positive and negative peaks are flattening out. As more harmonics are added, this process will continue until a nearly perfect square wave is produced. The **harmonic** content of a square wave, or any non-sinusoidal wave, may be viewed on an instrument known as a **spectrum analyzer**. The trace of a spectrum analyzer represents frequency. Its presentation is in the frequency domain. The graph of a sine wave is in the time domain as would be displayed on an **oscilloscope**.

Pulse Waveforms

A **pulse** is a voltage or current that momentarily makes a sharp change in amplitude. It remains at this value for a time and then returns to its original value. It is the type of waveform which can be created by connecting and

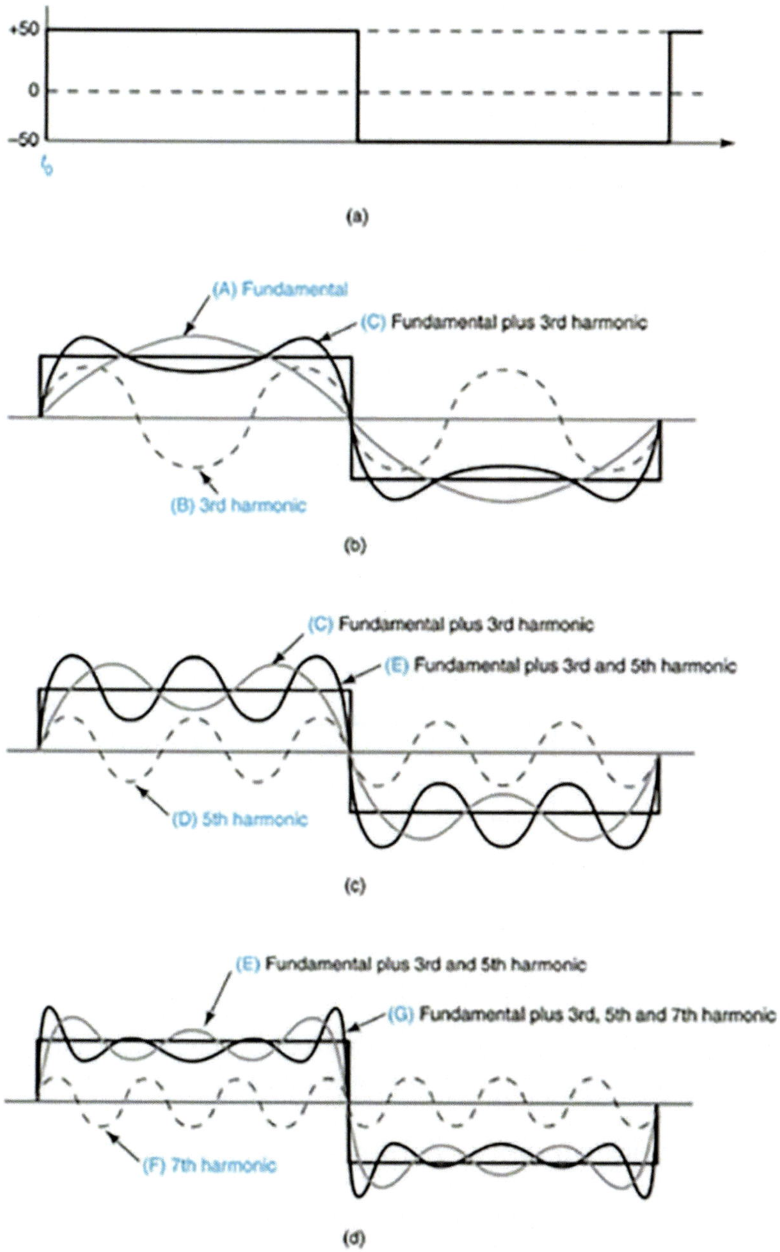

Figure 9-8 Harmonics: (a) square wave to be produced; (b) fundamental plus third harmonic; (c) fundamental plus third and fifth harmonics; (d) fundamental plus third, fifth, and seventh harmonics.

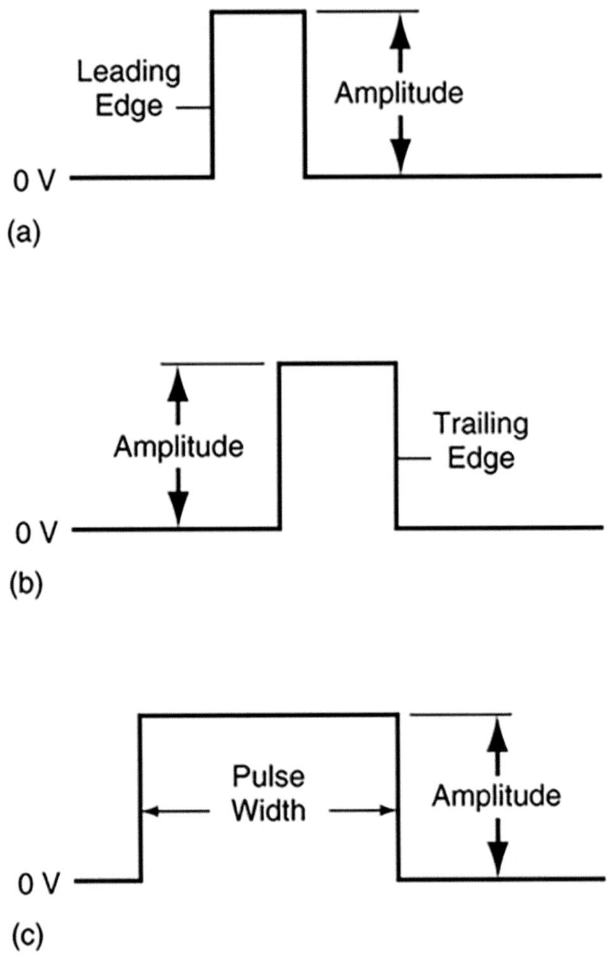

Figure 9-9 Pulses: (a) short duration; (b) medium duration; (c) long duration.

disconnecting a DC source to a load. The voltage across the load changes from 0 V to the positive supply voltage when the switch is closed and drops to 0 V when the switch is open.

As shown in **Figure 9-9**, a **pulse** may be of a short, medium, or relatively long duration. The difference between the lower and upper voltage levels of the pulse is known as its amplitude. The waveforms in this figure are shown in the time domain, with time increasing from left to right. Thus, the left rise of the pulse is known as the **leading edge**, while the right rise is known as the **trailing edge**. The pulse width is the time between the leading and trailing edges. The pulse may repeat itself over a definite period. The **pulse**

repetition time (PRT) is the time from the leading edge of one pulse to the leading edge of the next. The rate at which the pulses occur is known as the **pulse repetition frequency (PRF)** and is equal to the reciprocal of the PRT. The PRF is also referred to as the pulse repetition rate (PRR). PRF and PRT are related as follows:

$$PRF = \frac{1}{PRT}.$$

9.3 Single- and Three-phase AC

Electrical **single-phase AC voltage** is produced by single-phase AC generators, or it can be obtained across two power lines of a three-phase system. A single-phase AC source has a hot wire and a neutral wire to carry electrical current. The **neutral** is grounded to help prevent electrical shocks. Single-phase power is the type of power distributed to our homes. A **three-phase AC** source has three power lines that carry electrical current. Three-phase voltage is produced by three-phase generators at electrical power plants. Three-phase voltage is a combination of three single-phase voltages that are electrically connected. This voltage is similar to three single-phase AC sine waves separated in phase by 120°.

Three-phase AC is used to power large equipment in industry and commercial buildings. It is not distributed to homes. There are three current-carrying power lines on a three-phase system. Two common types of configurations are possible in three-phase systems. These are the three-phase wye and delta systems. A wye configuration has a common neutral connection, and a three-phase delta system does not.

Remember that **single-phase AC** voltage is in the form of a **sine wave**. Single-phase AC voltage is used for low power applications, primarily in the home. Almost all electrical power is generated and transmitted over long distances as three-phase AC. Three coils are placed 120° apart in a generator to produce three-phase AC voltage. Most AC motors over 1 hp in size operate with three-phase AC power applied. Most industries and commercial buildings have three-phase equipment.

Three-phase AC systems have several **advantages** over single-phase systems. In a single-phase system, the power is said to be "pulsating." The peak values along a single-phase AC sine wave are separated by 360°, as shown in **Figure 9-10(a)**. This is similar to a one-cylinder gas engine. A three-phase system is somewhat like a multi-cylinder gas engine. The power is steadier than in a single-phase system. One cylinder is compressing when

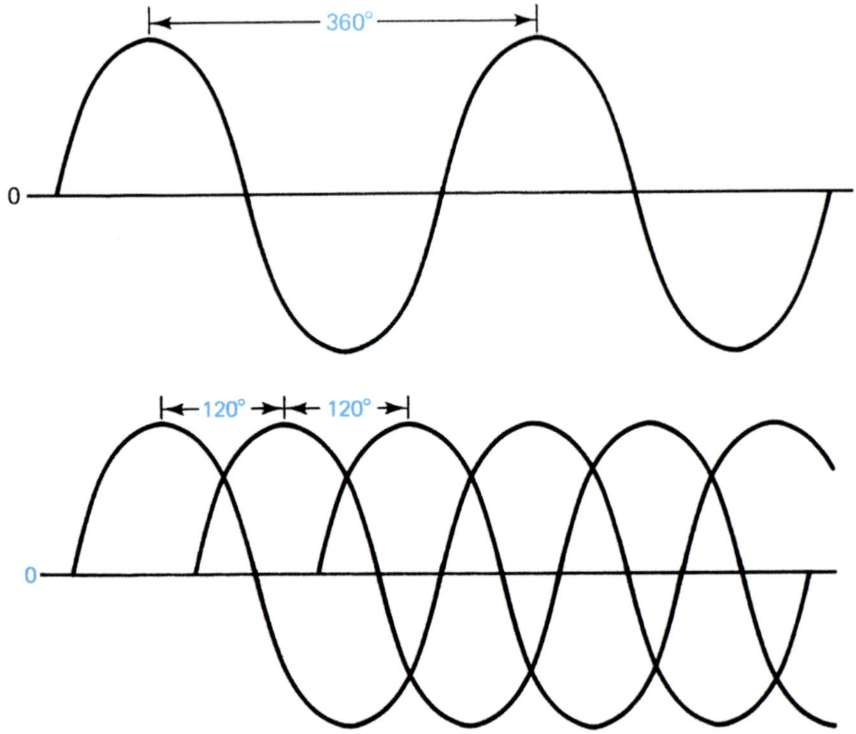

Figure 9-10 Comparison of (a) single-phase and (b) three-phase AC voltages.

the others are not, which is similar to the voltages in three-phase AC systems. The power of one separate phase is pulsating, but the total power is more constant. The peak values of three-phase AC are separated by 120°, as shown in **Figure 9-10(b)**. This makes three-phase AC power more desirable to use. The power ratings of motors and generators are greater when three-phase AC power is used. For a certain frame size, the rating of a three-phase AC motor is almost 50% larger than a similar single-phase AC motor.

9.4 Resistive AC Circuits

Ohm's law can be used with AC circuits containing resistors. **Kirchhoff's voltage and current laws** also apply to AC circuits containing resistors. The effect of applied voltage and current in an AC resistive circuit can be observed by connecting an AC source to a resistor.

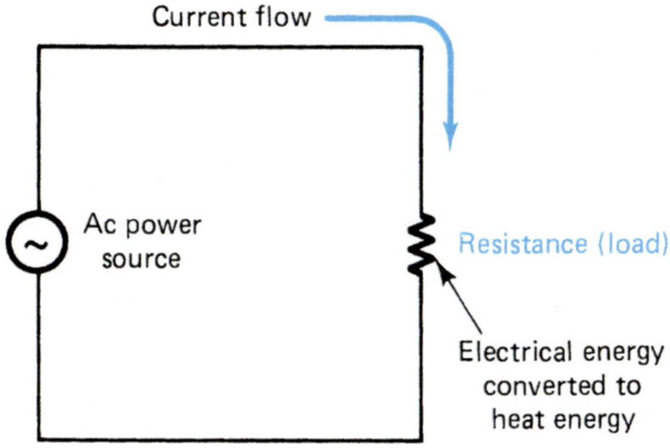

Figure 9-11 Resistive AC circuit.

The use of Ohm's and Kirchhoff's laws for AC circuits is influenced by the effects of two circuit elements which are normally not present in DC circuits – inductors and capacitors. An **inductor** is a coil of wire that has the property to oppose changes in current flowing through it due to energy stored in a magnetic field. A **capacitor** is a device that opposes changes in voltage due to energy stored in its electrostatic field. The **electrostatic field** is created between two oppositely charged metal plates separated by a dielectric material (insulator). The circuit properties of inductors and capacitors will be discussed in subsequent chapters.

AC circuits are similar in many ways to DC circuits. They have **a source**, **a load**, **a path**, and usually **controls** and **indicators**. AC circuits are classified by their electrical characteristics – resistive, inductive, or capacitive. All AC circuits are resistive, inductive, capacitive, or have a combination. The operation of each type of electrical circuit is different. The nature of alternating current causes certain electrical circuit properties to exist.

The simplest type of AC circuit is a resistive circuit, such as that illustrated in **Figure 9-11**. A **resistive circuit** is the same with AC applied as it is with DC applied. In DC circuits, the following formulas are used:

$$V = I \times R$$

$$I = \frac{V}{R}$$

$$R = \frac{V}{I}.$$

These show that when the voltage applied to a circuit is increased, the current will increase. Also, when the resistance of a circuit is increased, the current decreases. The waveforms of **Figure 9-12** show the relationship of the voltage and current in a resistive AC circuit. Voltage and current are in phase. In phase means that the minimum and maximum values of voltage and current occur at the samc time. The power converted by the resistance is found by multiplying voltage times current ($P = V \times I$). Thus, when an AC circuit has only resistance, it is similar to a DC circuit.

9.5 Measuring AC Voltages

Using a Digital or Analog Meter

A **multimeter** or **volt-ohm-meter (VOM)** may be used to measure AC voltage as shown. AC voltage is measured in the same way as DC voltage with the exception that when measuring AC voltage, proper polarity does not have to be observed. When measuring AC voltage, an appropriate AC voltage range of the meter must be selected. When measuring unknown AC voltage values, the highest meter range should be selected first and then reduced if needed.

A **digital multimeter** with a numerical readout, discussed in **Chapter 3**, can be used to measure AC voltage. Also, AC voltage measurement can be

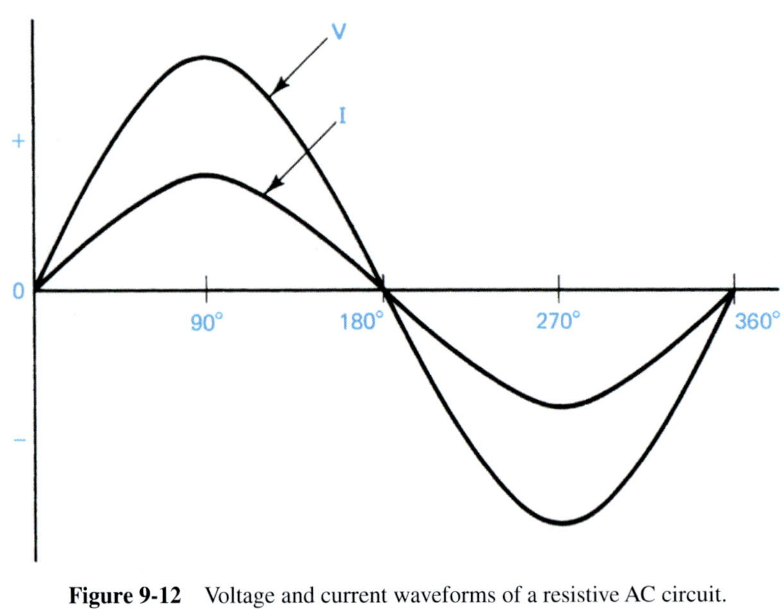

Figure 9-12 Voltage and current waveforms of a resistive AC circuit.

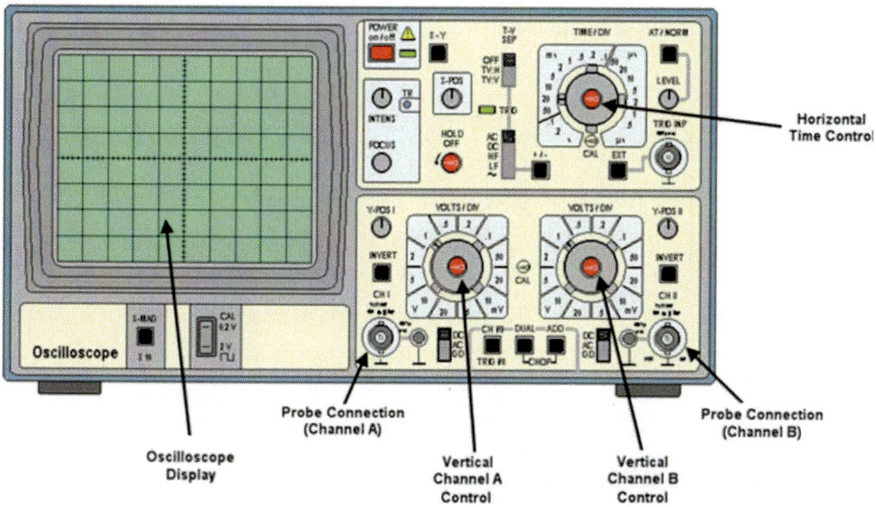

Figure 9-13 Oscilloscope screen and controls used to measure AC voltage.

made with an **analog multimeter**, having an indicating pointer that deflects on a graduated scale. Both analog and digital meters measure the **RMS** or **effective value** of an AC waveform.

Using an Oscilloscope

Another way to measure AC voltage is with an **oscilloscope**, such as the one shown in **Figure 9-13(a)**. Oscilloscopes provide a visual display of waveforms on their screen. They are also used to measure a wide range of frequencies with precision.

Oscilloscopes or "scopes" are used to examine wave shapes. For electronic servicing, it is necessary to be able to observe the voltage waveform while troubleshooting. An oscilloscope permits various voltage waveforms to be visually analyzed. It produces an image on its screen. The controls must be properly adjusted. The image, called a **trace**, is usually a line on the screen or cathode-ray tube (CRT). A stream of electrons strikes the phosphorescent coating on the inside of the screen, causing the screen to produce light.

The oscilloscope displays voltage waveforms on two axes, like a graph. The **horizontal axis** on the screen is the time axis. The **vertical axis** is the voltage axis. An AC waveform is displayed on the CRT as shown in **Figure 9-14**. For the CRT to display a trace properly, the internal circuits of the scope must be properly adjusted. These adjustments are made by controls

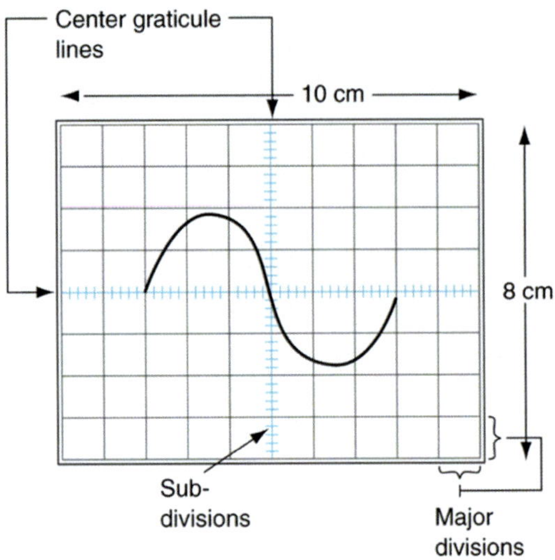

Figure 9-14 AC waveform displayed on the screen of an oscilloscope.

on the front of the oscilloscope. Oscilloscopes are slightly different, but most scopes have some of the following controls:

- **Intensity**: Controls the brightness of the trace, sometimes the on–off control.
- **Focus**: Adjusts the thickness of the trace so that it is clear and sharp.
- **Vertical position**: Adjusts the entire trace up or down.
- **Horizontal position**: Adjusts the entire trace to the left or right.
- **Vertical gain:** Controls the height of the trace.
- **Horizontal gain**: Controls the horizontal size of the trace.
- **Vertical attenuation** *or* **variable volts/cm**: Acts as a "coarse" adjustment to reduce the trace vertically.
- **Horizontal sweep** *or* **variable time/cm**: Controls the speed at which the trace moves across (sweeps) the CRT horizontally. This control determines the number of waveforms displayed on the screen.
- **Trigger**: Controls how the horizontal sweep is "locked in" with the circuitry of the scope.
- **Vertical input**: External connections used to apply an input to the vertical circuits of the scope.
- **Horizontal input**: External connections used to apply an input to the horizontal circuits of the scope.

The following procedure is used to adjust the oscilloscope controls to measure AC voltage. The names of some controls vary on different types of oscilloscopes.

1. Turn on the oscilloscope.
2. Adjust the intensity and focus controls until a bright, narrow, straight-line trace appears on the screen.
3. Use the horizontal position and vertical position controls to position the trace in the center of the screen.
4. Adjust the horizontal gain and the variable time/cm until the trace extends from the left of the screen to the right side of the screen. This allows the entire waveform to be displayed.
5. Connect the proper test probes into the oscilloscope's vertical input connections.
7. The scope is now ready to measure AC voltage.
8. After a waveform is displayed, adjust the vertical attenuation (volts/cm) and vertical gain controls until the height of the trace equals about 2 in., or 4 cm. Most scopes have scales that are marked in centimeters.
9. Adjust the Vernier or stability control until the trace becomes stable. One or more AC waveforms should appear on the screen of the scope.

An AC **resistive circuit** can be energized by a **function generator** or other AC power source (see **Figure 9-11)**. A multimeter can be used to measure the AC voltage developed across the resistor or an oscilloscope could be used to observe the sinusoidal AC waveform.

Summary

- An AC sinusoidal waveform changes in amplitude and direction periodically.
- The time taken to complete one complete AC waveform is called a time period or cycle time.
- The number of cycles per second is the frequency of a waveform.
- Frequency is measured in hertz (Hz).
- The instantaneous value of a waveform is its amplitude at any given time.
- The peak amplitude is the largest (most positive) value of an AC waveform.
- The peak-to-peak amplitude is the difference between the largest and smallest (most negative) voltage value.

- The effective or RMS AC voltage is the value of the AC voltage which will convert the same amount of energy as an equivalent DC voltage source.
- The phase difference between two waveforms refers to difference in the angular position between corresponding points on both waveforms.
- Phase difference is measured in degrees or radians.
- A sinusoidal waveform can be developed by rotating a vector.
- A harmonic is a sine wave whose frequency is a (whole number) multiple of the fundamental frequency.
- A square wave commonly has two voltage levels, a minimum and a maximum, with instantaneous changes in amplitude between the two levels.
- Single-phase power is the type of power distributed to our homes.
- Three-phase voltage is produced by three-phase generators at electrical power plants.
- A multimeter or volt-ohm-meter (VOM) may be used to measure AC voltage.
- Oscilloscopes provide a visual display of waveforms on their screen and are also used to measure a wide range of frequencies.

Formulas

$$f = \frac{1}{T}$$

$$T = \frac{1}{f}$$

Average value = 0.636 × Peak value
Effective value = 0.707 × Peak value
Peak value = 1.41 × Effective value

$$\text{radians} = \left(\frac{\pi}{180^\circ}\right) \times \text{degrees}$$

$$\text{degrees} = \left(\frac{180^\circ}{\pi}\right) \times \text{radians}$$

$$\text{PRF} = \frac{1}{\text{PRT}}$$

$$V = I \times R$$

$$I = \frac{V}{R}$$

$$R = \frac{V}{I}$$

Problems

1. An AC waveform completes two cycles in 10 seconds. Determine its time period for one cycle. (Two cycles are completed in 10 seconds; so one cycle is completed in $\frac{10}{2}$ = 5 seconds. Hence, T = 5 s.]
2. Determine the V_{+P}, V_{-P}, and $V_{P\text{-}P}$ of an AC waveform which has an RMS value of 6 V. [$V_P = 1.414 \times V_{RMS} = 1.414 \times 6 = 8.484$ V; $V_{-P} = -8.484$ V; $V_{P\text{-}P} = 2 \times V_{+P} = 2 \times 8.484 = 16.968$ V.]
3. An AC voltage with a peak value of 12 V produces an RMS value of ____________. [$V_{RMS} = 0.707 \times V_P = 0.707 \times 12 = 8.484$ V.]
4. A square wave has a frequency of 15,000 Hz. Determine the time that it takes to complete one complete square wave. [$T = \frac{1}{f} = \frac{1}{16{,}000} = 6.25\times 10^{-5}\ \text{s} = 62.5\mu\text{s}$]
5. Referring to **Figure 9-7**, determine the angle in degrees corresponding to $\frac{2\pi}{3}$ radians. [degrees $= \left(\frac{180°}{\pi}\right) \times$ radians $= \left(\frac{180°}{\pi}\right) \times \frac{2\pi}{3} = \left(\frac{360°}{3}\right) = 120°$.]

Self-examination/Answers

1. If it takes 4 seconds to complete 1 sine wave, determine the frequency of the waveform.
2. How long does it take to produce a sine wave of 50 Hz?
3. The peak value of a 25-V RMS AC voltage is _______________.
4. The length of a vector line represents its ____________.
5. If it takes 4 seconds to complete 1 square wave, determine the frequency of the waveform.

6. Square and triangular waveforms can be classified as ______________waveforms.
7. A square wave can be created by adding together (odd, even) harmonics to a fundamental sinusoidal waveform.
8. A single-phase AC source has a ________wire and a ________wire to carry electric current.
9. The number of hot or powered lines in a three-phase AC power distribution system is _______.
10. ________phase AC is distributed to homes, and ________phase AC is distributed to industrial facilities.
11. Each phase of a three-phase AC system has a phase difference of ________.
12. When AC is applied to a resistive circuit, the phase difference between the voltage and current is ____.
13. When the AC voltage applied to a resistive circuit is increased, the current __________.
14. When the AC voltage applied to a resistive circuit reverses in polarity, the polarity of the current (remains the same, becomes zero, reverses).
15. The (average, peak, RMS) voltage of AC is indicated by a multimeter.
16. The ____________of the meter is not important while measuring AC voltages.
17. While measuring an unknown AC voltage value, the range of the meter should first be set to its (lowest, highest) value.
18. A(n) ________________is commonly used to display the value and the shape of an AC waveform.
19. The (horizontal, vertical) controls of an oscilloscope determine the speed with which the display occurs.

Answers

1. $f = \frac{1}{T} = \frac{1}{4} = 0.25\ \text{Hz}$
2. $T = \frac{1}{f} = \frac{1}{50} = 0.02\ \text{s}$
3. $V_P = 1.414 \times V_{RMS} = 1.414 \times 25 = 35.35\ \text{V}$
4. magnitude
5. $f = \frac{1}{T} = \frac{1}{4} = 0.25\ \text{Hz}$
6. non-sinusoidal or complex
7. odd

8. hot, neutral
9. three
10. 1, 3
11. 120
12. 0
13. increases
14. reverses
15. RMS
16. polarity
17. highest
18. oscilloscope
19. horizontal

Glossary

AC
An abbreviation for alternating current.

Amplitude
The vertical height of an AC waveform.

Angle of lead or lag
The angle between applied voltage and current flow in an AC circuit, in degrees; in an inductive (*L*) circuit, voltage (*V*) leads current (*I*); in a capacitive (*C*) circuit, current (*I*) leads voltage (*V*).

Apparent power (volt-amperes)
The applied voltage times current delivered to an AC circuit.

Average voltage (V_{avg})
The value of an AC sine-wave voltage which is found by the formula: $V_{avg} = V_{peak} \times 0.637$.

Capacitance (C)
The property of a device to oppose changes in voltage due to energy stored in its electrostatic field.

Capacitor
A device that has capacitance and is usually made of two metal plate materials separated by a dielectric material (insulator).

Cycle
A sequence of events that causes one complete pattern of alternating current from a zero reference, in a positive direction, back to zero, then in a negative direction, and back to zero.

Effective voltage (V_{eff})
The value of an AC sine-wave voltage which has the same heating effect as an equal value of DC voltage. $V_{eff} = V_{peak} \times 0.707$

Electrostatic field
The space or area around a charged body in which the influence of an electrical charge is experienced.

Frequency response
A circuit's ability to operate over a range of frequencies.

Hertz (Hz)
The international unit of measurement of frequency equal to one cycle per second.

Inductance (L)
The property of a circuit to oppose changes in current due to energy stored in a magnetic field.

Inductor
A coil of wire that has the property of inductance and is used in a circuit for that purpose.

In phase
Two waveforms of the same frequency which pass through their minimum and maximum values at the same time and polarity.

Instantaneous voltage (V_i)
A value of AC voltage at any instant (time) along a waveform.

Lagging phase angle
The angle by which current lags voltage (or voltage leads current) in an inductive circuit.

Leading phase angle
The angle by which current leads voltage (or voltage lags current) in a capacitive circuit.

Peak-to-peak voltage ($V_{p\text{-}p}$)
The value of AC sine-wave voltage from positive peak to negative peak.

Peak voltage (V_{peak})
The maximum positive or negative value of AC sine-wave voltage. $V_{peak} = V_{eff} \times 1.41$

Period (time)
The time required to complete one AC cycle; Time $(T) = \dfrac{1}{\text{frequency}\ (f)}$

Phase angle (θ)
The angular displacement between applied voltage and current flow in an AC circuit.

Power (P)
The rate of doing work in electrical circuits, found by using the equation: $P = I \times V$

Power factor (PF)
The ratio of true power in an AC circuit and apparent power:

$$\text{PF} = \frac{\text{True power (in W)}}{\text{Apparent power (in VA)}}$$

Reactive circuit
An AC circuit that has the property of inductance or capacitance.

Reactive power (VAR)
The "unused" power of an AC circuit has inductance or capacitance, which is absorbed by the magnetic or electrostatic field of a reactive circuit.

Resistance (R)
Opposition to the flow of current in an electrical circuit; its unit of measurement is the ohm (Ω).

Resistive circuit
A circuit whose only opposition to current flow is resistance; a non-reactive circuit.

Root mean square (RMS) voltage
See **Effective voltage.**

Sawtooth waveform
A triangular-shaped AC waveform that gradually rises and falls.

Signal
An electrical waveform of varying value which is applied to a circuit.

Sine wave
The waveform of AC voltage.

Theta (θ)
The Greek letter used to represent the phase angle of an AC circuit.

Transformer
An AC power control device that transfers energy from its primary winding to its secondary winding by mutual inductance and is ordinarily used to increase or decrease voltage.

True power (W)
The power actually converted by an AC circuit, as measured with a wattmeter.

Volt-ampere (VA)
The unit of measurement of apparent power.

Volt-amperes reactive (VAR)
The unit of measurement of reactive power.

Waveform
The pattern of an AC frequency derived by looking at instantaneous voltage values that occur over time; on a graph, a waveform is plotted with instantaneous voltages on the vertical axis and time on the horizontal axis.

Wavelength
The distance between two corresponding points that represents one complete wave.

10

Sources of AC Electrical Energy

Many electrical systems require **alternating current (AC)** to function. AC is undoubtedly the most significant source of electrical energy today. AC is produced by converting mechanical energy into electrical energy through the use of a generator. The mechanical energy is used to move electrical conductors (turns of wire) through a magnetic field inside the generator. **Generators** rely on the principle of electromagnetic induction to convert mechanical energy into AC electrical energy. This type of electrical energy is available as single- or three-phase alternating current.

Objectives

1. Explain how mechanical energy can be transformed into electrical energy
2. Explain how electromagnetic induction is used to generate AC voltage
3. Explain the factors that affect the generation of voltage
4. Describe the operation of a single-phase AC generator
5. Describe the operation of three-phase AC generator

Chapter Outline

10.1 AC Electrical Generators Basics

AC electrical energy is produced by placing a conductor or group of conductors in a strong magnetic field. When there is relative motion between the conductor and the magnetic field, the conductors will cut across the magnetic lines of force. This causes a voltage to be induced in the conductor. If a complete electrical path is formed between the terminals of conductor,

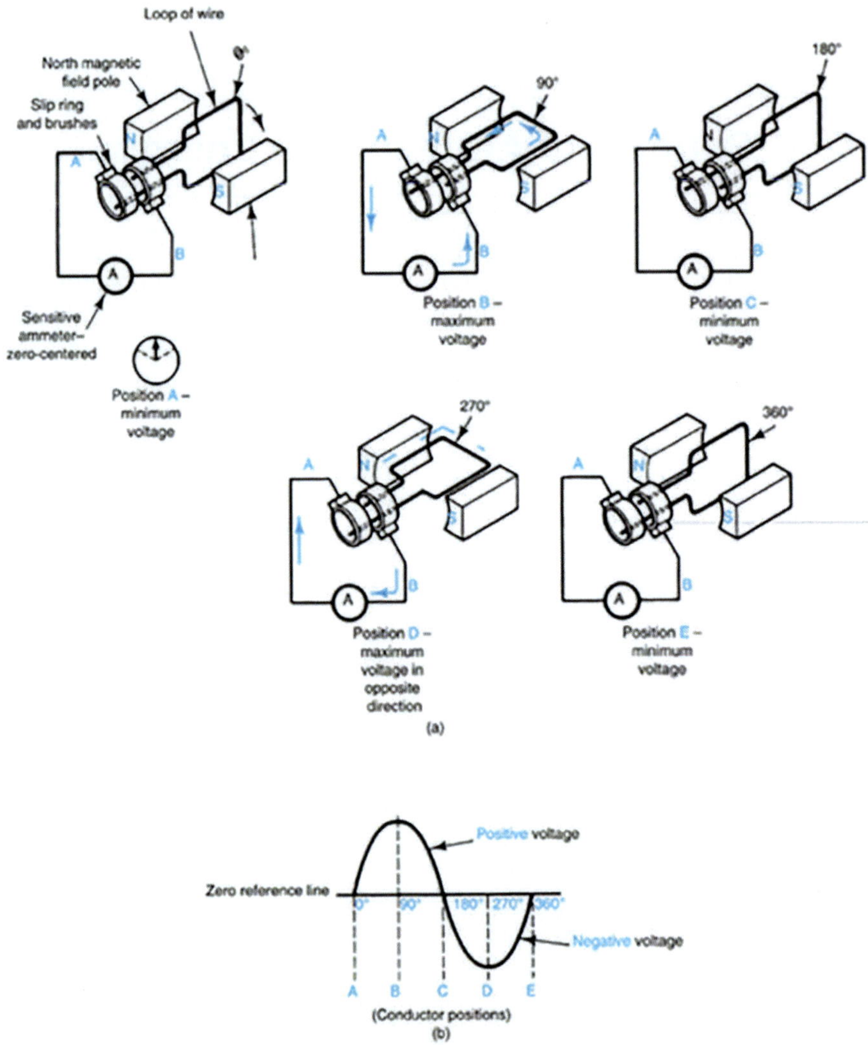

Figure 10-1 Voltage generation: (a) loop of wire rotated through a magnetic field; (b) voltage produced by the movement of the loop of wire.

the induced voltage will cause a current to flow through the conductor. The complete path will serve as an electrical load for the system.

A **generator** has a stationary part and a rotating part housed inside a machine assembly. The stationary part is called the **stator** and the rotating part is called the **rotor**. The generator has magnetic **field poles** of north and south polarities. Generators must have a method of producing rotary motion (mechanical energy). This system is called a **prime mover** and is connected to the generator shaft.

Figure 10-1 shows a loop of wire rotated through a magnetic field.

The position of the loop inside the magnetic field determines the amount of **induced voltage** and the resulting current flow. Each side of the coil moves across the magnetic lines of force in opposite directions. This movement causes an equal amount of electrical current to flow in opposite directions through the two sides of the loop. Note each position of the loop and the resulting output voltage in **Figure 10-1**. With every complete revolution of the conductor, the electrical current flows in one direction and then in the opposite direction. This method produces alternating current (AC). One complete rotation is called a **cycle**. The number of cycles per second is known as the frequency. Most AC generators produce 60 cycles per second (Hz).

The ends of each conductor that move across the magnetic field of the generator are connected to continuous metal ring as shown in **Figure 10-1**. This ring is called a **slip ring commutator** and is mounted on the same shaft as the conductors. Carbon **brushes** are used to make contact with the slip rings. The electrical current induced into each conductor flows through the slip rings to the brushes. When the conductor turns half a revolution, it causes current to flow in one direction through the slip ring assembly. During the next half-revolution of the coil, the positions of the two sides of the conductor are opposite. As a result of this, the direction of the **induced current** is reversed. Current now flows in the opposite direction. With all the generator parts working together, electrical power is produced.

The conductors that make up the rotor of a generator have many turns. The generated voltage is determined by the number of turns of wire used, the strength of the magnetic field, and the speed of the prime mover used to rotate the machine. **Electromagnetic induction** occurs when a conductor passes through a magnetic field and cuts across lines of force. As a conductor passes through a magnetic field, it cuts across the **magnetic flux lines**. As the conductor cuts across the flux lines, the magnetic field develops a force on the electrons of the conductor. The direction of the electron movement determines the polarity of the induced voltage.

The **left-hand rule** is used to determine the direction of electron flow. This rule for generators is stated as follows: Hold the thumb, forefinger, and middle finger of the left hand perpendicular to each point. Point the forefinger in the direction of the magnetic field from north to south. Point the thumb in the direction of the motion of the conductor. The middle finger will then point in the direction of electron current flow.

The amount of voltage induced into a conductor cutting across a magnetic field depends on the number of lines of force cut in a given time. This value is determined by the following three factors.

1. The speed of the relative motion between the magnetic field and the conductor
2. The strength of the magnetic field
3. The length of the conductor passed through the magnetic field

If the speed of the conductor cutting the magnetic lines of force is **increased**, the generated voltage increases. If the strength of the magnetic field is increased, the induced voltage also increases. A longer conductor allows the magnetic field to induce more voltage into the conductor. The induced voltage increases when each of the three quantities listed is increased.

In electrical generators, the coils move with respect to a magnetic field or flux. Electromagnetic induction occurs in accordance with **Faraday's law**. This law states that (1) if a magnetic flux that links a conductor loop has relative motion, a voltage is induced; and (2) the value of the induced voltage is proportional to the rate of change of flux.

The voltage induced in a conductor of a generator is defined by **Faraday's law** as follows:

$$V_i = B \times L \times v$$

where
V_i = induced voltage in volts
B = magnetic flux in teslas
L = length of conductor within the magnetic flux in
v = relative speed of the conductor in meters per second

Given: The conductors of the stator of a generator have a length of 0.5 M. The conductors move through a magnetic field of 0.8 tesla at a rate of 60 m/s.

Find: The amount of induced voltage in each conductor.

Solution:

$$V_i = B \times L \times v$$
$$= 0.8 \times 0.5 \times 60 = 24\,\text{V}.$$

10.2 Single-phase AC Generators

Single-phase electrical power is often used, particularly in homes. However, little electrical power is produced by single-phase generators. **Alternating**

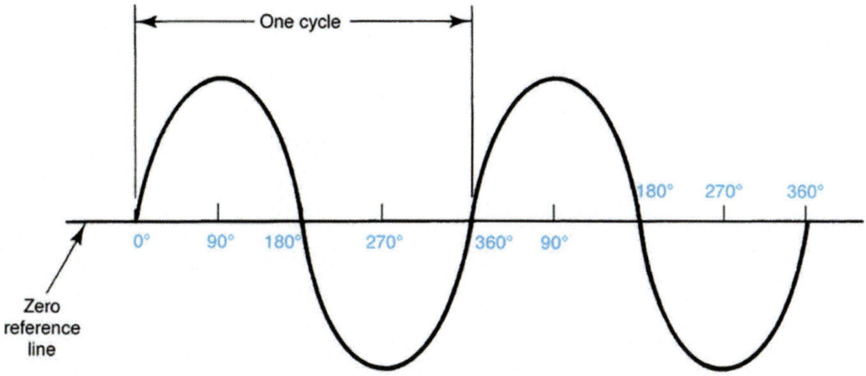

Figure 10-2 Two cycles of an alternating current (AC) sine wave.

current generators are usually referred to as "**alternators**." Single-phase electrical power used in homes is usually produced by three-phase generators at power plants. It is then converted to single-phase electrical energy before it is distributed to homes. Single-phase generators have several uses. The current produced by single-phase generators is in the form of a **sine wave**; it is so called due to its mathematical origin. It is based on the trigonometric sine function used in mathematics. Two cycles of single-phase AC voltage are shown in **Figure 10-2**. This voltage is known as a sine-wave voltage.

The voltage induced into the conductors of the armature varies as the **sine** of the angle of rotation between the conductors and the magnetic field (see **Figure 10-3**). The voltage induced at a specific time is called **instantaneous voltage** ($\mathbf{V_i}$). Voltage induced into an armature conductor at a specific time is found by using the following formula:

$$V_i = V_{\max} \times \sin\theta$$

where $\mathbf{V}_{max}$ is the maximum voltage induced into the conductor. The symbol theta (θ) is the angle of conductor rotation.

For example, at the 60° position, assume that the maximum voltage output is 100 V. The instantaneous voltage induced at 60° is $V_i = 100 \times \sin 60$. The **frequency** of the voltage produced by alternators is usually 60 hertz (Hz). A cycle of AC is generated when the rotor moves one complete revolution (360°). Cycles per second or **hertz** refers to the number of revolutions per second. For example, a speed of 60 revolutions per second (3600 rpm) produces a frequency of 60 Hz. The frequency of the AC generated by an alternator is found by using the following formula:

Degrees	Sine
0/360	0.000
30	0.500
60	0.866
90	1.000
120	0.866
150	0.500
180	0.000
210	−0.500
240	−0.866
270	−1.000
300	−0.866
330	−0.500

(a)

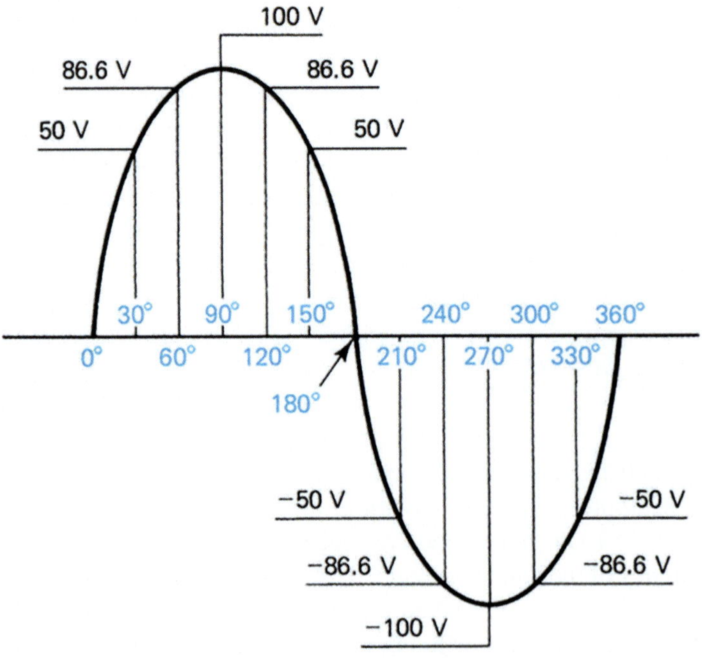

(b)

Figure 10-3 Generation of an AC sine wave: (a) sine values of angles from 0 to 360; (b) sine wave produced.

$$f = \frac{\text{number of magnetic pole pairs per phase} \times \text{speed of rotation (rpm)}}{120}.$$

The frequency is measured in hertz. If the number of poles (field coils) is increased, the speed of rotation can be reduced and still produce a 60-Hz frequency. For a generator to convert mechanical energy into electrical energy, three conditions must exist:

1. There must be a magnetic field.
2. Conductors must be placed adjacent to the magnetic field.
3. There must be relative motion between the magnetic field and conductors.

The two methods used to accomplish these conditions are the rotating-armature method and the rotating-field method, both shown in **Figure 10-4**. In the **rotating-armature method**, shown in **Figure 10-4(a)**, AC voltage is induced into the rotor conductors. The magnetic field is developed by a set of stationary field poles. Relative motion between the conductors and the magnetic field comes from a prime mover connected to the generator shaft. Prime movers can be gasoline engines, diesel engines, steam turbines, or electric motors. Remember that generators convert mechanical energy into electrical energy. The rotating-armature method can only be used to produce small amounts of electrical power. The major disadvantage is that the AC voltage passes through the slip-ring/brush assembly. High voltages would cause sparking or arc-over between the brushes and slip rings. The maintenance involved in replacing brushes and repairing slip-ring assemblies would be expensive. This method is used for alternators with low power outputs.

The **rotating-field method** shown in **Figure 10-4(b)** is used for alternators with larger power outputs. The DC excitation voltage is used to develop the magnetic field. DC voltage is applied to the rotor of the generator. The AC voltage output is induced into the stationary conductors of the machine. Because the **DC excitation voltage** is much smaller than the AC voltage output, maintenance problems are reduced. The conductors of the stationary part of the machine can be made larger. They will carry more current because they do not rotate.

Given: a six-pole three-phase alternator rotates at a speed of 3600 rpm.

Find: The frequency of the alternator.

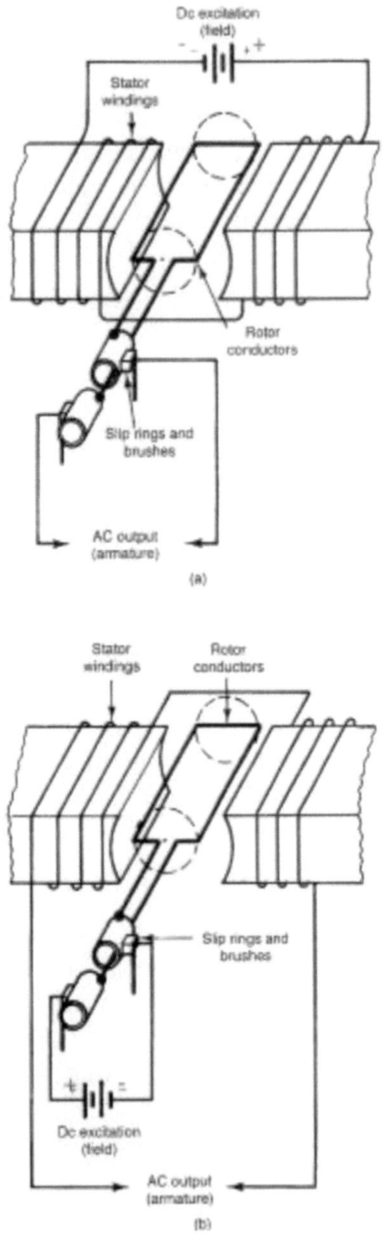

Figure 10-4 Generating voltage: (a) rotating-armature method; (b) rotating-field method.

Solution:

$$f = \frac{N/3 \times 3600}{120}$$
$$= 60\,\text{Hz}.$$

Note that if the number of poles is increased, the speed of rotation may be reduced while still maintaining a 60-Hz frequency.

10.3 Three-phase AC Generators

Most of the electrical power produced is **three-phase AC** produced at power plants. Power distribution systems use many three-phase generators (**alternators**) connected in parallel.

A simple diagram of a three-phase alternator and a three-phase voltage diagram are shown in **Figure 10-5**. The alternator output windings are connected in either of two ways – the wye connection and the delta connection.

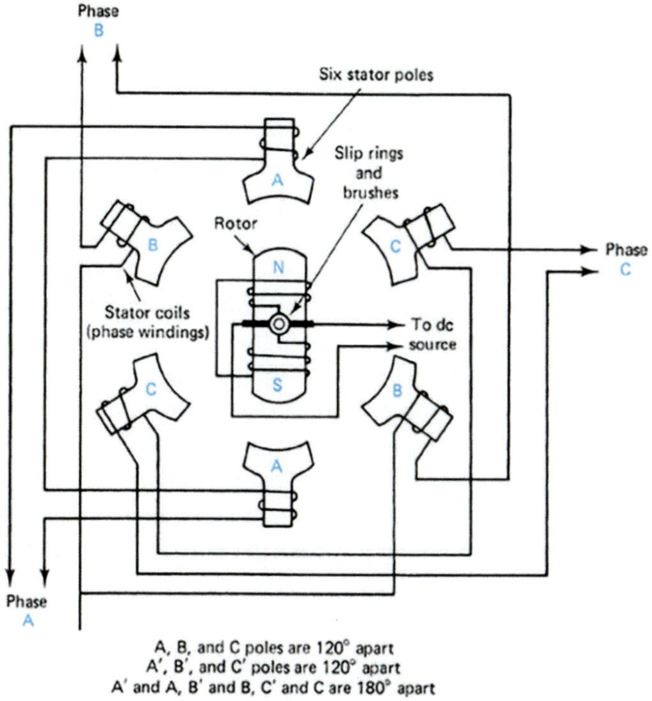

(a)

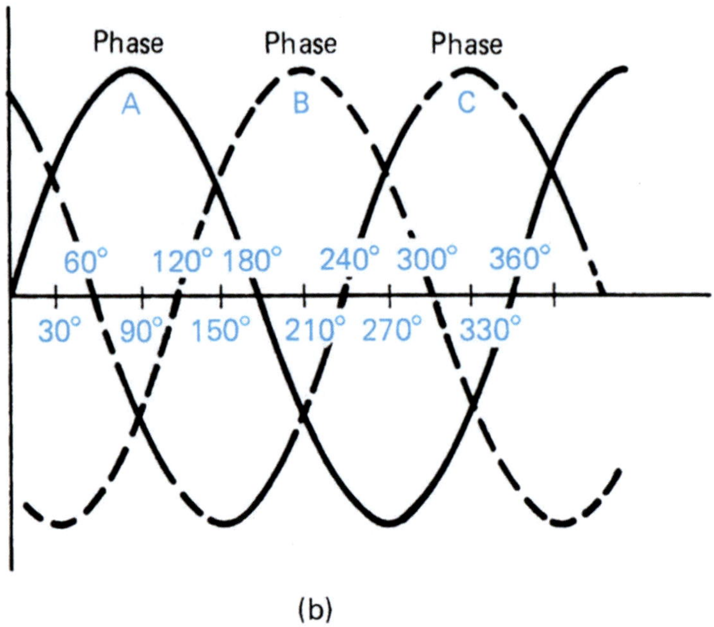

Figure 10-5 Three-phase system: (a) construction features; (b) diagram of three-phase voltage.

These three-phase connections are shown with schematics in **Figure 10-6**. These methods are also used for connecting the windings of three-phase transformers, three-phase motors, and other three-phase equipment. In the **wye connection** of **Figure 10-6(a)**, the beginnings or ends of each winding are connected. The other sides of the windings are the ac lines which extend from the alternator. The voltage across the power lines is called **line voltage (V_L)**. Line voltage is higher than the voltage across each phase. Line voltage (V_L) is equal to the square root of 3 (1.73) multiplied by the voltage across the phase windings (V_P), or $V_L = V_P \times 1.73$. The **line current (I_L)** is equal to the phase current (I_P), or $I_L = I_P$.

In the **delta connection** shown in **Figure 10-6(b)**, the end of one phase winding is connected to the beginning of the next phase winding. Line voltage (V_L) is equal **to phase voltage (V_p). Line current (I_L)** is equal to the **phase current (I_P)** multiplied by 1.73, or $I_L = I_P \times 1.73$. The differences between voltages and currents in wye and delta systems are as follows:

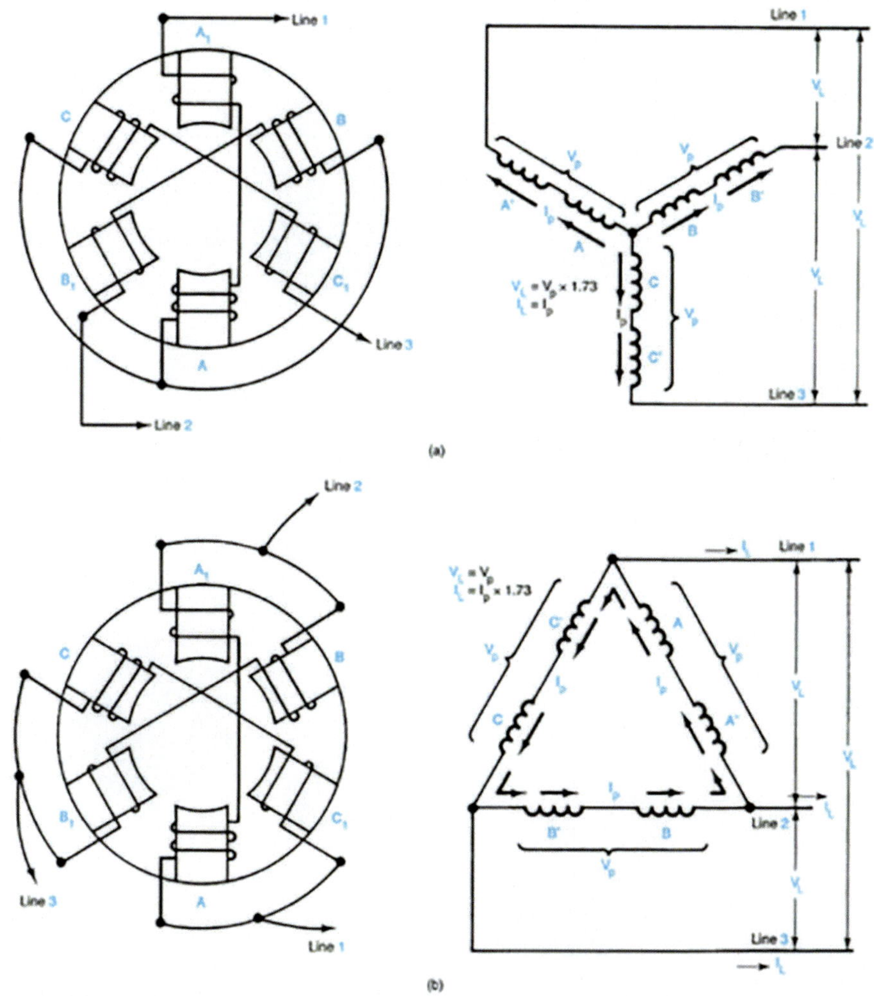

Figure 10-6 Three-phase connections: (a) wye connection, sometimes called a star connection; (b) delta connection.

Wye system	Delta system
$V_L = V_P \times 1.73$	$V_L = V_P$
$I_L = I_P$	$V_L = V_P \times 1.73$

Determine the line voltage developed by a wye-connected three-phase AC generator if the phase voltage developed across the windings is 100 V.

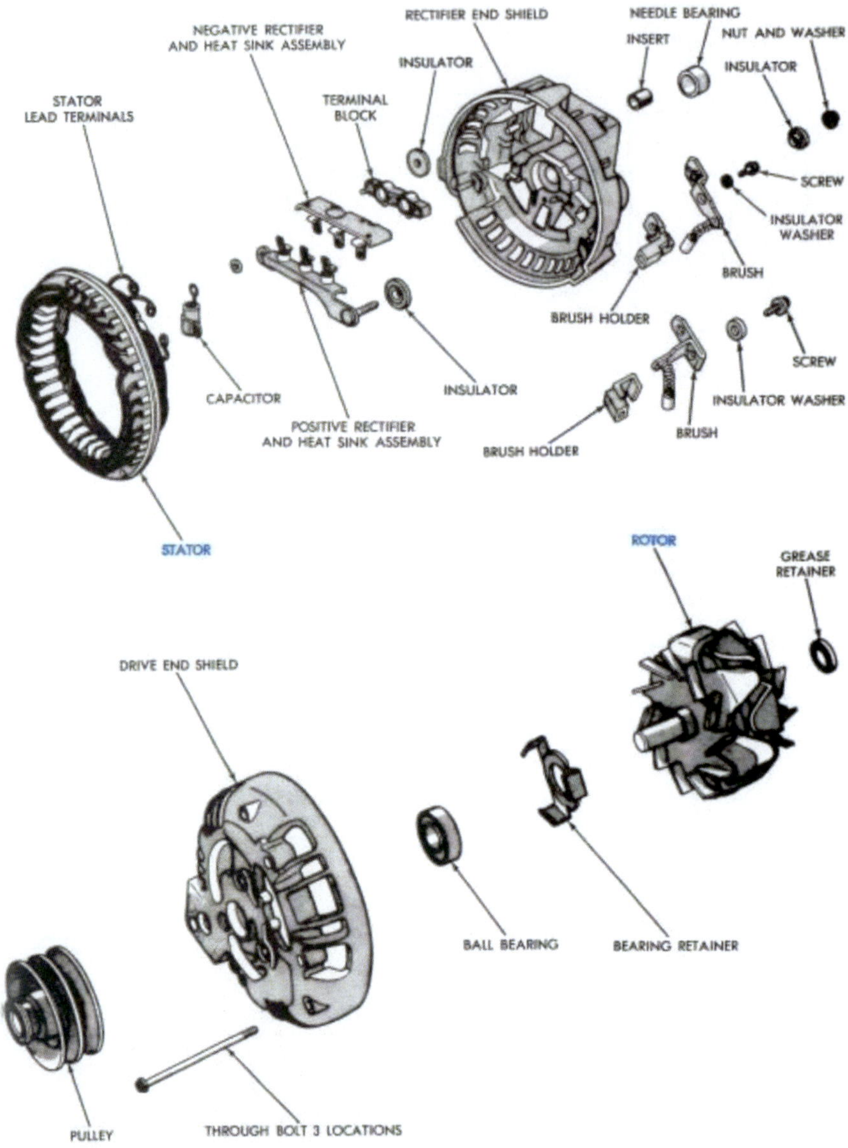

Figure 10-7 Parts of an automobile alternator (courtesy of Chrysler Corp.).

In a **wye system**: $V_L = V_P \times 1.73$ = 100 × 1.73 = 173 V.

Determine the line voltage developed by a delta-connected three-phase AC generator if the phase voltage developed across the windings is 100 V.

In a **Delta system**: $V_L = V_P$ = 100 V.

Three-phase power is used mainly for high-power industrial and commercial equipment. The power produced by three-phase generators is a more constant output than that of single-phase power. Three-phase power is more economical to supply energy to the large motors that are often used in industries. Three separate single-phase voltages can be delivered from a three-phase power system. It is more economical to distribute three-phase power from power plants to homes, cities, and industries. Three conductors are needed to distribute three-phase voltage. Six conductors would be necessary for three separate single-phase systems. Equipment that uses three-phase power is smaller in size than similar single-phase equipment. It saves energy to use three-phase power when possible.

One type of **three-phase alternator** is used in automobiles. The three-phase AC it produces is converted to DC by a rectifier circuit. The DC voltage is then used to charge the automobile battery. The charging time and voltage are controlled by a voltage regulator circuit. The parts of an automobile alternator are shown in **Figure 10-7**.

Analysis and Troubleshooting

Generators are used to convert mechanical energy into electrical energy. For effective troubleshooting of generators, one should have a good understanding of the type of generator being evaluated. With this knowledge of operation and test equipment, generator repair can be accomplished. Any moving part of a generator is subject to wear. This includes carbon brushes, commutator assembly, and the bearings. Inspection of the generator consists of observing the condition of components and noting excessive vibration or noise during its operation. Large electrical loads connected to a generator can cause excessive heating in the windings and can cause failure of the winding insulation. Adequate cooling of the generator should be provided, and care should be taken not to exceed its power ratings. Electrical tests include evaluation of the winding resistance, continuity, and its insulation resistance with respect to the metal housing of the generator. The winding resistance is typically very low, whereas the insulation resistance should be extremely high.

Summary

Through electromagnetic induction, voltage is developed in a conductor when it moves through a magnetic field.

- When voltage is induced in a conductor, current will flow through it only when there is a closed or complete path.
- The amount of voltage induced in a conductor moving through a magnetic field depends on the number of lines of force being cut within a given unit of time.
- In an AC generator, voltage is induced into a conductor when it moves through a stationary or fixed magnetic field.
- An AC generator uses slip rings to complete a path for current flow through an external electrical load.
- The frequency of the voltage developed by an AC generator depends on the number of magnetic poles pairs and the speed of rotation of the conductors.
- Two methods of generating single-phase AC are the rotating-armature method and the rotating-field method. Here, the AC output derived is the difference between these two methods of generation.
- In a three-phase AC generator, the stator is composed of three separate field windings which are displaced by 120°.
- In a three-phase AC generator, the rotor consists of a magnetic field that is rotated by a prime mover.
- The stationary field windings of a three-phase AC generator are connected in either a wye or a delta configuration. These configurations are based on how the individual windings are connected together.

Formulas

$$V_i = B \times L \times v$$

$$V_i = V_{\max} \times \sin\theta$$

$$f = \frac{\text{number of magnetic pole pairs per phase} \times \text{speed of rotation (rpm)}}{120}$$

Wye system: $V_L = V_P \times 1.73$, $I_L = I_P$

Delta system: $V_L = V_P$, $V_L = V_P \times 1.73$

Problems

Determine the strength of the magnetic field needed for inducing a voltage of 48 V in a conductor of length 1 m, rotating within the magnetic field at 60 m/s.

1. $[V_i = B \times L \times v$

$$B = \frac{V_i}{L \times v} = \frac{48}{1 \times 60} = 0.8T\ ].$$

2. Determine the value of the instantaneous voltage induced in a conductor which is rotated 30°, if the maximum voltage (V_{max}) developed at 90° is 25 V.
$[V_i = V_{max} \times \sin\theta = 25 \times 0.5 = 12.5\text{V}]$.

3. Determine the speed of rotation of a three-phase alternator having six poles, which is generating a voltage of 60 Hz.

$$[f = \frac{\text{number of magnetic pole pairs per phase} \times \text{speed of rotation (rpm)}}{120}$$

$$\text{speed of rotation (rpm)} = \frac{f \times 120}{\text{number of magnetic pole pair per phase}}$$

$$= \frac{60 \times 120}{6/3} = 3600 \text{ rpm }].$$

4. Determine the line voltage developed by a wye-connected three-phase AC generator if the phase voltage developed across the windings is 120 V.
[In a Wye system: $V_L = V_P \times 1.73 = 120 \times 1.73 = 207.6$ V].
5. Determine the line voltage developed by a delta-connected three-phase AC generator if the phase voltage developed across the windings is 120 V.
[In a Delta system: $V_L = V_P = 120$ V].

Self-examination/Answers

1. The amount of voltage generated due to electromagnetic induction depends upon ____________________, ____________________, and ____________________.
2. The stationary field of an AC generator is called the ____________________, while the rotating field is called the ____________________.
3. The direction of current flow in an AC generator is determined by the ____________________.

4. The ____________________ commutator is used with an AC generator.
5. The three conditions for converting mechanical energy into electrical energy are ____________ ______________________________, ______________________________, and __________________.
6. The frequency of AC generated depends on ______________________ and __________________.
7. Two methods for connecting the stator windings of a three-phase AC generator are ___________ and _____________.
8. In a _______________ connection of a three-phase AC generator, the beginning or ending of each winding are connected to a common point.
9. The ______________ voltage measures the voltage across the winding of a three-phase AC generator.
10. In a delta-connected three-phase AC generator, the ___________ voltage is larger than the ___________voltage.

Answers

1. strength of magnetic field, length of conductor, velocity
2. stator, rotor
3. position of the rotor coils with respect to the stator
4. slip-ring
5. magnetic field, conductors much be adjacent to the magnetic field, relative motive between the field and the conductors
6. number of magnetic poles, speed of rotation
7. delta, wye
8. wye
9. phase
10. line, phase

Glossary

AC
An abbreviation for alternating current.

Amplitude
The vertical height of an AC waveform.

Angle of lead or lag
The angle between applied voltage and current flow in an AC circuit, in degrees; in an inductive (*L*) circuit, voltage (*V*) leads current (*I*); in a capacitive (*C*) circuit, current (*I*) leads voltage (*V*).

Apparent power (volt-amperes)
The applied voltage times current delivered to an AC circuit.

Average voltage (V_{avg})
The value of an AC sine-wave voltage which is found by the formula: $V_{avg} = V_{peak} \times 0.637$.

Capacitance (C)
The property of a device to oppose changes in voltage due to energy stored in its electrostatic field.

Capacitor
A device that has capacitance and is usually made of two metal plate materials separated by a dielectric material (insulator).

Cycle
A sequence of events that causes one complete pattern of alternating current from a zero reference, in a positive direction, back to zero, and then in a negative direction, and back to zero.

Effective voltage (V_{eff})
The value of an AC sine-wave voltage which has the same heating effect as an equal value of DC voltage. $V_{eff} = V_{peak} \times 0.707$.

Electrostatic field
The space or area around a charged body in which the influence of an electrical charge is experienced.

Hertz (Hz)
The international unit of measurement of frequency equal to one cycle per second.

Inductance (L)
The property of a circuit to oppose changes in current due to energy stored in a magnetic field.

Inductor
A coil of wire that has the property of inductance and is used in a circuit for that purpose.

In phase
Two waveforms of the same frequency which pass through their minimum and maximum values at the same time and polarity.

Instantaneous voltage (V_i)
A value of AC voltage at any instant (time) along a waveform.

Lagging phase angle
The angle by which current lags voltage (or voltage leads current) in an inductive circuit.

Leading phase angle
The angle by which current leads voltage (or voltage lags current) in a capacitive circuit.

Peak-to-peak voltage ($V_{p\text{-}p}$)
The value of AC sine-wave voltage from positive peak to negative peak.

Peak voltage (V_{peak})
The maximum positive or negative value of AC sine-wave voltage. $V_{peak} = V_{eff} \times 1.41$.

Period (time)
The time required to complete one AC cycle; Time $(T) = \frac{1}{\text{frequency } (f)}$.

Phase angle (θ)
The angular displacement between applied voltage and current flow in an AC circuit.

Power (P)
The rate of doing work in electrical circuits, found by using the equation: $P = I \times V$.

Power factor (PF)
The ratio of true power in an AC circuit and apparent power:

$$\text{PF} = \frac{\text{True power (in W)}}{\text{Apparent power (in VA)}}.$$

Reactive circuit
An AC circuit that has the property of inductance or capacitance.

Reactive power (VAR)
The "unused" power of an AC circuit has inductance or capacitance, which is absorbed by the magnetic or electrostatic field of a reactive circuit.

Resistance (R)
Opposition to the flow of current in an electrical circuit; its unit of measurement is the ohm (Ω).

Resistive circuit
A circuit whose only opposition to current flow is resistance, a nonreactive circuit.

Root mean square (RMS) voltage
See **Effective voltage**.

Sine wave
The waveform of AC voltage.

Theta (θ)
The Greek letter used to represent the phase angle of an AC circuit.

Transformer
An AC power control device that transfers energy from its primary winding to its secondary winding by mutual inductance and is ordinarily used to increase or decrease voltage.

True power (W)
The power actually converted by an AC circuit, as measured with a wattmeter.

Volt-ampere (VA)
The unit of measurement of apparent power.

Volt-amperes reactive (VAR)
The unit of measurement of reactive power.

Waveform
The pattern of an AC frequency derived by looking at instantaneous voltage values that occur over time; on a graph, a waveform is plotted with instantaneous voltages on the vertical axis and time on the horizontal axis.

11

Capacitance and Capacitive Reactance

The response of an electrical circuit depends on the type of energy being applied and the components being used. For electrical circuits that consist of only resistors, the application of AC or DC energy will cause a similar response, resulting in the dissipation of heat energy. However, other circuit components, such as **capacitors**, will respond differently. Capacitors have a tendency to oppose changes in the voltage applied to it and are commonly used in AC circuits.

A capacitor consists of two conductive **plates** separated by an **insulator**. When voltage is applied to the plates, an electrical charge is transferred from the source to the plates. This causes an electrostatic charge to be developed on the conductors. The conductors will then store this charge even when the source is removed. The applied energy is stored in an electrical field which is developed across the plates.

The amount of **charge** that can be developed on the conductors depends on the applied voltage, the surface area of the conductors (called plates), the quality of the insulating material, and the distance between the plates. The closer the two plates are placed together without being in contact with each other and the larger the surface area, the more the charge that can be stored on the plates. The term capacitance (C) refers to the ability of a device to store a difference in potential charge or voltage. The fundamental unit of capacitance is a **farad (F or Fd)**.

When **DC voltage** is applied to the plates of a capacitor, it will charge to the value of the source voltage. The insulating material that is placed between the plates obstructs the current flow. The capacitor then remains charged to the value of the DC voltage source. When the voltage source is removed, the charge of the capacitor may be depleted over a period of time. DC current flows to or from a capacitor connected to a DC voltage source only when the source is turned on or off. During the charging process, an electrostatic change is developed on the plates. Once a capacitor is charged to the source voltage no further charging action will take place. When this occurs, no DC current can flow.

If the same capacitor is connected to an **AC voltage source**, the voltage applied to the plates constantly changes. The capacitor receives energy from the source during the first part of each **alternation** and returns the energy to the source during the second part of each alternation. A capacitor, therefore, charges and discharges during each alternation. For the next alternation, the process is reversed. Thus, AC voltage applied to a capacitor causes a constant change in the amount of charge developed. Capacitors have the ability to pass AC current because of this **charging** and **discharging** action. They can, therefore, be used to pass AC current and block DC current flow. The opposition to a change in voltage is called **capacitive reactance (X_C)** which is measured in ohms. The frequency of the applied AC also influences the response of the capacitor. When high frequency AC is applied to a capacitor, charging and discharging of the capacitor occur at the same rate. As a result of this, it is easier for a capacitor to respond to higher frequencies better than lower frequencies.

Objectives

1. Define capacitance
2. Describe the construction of a capacitor
3. Explain the response of a capacitor when energy is applied
4. List the factors that determine the value of a capacitor
5. Identify different types of capacitors
6. Determine the total capacitance of a circuit with capacitors in series or parallel
7. Explain the response of a capacitive voltage divider circuit
8. Determine capacitive reactance
9. Examine the response of resistor–capacitor circuits

Chapter Outline

11.1 Capacitor Construction

Figure 11-1 shows the construction of a **capacitor**. This includes two or more conductive plates and a piece of insulating material. The **plate** is often made of a foil-type material, similar to aluminum foil used in cooking. The function of the conductive plates is to accumulate an **electrostatic charge** when energy is applied to it. One of the plates will develop a positive change and the other plate a negative charge depending on the polarity of the energizing source connected to it. These plates may be designated as

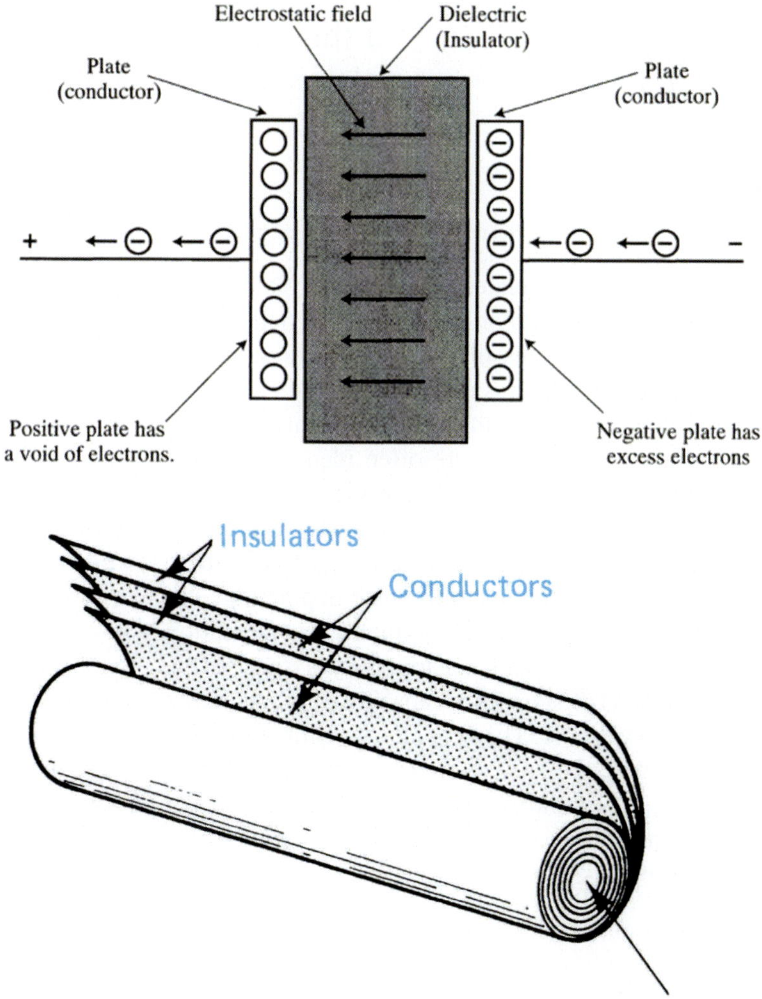

Figure 11-1 Capacitors are two conductors separated by an insulator.

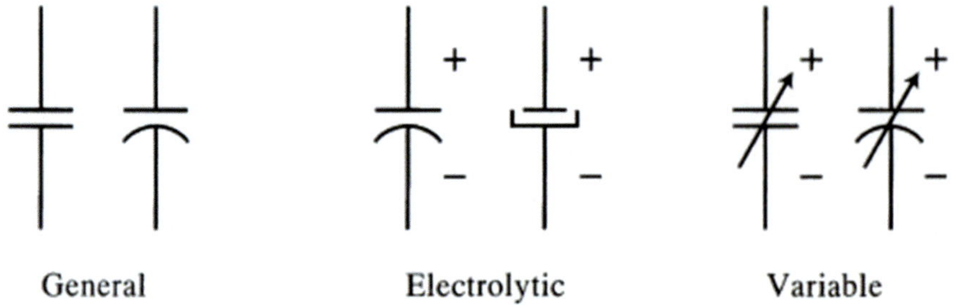

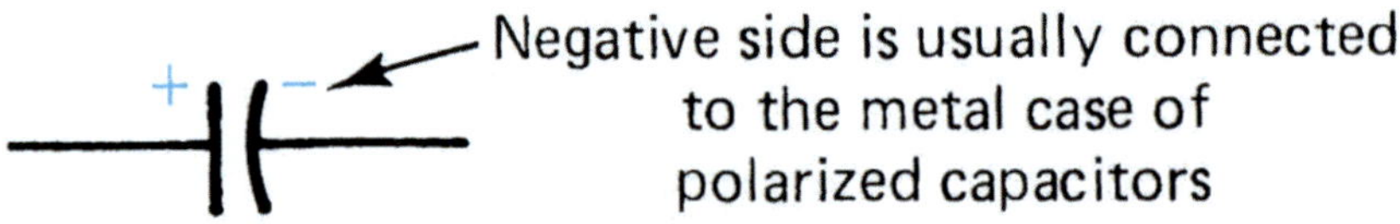

(a)

Figure 11-2 Schematic symbols representing capacitors.

the positive and negative plates, respectively. Note that when the energizing source is AC, the charge polarity of the plates will alternate as well.

The conductive plates of a capacitor are separated by some form of **insulating material**, which is referred to as a dielectric. Some capacitors use air as the dielectric, while others use materials such as electrolytes, paper, mica, or ceramics. Capacitors are often identified by the type of **dielectric material** used its construction. The value of the capacitance is influenced by the type of dielectric used in its construction. An electrostatic field is developed across the dielectric, from the **negative plate** to the **positive plate**. This field is considered to be a difference in potential charge or voltage.

The **symbol** of a capacitor is a graphic representation of its construction. It has two conductive plates separated by a space which is the insulating material. **Figure 11-2** shows six capacitor symbols in three groups: general, electrolytic, and variable. The general capacitor symbol is used for capacitors that are not polarized, such as those used with AC. The electrolytic-type capacitor must show polarity signs. Variable capacitors have an arrow through the symbol which indicates value changes. Various types of capacitors are shown in **Figure 11-3**.

(a)

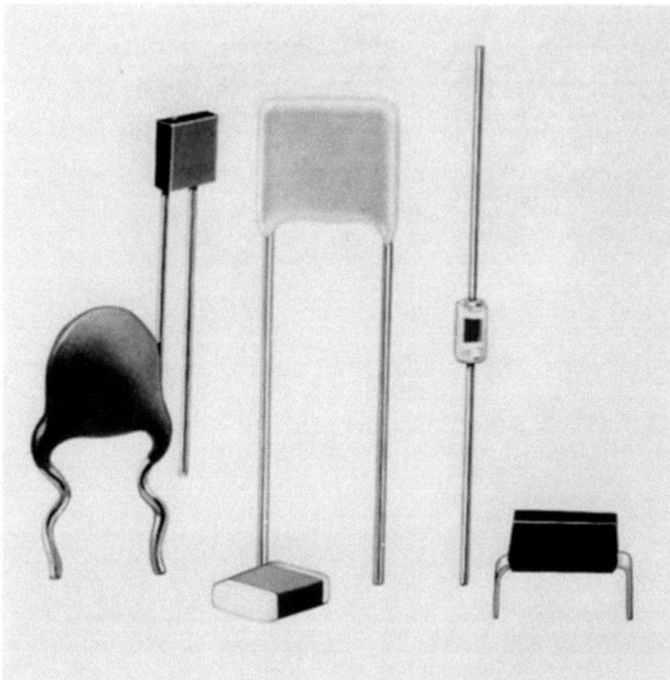

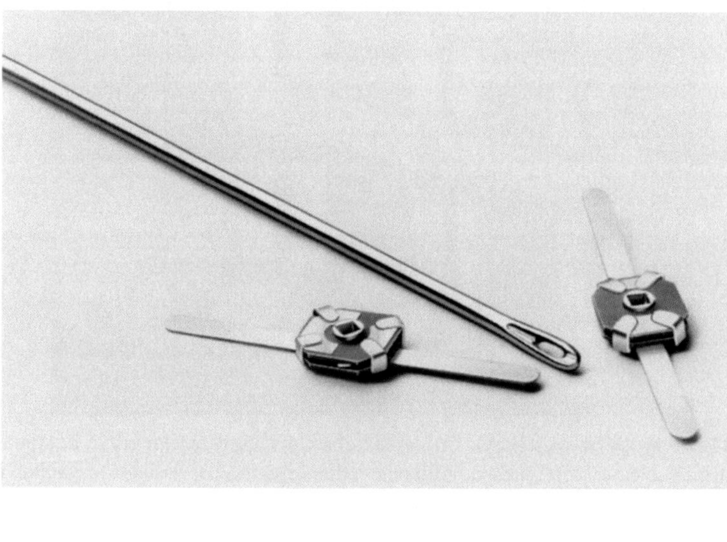

(c)

(d)

Figure 11-3 Types of capacitors: (a) small capacitors used in electronic equipment; (b) variable trimmer capacitors used in communication circuits; (c) variable capacitor used in tuning circuits; (d) capacitors used for power factor correction on electrical power systems [(a) courtesy of Centralab, Inc., a North American Philips Co.; (b) courtesy of Johanson Manufacturing Corp.; (c) courtesy of J.W. Miller Division/Bell Industries; (d) courtesy of McGraw-Edison Co.].

11.2 Capacitor Operation

In order for a **capacitor** to function, it must have some form of energy applied to it. This energy is stored by a capacitor in its **electrostatic field** when voltage is applied. When this occurs, the amount of energy being stored is determined by the following equation:

$$W = \frac{1}{2} \times C \times V^2$$

where
W = energy stored in joules
C = capacitance in farads
V = voltage applied across the capacitor plates in volts

A 100-μF capacitor has 120 V applied to it. Determine the amount of energy stored by the capacitor.

$$W = \frac{1}{2} \times C \times V^2 = \frac{1}{2} \times 100 \times 10^{-6} \times 120^2 = 0.72 \text{ joules.}$$

DC Capacitor Operation

The energy applied to a **capacitor** can be either DC or AC. Refer to the circuit of **Figure 11-4**, which shows a DC source connected to a capacitor–resistor network using a three-position switch.

When the switch is moved to the **charging** position (1), the capacitor begins charging through the series resistor *R*. When a voltage is placed across the capacitor, an electrostatic field or charge is developed across its plates. This places the capacitor across the 6-V source voltage, which causes a displacement of electrons in the circuit. Electrons move away from the negative polarity and toward the positive polarity of the battery. A surge of current flows as the capacitor charges to the value of the source voltage. Electrons accumulate on the **negative plate** of the capacitor. At the same time, electrons leave the **positive plate**, which causes a difference in potential charge to develop across the capacitor. When electrons move onto the negative plate, it becomes more negative. Also, as electrons leave the positive plate, it becomes more positive.

The **polarity** of the potential that exists across the capacitor opposes the source voltage. As the capacitor continues to charge, the voltage across the capacitor continues to increase as well. Current flow stops when the voltage across the capacitor is equal to the source voltage. At this time, the two

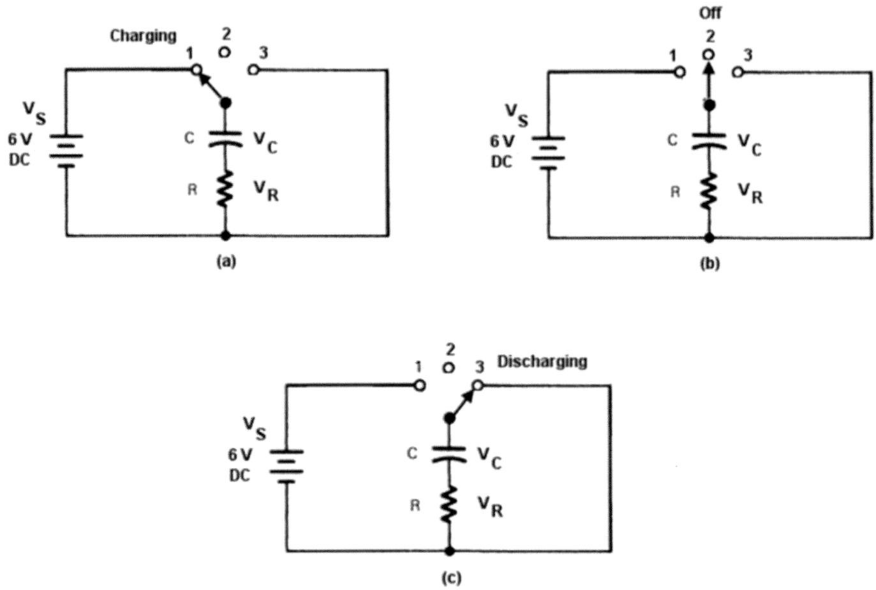

Figure 11-4 Capacitor charge and discharge circuit.

voltages cancel each other. No current actually flows through the capacitor because the material between the plates of a capacitor is an insulator.

An important safety factor to remember is that a capacitor may hold a charge for a long time. Capacitors can be an electrical shock hazard if not handled properly. This can be observed by placing the switch in the off position, where it is disconnected from the source voltage.

When a capacitor is **discharged**, the charge on the plates becomes neutralized. A current path between the two plates must be developed. When the switch in **Figure 11-4** is moved to discharging position (3), the electrons on the **negative plate** move to the **positive plate** and neutralize the charge. The capacitor releases the energy it has absorbed during its charging as it discharges through the resistor.

Figure 11-5 shows the timing diagram of the capacitor charging and discharging circuits shown in **Figure 11-4**. Supply voltage V_S is applied to the circuit. V_C is the voltage across the capacitor and V_R is the voltage across the resistor. When the switch is placed in the **charging** position, the capacitor voltage rises quickly toward the applied voltage. This is shown in the voltage timing diagram of V_C. While the capacitor voltage is increasing, the voltage across the resistor decreases as shown in the timing diagram of V_R. Note that the sum of voltages V_C and V_R equal the supply voltage at every instant. When the capacitor is fully charged, the resistor voltage is zero.

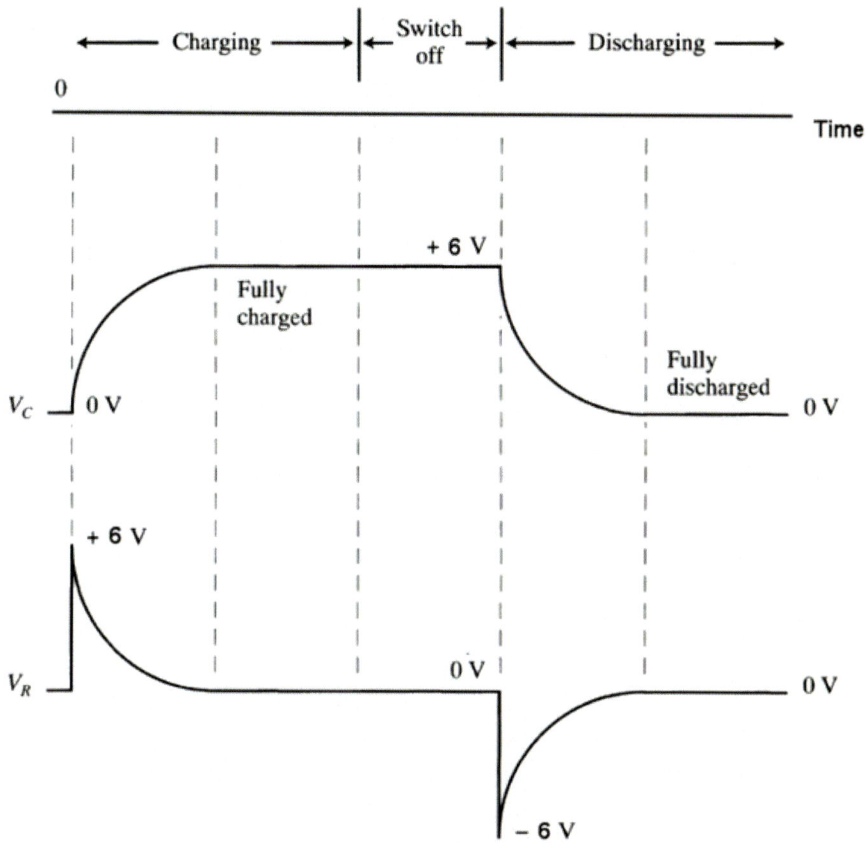

Figure 11-5 Timing diagram of one charge/discharge cycle of the capacitor circuit in **Figure 14-4**.

When the switch is moved to the off position, the capacitor remains fully charged. This condition can be maintained for long periods of time. It could last indefinitely if the dielectric of the capacitor were perfect and had no leakage current. When the switch is moved to the **discharging** position, the capacitor discharges through the resistor. During the discharge time, the resistor instantly switches its polarity. It falls from 0 to –6 V instantaneously and then follows the capacitor as it discharges.

AC Capacitor Operation

When AC voltage is applied to a pure resistive circuit, the voltage and current are in step or "in phase" with each other. On the other hand, when AC is

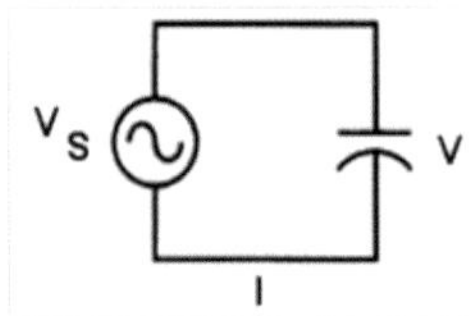

(a)

Capacitor voltage and current phase relationships

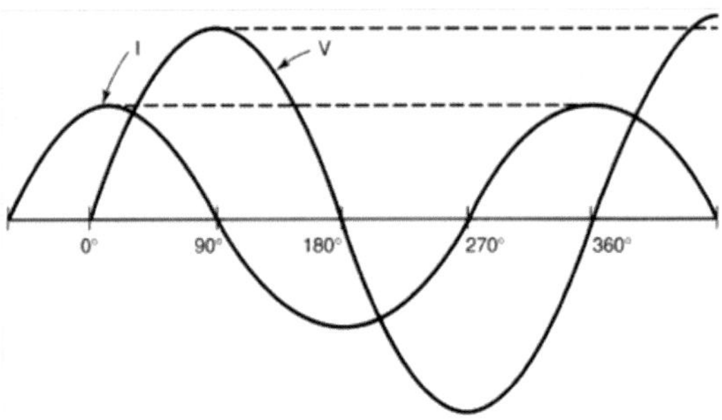

(b)

Figure 11-6 Capacitor voltage and current phase relationships.

applied to a pure capacitive circuit, the current and voltage will respond differently. **Figure 11-6(a)** shows AC voltage (V_S) applied to a capacitor. When this occurs, charging current (I) flows and causes a voltage (V) to be developed across the plates of the capacitor. The resulting current and voltage waveforms are shown in **Figure 11-6(b)**.

When AC energy is first applied to an **uncharged** capacitor, the initial voltage developed across its plates is zero. **Charging current** flows in the circuit and is at a maximum value, even though the voltage developed across the capacitor plates is zero. The current in a capacitive circuit thus leads or occurs before the voltage. As the capacitor begins to charge, the voltage developed across its plates increases, and the value of the current flow decreases. When the capacitor charges to its maximum value, which is the same as the supply voltage for an alternation, the current becomes zero. The voltage changes from zero to maximum in 90°, whereas the current decreases from maximum to zero in 90°. Thus, the voltage and current are considered to be **out of phase** by 90°. This process is reversed in the next alternation.

The 90° phase difference between current and voltage of a pure capacitor circuit is the maximum possible value that can occur. In an actual

AC circuit, it is not possible to have a pure capacitor. Some resistance is always present in actual circuits. Combinations of resistance and capacitance cause the phase difference between the current and the voltage in the circuit to vary between 0° and 90°.

11.3 Factors that Affect Capacitance

Capacitance is the property of a circuit to oppose changes in voltage. The three factors that determine the capacitance of a capacitor are as follows and are shown in **Figure 11-7**:

1. Plate area: Increasing plate area increases capacitance.
2. Distance between plates: Capacitance is decreased when the distance between the plates increases.

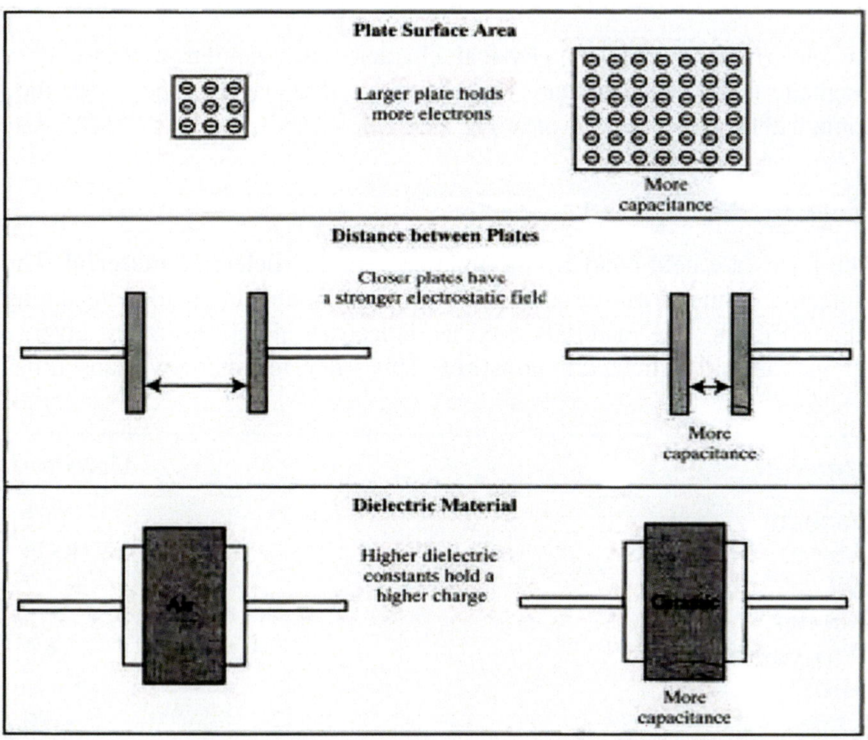

Figure 11-7 Factors affecting capacitance.

3. Dielectric material: Dielectrics, including air, are used as insulators between capacitor plates. They are rated with a dielectric constant (K). Capacitors with higher dielectric constants have higher capacitance.

Plate Surface Area

The surface of the plates is used to develop an **electrostatic charge**. A larger surface area holds more electrons and holes; therefore, it holds more electric charge. A capacitor charges to the applied voltage. A larger capacitance takes longer to reach full charge and longer to discharge. In the construction of some types of capacitors, the **plates** are made of a foil material. Many layers of this foil can be wrapped together with an insulator in between. This produces the maximum surface area in the smallest space possible.

Distance between Plates

An electric field displays physical characteristics similar to those of a magnetic field. The plates are like the poles of a magnet. The closer they come without touching, the stronger the field.

Dielectric Material

The plates are held apart by the thickness of the **dielectric material**. The dielectric should be as thin as possible while still maintaining good insulating characteristics. The materials used as dielectrics in capacitors are given a rating, called the **dielectric constant**. This value reflects how many times

Material	*Constant*
Vacuum	1
Air	1.0006
Oil	2
Rubber	2.5—35
Wax paper	3.5
Mica	5—6
Glass	5—10
Pure water	81
Ceramic	up to 7500

Figure 11-8 Dielectric constants of selected materials.

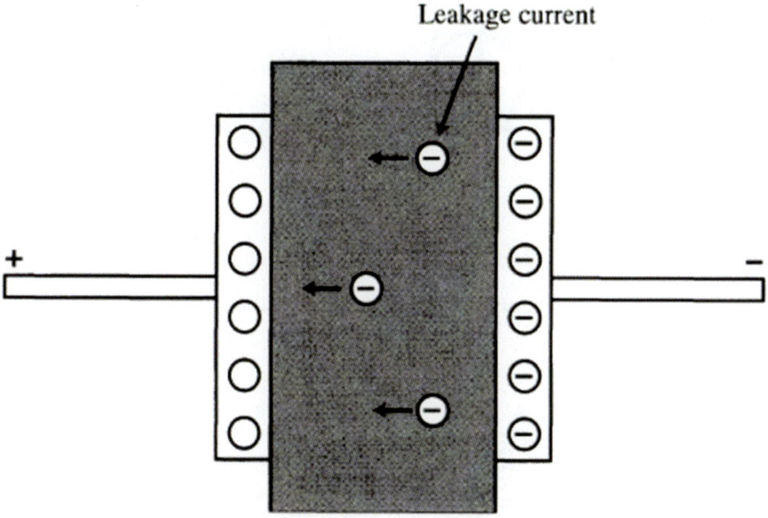

Figure 11-9 Capacitors have a leakage current between their plates. A stronger dielectric produces a smaller leakage.

better it is than a vacuum. For example, as shown in **Figure 11-8**, air has a dielectric constant value of 1.0006. This is effectively the same as a vacuum. Mica has a dielectric constant of five. Therefore, mica increases the capacitance five times larger than air or a vacuum.

Along with the values given for dielectric constants, consideration must be given to the quality of the dielectric material. As seen in **Figure 11-9**, leakage current is a result of electrons going across the dielectric. Any leakage decreases the electric field and decreases the capacitance.

11.4 Capacitor Types

Capacitors are classified as either fixed or variable. **Fixed capacitors** have one value of capacitance. Variable capacitors are constructed to allow capacitance to be varied over a range of values. **Variable capacitors** often use air as the dielectric. The capacitance is varied by changing the position of the movable plates. This changes the plate area of the capacitor. The distance between the plates is adjusted to vary the capacitance. When the movable plates are fully meshed together with the stationary plates, the capacitance is maximum.

Figure 11-10 shows several variable capacitors. These are used in tuning filters. One use of the variable capacitor is in the tuning dial on a portable AM/FM radio. Air dielectric capacitors are usually used as variables.

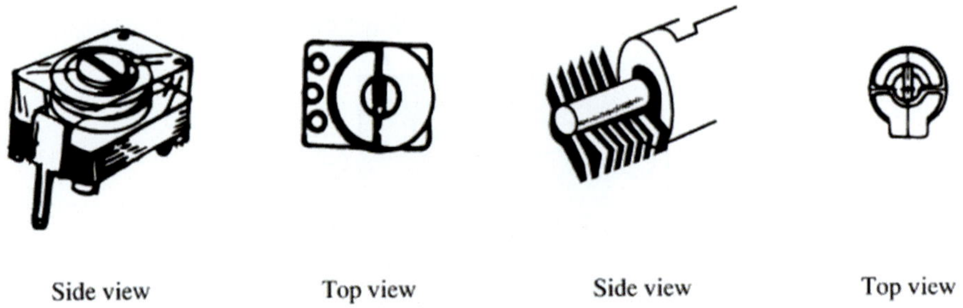

Figure 11-10 Variable capacitors.

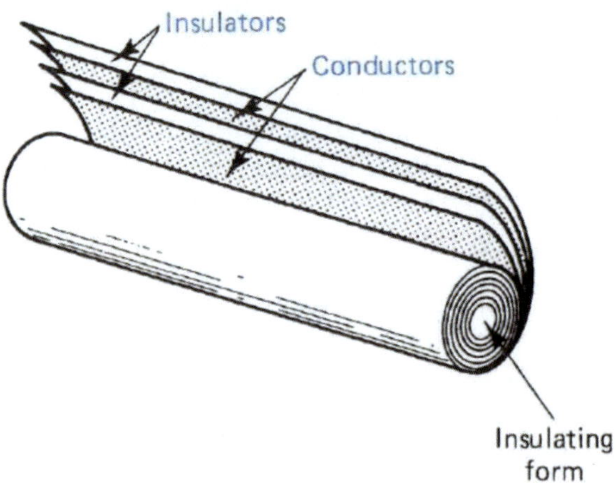

(a) Paper capacitor construction

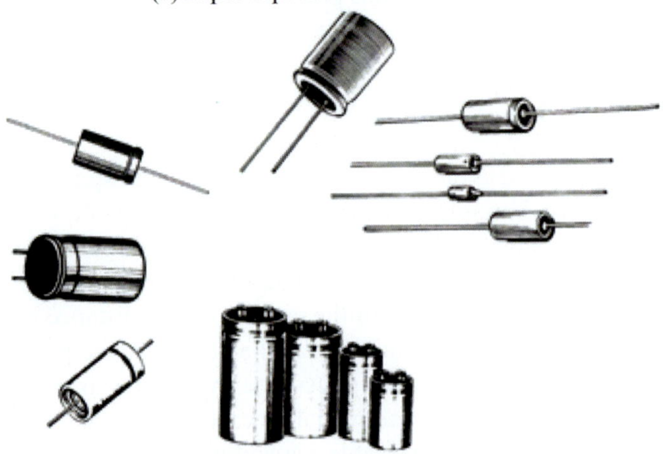

(b) Polarized electrolytic capacitors

Figure 11-11 Fixed capacitors.

Several types of capacitors are shown in **Figure 11-3**. Capacitor values in **microfarads (μF)** and **voltage ratings (DCWV)** are usually marked on the body of the capacitor.

Fixed capacitors come in many types. Some types of fixed capacitors are as follows:

1. **Paper capacitors**: Paper capacitors use paper as their dielectric. As shown in **Figure 11-11(a)**, they are made of flat strips of metal foil plates separated by a dielectric, which is usually waxed paper. Paper capacitors have values in the picofarad and low microfarad ranges. The voltage ratings are usually less than 600 V. Paper capacitors are usually sealed with wax to prevent moisture problems. The **voltage rating** of capacitors is important to note. A typical set of values marked on a capacitor might be "10 μF, 50 DCWV." This capacitor would have a capacitance of 10 μF and a "**DC working voltage**" of 50 V. This means that a voltage in excess of 50 V could damage the dielectric of the capacitors.
2. **Mica capacitors**: Mica capacitors have a layer of mica and then a layer of plate material. Their capacitance is usually small (in the picofarad range). They are small in physical size but have high voltage ratings.
3. **Oil-filled capacitors**: Oil-filled capacitors are used when high capacitance and high voltage ratings are needed. They are like paper capacitors immersed in oil. The paper, when soaked in oil, has a high dielectric constant.
4. **Ceramic capacitors:** Ceramic capacitors use a ceramic dielectric. The plates are thin films of metal deposited on ceramic material or made in the shape of a disk. They are covered with a moisture-proof coating and have high voltage ratings.
5. **Electrolytic capacitors**: Electrolytic capacitors are used when very high capacitance is needed. Representative electrolytic capacitors are shown in **Figure 11-11(b)**. Electrolytic capacitors contain a paste electrolyte. They have two metal plates with the **electrolyte** between them and are usually housed in a cylindrical aluminum can. The aluminum can is the negative terminal of the capacitor. The positive terminal (or terminals) is brought out of the can at the bottom. The size and voltage rating are usually printed on the capacitor. Electrolytic capacitors often have two or more capacitors housed in one unit. They are called **multisection capacitors**. The **positive plate** of an electrolytic capacitor is aluminum foil covered with a thin

oxide film. The film is formed by an electrochemical reaction and acts as the dielectric. A strip of paper that contains a paste electrolyte is placed next to the positive plate. The electrolyte is the **negative plate** of the capacitor. Another strip of aluminum foil is placed next to the electrolyte. These three layers are then coiled up and placed into a cylinder. These capacitors have a lower leakage resistance as compared to other capacitor types and do not retain the voltage developed on the plates for long periods of time.

Electrolytic capacitors are said to be "**polarized**." This means that each plate of the capacitor has a specific polarity. When used in a DC circuit, the positive terminal of the voltage supply should be connected to positive plate of the capacitor. If the positive plate is connected to the negative terminal of the source, the **dielectric** oxide film will break down, and the dielectric material becomes conductive. If connected in **reverse polarity**, the capacitor may explode due to chemical reaction of the electrolyte. Thus, it is critical to connect a polarized capacitor with the proper polarity. Capacitor polarity is marked on the capacitor for identifying the polarity of the plates. Generally, electrolytic capacitors have high capacitance values. This permits a capacitor to be used in applications that require the voltage to be maintained at a relatively constant value, such as a DC power supply. Some special high-value electrolytic capacitors can be used in AC circuitry.

6. **Ultracapacitors**: A capacitor is capable of storing electrical energy for long periods. When energy is needed, the capacitor can be made to act as a source and provide its stored energy to some load. To store a large amount of energy, however, the capacitor must be quite large in value. Large-value capacitors are of the electrolytic variety and are composed of metals and chemicals. Problems involved with electrolytic capacitors include (1) high **leakage currents** and (2) **degradation** of the dielectric material, which causes it to have a limited shelf life, beyond which it is no longer usable.

   An ultracapacitor will store electrical energy indefinitely with no need for recharging and no deterioration in its physical characteristics. These use either a **liquid** or a **dry type** of dielectric. The charge in an ultracapacitor is in the form of ions, which are trapped in tiny grooves within a coating of titanium. The ions can be released upon demand to produce a flow of current, similar to the discharging of a conventional capacitor.

The ultracapacitor has several advantages over other electrical energy sources. Unlike batteries, the ultracapacitor can be **recharged** any number of times and has an unlimited shelf life. Unlike the conventional capacitor, it has little or no leakage current, which enables it to retain a charge for a very long period.

11.5 Capacitor Ratings

Capacitors have two ratings: the **dielectric strength** and the **capacitance** value. Usually, these two values are stamped on the body of the capacitor. Most electronic components have ratings for current and/or power. However, neither of these is necessary for a capacitor, because current does not travel through it.

Dielectric Strength

The **dielectric strength** states the maximum voltage that can be applied to the capacitor without a destructive breakdown. The dielectric strength is given as a voltage rating. Many small **disc capacitors** have voltage ratings of over 600 V. Electrolytic capacitors have much smaller voltage ratings. Some ratings are as low as 10 V.

Voltage ratings are usually stated as a DC voltage or sometimes as **DCWV**, standing for DC working voltage. In some cases, ratings are given for a surge voltage. This is a voltage that the capacitor can survive but only if it lasts for a brief period of time. This rating is usually much higher. In addition, a rating may be given for AC voltages.

Unit of Measure of Capacitance

A capacitor is measured by a unit called the **farad**. One farad is the amount of capacitance that permits a current of 1 A to flow when the voltage change across the plates of a capacitor is 1 V per second. A 1-farad capacitor will store a **coulomb** of charge (equivalent to 6.24×10^{18} electrons) when the applied voltage across the terminals is 1 volt. The farad is too large for practical use. The **microfarad** (one-millionth of a farad, abbreviated as **μF** or **MFD**) is the most common subunit of capacitance. For high-frequency circuits, the microfarad is also too large. The subunit **micro-microfarad** (one-millionth of a microfarad, abbreviated as **μμF** or MMFD) is then used. This subunit is called the **picofarad (pF)** to avoid confusion. The multipliers milli- (10^{-3}) and nano- (10^{-9}) are not commonly used.

11.6 Capacitor Connections

When two or more capacitors are connected in a circuit, the total or effective capacitance will be a combination of the individual capacitance values. The type of connection involved will determine whether the value of the total capacitance will increase or decrease in value. In this regard, capacitors can be connected in series and parallel, similar to connection of other circuit components.

Capacitors in Series

When capacitors are connected in **series**, the effect is the same as increasing the thickness of the dielectric, as shown in **Figure 11-12**. A thicker dielectric inversely affects the net capacitance. As the thickness of the dielectric material increases, the value of the working voltage which can be applied across the plates increases as well. The **reciprocal formula** is used for total capacitance in series. It is the same type of formula used for resistors in parallel.

The formulas for capacitors in series are as follows:

$$\frac{1}{C_T} = \frac{1}{C_1} + \frac{1}{C_2} + \frac{1}{C_3} + \ldots \frac{1}{C_N}$$

C_T is the total capacitance, measured in farads.

C_1 through C_N are the individual capacitances, measured in farads.

N is the number of capacitors in series.

$$C_T = \frac{C}{N}$$

C_T is the total capacitance, measured in farads.

C is the capacitance value, measured in farads.

N is the number of capacitors in series.

$$C_T = \frac{C_1 \times C_2}{C_1 + C_2}$$

C_T is the total capacitance, measured in farads.

C_1 and C_2 are the individual capacitances, measured in farads.

Determine the **total capacitance** of two capacitors connected in series with values of 250 and 125 μF.

$$C_T = \frac{C_1 \times C_2}{C_1 + C_2}$$

$$C_T = \frac{250\ \mu F \times 125\ \mu F}{250\ \mu F + 125\ \mu F}$$

$$C_T = 83.3\ \mu F$$

Capacitors in Parallel

Capacitors connected in **parallel** have the effect of increasing the plate area, as shown in **Figure 11-13**. An increased plate area increases the capacitance, with the resultant being the sum of the capacitors connected in parallel. This formula is of the same type as the formula used for resistors and inductors in series.

The formula for capacitors in parallel is as follows:

$$C_T = C_1 + C_2 + C_3 + \cdots + C_N$$

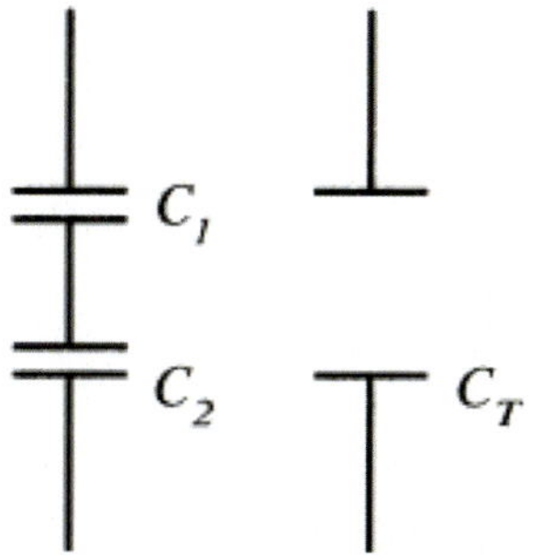

$$\frac{1}{C_T} = \frac{1}{C_1} + \frac{1}{C_2}$$

Figure 11-12 Capacitors connected in series.

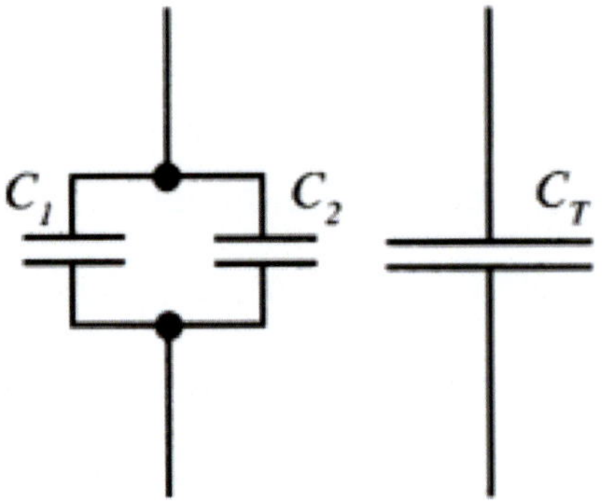

$$C_T = C_1 + C_2$$

Figure 11-13 Capacitors connected in parallel.

where C_T is the total capacitance, measured in Farads, C_1–C_N are individual capacitance, measured in Farads, and N is the number of capacitors connected in parallel.

Four 47-μF capacitors are connected in parallel. What is the total capacitance?

$$C_T = C_1 + C_2 + C_3 + \ldots C_N$$

$$C_T = 47\ \mu\text{F} + 47\ \mu\text{F} + 47\ \mu\text{F} + 47\ \mu\text{F}$$

$$C_T = 188\ \mu\text{F}$$

11.7 Capacitive Voltage Dividers

A **capacitive voltage divider** results from capacitors being connected in series; see **Figure 11-14**. The amount of voltage across the capacitors is a ratio of the opposite capacitance to the total multiplied by the applied voltage.

In equation form:

Voltage across C_1:

$$V_{C_1} = \frac{C_2}{C_1 + C_2} \times V_A$$

V_{C_1} is the voltage across C_1, measured in volts.
C_1 and C_2 are individual capacitances, measured in farads.
V_A is the applied voltage, measured in volts.

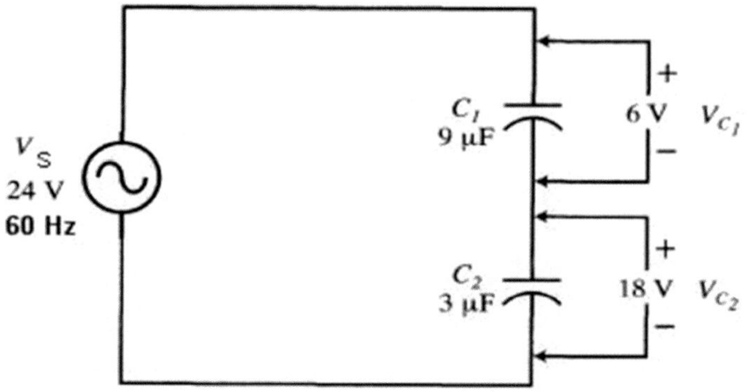

Figure 11-14 In a capacitive voltage divider, the smaller capacitance value has the largest voltage drop.

Voltage across C_2:

$$V_{C_2} = \frac{C_1}{C_1 + C_2} \times V_A$$

V_{C_2} is the voltage across C_2, measured in volts.
C_1 and C_2 are individual capacitances, measured in farads.
V_A is the applied voltage, measured in volts.

Verify the voltage readings in **Figure 11-14**.

$$V_{C_1} = \frac{C_2}{C_1 + C_2} \times V_A$$

$$V_{C_1} = \frac{3\ \mu F}{9\ \mu F + 3\ \mu F} \times 24\ V$$

$$V_{C_1} = 6\ V$$

$$V_{C_2} = \frac{C_1}{C_1 + C_2} \times V_A$$

$$V_{C_2} = \frac{9\ \mu F}{9\ \mu F + 3\ \mu F} \times 24\ V$$

$$V_{C_2} = 18\ V$$

11.8 Capacitive Reactance

An opposition to the flow of AC current caused by the electrostatic field of a capacitor is due to a property called **capacitive reactance (X_C)**, which is measured in ohms (Ω). It may be regarded as the AC resistance of a capacitor. The capacitive reactance varies inversely with the applied frequency and the capacitance. This can be determined by the formula:

$$X_C = \frac{1}{2 \times \pi \times f \times C}$$

where
X_C = capacitive reactance in ohms
2π = mathematical expression of one sine wave (0°–360°)
f = frequency of the source in Hz
C = capacitance in farads

- Determine the capacitive reactance of a 200-μF capacitor which is connected to an AC source that has a frequency of:
- (a) 60 Hz

$$X_C = \frac{1}{2 \times \pi \times f \times C} = \frac{1}{2 \times \pi \times 60 \times 200 \times 10^{-6}} = \frac{1}{0.0754} = 13.26\ \Omega$$

(b) 5000 Hz

$$X_C = \frac{1}{2 \times \pi \times f \times C} = \frac{1}{2 \times \pi \times 6000 \times 200 \times 10^{-6}} = \frac{1}{0.000754} = 0.133\ \Omega.$$

Note that these calculations have shown that when the frequency of the AC source **increases**, it causes a corresponding decrease in the value of the capacitive reactance of a circuit. At very high frequencies this value is very small, while at very low frequencies (DC level), this value is very high. Hence, the capacitance reactance is a variable factor, which depends on the frequency of the AC source voltage. When energized by a DC source, its capacitive reactance is infinite.

The total opposition of a circuit to current flow is called impedance. For a circuit that contains only a capacitor, the impedance (*Z*) is the same as the capacitive reactance (X_C). The calculation of the capacitive reactance of a pure capacitive circuit can be determined by using **Ohm's law**. This is given as follows:

$$X_C = \frac{V_C}{I_C}.$$

- Determine the capacitive reactance of the circuit shown in **Figure 11-15**. The circuit is energized by a 40-V_{RMS} source, which causes a current of 2 A.

$$X_C = \frac{40}{2} = 20\,\Omega.$$

Capacitive Reactance Connections

The **total reactance** of AC components connected in series or parallel is determined by the same formulas that are used for determining the total resistance of a circuit.

Series Capacitive Reactance

The total reactance of series-connected capacitors is determined by adding the individual capacitive reactance values. This is similar to that of series-connected resistors.

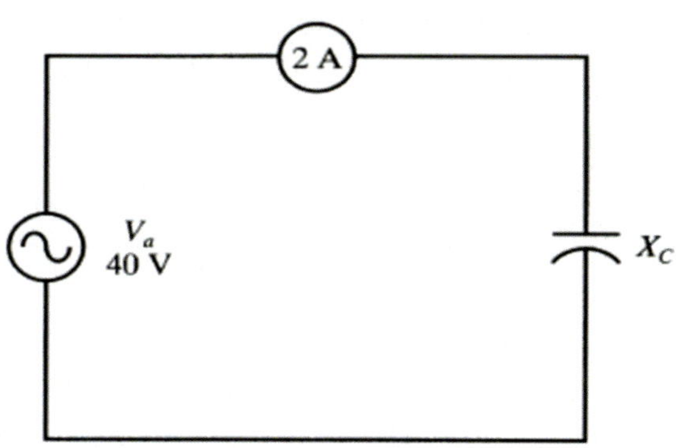

Figure 11-15 Determining the capacitive reactance of a circuit.

$$X_{C_T} = X_{C_1} + X_{C_2} + X_{C_3} + \ldots X_{C_N}$$

X_{C_T} is the total capacitive reactance, measured in ohms.

X_{C_1} through X_{C_N} are the individual capacitive reactances, measured in ohms.

The **total capacitance** of series-connected capacitors in terms of the frequency and individual capacitor values is given by

$$X_{CT} = X_{C1} + X_{C2} + \cdots + X_{CN}$$
$$= \frac{1}{2\pi \times f \times C_1} + \frac{1}{2\pi \times f \times C_2} + \cdots + \frac{1}{2\pi \times f \times C_N}.$$

Note that when the value of the capacitors in a series configuration is given, the reciprocal of the total capacitance equals the sum of the individual capacitor reciprocals. This is given as follows:

$$\frac{1}{C_T} = \frac{1}{C_1} + \frac{1}{C_2} + \cdots + \frac{1}{C_N}.$$

Note the difference in the formulas used for determining the total capacitive reactance and that for determining the total capacitance of a series-connected capacitors.

Determine the total capacitive reactance of the series-connected capacitors as shown in **Figure 11-16**. It consists of 20-, 30-, and 50-Ω capacitors connected in series.

$$X_{CT} = X_{C1} + X_{C2} + X_{C3} = 20 + 30 + 50 = 100\,\Omega.$$

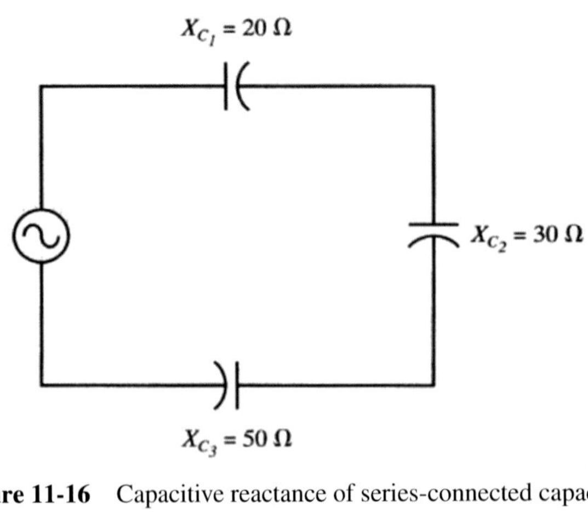

Figure 11-16 Capacitive reactance of series-connected capacitors.

Parallel Capacitive Reactance

The reciprocal of the **total reactance** of parallel-connected capacitors is determined by adding the individual capacitive reactance reciprocal values. This is similar to that of parallel-connected resistors.

$$\frac{1}{X_{C_T}} = \frac{1}{X_{C_1}} + \frac{1}{X_{C_2}} + \frac{1}{X_{C_3}} + \ldots \frac{1}{X_{C_N}}$$

X_{C_T} is the total capacitive reactance, measured in ohms.

X_{C_1} through X_{C_N} are the individual capacitive reactances, measured in ohms.

The total capacitance of parallel-connected capacitors in terms of the frequency and individual capacitor values is given:

$$\frac{1}{X_{CT}} = \frac{1}{X_{C1}} + \frac{1}{X_{C2}} + \cdots + \frac{1}{X_{CN}} = \frac{1}{\frac{1}{2\pi \times f \times C_1}} + \frac{1}{\frac{1}{2\pi \times f \times C_2}} + \cdots + \frac{1}{\frac{1}{2\pi \times f \times C_N}}$$
$$= 2\pi \times f \times C_1 + 2\pi \times f \times C_2 + \cdots + 2\pi \times f \times C_N.$$

Note that when the value of the capacitors in a parallel configuration is given, the total capacitance equals the sum of the individual capacitor values. This is given as follows:

$$C_T = C_1 + C_2 + C_3 + \cdots + C_N.$$

Note the difference in the formulas used for determining the total capacitive reactance and that for determining the total capacitance of a parallel-connected capacitors.

- Determine the total capacitive reactance of parallel-connected capacitors as shown in **Figure 11-17**. It consists of a 20-, 30-, and 50-Ω capacitors connected in parallel.

$$\frac{1}{X_{CT}} = \frac{1}{X_{C1}} + \frac{1}{X_{C2}} + \frac{1}{X_{C3}} = \frac{1}{20} + \frac{1}{30} + \frac{1}{50} = 0.05 + 0.033 + 0.2 = 0.283$$

$$X_{CT} = \frac{1}{0.283} = 3.53\,\Omega.$$

When the value of the total capacitive reactance (X_{CT}) is known, the value of the total capacitance C_T can be obtained by

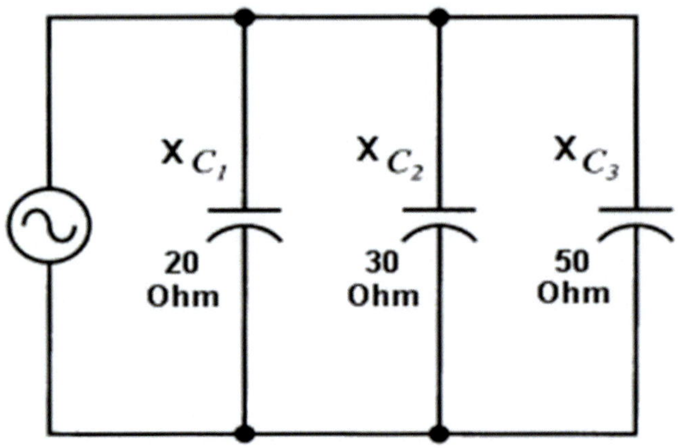

Figure 11-17 Capacitive reactance of series-connected capacitors.

$$X_{CT} = \frac{1}{2 \times \pi \times f \times C_T}$$

$$C_T = \frac{1}{2 \times \pi \times f \times X_{CT}}.$$

11.9 Resistor and Capacitor Circuits

The operation of an AC electrical circuit depends on the specific component being used. The simplest type of AC circuit consists of only one type of component. **Figure 11-18(a)** shows a circuit that has a resistor connected to an AC source of energy. When the voltage applied to a circuit is **increased**, the current will increase, and when the voltage is **decreased**, the current will decrease. The waveforms of **Figure 11-18(b)** show the relationship between the voltage and current in a **resistive AC circuit**. Note that the voltage and current waveforms are in phase. This means that a change in the value of the applied voltage causes a corresponding change in current flow. The resulting power converted by the resistance is found by multiplying voltage times current (P = V × I). During the 0°–180° interval, both the voltage and current are positive, resulting in a positive power value (+P = +V × +I). During the 180°–360° interval, both the voltage and current are negative, resulting in a positive power value (+P = –V × –I). A resulting **power curve** for a resistive AC circuit is shown in **Figure 11-18(b)**.

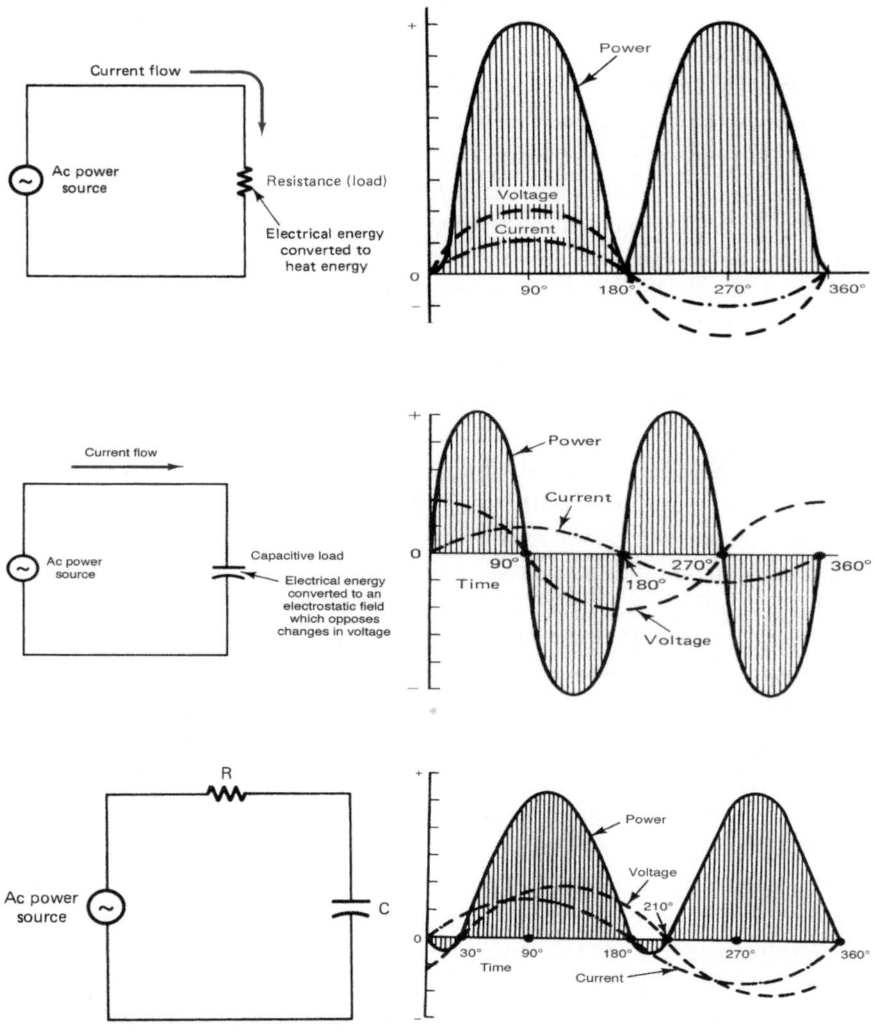

Figure 11-18 AC circuits and waveforms: (a) AC resistive circuit and (b) waveforms. (c) AC capacitive circuit and (d) waveforms; (e) AC resistive–capacitive (RC) circuit and (f) waveforms.

Figure 11-18(c) shows a circuit that has a capacitor connected to an AC source of energy. Capacitors have the ability to store an electrical charge.

When AC is applied to a capacitor, the **changing** value of current will cause the capacitor to charge and discharge. The voltage and current waveforms of a purely capacitive circuit (no resistance) are shown in **Figure 11-18(d)**. The value of the current flowing in a capacitive circuit depends on the rate of change of the applied voltage. The most rapid change in voltage occurs at the 0° and 180° positions as the polarity of the source changes. At these positions, maximum current is developed in the circuit. The rate of change of the voltage is very slow near the 90° and 270° positions and only a small amount of current flows. Thus, **current leads voltage** by 90° in a **purely capacitive circuit**. The angular separation between voltage and current waveforms is called the **phase angle**. During the 0°–90° interval, the current and voltage are both positive, resulting in a positive power value (+P = +V × +I). During the 90°–180° interval, the voltage is negative, and the current is positive, resulting in a **negative power** value (–P = –V × +I). During the 180°–270° interval, the current and voltage are both negative, resulting in a positive power value (+P = –V × –I). During the 270°–360° interval, the voltage is positive, and the current is negative, resulting in a negative power value (–P = +V × –I). A resulting power curve for a capacitive AC circuit is shown in **Figure 11-18(d)**. It should be noted that during an operational cycle, the positive and negative power waveforms cancel each other. As a result of this, no power is converted by a pure capacitive AC circuit.

In practice, AC circuits may contain a combination of two or more components. In this regard, a circuit could have resistance and capacitance in its construction. A sample **resistive–capacitive (RC) circuit** is shown in **Figure 11-18(e)**. In an *RC* circuit, the **current leads the voltage** by a phase angle between 0° and 90°. If capacitance in a circuit increases, the **phase angle** increases. The waveforms of **Figure 11-18(f)** show an *RC* circuit in which **the current leads the voltage** by 30°. During the 0°–30° interval, the voltage is negative, and the current is positive, resulting in a **negative power** value (–P = –V × +I). During the 30°–180° interval, the voltage is positive, and the current is positive, resulting in a **positive power** value (+P = +V × +I). During the 180°–210° interval, the voltage is positive, and the current is negative, resulting in a negative power value (–P = +V × –I). During the 210°–360° interval, the voltage is negative, and the current is negative, resulting in a positive power value (+P = –V × –I). No power is converted in the circuit during the 0°–30° and the 180°–210° intervals. In this *RC* circuit, most of the electrical energy supplied by the source is converted to another form of energy. Over a complete operational cycle, the circuit has net positive power. Thus, in practical RC circuits, a large part of the electrical power supplied by the source is utilized by the circuit components.

Series RC Circuits

In any **series AC circuit**, the current (I) is the same in all parts of the circuit. In a series *RC* circuit, the value and phase of the current flowing through the resistor (I_R) and the capacitor (I_C) is in phase. Recall that the voltage across a resistor (V_R) is in phase with the current. However, the voltage across a capacitor (V_C) leads the current by 90°. The voltage drop across each component in a series circuit when added together equals the source voltage (V_A). In a series *RC* circuit, the addition of voltages must take into account the phase relationship of V_C and V_R. This relationship can be conveniently represented using **vectors or phasors** (rotating vectors). The length of the vector corresponds to the value of the voltage, and the direction of the vector corresponds to the **phase angle**. In vector diagrams, a leading phase angle θ is indicated by drawing the vector in the counter-clockwise direction, making an angle θ from a given reference (usually horizontal) line. Similarly, a lagging phase angle is indicated by showing the vector in the clockwise direction, making an angle θ from a given reference line.

Since the phase difference between V_C and V_R is 90°, a right-angle triangle is used to represent these values. The two voltages, V_C and V_R, form the sides of the right-triangle and the applied voltage (V_A) is represented by the hypotenuse. This triangle is referred to as a voltage triangle.

The values of the voltage (V_A) and current (I) in a series circuit are used to determine the impedance (Z) offered by the circuit. This is given as $Z = \frac{V_A}{I}$. The capacitive reactance (X_C) is obtained by using $X_C = \frac{V_C}{I}$, and the resistance (R) is obtained by using $R = \frac{V_R}{I}$. This results in an impedance triangle which is similar in shape to the voltage triangle. The capacitive reactance (X_C) and the resistance (R) form the sides of the triangle and the impedance (Z) is represented by the hypotenuse.

If the values of the **resistance** and **capacitive reactance** of an AC series-connected RC circuit are known, the impedance triangle can be constructed first. These values are used to determine the impedance of the circuit. Next, by using the values of the applied voltage and the impedance, the current in the circuit can be evaluated. The value of the series current and the resistance or reactance values are used to determine the voltage drop across the circuit components. A voltage triangle can then be constructed from the values of the voltage drops and the applied voltage.

General Procedure for Solving AC Series RC Circuit Problems

1. Draw the circuit diagram. Label with known values.
2. If the capacitive reactance is not given, it can be determined by using:

$$X_C = \frac{1}{2 \times \pi \times f \times C}.$$

3. Draw an impedance triangle to calculate Z.
4. Draw the R "vector" horizontally. From the tip of the R vector, draw the X_C vector vertically downward.
5. Vectorially add the vectors and then draw Z.
6. Use the Pythagorean theorem to calculate the value of Z by using the following formula:

$$Z = \sqrt{R^2 + X_C^{\ 2}}.$$

7. Calculate the current in the circuit.

$$I = \frac{V}{Z}.$$

8. Calculate the voltage drop across each component.
9. Voltage across resistor: $V_R = I \times R.$
10. Voltage across capacitor: $V_C = I \times X_C.$
11. Draw a vector diagram of the voltages. Use I as the horizontal reference. Draw V_R in phase with the current. Draw V_C vertically downward as the capacitor voltage lags behind the current by 90°.
12. Vectorially add the voltages to obtain the applied voltage:

$$V_A = \sqrt{V_R^{\ 2} + V_C^{\ 2}}.$$

13. Use the Pythagorean theorem to calculate the value of the applied voltage to verify the solution. Using the proper scale while drawing the vectors makes it possible to spot any errors in the calculations.
14. Calculate the phase angle. This requires use of the trigonometric identity, cos θ, which is given as follows:

$$\cos\theta = \left(\frac{\text{opposite}}{\text{hypotenuse}}\right) = \left(\frac{V_R}{V_A}\right)$$

15. The phase angle can be obtained by taking the inverse of the cosine. This is indicated as $\cos^{-1}$:

$$\theta = \cos^{-1}\left(\frac{V_R}{V_A}\right)$$

16. A solved example of an AC series *RC* circuit, including the voltage and impedance triangles, is shown in **Figure 11-19**.

Parallel RC circuits

The basic formulas used with **parallel AC circuits** are different from those of series circuits. The impedance (Z) of a parallel circuit is less than individual

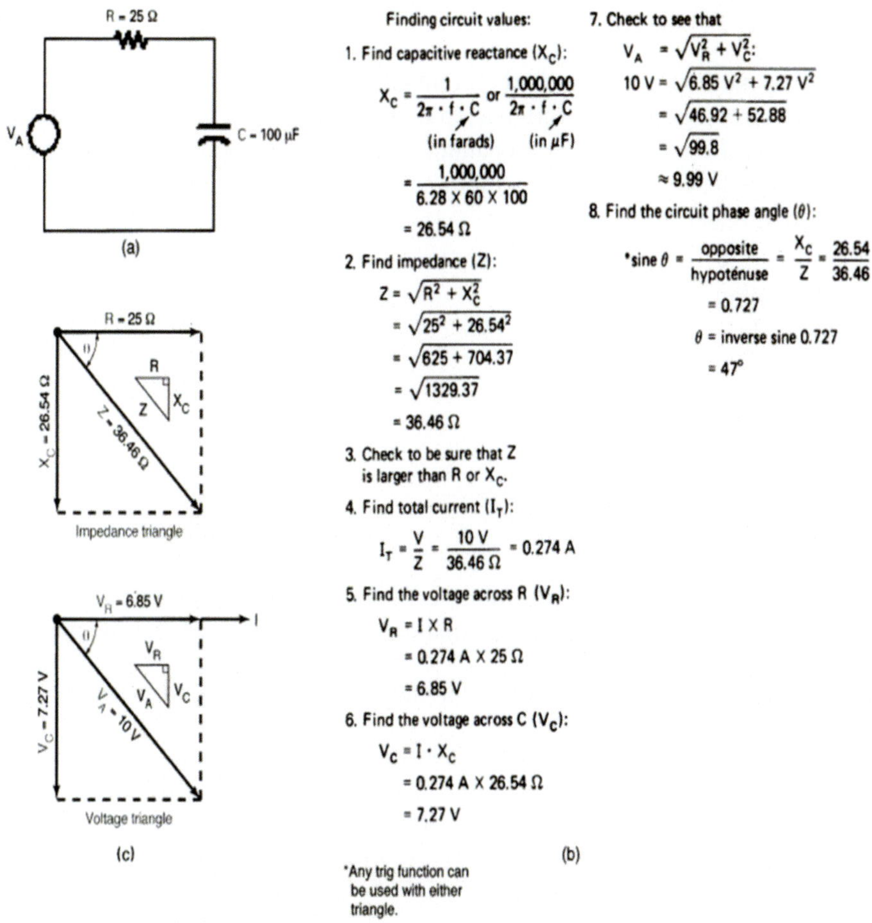

Figure 11-19 AC series *RC* circuit (a) diagram; (b) procedure for finding circuit values; (c) impedance and voltage triangles.

branch values of resistance and capacitive reactance. In practice, an impedance triangle is not used to show the operation of a parallel circuit. Instead, an admittance triangle is constructed. This shows the reciprocals of the circuit **impedance (Z)**, the **capacitive reactance (X_C)**, and the **resistance (R)**.

The voltage of a parallel AC circuit is the same across each branch. A **right triangle** is drawn to show the currents in the capacitive and resistive branches of the parallel *RC* circuit. The currents through the branches of a parallel AC circuit are shown by a right triangle called a current triangle.

The current through the capacitor (I_C) is shown leading the current through the resistor (I_R) by 90°. Because these values form a right triangle, the total current may be found by using the following formula:

$$I_T = \sqrt{I_R^{\ 2} + I_C^{\ 2}}.$$

This method is used to find currents in parallel RC circuits.

When components are connected in parallel, finding impedance is more difficult. An impedance triangle cannot be used. A method that can be used to find impedance is to use an admittance triangle. The following quantities are plotted on the triangle:

admittance: $Y = \dfrac{1}{Z}$;

conductance: $G = \dfrac{1}{R}$;

capacitive susceptance: $B_C = \dfrac{1}{X_C}$.

These quantities are the **inverse** of each type of opposition to AC. Because total impedance (Z) is the smallest quantity in a parallel AC circuit, its reciprocal ($Y = \dfrac{1}{Z}$) becomes the largest quantity on the admittance triangle (just as ½ is larger than ¼). The values are in **siemens**.

Parallel AC circuits are similar in several ways to series AC circuits except that the applied voltage is used as the reference, instead of the current that is used as a reference in series AC circuits.

General Procedure to Solve AC Parallel RC Circuit Problems

1. Draw the circuit diagram.
2. Label the known values.
3. Label the branch currents as I_R, I_C, and I_T depending on the kind of component in each branch.
4. If the capacitive reactance is not given, solve for it using:

$$X_C = \frac{1}{2 \times \pi \times f \times C}.$$

5. Use Ohm's law to calculate the current in each branch.
6. Draw a **vector diagram** of the currents. Use the applied voltage as the horizontal reference. Draw the total I_R in phase with the voltage. Draw the I_C vertically upward as the current flowing in the capacitor leads the current flowing in the resistor.
7. Vectorially add the currents to obtain the total current, I_T, delivered by the source.
8. Use the **Pythagorean theorem** to calculate the value of I_T. Drawing the vectors approximately to scale will allow you to pick up any gross errors you may have made in calculations.
9. Solve for the total impedance of the parallel circuit using:

$$Z_T = \frac{V}{I_T}.$$

A solved example of an AC parallel RC circuit, including voltage and admittance triangles, is shown in **Figure 11-20**.

Troubleshooting Capacitors

Capacitor problems are of three types: **opens**, **shorts**, or **leaks**. An **open capacitor** is often caused by one of the connecting wires becoming open where it connects to the plate. **Shorts** occur when current arcs through the dielectric, leaving a current path that allows excess current to flow. A **leaky capacitor** is one that is not shorted but does allow current through the dielectric. Thus, it never fully charges. Leaky capacitors are more likely to occur as equipment ages. An **analog scale** ohmmeter can be used to test capacitors for open, short, and leakage. An analog ohmmeter should be used for these measurements. Different ohms scales should be tried to find which works the best with a particular meter.

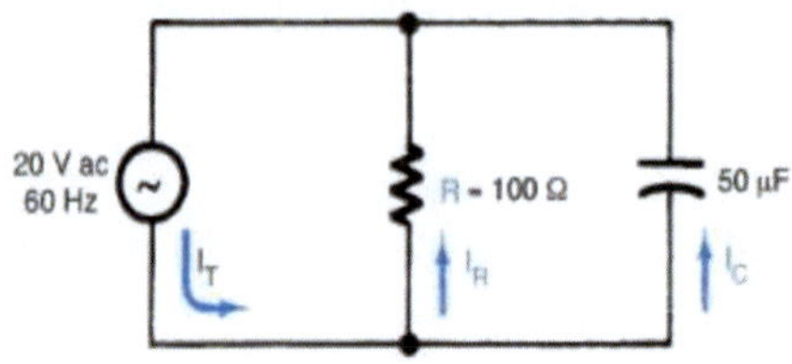

(a)

Finding circuit values:

1. Find capacitive reactance (X_C):

$$X_C = \frac{1,000,000}{2\pi \cdot f \cdot C\,(\mu F)}$$

$$= \frac{1,000,000}{6.28 \times 60 \times 50}$$

$$= 53\ \Omega$$

2. Find current through R (I_R):

$$I_R = \frac{V}{R} = \frac{20\ V}{100\ \Omega} = 0.2\ A$$

3. Find current through C (I_C):

$$I_C = \frac{V}{X_C} = \frac{20\ V}{53\ \Omega} = 0.377\ A$$

4. Find total current (I_T):

$$I_T = \sqrt{I_R^2 + I_C^2}$$

$$= \sqrt{0.2^2 + 0.377^2}$$

$$= \sqrt{0.04 + 0.142}$$

$$= \sqrt{0.182}$$

$$= 0.427\ A$$

5. Check to see that I_T is larger than I_R or I_C.
6. Find impedance (Z):

$$Z = \frac{V}{I_T} = \frac{20\ V}{0.427\ A} = 46.84\ \Omega$$

7. Check to see that Z is less than R or X_C.
8. Find circuit phase angle (θ):

$$\text{cosine } \theta = \frac{\text{adjacent}}{\text{hypotenuse}} = \frac{G}{Y} = \frac{10}{21.3}$$

$$= 0.469$$

$$\theta = \text{inverse cosine } 0.469$$

$$= 62^\circ$$

(b)

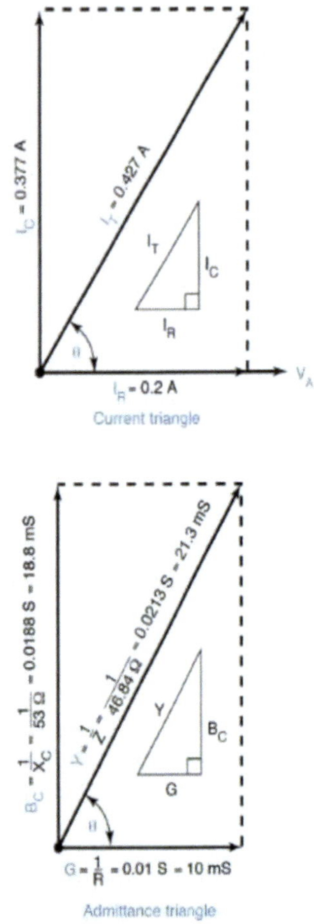

(c)

Figure 11-20 RC parallel circuit (a) diagram; (b) procedure for finding circuit values; (c) current and admittance triangles.

First, short the leads of the capacitor together. This eliminates any charge that may already be present. If the testing is repeated, be sure to repeat discharging the capacitor. Next, connect the ohmmeter across the leads of the capacitor. A normal capacitor, when connected across the ohmmeter, initially shows a low resistance. Then as the capacitor charges, the needle moves toward infinity. If the capacitor is shorted or leaky, the resistance reading stays low and does not move toward infinity. The normal resistance should be between 500 kΩ and 1 MΩ for capacitors above 1 μF. Smaller capacitors charge too quickly for the ohmmeter to respond and should read infinite on all scales. When checking **electrolytic capacitors**, the polarity of the meter must match the polarity of the capacitor leads. A better check of a capacitor is made with a **capacitor analyzer**. This instrument checks the value of the capacitor and measures the leakage current under normal working voltage of the capacitor.

Review Questions

1. The resistance value of a shorted capacitor is (low/high). [low]
2. The resistance value of an open capacitor is (low/high). [high]
3. A capacitor that initially shows a low resistance when an ohmmeter is connected, but changes to a high resistance value in a short time, indicates that the capacitor is (shorted, good, open).

Summary

A capacitor is a device that has the ability to store electric charge.

- Capacitance is measured in farads (F), microfarads (μF), and picofarads (pF).
- A capacitor consists of two or more conductive plates separated by an insulating material.
- Capacitance is influenced by the area of the plates, the distance between the plates, and the type of dielectric material.
- When voltage is initially applied to an uncharged capacitor, it causes a charging current to flow.
- The charge across the plates of a capacitor builds up very quickly during the charging process.
- When the charge voltage developed on the plates of a capacitor equals the supply voltage, no further charging action occurs.

- When discharging a capacitor, the charge voltage across the plates will decrease in value with respect to time.
- When capacitors are connected in series, the resulting capacitance decreases in value because the addition of each capacitor increases the distance between the plates.
- When capacitors are connected in parallel, the resulting capacitance increases in value because the addition of each capacitor increases the area of the plates where charges are stored.
- Capacitive reactance (X_C) that is measured in ohms is the AC resistance offered by a capacitor.
- Capacitive reactance is inversely proportional to the value of the capacitance and the frequency applied to it.
- In a pure resistive AC circuit, the voltage and current are in phase.
- In a pure capacitive circuit, the current leads the voltage by 90°.
- Vector or phasor diagrams are commonly used to represent the phase relationship between voltages and currents in AC circuits.
- In a series *RC* circuit, the current flowing through circuit components is the same.
- In a parallel *RC* circuit, the voltage across the circuit components is the same.
- No power is consumed by a purely capacitive AC circuit.
- Practical capacitor circuits are very efficient in converting power.
- Capacitors are susceptible to open, short, and leakage faults.

Formulas

- Capacitors in series:

$$\frac{1}{C_T} = \frac{1}{C_1} + \frac{1}{C_2} + \frac{1}{C_3} + \ldots \frac{1}{C_N}$$

C_T is the total capacitance, measured in farads.

C_1 through C_N are the individual capacitances, measured in farads.

N is the number of capacitors in series.

$$C_T = \frac{C}{N}$$

C_T is the total capacitance, measured in farads.

C is the capacitance value, measured in farads.

N is the number of capacitors in series.

$$C_T = \frac{C_1 \times C_2}{C_1 + C_2}$$

C_T is the total capacitance, measured in farads.

C_1 and C_2 are the individual capacitances, measured in farads.

- Capacitors in parallel:

$$C_T = C_1 + C_2 + C_3 + \cdots + C_N.$$

- Voltage divider formed by two series capacitors C_1 and C_2:

Voltage across C_1:

$$V_{C_1} = \frac{C_2}{C_1 + C_2} \times V_A$$

V_{C_1} is the voltage across C_1, measured in volts.
C_1 and C_2 are individual capacitances, measured in farads.
V_A is the applied voltage, measured in volts.

Voltage across C_2:

$$V_{C_2} = \frac{C_1}{C_1 + C_2} \times V_A$$

V_{C_2} is the voltage across C_2, measured in volts.
C_1 and C_2 are individual capacitances, measured in farads.
V_A is the applied voltage, measured in volts.

- Capacitive reactance:

$$X_C = \frac{1}{2 \times \pi \times f \times C}$$

$$X_C = \frac{V_C}{I_C}.$$

- Series capacitive reactance

$$X_{C_T} = X_{C_1} + X_{C_2} + X_{C_3} + \ldots X_{C_N}$$

X_{C_T} is the total capacitive reactance, measured in ohms.

X_{C_1} through X_{C_N} are the individual capacitive reactances, measured in ohms.

$$\begin{aligned} X_{CT} &= X_{C1} + X_{C2} + \cdots + X_{CN} \\ &= \frac{1}{2\pi \times f \times C_1} + \frac{1}{2\pi \times f \times C_2} + \cdots + \frac{1}{2\pi \times f \times C_N} \end{aligned}$$

- Parallel capacitive reactance:

$$\frac{1}{X_{C_T}} = \frac{1}{X_{C_1}} + \frac{1}{X_{C_2}} + \frac{1}{X_{C_3}} + \ldots \frac{1}{X_{C_N}}$$

X_{C_T} is the total capacitive reactance, measured in ohms.

X_{C_1} through X_{C_N} are the individual capacitive reactances, measured in ohms.

- Total capacitive reactance (X_{CT}) terms of the total capacitor value (C_T)

$$X_{CT} = \frac{1}{2 \times \pi \times f \times C_T}$$

$$C_T = \frac{1}{2 \times \pi \times f \times X_{CT}}.$$

In an *RC* series circuit:

Impedance: $Z = \sqrt{R^2 + X_C^{\,2}}$

Current: $I = \frac{V}{Z}$

Voltage drop across resistor: $V_R = I \times R$

Voltage drop across capacitor: $V_C = I \times X_C$

Applied voltage: $V_A = \sqrt{V_R^{\ 2} + V_C^{\ 2}}$

Phase angle by which current leads voltage: $\theta = \cos^{-1}\left(\frac{V_R}{V_A}\right)$

- In an *RC* parallel circuit:

  $I_T = \sqrt{I_R^{\ 2} + I_C^{\ 2}}.$

Problems

Three capacitors of 30, 60, and 90-μF are connected in series with a 50-Hz AC source. Determine the total capacitance and capacitive reactance. [16.364 μF, $X_{CT} = X_{C1} + X_{C2} + X_{C3}$].

$$X_{CT} = X_{C1} + X_{C2} + X_{C3} = \frac{1}{2\pi\times50\times30\times10^{-6}} + \frac{1}{2\pi\times50\times60\times10^{-6}} + \frac{1}{2\pi\times50\times90\times10^{-6}}$$
$$= 106.1 + 53 + 35.4 = 194.4\,\Omega.$$

To verify the value of total capacitance: $X_C = \frac{1}{2\times\pi\times f\times C_T}$

$$C_T = \frac{1}{2\times\pi\times50\times X_{CT}}$$

$$C_T = \frac{1}{2\times\pi\times50\times194.4} = 1.637\times10^{-5} = 16.37\mu\text{F}.$$

- Three capacitors of 30, 60, and 90 μF are connected in parallel with a 50-Hz AC source. Determine the total capacitance and capacitive reactance. [180 μF,

$$\frac{1}{X_{CT}} = \frac{1}{X_{C1}} + \frac{1}{X_{C2}} + \frac{1}{X_{C3}} = \frac{1}{\frac{1}{2\pi\times50\times30\times10^{-6}}} + \frac{1}{\frac{1}{2\pi\times50\times60\times10^{-6}}} + \frac{1}{\frac{1}{2\pi\times50\times90\times10^{-6}}}$$
$$= 2\pi\times50\times30\times10^{-6} + 2\pi\times50\times60\times10^{-6} + 2\pi\times50\times90\times10^{-6}$$
$$= 0.0094 + 0.0188 + 0.0283 = 0.0565$$

$$X_{CT} = \frac{1}{0.0565} = 17.68\,\Omega.$$

To verify the value of total capacitance: $X_{CT} = \dfrac{1}{2\times\pi\times f\times C_T}$

$$C_T = \frac{1}{2\times\pi\times 50\times X_{CT}}$$

$$C_T = \frac{1}{2\times\pi\times 50\times 17.68} = 1.8\times10^{-4} = 180\mu F\,].$$

- Three 30-μF capacitors are connected in series. Determine the total capacitance. [10 μF]
- Three 30-μF capacitors are connected in parallel. Determine the total capacitance. [90 μF]
- If a 12-V DC source is connected across two 30-μF capacitors connected in series, determine the voltage across each capacitor. [6 V across each capacitor]
- Refer to **Figure 11-19**. If the value of R = 100 Ω and L = 20 μF, determine the value of voltages developed across the circuit components, and the phase angle, θ, between the current and voltage in the circuit.
- [Capacitive reactance:

$$X_C = \frac{1}{2\times\pi\times f\times C} = \frac{1}{2\times\pi\times 60\times 20\times10^{-6}} = \frac{10^6}{376.99\times 20} = 132.63\,\Omega.$$

Impedance:

$$Z = \sqrt{R^2 + X_C^{\ 2}} = \sqrt{100^2 + 753.98^2} = \sqrt{27590.587} = 166.1\Omega.$$

Series current: $I = \dfrac{V}{Z} = \dfrac{10}{166.1} = 0.0602\,\text{A}.$

$$V_R = I\times R = 0.0602\times 100 = 6.02\ \text{V}$$
$$V_C = I\times X_C = 0.0602\times 166.1 = 7.985\,\text{V}$$

Verify voltages add up to the supply voltage:

$$V_A = \sqrt{V_R^{\ 2} + V_C^{\ 2}} = \sqrt{6.02^2 + 7.985^2} = \sqrt{99.9966} = 9.999\,\text{V}.$$

Phase angle: leading (current leads voltage)].

$$\theta = \cos^{-1}\left(\frac{V_R}{V_A}\right) = \cos^{-1}\left(\frac{6.02}{10}\right) = \cos^{-1} 0.602 = 58.87^\circ$$

Answers

$$\frac{1}{C_T} = \frac{1}{C_1} + \frac{1}{C_2} + \frac{1}{C_3} = \frac{1}{30\mu} + \frac{1}{60\mu} + \frac{1}{90\mu} = 0.0611 \times 10^6;$$

$C_T = 16.364\mu\text{F}$,

$$X_{CT} = X_{C1} + X_{C2} + X_{C3} = \frac{1}{2\pi \times 50 \times 30 \times 10^{-6}} + \frac{1}{2\pi \times 50 \times 60 \times 10^{-6}} + \frac{1}{2\pi \times 50 \times 90 \times 10^{-6}}$$
$$= 106.1 + 53 + 35.4 = 194.4\,\Omega.$$

To verify the value of total capacitance: $X_C = \dfrac{1}{2 \times \pi \times f \times C_T}$

$$C_T = \frac{1}{2 \times \pi \times 50 \times X_{CT}}$$

$$C_T = \frac{1}{2 \times \pi \times 50 \times 194.4} = 1.637 \times 10^{-5} = 16.37\mu\text{F}\ ]$$

$C_T = 30\ \mu\text{F} + 60\ \mu\text{F} + 90\ \mu\text{F} = 180\ \mu\text{F}$,

$$\frac{1}{X_{CT}} = \frac{1}{X_{C1}} + \frac{1}{X_{C2}} + \frac{1}{X_{C3}} = \frac{1}{\dfrac{1}{2\pi \times 50 \times 30 \times 10^{-6}}} + \frac{1}{\dfrac{1}{2\pi \times 50 \times 60 \times 10^{-6}}} + \frac{1}{\dfrac{1}{2\pi \times 50 \times 90 \times 10^{-6}}}$$
$$= 2\pi \times 50 \times 30 \times 10^{-6} + 2\pi \times 50 \times 60 \times 10^{-6} + 2\pi \times 50 \times 90 \times 10^{-6}$$
$$= 0.0094 + 0.0188 + 0.0283 = 0.0565$$

$$X_{CT} = \frac{1}{0.0565} = 17.68\,\Omega.$$

To verify the value of total capacitance: $X_C = \dfrac{1}{2 \times \pi \times f \times C_T}$

$$C_T = \frac{1}{2 \times \pi \times 50 \times X_{CT}}$$

$$C_T = \frac{1}{2 \times \pi \times 50 \times 17.68} = 1.8 \times 10^{-4} = 180\mu\text{F}$$

$C_T = 30\ \mu/3 = 10\ \mu\text{F}$

$C_T = 3 \times 30\ \mu F = 90\ \mu F$

$$V_{C_1} = \frac{C_2}{C_1 + C_2} \times V = \frac{30\mu}{60\mu} \times 12 = 6\,V; V_{C_1} = 6\,V.$$

Capacitive reactance:

$$X_C = \frac{1}{2 \times \pi \times f \times C} = \frac{1}{2 \times \pi \times 60 \times 20 \times 10^{-6}} = \frac{10^6}{376.99 \times 20} = 132.63\,\Omega.$$

Impedance: $Z = \sqrt{R^2 + X_C{}^2} = \sqrt{100^2 + 753.98^2} = \sqrt{27590.587} = 166.1\,\Omega.$

Series current: $I = \frac{V}{Z} = \frac{10}{166.1} = 0.0602\,A.$

$V_R = I \times R = 0.0602 \times 100 = 6.02\,V$

$V_C = I \times X_C = 0.0602 \times 166.1 = 7.985\,V.$

Verify voltages add up to the supply voltage:

$V_A = \sqrt{V_R{}^2 + V_C{}^2} = \sqrt{6.02^2 + 7.985^2} = \sqrt{99.9966} = 9.999\,V.$

Phase angle: $\theta = \cos^{-1}\left(\frac{V_R}{V_A}\right) = \cos^{-1}\left(\frac{6.02}{10}\right) = \cos^{-1} 0.602 = 58.87^\circ$ leading (current leads voltage).

Self-examination/Answers

1. The material that separates the plates of a capacitor is a(n) ________________.
2. A(n) ________________ field is developed across the plates of a capacitor when it is being charged.
3. The distance between the plates of a capacitor is determined by the thickness of the ________________.
4. The plates of a capacitor are made up of ________________ material.
5. The charging current of a capacitor (increases/decreases) when the voltage developed across its plates equals the source voltage.

6. If a charged capacitor is disconnected from the supply or load, the voltage across the plates will _______________.
7. When AC voltage is applied across the plates of a capacitor, the phase difference between the charging current and the voltage developed across the plates of a capacitor is _________.
8. As the area of the capacitor plates increases, the capacitance _______________.
9. As the distance between the plates of a capacitor increases, the capacitance _____________.
10. As the value of the dielectric constant of the material between the plates increases, the capacitance _____________.
11. A polarized capacitor uses which kind of dielectric material?
12. What will happen if the working voltage rating of a capacitor is exceeded?
13. The leakage current of an ultracapacitor is very (small/high).
14. As the value of a capacitor increases, its ability to store charges _______________.
15. ______________________refers to the voltage that can be applied to a capacitor without causing a destructive breakdown of the dielectric material.
16. A picofarad capacitor is a thousand times (smaller/larger) than the value of a microfarad capacitor.
17. A 10- and a 30-μF capacitor are connected in series. Determine the total capacitance.
18. A 10- and a 30-*1*F capacitor are connected in parallel. Determine the total capacitance.
19. When capacitors are connected in series, the value of the working voltage that can be applied across the plates (increases/remains constant/decreases).
20. Referring to **Figure 11-14**, if $C_1 = 10$ μF and $C_2 = 30$ μF, determine the voltage across C_1.
21. Referring to **Figure 11-14**, if $C_1 = 10$ μF and $C_2 = 30$ μF, determine the voltage across C_2.
22. Referring to **Figure 11-14,** if $C_1 = 10$ μF and $C_2 = 30$ μF, determine the sum of the voltages across C_1 and C_2. How does this compare with the source voltage?
23. When the polarity of the capacitor voltage and current is negative, the polarity of the power developed by the capacitor is (positive, negative).

24. In a practical *RC* circuit, the voltage (leads, lags) behind the current by a phase angle between ______° and ______°.
25. In a series *RC* circuit, the (voltage across, current through) the components is the same, whereas in a parallel *RC* circuit, the (voltage across, current through) the components is the same.
26. In a series *RC* circuit, the impedance (Z) of the circuit can be determined graphically by using the ______________triangle.
27. In a series *RC* circuit, the value of the voltages developed across various components in the circuit can be determined graphically by using the ______________triangle.
28. In a parallel *RC* circuit, the admittance (Y) of the circuit can be determined graphically by using the ______________triangle.
29. The resistance value of a shorted capacitor is (low/high).
30. The resistance value of an open capacitor is (low/high).
31. A capacitor that initially shows a low resistance when an ohmmeter is connected, but changes to a high resistance value in a short time, indicates that the capacitor is (shorted, good, open).

Answers

1. insulator or dielectric
2. electrostatic
3. insulator or dielectric
4. conductive
5. decreases
6. remain constant
7. 90
8. increases
9. decreases
10. increases
11. Electrolytic
12. The capacitor will be destroyed due to the breakdown of the dielectric material
13. small
14. increases
15. Dielectric strength or working voltage
16. smaller
17. $\frac{1}{C_T} = \frac{1}{C_1} + \frac{1}{C_2}; C_T = \frac{C_1 \times C_2}{C_1 + C_2} = \frac{10\mu \times 30\mu}{10\mu + 30\mu} = 7.5\mu\text{F}$

18. $C_T = C_1 + C_2 = 10\ \mu F + 30\ \mu F = 40\ \mu F$
19. increases
20. $V_{C_1} = \dfrac{C_2}{C_1 + C_2} \times V = \dfrac{30\mu}{40\mu} \times 24 = 18\,V$
21. $V_{C_2} = \dfrac{C_1}{C_1 + C_2} \times V = \dfrac{10\mu}{40\mu} \times 24 = 6\,V$
22. V + 6 V = 24 V; same as supply voltage
23. positive
24. lags, 0, 90
25. current through, voltage across
26. impedance
27. voltage
28. admittance
29. low
30. high
31. good

Glossary

Admittance (Y)
The ease with which AC current flows through an impedance and is expressed as: $Y = \dfrac{1}{Z}$. It is measured in Siemens.

Capacitance (C)
The property of a device to oppose changes in voltage due to energy stored in its electrostatic field.

Capacitive reactance (X_C)
The opposition to the flow of AC current caused by a capacitive device (measured in ohms): $X_C = \dfrac{1}{2 \times \pi \times f \times C}$.

Capacitor
A device that has capacitance and is usually made of two metal plate materials separated by a dielectric material (insulator).

Conductance (G)

The ease with which current flows through a resistance and is expressed as: $G = \frac{1}{R}$. It is measured in Siemens.

Coulomb (Q)

6.24×10^{18} electrons represent one coulomb of charge.

Decay

A term used for a gradual reduction in value of a voltage or current.

Decay time

The time required for a capacitor to discharge to a certain percentage of its original charge or the time required for current through an inductor to reduce to a percentage of its maximum value.

Dielectric

An insulating material placed between the metal plates of a capacitor.

Dielectric constant

A number that represents the ability of a dielectric to develop an electrostatic field, as compared with air, which has a value of 1.0.

Dielectric strength

States the maximum voltage that can be applied to a capacitor before breakdown occurs.

Electrolytic Capacitor

A type of capacitor that is polarized.

Electrostatic Field

The attraction between negative and positive voltages.

Farad (F)

One farad is the amount of capacitance that permits a current of 1 A to flow when the voltage change across the plates of a capacitor is 1 V per second. A 1 farad capacitor will store a coulomb of charge (equivalent to 6.24×10^{18} electrons) when the applied voltage across the terminals is 1 V.

Filter
A circuit used to pass certain frequencies and attenuate all other frequencies.

Frequency
The number of AC cycles per second, measured in hertz (Hz).

Frequency response
A circuit's ability to operate over a range of frequencies.

Mica capacitor
A capacitor made of metal foil plates separated by a mica dielectric.

Parallel resonant circuit
A circuit that has an inductor and capacitor connected in parallel to cause response to frequencies applied to the circuit.

Plate
In reference to capacitors, it is a conductive surface which collects electrons on the negative side of the capacitor and electrons on the positive side.

Reactance (X)
The opposition to AC current flow due to inductance (X_L) or capacitance (X_C).

Reactive circuit
An AC circuit that has the property of inductance or capacitance.

Resonant circuit
See parallel resonant circuit and series resonant circuit.

Resonant frequency (f_r)
The frequency that passes most easily through a frequency-sensitive circuit when $X_L = X_C$ in the circuit:

$$f_r = \frac{1}{\sqrt{L \times C}}.$$

Selectivity
The ability of a resonant circuit to select a specific frequency and reject all other frequencies.

Susceptance (B)

The ease with which a capacitor passes AC current and is expressed as: $B_C = \frac{1}{X_C}$. It is measured in Siemens.

Working voltage

A rating of capacitors which is the maximum voltage that can be placed across the plates of a capacitor without damage occurring.

12

Inductance and Inductive Reactance

The response of an electrical circuit depends on the type of energy being applied and the components being used. For electrical circuits that consist of only resistors, the application of AC or DC energy will cause a similar response, resulting in the **dissipation** of heat energy. However, other circuit components, such as **inductors**, which consist of a number of conductors that form a coil, have a different response. When electrical energy is applied to the coils of an inductor, the resulting current flow produces a **magnetic field** around each coil. This causes the applied energy to be stored in the resulting electromagnetic field. The term **inductance (L)** refers to the ability of a device to oppose changes in the current due to energy stored in an electromagnetic field. The fundamental unit of inductance is a **henry (H)**.

When electrical energy is applied to a conductor, it causes a magnetic field to be produced. If the electrical energy applied is DC, the resulting magnetic field will be of a constant value and fixed polarity. The only change in the magnetic field occurs when the energy is initially applied and when it is removed or changed. This change in the magnetic field develops a voltage that is called **counter electromotive force (CEMF)**. The CEMF represents energy stored in the magnetic field. The developed CEMF then opposes any change in the current produced by the energy source.

If the electrical energy applied is AC, the value and polarity of the CEMF will be changing constantly as well. This will cause a continuous opposition to the current flow. This opposition is called **inductive reactance** **(X_L)** which is measured in **ohms**. The frequency of the applied AC also influences the response of the inductor. If the frequency of the applied AC changes rapidly, the opposition to the change will also occur at the same rate. As a result, higher frequency AC current encounters more opposition than lower frequency AC. **Inductors**, thus, have a tendency to oppose changes in the current applied to it and are commonly used in AC circuits.

Objectives

1. Define inductance
2. Describe the construction of an inductor
3. Explain the response of an inductor when energy is applied
4. List the factors that determine the value of an inductor
5. Identify different types of inductors
6. Determine the total inductance of a circuit with inductors in series or parallel
7. Explain the response of an inductive voltage divider circuit
8. Determine the inductive reactance of a circuit
9. Examine the response of resistor–inductor circuits

Chapter Outline

12.1 Inductor Construction

An **inductor** consists of several coils or turns of wire wound around some type of core material. When current flows through the coil, it develops a magnetic field. The core material which can be either air or metal is responsible for concentrating the **magnetic field**. The physical construction of an inductor determines how much magnetic flux will be produced by the inductor. Some types of inductor coils are shown in **Figure 12-1**.

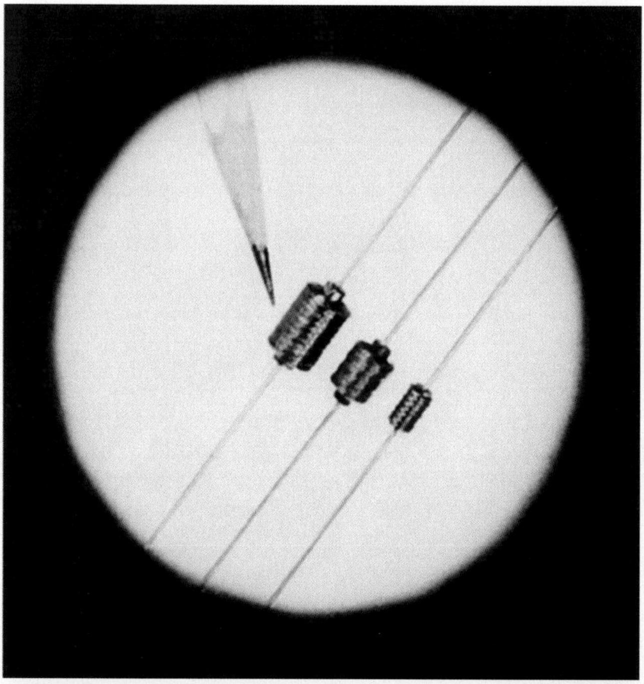

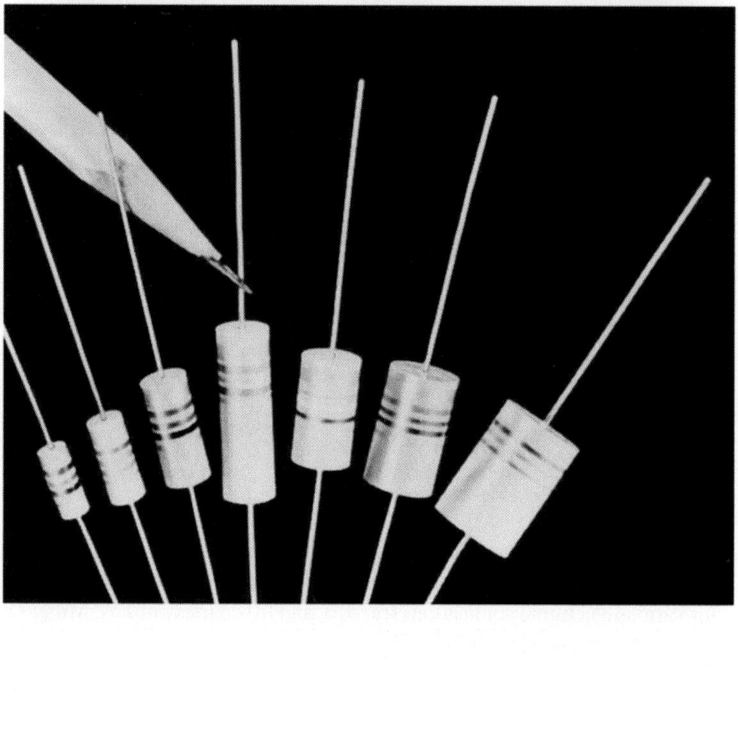

Figure 12-1 Some types of inductors: (a) subminiature inductors used with high frequencies; (b) inductors that use a color-coded value; (c) inductors used in equipment power supplies; (d) inductors for various electronic applications [(a), (b), and (d) courtesy of J.W. Miller Division/ Bell Industries; (c) courtesy of TRW/UTC Transformers].

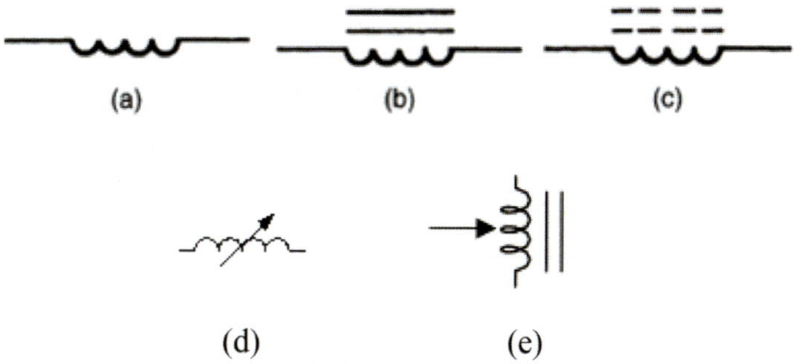

Figure 12-2 Schematic symbols used for representing inductors: (a) air core; (b) iron core; (c) powdered metal core; (d) variable air core; (e) variable iron core.

A **schematic symbols** of inductors a graphical representation of its construction. This symbol shows a coil of wire and some type of core material. **Figure 12-2** shows schematic symbols of five different inductor types that are in common use. The inductance of a coil can be either a fixed or variable value. A variable value is indicated by an arrow through the coil or pointing at it. The symbol of an inductor also indicates the type of core material used in its construction. A metal core is shown using two parallel lines drawn alongside the core. Broken parallel lines show that the core is made of some form of powdered metal. An absence of parallel lines alongside the coil indicates an air core.

12.2 Inductor Operation

When an external voltage source is first applied to the coil of an **inductor**, it causes a current to flow through the coil. This current establishes an **electromagnetic field** around the coil. As the value of the current **increases**, the electromagnetic field expands and increases its value. If the value of this current changes, it will cause a corresponding change in the magnetic field. A **decrease** in current causes the electromagnetic field to collapse and decreases its value. When the current is constant, the magnetic field remains stationary and constant in value as well.

Whenever there is relative motion between a magnetic field and a coil, a voltage will be developed or induced in the coil. This voltage will oppose the initial energizing voltage applied to the coil. As a result, the **induced voltage** will have a reversed polarity. Its value will depend on the speed at which the magnetic field changes. The induced voltage is called **counter electromotive force (CEMF)** or **back EMF**. The ability or property of a coil to generate

CEMF is called **self-inductance** or simply inductance. **Inductance** is represented by the letter "L." The unit of inductance is the **henry (H)**. A 1-H inductor will cause a CEMF of 1 V to be developed when the current changes at 1 A per second.

In order for an **inductor** to function, it must have some form of energy applied to it. When this energy is applied, it causes a current flow in the coil which produces a magnetic field. The coil then has the ability to store the applied energy in its magnetic field. The amount of energy being stored is defined by the following equation:

$$W = \frac{1}{2} \times L \times I^2$$

where
W = energy stored in joules
L = coil inductance in henries
I = current flowing through the coil in amperes

A l-H coil has 5.8 A of current flowing through it. Determine the amount of energy stored in the coil.

$$W = \frac{1}{2} \times L \times I^2 = \frac{1}{2} \times 1 \times 5.8^2 = 168.2 \text{ joules.}$$

The energy applied to an **inductor** can be either DC or AC. The response of the inductor will depend on the type of input energy. DC energy is applied to an inductor shown in **Figure 12-3**. In **Figure 12-3(a)**, when no energy is applied, the inductor is inoperative.

In **Figure 12-3(b)**, energy is applied to the inductor by closing the switch (Sw). This causes an expanding **magnetic field** to be developed around the coil as indicated. Note the polarity of the resulting magnetic field which travels externally from the north to the **south pole** of the coil. As the magnetic field expands, it cuts across the coils of the inductor, which develops a CEMF. This CEMF opposes the applied energy source (V_S) only when the magnetic field is changing in value. After energy has been applied to the coil for a period of time, the magnetic field becomes stationary and no longer increases in value. A stationary magnetic field will maintain or store the energy that has been applied to it from the source.

In **Figure 12-3(c)**, the energy source is disconnected from the **inductor** by opening the switch (Sw). This causes the stationary magnetic field to collapse inward and decrease in value. This changing magnetic field will cut across the coils of the inductor, which develops a **CEMF**. This CEMF

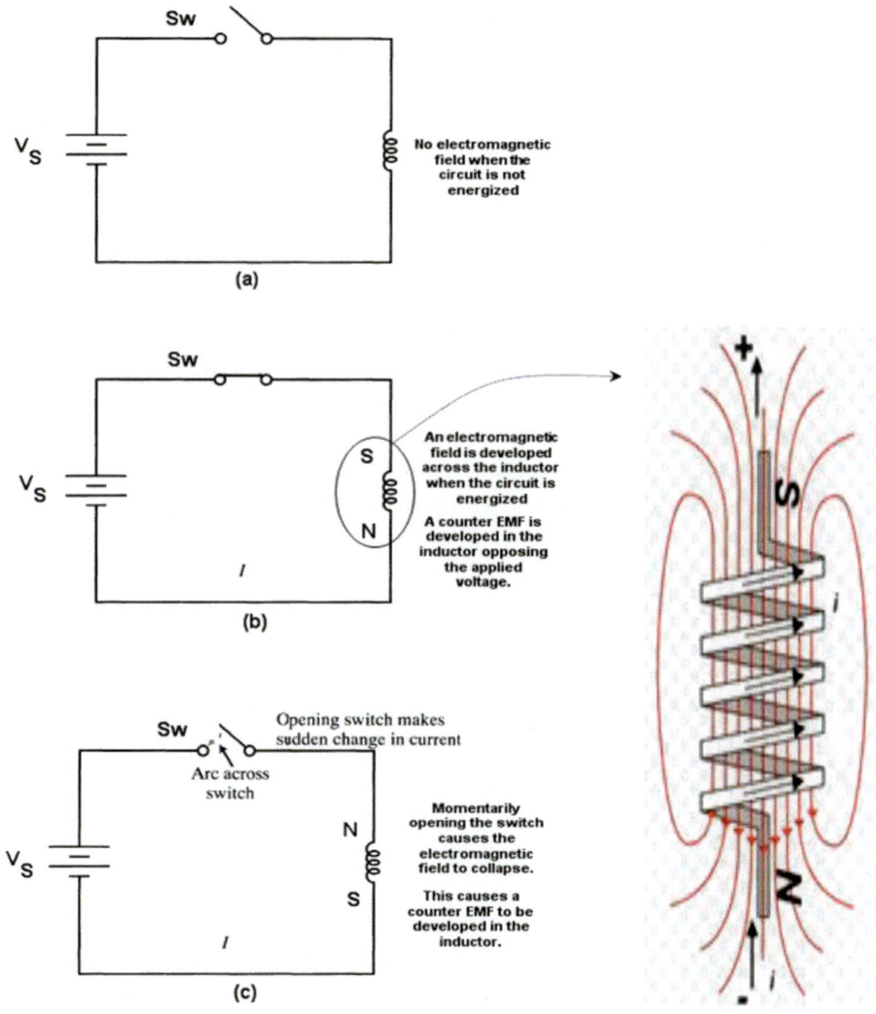

Figure 12-3 DC inductor circuit.

causes a continuation of current flow in the circuit even though the source (V_S) has been removed. This change in the magnetic field occurs almost instantaneously and results in a very high CEMF being developed across the contacts of the open switch. This could cause some arcing between the contacts of the switch while it is being opened. Note also that the polarity of the inductor momentarily changes as a result of the induced CEMF. After energy has been removed from the coil for a period of time, the magnetic field no longer exists.

When an inductor is powered by a DC source, it develops **CEMF** only when the circuit is initially energized and when the energy is removed from the inductor. If the value of the energy changes continuously, the inductor will also continuously develop a CEMF. This occurs when an inductor is energized by an AC source.

Phase Relationship in AC Inductor Circuits

A change in current through a coil causes a change in the **magnetic flux** around the coil. Voltage changes at the maximum rate when passing through its zero value at 0°, 180°, and 360°, as shown in **Figure 12-4**. The magnetic flux changes are also greatest at these times. The CEMF of the coil is at maximum value at these times also. Because CEMF opposes source voltage, it is directly opposite to the applied voltage in the diagram.

Lenz's law states that the **counter voltage (CEMF)** of an inductor always opposes a change in current. At 90°, the counter voltage is opposite in polarity to the applied voltage. This opposes the rise in current and causes current to equal zero. When the values of the counter voltage and the applied voltage are at 0°, the current is at a maximum value. This causes the current flow to lag behind the applied voltage by 90°. The DC resistance of a coil is generally a very small value. The largest opposition to AC current flow through a coil is the counter voltage (CEMF). The applied voltage is slightly higher than the CEMF. If they are equal, no current could flow. The resistance of the coil is usually not taken into account because its value is so small.

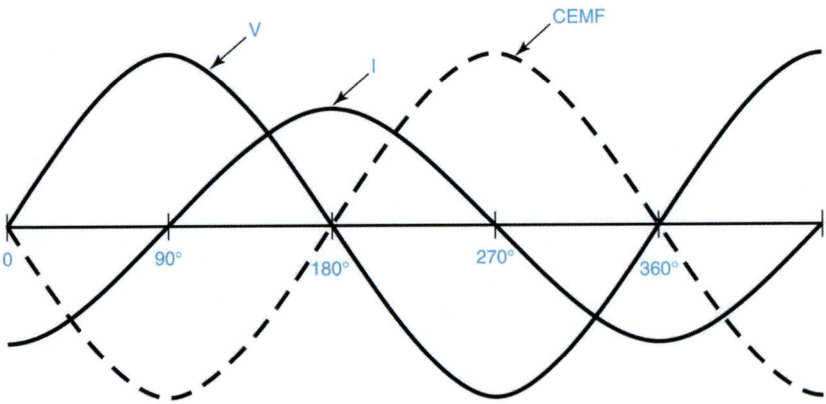

Figure 12-4 Inductive phase relationships.

12.3 Factors that Affect Inductance

The **inductance** of a coil depends on a number of factors based on its construction, which are shown in **Figure 12-5**.

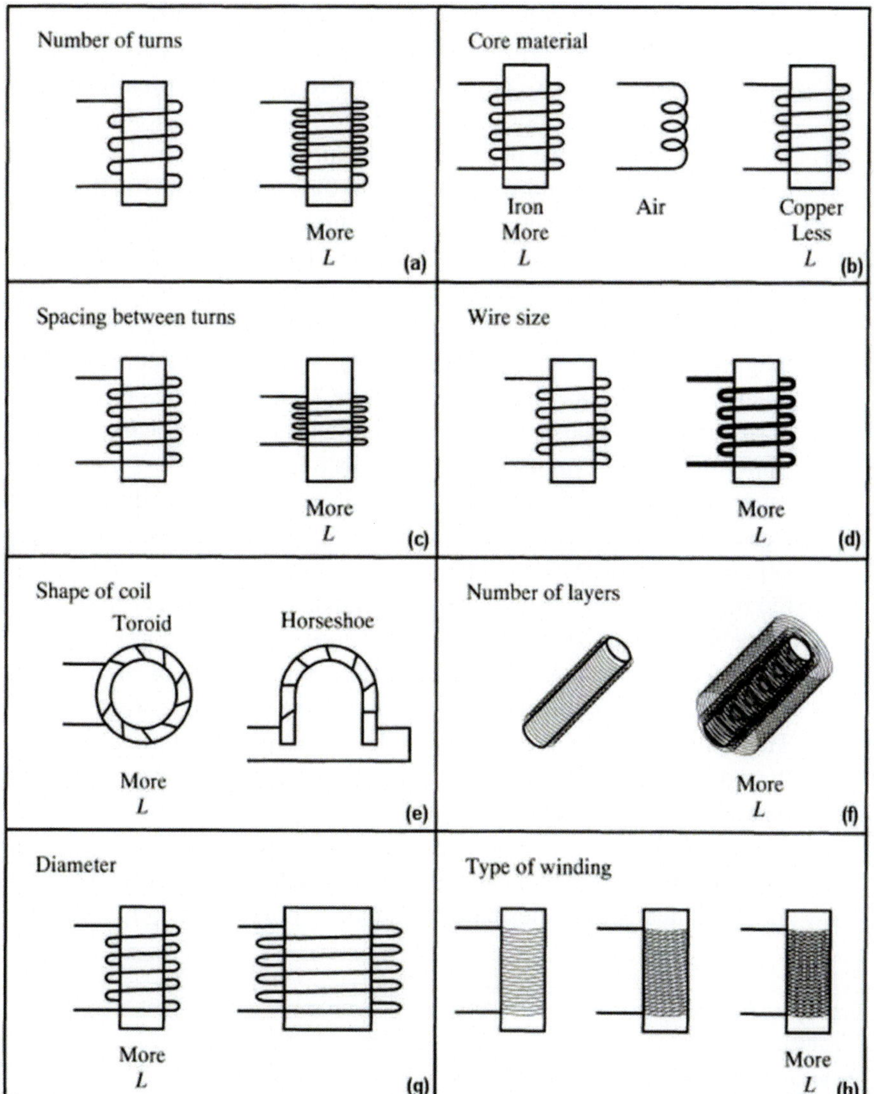

Figure 12-5 Factors affecting inductance.

Number of turns (N): As the number of turns increase, so does the inductance. Inductance is proportional to the square of the turns used in its construction. So, if the number of turns **increases** by a factor of 2, the inductance increases four times.

1. **Type of core material**: The magnetic flux passes through the core of an inductor. The type of core material determines its ability to conduct magnetic flux. This is referred to as the **permeability** of a material. Usually, the permeability of the core material is compared to that of air. The relative permeability of air is regarded as 1, and other materials are compared with this value. Certain materials such as soft iron will permit magnetic flux to pass through easily and concentrate the flux within the material. This type of material is considered to have a high permeability, and when used in the core, it will increase inductance. A diamagnetic core, such as copper, will weaken the magnetic field, lowering its permeability. When used as core material, this will decrease the inductance. **Figure 12-6** shows the permeability of different materials used in the construction of an inductor.
2. **Spacing between turns**: When the turns of a coil are close together, it causes the magnetic field to be more concentrated around the coil. This causes an increase in the flux density, which increases the inductance. When the spacing between the coils increases, it causes a reduction of the magnetic flux, which lowers the inductance.
3. **Wire size**: The size of the wire used in the construction of the coil influences the amount of current which can be carried by an inductor. A larger wire has lower resistance and permits more current to flow. This will produce a stronger magnetic field. A larger wire also makes the spacing between the turns closer. Therefore, a larger wire size results in more inductance.
4. **Shape of the core**: The shape of the core around which the coil is wound influences its inductance. The best shape for a coil is one where the north and south poles have the smallest possible **air gap**, allowing

Material	Permeability (N/A^2)
Free space (vacuum) Magnetic iron	$1 \times 4\pi \text{ x } 10^{-7} = 1.2566371 \times 10^{-6}$ $200 \times 4\pi \text{ x } 10^{-7} = 2.513274 \times 10^{-4}$
Permalloy (78.5% nickel, 21.5% iron)	$8{,}000 \times 4\pi \text{ x } 10^{-7} = .0251327$
Mumetal (75% nickel, 18% iron, 5% copper, 2% chromium)	$20{,}000 \times 4\pi \text{ x } 10^{-7} = 2.513274 \times 10^{-3}$

Figure 12-6 Permeability of selected materials.

a circulation of flux within the coil. The U-shaped or horseshoe core provides an excellent design and is used in many applications. However, the best shape for an inductor is the toroid, which looks like a doughnut. The **toroid** has the highest concentration of magnetic flux and the least leakage. The toroid coil is most popular when large values of inductance are needed for filters. In contrast, a bar-shaped construction of the core causes its magnetic field to be dispersed over a large area. This, in general, reduces the inductance.

5. **Number of layers**: When a coil is wound in multiple layers, it causes the magnetic field to be more concentrated. Consequently, a higher value of inductance will occur with increased layers.
6. **Diameter**: The diameter of the coil is closely related to the spacing of the turns. Inductance is increased with a more concentrated magnetic field.
7. **Type of winding**: Crisscrossing the windings improves the angle of the magnetic field. A right angle, 90°, induces the highest voltage. The crisscrossed coil is popular with miniature inductors.

12.4 Inductor Types

Figure 12-1 shows various types of **inductors**. Inductor types are based on their application. When used in high-frequency AC circuits, they are exclusively air core inductors for tuning purposes. The value of inductance can be adjusted to select or pass a specific frequency. Inductors are also used in low frequency circuitry. These include signal and high power applications. The inductors used in such applications are predominantly wire wound coils with a metal core. Evolving developments in inductors include thin-film construction techniques which are used for noise suppression filtering applications in integrated circuits. Multi-layer inductor structures that have a very small resistance are used for low transmission loss applications in mobile communication systems.

12.5 Inductor Ratings

Inductors have applications in both DC and AC circuits. A typical application of an inductor in a DC circuit is as a filter to help regulate fluctuations in the current. With some DC power supplies, such as the supplies for computer circuits, regulation is critical. A radio frequency filter circuit is a typical AC application. The inductor is used to select a specific frequency or to select a group of frequencies. Inductors have four basic ratings:

1. **Inductance value** measured in henrys (H) is usually specified in mH or μH.
2. **DC resistance** measured in ohms. Generally speaking, the DC resistance should be as low as possible, with typical values of 1 Ω or less.
3. **Maximum current** flow measured in amperes.
4. **Quality factor (Q)** of an inductor is a ratio of AC resistance to DC resistance. The AC resistance of a coil is called its inductive reactance, X_L. It depends on the value of the inductance L and the applied frequency. Q is measured at a particular frequency.
5. An inductor is often referred to as a **choke coil** because it has a high AC resistance and a low DC resistance. Thus, a choke coil blocks high-frequency AC and passes DC energy.

12.6 Inductor Connections

Similar to other electrical components, inductors may be connected together in various configurations. These include **series**, **parallel**, or **combination** arrangements of inductors. Since each inductor responds to magnetic field changes, there may be an inter-relationship between inductors due to field changes, which is referred to as mutual inductance. This must be taken into account when determining the total inductance of a configuration. **Inductors** can be connected so that their magnetic fields do or do not interact with each other. Formulas for calculating the **total inductance (L_T)** of a configuration must take into account any developed mutual inductance.

Inductors in Series

When inductors are connected in a **series** configuration, the end of one coil is connected to the beginning of the subsequent coil, so that there is only one path for the current. This is similar to a series resistor configuration. The **total inductance (L_T)** is determined by the sum of the individual inductances, when there is no interaction between the magnetic fields of individual inductors.

This is given by the following expression:

$$L_T = L_1 + L_2 + L_3 + \cdots + L_N$$

where L_T is the total inductance, L_1, …, L_N are individual inductances, and N is the number of inductors.

- Determine the total inductance of three inductors connected in series with the values: 35, 50, and 75 mH
  L_T = 35 mH + 50 mH + 75 mH = 160 mH.

Inductors in Parallel

When inductors are connected in a **parallel** configuration, the beginning of each coil is connected to a common connection point, and the ending of each coil is connected to another common connection point. Thus, there are as many paths for current flow as there are inductors in the circuit. This is similar to a parallel resistor configuration. The reciprocal of the total inductance ($1/L_T$) is determined by the sum of the reciprocals of individual inductances, when there is no interaction between the magnetic fields of individual inductors.

This is given by the following expression:

$$\frac{1}{L_T} = \frac{1}{L_1} + \frac{1}{L_2} + \frac{1}{L_3} + \cdots + \frac{1}{L_N}$$

where L_T is the total inductance, L_1, …, L_N are individual inductances, and N is the number of inductors.

In a special case with only two inductors L_1 and L_2, the total inductance L_T can be expressed as follows:

$$\frac{1}{L_T} = \frac{1}{L_1} + \frac{1}{L_2}$$

$$L_T = \frac{L_1 \times L_2}{L_1 + L_2}.$$

- A 40-μH inductor is connected in parallel with a 60-μH inductor. What is their combined inductance?

$$\frac{1}{L_T} = \frac{1}{40\mu} + \frac{1}{60\mu} = 25 \times 10^{-3} + 16.7 \times 10^{-3}$$

L_T = 24 μH.

Mutual Inductance

When inductors are connected, a property called **mutual inductance (M)** must be considered. Mutual inductance is the magnetic field interaction or flux linkage between coils. The amount of flux linkage is called the **coefficient of coupling (k)**. If all the lines of force of one coil cut across a nearby coil, it is called unity coupling. There are many possibilities, determined by coil placement, of coupling between coils. The amount of mutual inductance between coils is found by using the following formula:

$$\text{Mutual inductance } (M) = k \times \sqrt{L_1 \times L_2}\,.$$

The term k is the coefficient of coupling, which gives the amount of coupling. L_1 and L_2 are the inductance values of the coils. Mutual inductance should also be considered when two or more coils are connected together. Depending on the locations of the north and south poles, one inductor can either **aid** the field of the other inductor or it can **oppose** the other inductor's field. Phasing dots are used to indicate the direction in which the coil is wound. Usually, a dot indicates the beginning of the coil, and the unmarked end indicates the ending of the coil. **Figure 12-7** shows the use of phasing

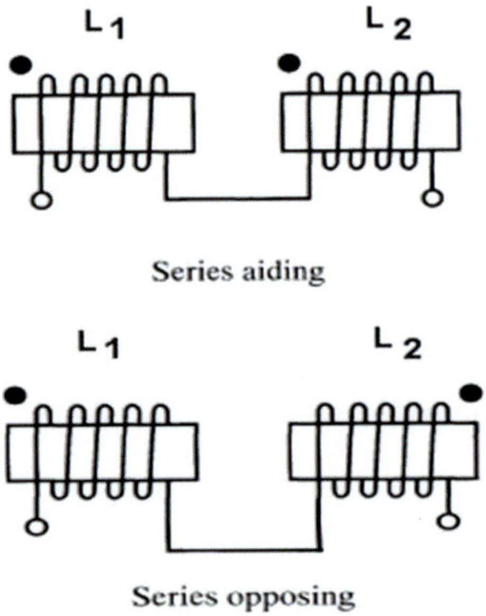

Figure 12-7 Inductors connected in series.

dots and two different wiring connections. Mutual inductance is also the fundamental principle used when designing and constructing transformers.

When inductors are connected so that the magnetic field of one inductor affects other inductors, the resulting mutual inductance changes the value of the total inductance. The effect of **mutual inductance** depends on the physical positioning of the inductors. Their distance apart, and the direction in which they are wound, affects mutual inductance. Inductors are connected in series or parallel with an aiding or opposing mutual inductance (M).

Series Inductor Connections with Mutual Induction

Aiding – A series aiding inductor circuit is shown in **Figure 12-7(a)**. Note that the end of the first coil (L_1) is connected at the beginning of the second coil (L_2). Both coils develop a mutually aiding magnetic field, which is expressed as +2M. The "+" indicates that the mutual inductance aids the self-inductances, "2" indicates the number of inductors, and "M" is the mutual inductance between L_1 and L_2. This is expressed as follows:

$$L_T = L_1 + L_2 + 2M.$$

Opposing – A series opposing inductor circuit is shown in **Figure 12-7(b)**. Note that the end of the first coil (L_1) is connected to the end of the second coil (L_2). Both coils develop a mutually opposing magnetic field, which is expressed as –2 M. The "–" indicates that the mutual inductance opposes the self-inductances, "2" indicates the number of inductors, and "M" is the mutual inductance between L_1 and L_2. This is expressed as follows:

$$L_T = L_1 + L_2 - 2M.$$

Parallel Inductor Connections with Mutual Induction

Aiding – A parallel aiding inductor circuit is shown in **Figure 12-9(a)**. Note that the beginning of both coils (L_1 and L_2) is connected to a common point and that the ending of both coils is connected to another common point. The "+M" term is identified with each coil. The "+" indicates an aiding mutual inductance of value "M." This is expressed as follows:

$$\frac{1}{L_T} = \frac{1}{L_1 + M} + \frac{1}{L_2 + M}.$$

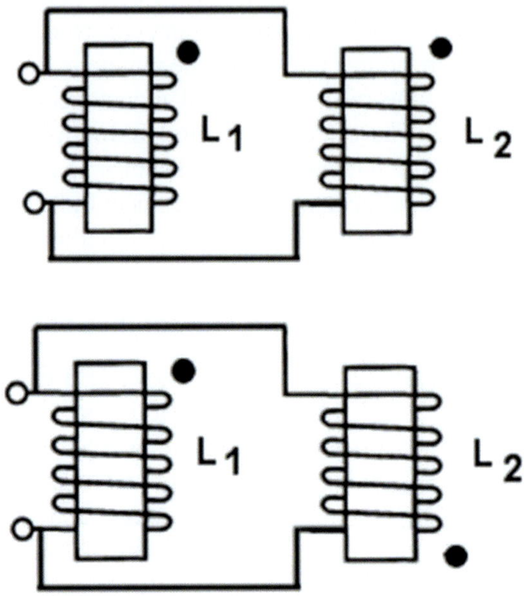

Figure 12-8 Inductors connected in parallel.

Opposing – A parallel opposing inductor circuit is shown in **Figure 12-9(b)**. Note that the beginning of the first coil (L_1) and the ending of the second coil (L_2) are connected to a common point, while the ending of the first coil and the beginning of the second coil is connected to another common point. The "–M" term is identified with each coil. The "–" indicates an opposing mutual inductance of value "M." This is expressed as follows:

$$\frac{1}{L_T} = \frac{1}{L_1 - M} + \frac{1}{L_2 - M}.$$

12.7 Inductor Voltage Dividers

An **inductive voltage divider** results from inductors being connected in **series**, as shown in **Figure 12-9**. The amount of voltage across the inductor is a ratio of its inductance to the total inductance multiplied by the applied AC voltage.

Voltage across L_1:

$$V_{L1} = \frac{L_1}{L_1 + L_2} \times V_A$$

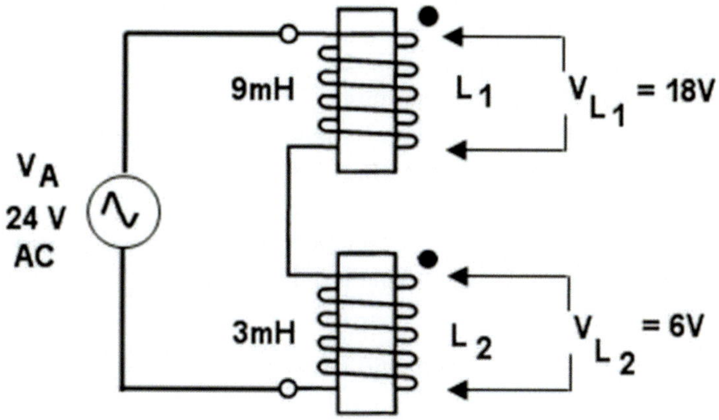

Figure 12-9 Inductive voltage divider.

where V_{L1} is the voltage across L_1, L_1 and L_2 are individual inductances, and V_A is the applied voltage.

Voltage across L_2:

$$V_{L2} = \frac{L_2}{L_1 + L_2} \times V_A$$

where V_{L2} is the voltage across L_2, L_1 and L_2 are individual inductances, and V_A is the applied voltage.

Verify the voltage readings in **Figure 12-10**.

The voltage across inductor L_1 is as follows:

$$V_{L1} = \frac{L_1}{L_1 + L_2} \times V_A = \frac{9m}{9m + 3m} \times 24 = 18\text{ V}.$$

The voltage across inductor L_2 is as follows:

$$V_{L2} = \frac{L_2}{L_1 + L_2} \times V_A = \frac{3m}{9m + 3m_2} \times 24 = 6\text{ V}.$$

12.8 Inductive Reactance

The opposition to current flow of an inductive device depends on the resistance of the wire and the magnetic properties of the circuit. This opposition due to the electromagnetic effect is called **inductive reactance (X_L)**, which is measured in ohms (Ω). It may be regarded as the AC resistance of an inductor. The inductive reactance varies directly with the applied frequency and the value of the inductance. This can be determined by the following formula:

$$X_L = 2 \times \pi \times f \times L$$

where
X_L = capacitive reactance in ohms
2π = mathematical expression of one sine wave (0°–360°)
f = frequency of the source in Hz
L = inductance in henrys

- Determine the inductive reactance of a 2-H inductor which is connected to an AC source that has a frequency of
- (a) 60 Hz.

$$X_L = 2 \times \pi \times f \times L = 2 \times \pi \times 60 \times 2 = 753.982\,\Omega.$$

(b) 6000 Hz

$$X_L = 2 \times \pi \times f \times L = 2 \times \pi \times 6000 \times 2 = 75{,}398.2 = 75.4\,\text{k}\Omega.$$

Note that these calculations show that when the **frequency** of the AC source **increases**, it causes a corresponding increase in the value of the inductive reactance of a circuit. At very high frequencies, this value is very large, while at very low frequencies (DC level), this value is very small. Hence, the **inductive reactance** is a variable factor, which depends on the frequency of the AC source voltage. At zero frequency (or DC), there is no opposition to current flow due to inductance. When this condition occurs, only the coil's DC resistance value limits current flow in the circuit. Since the DC resistance of a coil is usually very small, this current can become relatively large.

Many AC machines use magnetic circuits in one form or another. The **inductive reactance** of an AC circuit usually has more effect on current flow than resistance. An ohmmeter measures DC resistance only. Inductive reactance must be calculated or determined experimentally by a specialized meter. The total opposition of a circuit to current flow is called **impedance**. For a circuit that contains only an inductor, the impedance (Z) is the same as the inductive reactance (X_L). The calculation of the inductive reactance of a **pure inductive circuit** can be determined by using Ohm's law. This is given as follows:

$$X_L = \frac{V_L}{I_L}.$$

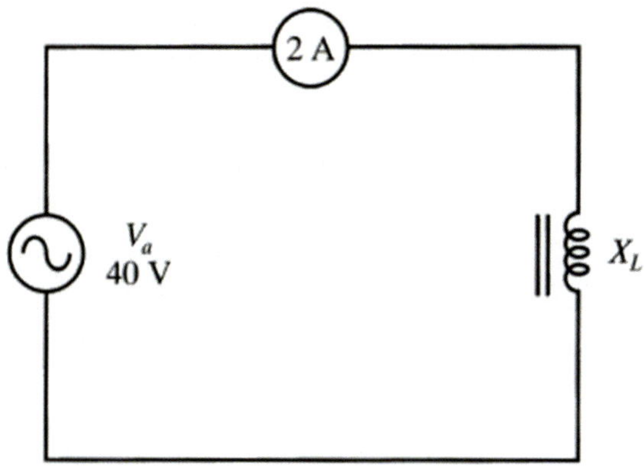

Figure 12-10 Determining the inductive reactance of a circuit.

- Determine the inductive reactance of the circuit shown in **Figure 12-10**. The circuit is energized by a $40V_{\text{RMS}}$ source, which causes a current of 2 A.

$$X_L = \frac{40}{2} = 20\Omega.$$

Inductive Reactance Connections

The total reactance of AC components connected in series or parallel is determined by the same formulas which are used for determining the total resistance of a circuit.

Series Inductive Reactance

The total reactance of series-connected inductors is determined by adding the individual inductive reactance values. This is similar to that of series-connected resistors.

$$X_{L_T} = X_{L_1} + X_{L_2} + X_{L_3} + \ldots X_{L_N}$$

X_{L_T} is the total inductive reactance, measured in ohms.

X_{L_1} through X_{L_N} are the individual inductive reactances, measured in ohms.

The total inductance of series-connected inductors in terms of the frequency and individual inductor values is given by

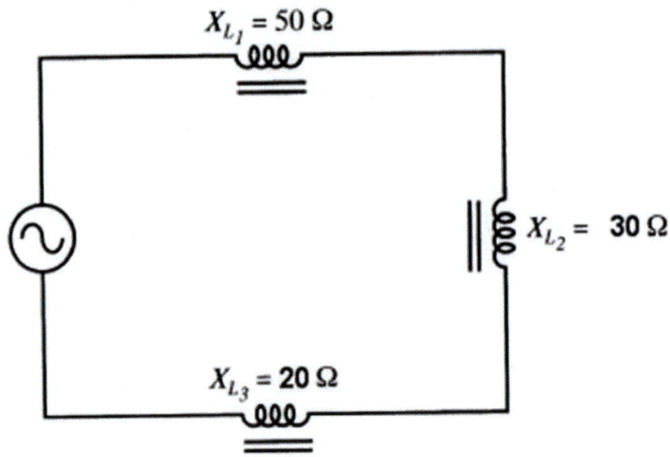

Figure 12-11 Inductive reactance of series-connected inductors.

$$X_{LT} = X_{L1} + X_{L2} + \cdots + X_{LN}$$
$$= 2\pi \times f \times L_1 + 2\pi \times f \times L_2 + \cdots + 2\pi \times f \times L_N.$$

Note that when the value of the inductors in a series configuration is given, total inductance is the sum of the individual inductances. This is given as follows:

$$L_T = L_1 + L_2 + L_3 + \cdots + L_N.$$

- Determine the total inductive reactance of the series-connected inductors as shown in **Figure 12-11**. It consists of 50-, 30-, and 20-Ω inductors connected in series.

  $$X_{LT} = X_{L1} + X_{L2} + X_{L3} = 50 + 30 + 20 = 100\,\Omega.$$

Parallel Inductive Reactance

The reciprocal of the total reactance of parallel-connected inductors is determined by adding the individual inductive reactance reciprocal values. This is similar to that of parallel-connected resistors.

$$\frac{1}{X_{L_T}} = \frac{1}{X_{L_1}} + \frac{1}{X_{L_2}} + \frac{1}{X_{L_3}} + \dots \frac{1}{X_{L_N}}$$

X_{L_T} is the total inductive reactance, measured in ohms.

X_{L_1} through X_{L_N} are the individual inductive reactances, measured in ohms.

The total inductance of parallel-connected inductors in terms of the frequency and individual inductor values is given:

$$\frac{1}{X_{LT}} = \frac{1}{X_{L1}} + \frac{1}{X_{L2}} + \cdots + \frac{1}{X_{LN}} = \frac{1}{2\pi \times f \times L_1} + \frac{1}{2\pi \times f \times L_2} + \cdots + \frac{1}{2\pi \times f \times L_N}.$$

Note that when the value of the inductors in a parallel configuration is given, the reciprocal of the total inductance equals the sum of the reciprocals of individual inductances. This is given as follows:

$$\frac{1}{L_T} = \frac{1}{L_1} + \frac{1}{L_2} + \frac{1}{L_3} + \cdots + \frac{1}{L_N}.$$

- Determine the total inductive reactance of parallel-connected inductors as shown in **Figure 12-12**. It consists of 50-, 30-, and 20-Ω inductors connected in parallel.

$$\frac{1}{X_{LT}} = \frac{1}{X_{L1}} + \frac{1}{X_{L2}} + \frac{1}{X_{L3}} = \frac{1}{50} + \frac{1}{30} + \frac{1}{20} = 0.2 + 0.033 + 0.05 = 0.283$$

$$X_{LT} = \frac{1}{0.283} = 3.53\,\Omega.$$

When the value of the total inductive reactance (X_{LT}) is known, the value of the total inductance L_T can be obtained by

$$X_{LT} = 2 \times \pi \times f \times L_T$$

$$L_T = \frac{X_{LT}}{2 \times \pi \times f}.$$

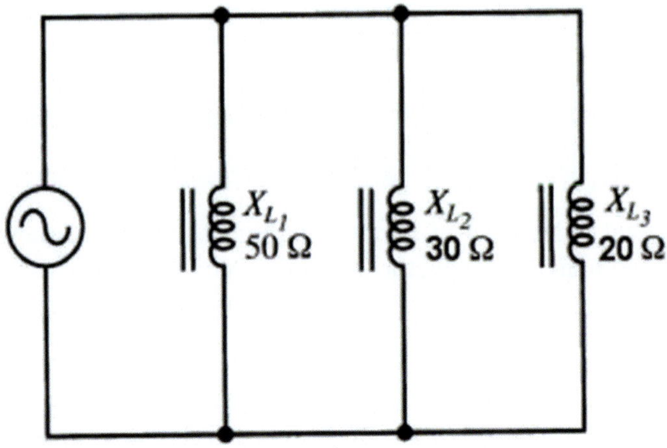

Figure 12-12 Inductive reactance of series-connected inductors.

12.9 Resistor and Inductor Circuits

The operation of an AC electrical circuit depends on the specific component being used. The simplest type of AC circuit consists of only one type of component. **Figure 12-13(a)** shows a circuit that has a resistor connected to an AC source of energy.

When the voltage applied to a circuit is **increased**, the current will increase, and when the voltage is decreased, the current will decrease. The waveforms of **Figure 12-13(b)** show the relationship between the voltage and current in a **resistive AC circuit**. Note that the voltage and current waveforms are in phase. This means that a change in the value of the applied voltage causes a corresponding change in current flow. The resulting power converted by the resistance is found by multiplying voltage times current (P = V × I). During the 0°–180° interval, both the voltage and current are positive, resulting in a **positive power** value (+P = +V × +I). During the 180°–360° interval, both the voltage and current are negative, resulting in a positive power value (+P = –V × –I). A resulting power curve for a resistive AC circuit is shown in **Figure 12-14(b)**.

Figure 12-13(c) shows a circuit that has an inductor connected to an AC source of energy. All motors, generators, and transformers have inductance. Inductance is due to the **counter electromotive force (CEMF)** produced when a magnetic field is developed around a coil of wire. The magnetic field produced around coils affects a circuit. Inductors store energy in their magnetic field. The CEMF produced by a magnetic field offers opposition to change in the current of a circuit. Thus, in an inductive circuit, voltage

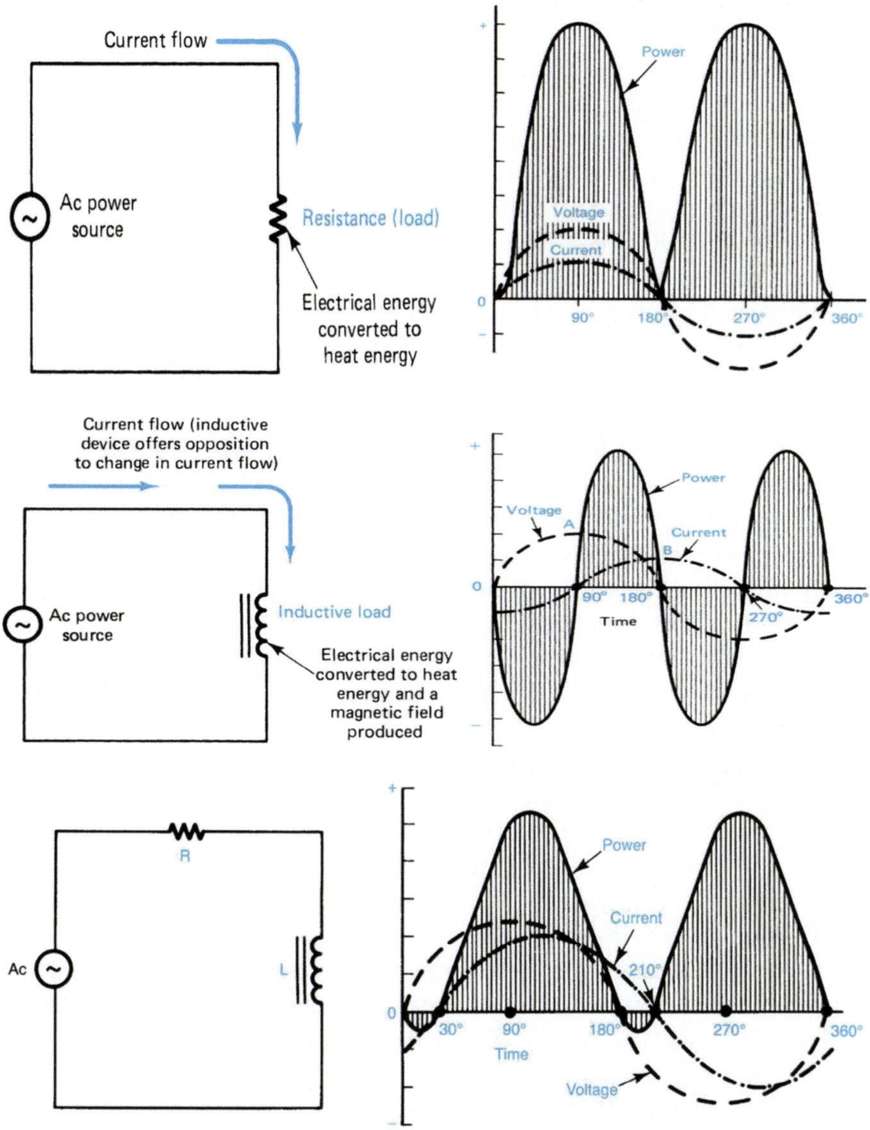

Figure 12-13 AC circuits and waveforms: (a) AC resistive circuit and (b) waveforms; (c) AC inductive circuit and (d) waveforms; (e) AC resistive–inductive (RL) circuit and (f) waveforms.

(V_L) leads the current (I_L). If the circuit was purely inductive (containing no resistance), the voltage leads the current by 90°. The voltage and current waveforms of a purely inductive circuit (no resistance) are shown in **Figure 12-13(d)**. The value of the voltage across an inductor depends on the rate of change of the current. The most rapid change in current occurs at the 90° and 270° positions as the polarity of the current changes. At these positions, maximum voltage is developed in the circuit. The rate of change of the current is very slow near the 0° and 180° positions and the voltage developed is very small. Thus, current leads voltage by 90° in a purely inductive circuit. During the 0°–90° interval, the voltage is positive, and the current is negative, resulting in a negative power value (–P = +V × –I).

During the 90°–180° interval, both the voltage and the current are positive, resulting in a positive power value (+P = +V × +I). During the 180°–270° interval, the voltage is negative and the current is positive, resulting in a negative power value (–P = –V × +I). During the 270°–360° interval, the current and voltage are both negative, resulting in a positive power value (+P = –V × –I). A resulting **power curve** for an **inductive AC circuit** is shown in **Figure 12-13(d). Negative power** means that electrical energy is returned from one load to the source without being converted to another form. It should be noted that during an operational cycle, the positive and negative power waveforms cancel each other. The power converted in a **purely inductive circuit** is equal to zero. As a result of this, no power is converted by a pure inductive AC circuit.

In practice, AC circuits may contain a combination of two or more components. In this regard, a circuit could have resistance and capacitance in its construction.

A sample **resistive–inductive (RL) circuit** is shown in **Figure 12-13(e)**. In an RL circuit, the current lags behind the voltage by a phase angle between 0° and 90°. If inductance in a circuit increases, the phase angle increases. The waveforms of **Figure 12-13(f)** show an *RL* circuit in which the current lags the voltage by 30°. During the 0°–30° interval, the voltage is positive, and the current is negative, resulting in a negative power value (–P = +V × –I). During the 30°–180° interval, the voltage is positive, and the current is positive, resulting in a positive power value (+P = +V × +I). During the 180°–210° interval, the voltage is negative, and the current is positive, resulting in a negative power value (–P = –V × +I). During the 210°–360° interval, the voltage is negative, and the current is negative, resulting in a positive power value (+P = –V × –I). No power is converted in the circuit during the 0°–30° and the 180°–210° intervals. In this *RL* circuit, most of the electrical energy supplied by the source is converted to another form of energy. Over a complete operational cycle, the circuit has net positive power. Thus, in practical *RL* circuits, a large part of the electrical power supplied by the source is utilized by the circuit components.

Series RL circuits

In any **series AC circuit**, the current (I) is the same in all parts of the circuit. In a **series RL circuit**, the value and phase of the current flowing through the resistor (I_R) and the inductor (I_L) is in phase. Recall that the voltage across a resistor (V_R) is in phase with the current. However, the voltage across an inductor (V_L) **leads** the current by 90°. The voltage drop across each component in a series circuit when added together equals the source voltage (V_A). In a series *RL* circuit, the addition of voltages must take into account the phase relationship between V_L and V_R. This relationship can be conveniently represented using **vectors or phasors** (rotating vectors). The length of the vector corresponds to the value of the voltage, and the direction of the vector corresponds to the **phase angle**. In vector diagrams, a leading phase angle is indicated by drawing the vector in the counter-clockwise direction, making an angle from a given reference (usually horizontal) line. Similarly, a lagging phase angle is indicated by showing the vector in the clockwise direction, making an angle from a given reference line. Since the phase difference between V_L and V_R is 90°, a **right-angle triangle** is used to represent these values. The two voltages, V_L and V_R, form the sides of the right-triangle and the **applied voltage (V_A)** is represented by the **hypotenuse**. This triangle is referred to as a **voltage triangle**.

The values of the voltage (V_A) and current (I) in a series circuit are used to determine the **impedance (Z)** offered by the circuit. This is given as $Z = \frac{V_A}{I}$. The **inductive reactance (X_L)** is obtained by using $X_L = \frac{V_L}{I}$, and the **resistance (R)** is obtained by using $R = \frac{V_R}{I}$. This results in an **impedance triangle** which is similar in shape to the voltage triangle. The inductive reactance (X_L) and the resistance (R) form the sides of the triangle and the impedance (Z) is represented by the hypotenuse.

If the values of the resistance and inductive reactance of an **AC series-connected RL circuit** are known, an impedance triangle can be constructed. These values are used to determine the **impedance** of the circuit. Next, by using the values of the applied voltage and the impedance, the current of the circuit can be evaluated. The value of the current and that of the resistance or reactance are used to determine the voltage drop across the circuit components. A **voltage triangle** can then be constructed from the values of the voltage drops and the applied voltage.

General Procedure for Solving AC Series-connected RL Circuit Problems

1. Draw the circuit diagram. Label with known values.
2. If the inductive reactance is not given, it can be determined by using $X_L = 2 \times \pi \times f \times L.$
3. Draw an impedance triangle to calculate Z.
4. Draw the R "vector" horizontally. From the tip of the R vector, draw the X_L vector vertically downward.
5. Vectorially add the vectors and then draw Z.
6. Use the Pythagorean theorem to calculate the value of Z by using the formula: $Z = \sqrt{R^2 + X_L{}^2}.$
7. Calculate the current in the circuit. $I = \dfrac{V}{Z}.$
8. Calculate the voltage drop across each component.
   Voltage across resistor: $V_R = I \times R$
   Voltage across inductor: $V_L = I \times X_L.$
9. Draw a vector diagram of the voltages. Use I as the horizontal reference. Draw V_R in phase with the current. Draw V_L vertically upward as the inductor voltage leads the current by 90°.
10. Vectorially add the voltages to obtain the applied voltage. $V_A = \sqrt{V_R{}^2 + V_L{}^2}.$
11. Use the Pythagorean theorem to calculate the value of the applied voltage as a check on the solution. Using the proper scale while drawing the vectors makes it possible to spot any errors in the calculations.
12. Calculate the phase angle. This requires use of the trigonometric identity, cos θ, which is given as follows:

    $$\cos\theta = \left(\frac{\text{opposite}}{\text{hypotenuse}}\right) = \left(\frac{V_R}{V_A}\right)$$

13. The phase angle can be obtained by taking the inverse of the cosine. This is indicated as $\cos^{-1}$:

    $$\theta = \cos^{-1}\left(\frac{V_R}{V_A}\right)$$

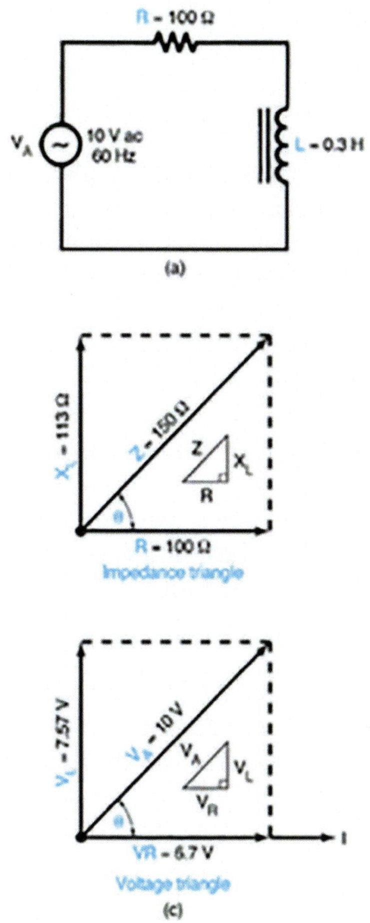

Finding circuit values:

1. Find inductive reactance (X_L):

$$X_L = 2\pi \cdot f \cdot L$$
$$= 6.28 \times 60 \times 0.3$$
$$= 113\ \Omega$$

2. Find impedance (Z):

$$Z = \sqrt{R^2 + X_L^2}$$
$$= \sqrt{100^2 + 113^2}$$
$$= \sqrt{10{,}000 + 12{,}769}$$
$$= \sqrt{22{,}769}$$
$$= 150\ \Omega$$

3. Check to be sure that Z is larger than R or X_L.

4. Find total current (I_T):

$$I_T = \frac{V}{Z} = \frac{10\text{ V}}{150\ \Omega} = 0.067\text{ A}$$

5. Find the voltage across R (V_R):

$$V_R = I \times R$$
$$= 0.067\text{ A} \times 100\ \Omega$$
$$= 6.7\text{ V}$$

6. Find the voltage across L (V_L):

$$V_L = I \cdot X_L$$
$$= 0.067\text{ A} \times 133\ \Omega$$
$$= 7.57\text{ V}$$

7. Check to see that

$$V_A = \sqrt{V_R^2 + V_L^2}$$
$$10\text{ V} = \sqrt{6.7\text{ V}^2 + 7.57\text{ V}^2}$$
$$= \sqrt{44.89 + 57.3}$$
$$= \sqrt{102.2}$$
$$\approx 10.1\text{ V}^*$$

($\approx$ approximately equal to)

8. Find circuit phase angle (θ):

$$\text{cosine } \theta = \frac{\text{adjacent}}{\text{hypotenuse}} = \frac{V_R}{V_A} = \frac{6.7}{10}$$
$$= 0.67$$
$$= \text{inverse cosine } 0.67$$
$$= 48^\circ$$

*Slight difference is due to rounding off of numbers as they are calculated.

(b)

Figure 12-14 AC series *RL* circuit (a) diagram; (b) procedure for finding circuit values; (c) impedance and voltage **triangles**.

A solved example of an AC series RL circuit, including the voltage and impedance triangles, is shown in **Figure 12-14**.

Parallel RL circuits

The basic formulas used with **parallel AC circuits** are different from those of series circuits. The **impedance (Z)** of a parallel circuit is less than individual branch values of resistance and capacitive reactance. There is no impedance triangle for parallel circuits; instead, an **admittance triangle** may be drawn which uses the reciprocal of the circuit impedance (Z), that of the **inductive reactance (X_L)** and the **resistance (R)**. The voltage of a parallel AC circuit is the same across each branch. A **right triangle** is drawn to show the currents in the capacitive and resistive branches of the parallel RL circuit. The currents through the branches of a parallel AC circuit are shown by a right triangle called a **current triangle**.

The current through the **inductor (I_L)** is shown **lagging** the current through the resistor (I_R) by 90°. Because these values form a right triangle, the total current may be found by using the following formula:

$$I_T = \sqrt{I_R^{\,2} + I_L^{\,2}}.$$

This method is used to find currents in parallel RL circuits. When components are connected in parallel, finding impedance is more difficult. An impedance triangle cannot be used. A method that can be used to find impedance is to use an **admittance triangle**. The following quantities are plotted on the triangle:

admittance: $Y = \dfrac{1}{Z}$;

conductance: $G = \dfrac{1}{R}$; $B_L = \dfrac{1}{X_L}$

capacitive susceptance: .

These quantities are the **inverse** of each type of opposition to AC. Because **total impedance (Z)** is the smallest quantity in a parallel AC circuit, its **reciprocal** ($Y = \dfrac{1}{Z}$) becomes the largest quantity on the admittance triangle (just as ½ is larger than ¼). The values are in Siemens. Parallel AC circuits are similar in several ways to series AC circuits except that the applied voltage is used as the reference, instead of the current that is used as a reference in series AC circuits.

General Procedure to Solve AC Parallel RL Circuit Problems

1. Draw the circuit diagram. Label the known values.
2. Label the branch currents as I_R, I_L, and I_T depending on the kind of component in each branch.
   If the inductive reactance is not given, solve for it using:
   $X_L = 2 \times \pi \times f \times L.$
3. Use Ohm's law to calculate the current in each branch.
4. Draw a vector diagram of the currents. Use the applied voltage as the horizontal reference. Draw the total I_R in phase with the voltage. Draw the I_L vertically upward as the current flowing in the capacitor leads the current flowing in the resistor.
5. Vectorially add the currents to obtain the total current, I_T, delivered by the source.
6. Use the Pythagorean theorem to calculate the value of I_T. Drawing the vectors approximately to scale will allow you to pick up any gross errors you may have made in calculations.
7. Solve for the total impedance of the parallel circuit using:

$$Z_T = \frac{V}{I_T}.$$

A solved example of an **AC parallel RL circuit**, including voltage and admittance triangles, is shown in **Figure 12-15**.

12.10 Troubleshooting Inductors

Inductors are constructed using several conductors or turns of wire wound around a common core. The individual turns are insulated from each other and the core material. **Metal core inductors** are more susceptible to faults as compared to air core inductors. The coil of an inductor generally has a low resistance value. Troubleshooting an inductor generally involves some evaluation of its resistance. The most significant **resistance measurement** is between the coil and the core. Ideally, this resistance should be infinite or an extremely high value. A lower resistance value indicates that insulation breakdown of the coil has occurred. The resistance between the terminals of the inductor should also be evaluated. As the resistance is tested using a

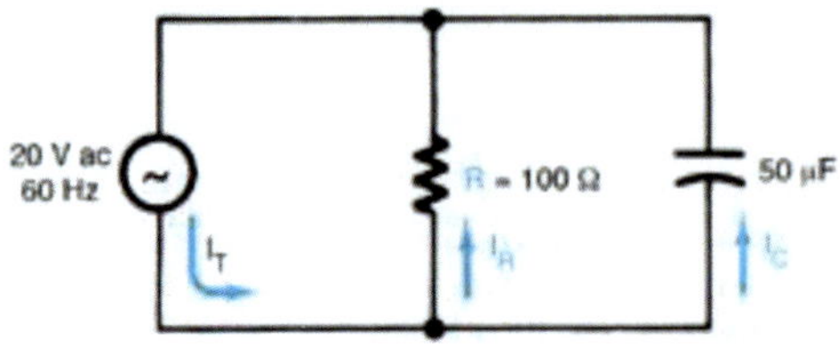

(a)

Finding circuit values:

1. Find capacitive reactance (X_C):

$$X_C = \frac{1{,}000{,}000}{2\pi \cdot f \cdot C\,(\mu F)}$$

$$= \frac{1{,}000{,}000}{6.28 \times 60 \times 50}$$

$$= 53\ \Omega$$

2. Find current through R (I_R):

$$I_R = \frac{V}{R} = \frac{20\ V}{100\ \Omega} = 0.2\ A$$

3. Find current through C (I_C):

$$I_C = \frac{V}{X_C} = \frac{20\ V}{53\ \Omega} = 0.377\ A$$

4. Find total current (I_T):

$$I_T = \sqrt{I_R^2 + I_C^2}$$

$$= \sqrt{0.2^2 + 0.377^2}$$

$$= \sqrt{0.04 + 0.142}$$

$$= \sqrt{0.182}$$

$$= 0.427\ A$$

5. Check to see that I_T is larger than I_R or I_C.

6. Find impedance (Z):

$$Z = \frac{V}{I_T} = \frac{20\ V}{0.427\ A} = 46.84\ \Omega$$

7. Check to see that Z is less than R or X_C.

8. Find circuit phase angle (θ):

$$\text{cosine}\ \theta = \frac{\text{adjacent}}{\text{hypotenuse}} = \frac{G}{Y} = \frac{10}{21.3}$$

$$= 0.469$$

$$\theta = \text{inverse cosine } 0.469$$

$$= 62^\circ$$

(b)

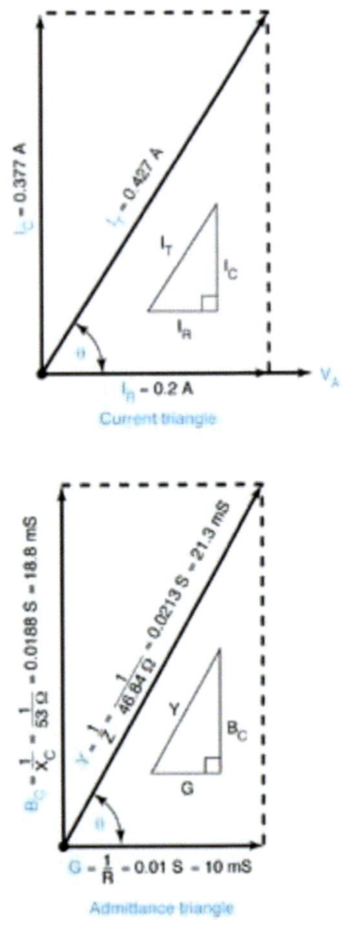

(c)

Figure 12-15 Parallel *RLC* circuit example: (a) circuit; (b) procedure for finding circuit values; (c) circuit triangles.

DC **ohmmeter**, which has a frequency of 0 Hz, the inductance value should be low or zero. Thus, the value of the resistance of the coil is usually very low. An **infinite** or extremely high value of resistance between the terminals indicates that the coil is open.

Since an inductor is primarily an AC component, its behavior should also be evaluated with AC energy applied. As the f**requency** of the AC is **increased**, the value of the AC resistance will increase proportionately. The resulting current of the circuit will show a corresponding **decrease** in value with an increase in the applied frequency.

Summary

- An inductor consists of several coils or turns of wire wound around some type of core material.
- When electrical energy is applied to the coils of an inductor, it will cause a resulting current flow which produces a magnetic field around each coil.
- Inductance (L) refers to the ability of a device to oppose changes in current flow due to energy stored in an electromagnetic field.
- The physical construction of an inductor determines how much magnetic flux is produced.
- The fundamental unit of inductance is the Henry (H), and common units include mH and μH.
- In an inductor, a changing magnetic field produces a counter electromotive force (CEMF).
- When DC is applied to an inductor, it develops a CEMF only when the circuit is initially energized and when the energy is removed or changed.
- When AC is applied to an inductor, it will continuously develop a CEMF in response to a continuously changing magnetic field.
- The main factors that influence the inductance of a coil are the number of turns, the type, and shape of the core material.
- When inductors are connected in series, the beginning of one coil is connected to the end of the subsequent coil; there is only one path for the current. The total inductance of the configuration is the sum of the individual inductances.
- When inductors are connected in parallel, the beginning of each coil is connected to a common connection point, and the ending of each coil is connected to another connection point. This provides multiple

paths for the current. The reciprocal of the total inductance ($1/L_T$) is equal to the sum of the individual reciprocal inductances.

- Mutual inductance (M) represents a magnetic flux linkage between coils which can be aiding or opposing.
- When inductors are connected in series and energized by an AC source, the voltage drop across each inductor depends on the ratio of the inductance values.
- Inductive reactance (X_L) which is measured in ohms is the AC resistance offered by an inductor.
- Inductive reactance is directly proportional to the value of the inductance and the frequency applied to it.
- In a pure resistive AC circuit, the voltage and current are in phase.
- In a pure inductive circuit, the voltage leads the current by 90°.
- Vector or phasor diagrams are commonly used to represent the phase relationship between voltages and currents in AC circuits.
- In a series RL circuit, the current flowing through circuit components is the same.
- In a parallel RL circuit, the voltage across the circuit components is the same.
- No power is consumed by a purely inductive circuit.
- Practical inductor circuits are very efficient in converting power.
- Inductors are susceptible to faults which include shorting of the coil to the core, or open coil condition between its terminals. These faults can be identified with an ohmmeter or a test circuit.

Formulas

Series inductors connection:

$$L_T = L_1 + L_2 + L_3 + \cdots + L_N.$$

Parallel inductor connection:

$$\frac{1}{L_T} = \frac{1}{L_1} + \frac{1}{L_2} + \frac{1}{L_3} + \cdots + \frac{1}{L_N}.$$

Parallel connection of two inductors:

$$\frac{1}{L_T} = \frac{1}{L_1} + \frac{1}{L_2}; \; L_T = \frac{L_1 \times L_2}{L_1 + L_2}.$$

Mutual inductance:

$$M = k \times \sqrt{L_1 \times L_2}.$$

Series inductor connection with aiding mutual inductance:

$$L_T = L_1 + L_2 + 2M.$$

Series inductor connection with opposing mutual inductance:

$$L_T = L_1 + L_2 - 2M.$$

Parallel inductor connection with aiding mutual inductance:

$$\frac{1}{L_T} = \frac{1}{L_1 + M} + \frac{1}{L_2 + M}.$$

Parallel inductor connection with opposing mutual inductance:

$$\frac{1}{L_T} = \frac{1}{L_1 - M} + \frac{1}{L_2 - M}.$$

Series inductor voltage divider formed by L_1 and L_2:

$$V_{L1} = \frac{L_1}{L_1 + L_2} \times V_A$$

$$V_{L2} = \frac{L_2}{L_1 + L_2} \times V_A.$$

Inductive reactance:

$$X_L = 2 \times \pi \times f \times L$$

$$X_L = \frac{V_L}{I_L}.$$

Series inductive reactance

$$X_{L_T} = X_{L_1} + X_{L_2} + X_{L_3} + \ldots X_{L_N}$$

X_{L_T} is the total inductive reactance, measured in ohms.

X_{L_1} through X_{L_N} are the individual inductive reactances, measured in ohms.

Parallel inductive reactance:

$$\frac{1}{X_{L_T}} = \frac{1}{X_{L_1}} + \frac{1}{X_{L_2}} + \frac{1}{X_{L_3}} + \ldots \frac{1}{X_{L_N}}$$

X_{L_T} is the total inductive reactance, measured in ohms.

X_{L_1} through X_{L_N} are the individual inductive reactances, measured in ohms.

Total inductive reactance:

$$X_{LT} = 2 \times \pi \times f \times L_T$$

$$L_T = \frac{X_{LT}}{2 \times \pi \times f}.$$

In an RL series circuit:

Impedance: $Z = \sqrt{R^2 + X_L{}^2}$

Current: $I = \frac{V}{Z}.$

Voltage drop across resistor: $V_R = I \times R.$

Voltage drop across inductor: $V_L = I \times X_L.$

Applied voltage: $V_A = \sqrt{V_R{}^2 + V_L{}^2}.$

Phase angle by which voltage leads current: $\theta = \cos^{-1}\left(\frac{V_R}{V_A}\right)$

In an RL parallel circuit:

$$I_T = \sqrt{I_R{}^2 + I_L{}^2}.$$

Problems

Determine the total inductance and inductive reactance of a circuit which has 4- and 16-H inductors connected in series with no mutual inductance. This circuit is energized with a 5-V, 50-Hz AC source.

[$L_T = L_1 + L_2 = 4 + 16 = 20\,\text{H}$

$$X_{LT} = X_{L1} + X_{L2} + \cdots + X_{LN}$$
$$= 2\pi \times f \times L_1 + 2\pi \times f \times L_2 + \cdots + 2\pi \times f \times L_N$$
$$= 2\pi \times f \times (L_1 + L_2) = 2 \times \pi \times 50 \times 20 = 6283.18\,\Omega.$$

To verify the value of the total inductance:

$$L_T = \frac{X_{LT}}{2 \times \pi \times f} = \frac{6283.18}{6.283 \times 50} = 20\,\text{H}\,].$$

- Determine the total inductance and inductive reactance of a circuit which has 4- and 16-H inductors connected in parallel with no mutual inductance. This circuit is energized with a 5-V, 50-Hz AC source:

$$[\frac{1}{L_T} = \frac{1}{L_1} + \frac{1}{L_2};$$

$$L_T = \frac{L_1 \times L_2}{L_1 + L_2} = \frac{4 \times 16}{4 + 16} = \frac{64}{20} = 3.2\,\text{H}$$

$$\frac{1}{X_{LT}} = \frac{1}{X_{L1}} + \frac{1}{X_{L2}} = \frac{1}{2\pi \times f \times L_1} + \frac{1}{2\pi \times f \times L_2} = \frac{1}{2\pi \times 50 \times 4} + \frac{1}{2\pi \times 50 \times 16}$$

$$\frac{1}{X_{LT}} = 7.957 \times 10^{-4} + 1.989 \times 10^{-4}$$

$$X_{LT} = \frac{1}{9.946 \times 10^{-4}} = 1005.38\,\Omega.$$

To verify the value of the total inductance:

$$L_T = \frac{X_{LT}}{2 \times \pi \times f} = \frac{1005.38}{6.283 \times 50} = 3.2\,\text{H}\,].$$

Determine the mutual inductance (M) of a 4- and a 16-H coil connected in a series aiding configuration, with the coefficient of coupling (k) of 0.5.

$$[\text{M} = k \times \sqrt{L_1 \times L_2} = 0.5 \times \sqrt{4 \times 16} = 0.5 \times \sqrt{64} = 0.5 \times 8 = 4\,\text{H}]$$

Determine the total inductance of the circuit which has a 4- and a 16-H coil connected in a series aiding configuration, with the coefficient of coupling (k) of 0.5:

[$L_T = L_1 + L_2 + 2M = 4 + 16 + 2 \times 4 = 20 + 8 = 28$ H].

Determine the total inductance of the circuit which has a 4- and a 16-H coil connected in a parallel aiding configuration, with the coefficient of coupling (k) of 0.5:

$$[\frac{1}{L_T} = \frac{1}{L_1 + M} + \frac{1}{L_2 + M} = \frac{1}{4+4} + \frac{1}{16+4} = \frac{1}{8} + \frac{1}{20} = 0.125 + 0.05 = 0.625$$

$$L_T = \frac{1}{0.625} = 1.6\,\text{H}]$$

Refer to **Figure 12-10**. If the value of the inductance L_1 = 25 mH and L_2 = 50 mH, determine the values of V_{L1} and V_{L2}:

$$[V_{L1} = \frac{L_1}{L_1 + L_2} \times V_A = \frac{25m}{25m + 50m} \times 24 = \frac{25m}{75m} \times 24 = 8\,\text{V}$$

$V_{L2} = V_A - V_{L1} = 24 - 8 = 16$ V].

Refer to **Figure 12-14**. If the value of R = 500 Ω and L = 2 H, determine the value of voltages developed across the circuit components, and the phase angle, , between the current and voltage in the circuit.

[Inductive reactance: $X_L = 2 \times \pi \times f \times L = 2 \times \pi \times 60 \times 2 = 753.98\,\Omega$.

Impedance: $Z = \sqrt{R^2 + X_L^{\,2}} = \sqrt{500^2 + 753.98^2} = \sqrt{818489.2} = 904.7\,\Omega$

Series current: $I = \frac{V}{Z} = \frac{10}{904.7} = 0.011\,\text{A}$.

$V_R = I \times R = 0.011 \times 500 = 5.5\,\text{V}$

$V_L = I \times X_L = 0.011 \times 753.98 = 8.4\,\text{V}$.

Verify voltages add up to the supply voltage:

$V_A = \sqrt{V_R^{\,2} + V_L^{\,2}} = \sqrt{5.5^2 + 8.4^2} = \sqrt{100.81} = 10.04\,\text{V}$.

Phase angle: $\theta = \cos^{-1}\left(\frac{V_R}{V_A}\right) = \cos^{-1}\left(\frac{5.5}{8.4}\right) = \cos^{-1} 0.6548 = 54.55^\circ$ lagging].

Self-examination/Answers

1. The broken parallel lines on an inductor symbol indicate that the core material is __________.
2. The absence of parallel lines on an inductor symbol indicates that the core material is _________.
3. In an inductor, coils of wire are wound around a common ____________.
4. When DC energy is first applied to an inductor, it causes a(n) ______________magnetic field to be developed.
5. A __________is developed when there is relative motion between a coil and a magnetic field.
6. _________________ developed in an inductor is considered to be a self-induced voltage.
7. When AC energy is applied to an inductor, a continuously changing ____________ field is developed.
8. In an inductor energized by AC energy, the voltage across the inductor ________the current by ________°.
9. When the number of turns of an inductor increases, the value of its inductance ____________.
10. The type of core material that has a relative permeability of 1 is __________.
11. The permeability of an iron core is (greater than, equal to, smaller than) the permeability of air.
12. An inductor in the form of a _____________ core has the greatest concentration of magnetic lines of force.
13. __________ core inductors are primarily used in high-frequency AC circuits.
14. Inductors with coils wound around __________ cores are primarily used in low frequency and high power circuit applications.
15. Which of the three inductors, $L_1 = 5$ H, $L_2 = 5$ mH, and $L_3 = 5$ μH, has the highest value?
16. The DC resistance of an inductor is typically very (small/large).
17. The AC resistance of an inductor is typically very (small/large).
18. An inductor is sometimes also referred to as a ________coil.
19. Determine the total inductance of a circuit which has three 5 mH connected in series with no mutual inductance.
20. Determine the total inductance of a circuit which has three 5 mH connected in parallel with no mutual inductance.

21. _______________ inductance is the magnetic field interaction or flux linkage between coils in a circuit.
22. The magnetic field of one coil may either ___________ or ___________ the magnetic field of another coil in a circuit and must be taken into account when determining the total inductance.
23. A (series, parallel) connection of inductances in an AC circuit will function as a voltage divider.
24. If two inductances of equal value are connected in series with an AC voltage source (V_S), the voltage developed across each inductor will be (0, V_S, $2V_S$, $\frac{V_S}{2}$).
25. Refer to **Figure 12-9**. If the supply voltage applied to the circuit is reduced to 12 V, the voltage developed across inductor L_1 will be _________ and across L_2 will be _________.
26. Determine the inductive reactance of a circuit which has a 100-Hz source connected to a 2-mH inductor.
27. Referring to **Figure 12-10**, if the value of the AC source voltage is 20 V and that of current is 3A, determine the value of X_L.
28. Determine the total inductive reactance of a circuit which has two inductors having an inductive reactance of 10 and 20 □ connected in:
    (a) series
    (b) parallel
29. When the polarity of the inductor voltage and current is negative, the polarity of the power developed by the inductor is (positive, negative).
30. In a practical RL circuit, the voltage (leads, lags) behind the current by a phase angle between ______° and ______.
31. In a series RL circuit, the (voltage across, current through) the components is the same, whereas in a parallel RL circuit, the (voltage across, current through) the components is the same.
32. In a series RL circuit, the impedance (Z) of the circuit can be determined graphically by using the ______________triangle.
33. The resistance between the coil and the core material of an inductor should ideally be __________.
34. The resistance between the two terminals of an inductor should be (low/medium/high).
35. If the resistance between the terminals of an inductor is infinite, this indicates that the coils of the inductor are (shorted/normal/open).

Answers

1. powdered metal
2. air
3. core
4. expanding
5. voltage
6. CEMF or counter electromotive force
7. magnetic
8. leads, 90
9. increases
10. air
11. greater than
12. toroid
13. air
14. metal
15. L_1
16. small
17. large
18. choke
19. 15 mH
20. 1.667 mH
21. Mutual
22. aid, oppose
23. series
24. $\frac{V_S}{2}$
25. 9 V, 3 V
26. $X_L = 2 \times \pi \times f \times L = 2 \times \pi \times 100 \times 2 \times 10^{-3} == 1.257\,\Omega$
27. $X_L = \frac{V_L}{I_L} = \frac{20}{3} = 6.66\,\Omega$
28. (a) series [$X_{LT} = X_{L1} + X_{L2} = 10 + 20 = 30\,\Omega$]

    (b) parallel [$\frac{1}{X_{LT}} = \frac{1}{X_{L1}} + \frac{1}{X_{L2}} = \frac{1}{10} + \frac{1}{20} = 0.1 + 0.05 = 0.15$

    $X_{LT} = \frac{1}{0.15} = 6.66\,\Omega$]

29. positive
30. leads, 0, 90
31. current through, voltage across
32. impedance
33. infinite
34. low
35. open

Problems Answers

$$L_T = L_1 + L_2 = 4 + 16 = 20\,\text{H}$$

$$\begin{aligned} X_{LT} &= X_{L1} + X_{L2} + \cdots + X_{LN} \\ &= 2\pi \times f \times L_1 + 2\pi \times f \times L_2 + \cdots + 2\pi \times f \times L_N \\ &= 2\pi \times f \times (L_1 + L_2) = 2 \times \pi \times 50 \times 20 = 6283.18\,\Omega. \end{aligned}$$

To verify the value of the total inductance:

$$L_T = \frac{X_{LT}}{2 \times \pi \times f} = \frac{6283.18}{6.283 \times 50} = 20\,\text{H}$$

$$\frac{1}{L_T} = \frac{1}{L_1} + \frac{1}{L_2};$$

$$L_T = \frac{L_1 \times L_2}{L_1 + L_2} = \frac{4 \times 16}{4 + 16} = \frac{64}{20} = 3.2\,\text{H}$$

$$\frac{1}{X_{LT}} = \frac{1}{X_{L1}} + \frac{1}{X_{L2}} = \frac{1}{2\pi \times f \times L_1} + \frac{1}{2\pi \times f \times L_2} = \frac{1}{2\pi \times 50 \times 4} + \frac{1}{2\pi \times 50 \times 16}$$

$$\frac{1}{X_{LT}} = 7.957 \times 10^{-4} + 1.989 \times 10^{-4}$$

$$X_{LT} = \frac{1}{9.946 \times 10^{-4}} = 1005.38\,\Omega.$$

To verify the value of the total inductance:

$$L_T = \frac{X_{LT}}{2\times\pi\times f} = \frac{1005.38}{6.283\times 50} = 3.2\,\mathrm{H}\,]$$

$$\mathrm{M} = k\times\sqrt{L_1\times L_2} = 0.5\times\sqrt{4\times 16} = 0.5\times\sqrt{64} = 0.5\times 8 = 4\,\mathrm{H}$$

$$L_T = L_1 + L_2 + 2M = 4 + 16 + 2\times 4 = 20 + 8 = 28\ \mathrm{H}$$

$$\frac{1}{L_T} = \frac{1}{L_1+M} + \frac{1}{L_2+M} = \frac{1}{4+4} + \frac{1}{16+4} = \frac{1}{8} + \frac{1}{20} = 0.125 + 0.05 = 0.625$$

$$L_T = \frac{1}{0.625} = 1.6\,\mathrm{H}$$

$$V_{L1} = \frac{L_1}{L_1+L_2}\times V_A = \frac{25m}{25m+50m}\times 24 = \frac{25m}{75m}\times 24 = 8\,\mathrm{V}$$

$$V_{L2} = V_A - V_{L1} = 24 - 8 = 16\ \mathrm{V}$$

Inductive reactance: $X_L = 2\times\pi\times f\times L = 2\times\pi\times 60\times 2 = 753.98\,\Omega$

Impedance: $Z = \sqrt{R^2 + X_L{}^2} = \sqrt{500^2 + 753.98^2} = \sqrt{818489.2} = 904.7\,\Omega$

Series current: $I = \dfrac{V}{Z} = \dfrac{10}{904.7} = 0.011\,\mathrm{A}$

$$V_R = I\times R = 0.011\times 500 = 5.5\,\mathrm{V}$$
$$V_L = I\times X_L = 0.011\times 753.98 = 8.4\,\mathrm{V}$$

Verify voltages add up to the supply voltage:

$$V_A = \sqrt{V_R{}^2 + V_L{}^2} = \sqrt{5.5^2 + 8.4^2} = \sqrt{100.81} = 10.04\,\mathrm{V}$$

Phase angle: lagging

$$\theta = \cos^{-1}\left(\frac{V_R}{V_A}\right) = \cos^{-1}\left(\frac{5.5}{8.4}\right) = \cos^{-1} 0.6548 = 54.55^\circ$$

Glossary

Admittance (Y)
The ease with which AC current flows through an impedance and is expressed as: $Y = \frac{1}{Z}$. It is measured in Siemens.

Air-core inductor
A coil wound on an insulated core or a coil of wire that does not have a metal core.

Angle of lead or lag
The angle between applied voltage and current flow in an AC circuit, in degrees; in an inductive (L) circuit, voltage (V) leads current (I); in a capacitive (C) circuit, current (I) leads voltage (V).

Center tap
An electrical connection point at the center of a wire coil or transformer winding.

Choke coil
An inductor coil used to block the flow of AC current and pass DC current.

Conductance (G)
The ease with which current flows through a resistance and is expressed as: $G = \frac{1}{R}$. It is measured in Siemens.

Electromagnet
A coil of wire wound on a metal core so that as current flows through the coil, it becomes magnetized.

Faraday's law
The value of the induced EMF is directly proportional to the rate of change of current through an inductor.

Flux (Φ)
Invisible lines of force that extend around a magnetic material.

Flux density
The number of lines of force per unit area of a magnetic material or circuit.

Henry
The unit of measurement of inductance that is produced when a voltage of 1 V is induced when the current through a coil is changing at a rate of 1 A per second.

Inductance (L)
The property of a circuit to oppose changes in current due to energy stored in a magnetic field.

Impedance (Z)
The total opposition to current flow in an AC circuit which is a combination of resistance (R) and reactance (X) in a circuit; measured in ohms.

$$Z = \sqrt{R^2 + X^2}.$$

Inductive circuit
A circuit that has one or more inductors or has the property of inductance, such as an electric motor circuit.

Inductive reactance (X_L)
The opposition to current flow in an AC circuit caused by an inductance (L), measured in ohms. $X_L = 2\pi fL$.

Inductor
A coil of wire that has the property of and is used in a circuit for that purpose.

Lagging phase angle
The angle by which current lags voltage (or voltage leads current) in an inductive circuit.

Lenz's law
The counter EMF developed by an inductor always opposes a change in the current. Thus, the current caused by the induced EMF will create a magnetic field that opposes the original field that produced it.

Magnetic field
Magnetic lines of force that extend from a north pole and enter a south pole to form a closed loop around the outside of a magnetic material.

Mutual inductance (M)
When two coils are located close together so that the magnetic flux of the coils affects one another in terms of their inductance properties.

Permeability (μ)
The ability of a material to conduct magnetic lines of force. The permeability of air (μ_0) is $4 \times \pi \times 10^{-7}$ or 1.26×10^{-6} N/A^2. The permeability of other materials is given relative to that of air, which is considered to be relative permeability of the material.

Quality factor (Q)
The "figure of merit" or ratio of inductive reactance and resistance in a frequency-sensitive circuit. $Q = \frac{X_L}{R} = \frac{2 \times \pi \times f \times L}{R}$.

Reactance (X_L)
The opposition to AC current flow due to inductance (X_L) or capacitance (X_C).

Reactive circuit
An AC circuit that has the property of inductance or capacitance.

Susceptance (B)
The ease with which an inductor passes AC current and is expressed as: $B_L = \frac{1}{X_L}$. It is measured in Siemens.

13

Transformers

Transformers are important electrical devices. They are used to either increase or decrease AC voltage and current. Transformers are made by using two separate sets of wire windings which are wound on a core. These are called the primary and the secondary windings. Energy is applied to the primary winding and the output is developed by the secondary winding.

A transformer uses electromagnetism to link the windings together When AC is applied to the **primary winding**, it develops a magnetic field that is continually changing. This **magnetic field** is used to link the individual windings together through **mutual inductance** (see **Figure 13-1**).

The changing magnetic field induces AC in the secondary winding. Thus, when AC energy is applied to the primary winding, it permits the transfer of energy to the windings that are commonly coupled.

Transformers have a wide range of applications. One high-power application is where the power companies use transformers to increase the voltage on **transmission lines**. This decreases the power lost during transmission. The power company then decreases the voltage for use by the consumer. A low-power application of transformers is in an AC adaptor used to power small electronic devices. The transformer reduces the 120 V coming from an outlet to a much lower voltage, such as 12-V AC. Transformers can

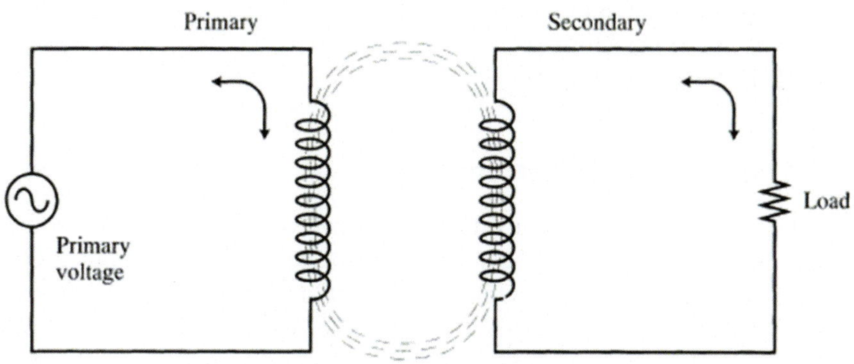

Figure 13-1 Mutual inductance results in transformer action.

also be found in computers, televisions, microwave ovens, and sound systems. A transformer that increases voltage is called a **step-up transformer** and one that decreases voltage is called a **step-down transformer**. Transformers are also used to match the resistance of a load to the internal resistance of the power source. This produces a maximum transfer of power.

Objectives

1. Describe how a transformer is constructed
2. Explain the operation of a transformer
3. Determine the turns ratio, voltage ratio, and current ratio of a transformer
4. Determine the power and operational efficiency of a transformer
5. Describe the losses that occur in the operation of a transformer and ways to reduce these losses
6. Calculate the reflected resistance of a transformer and perform impedance matching
7. Recognize the effects of transformer loading

Chapter Outline

13.1 Transformer Construction

A transformer is generally constructed with two or more coils or windings wound on a common core. The winding connected to the AC input energy source is called the primary. Other coils wound around the common core are regarded to be the secondary windings. The AC output of a transformer is developed across the secondary windings. There is generally no electrical connection between the primary and the secondary windings. However, the

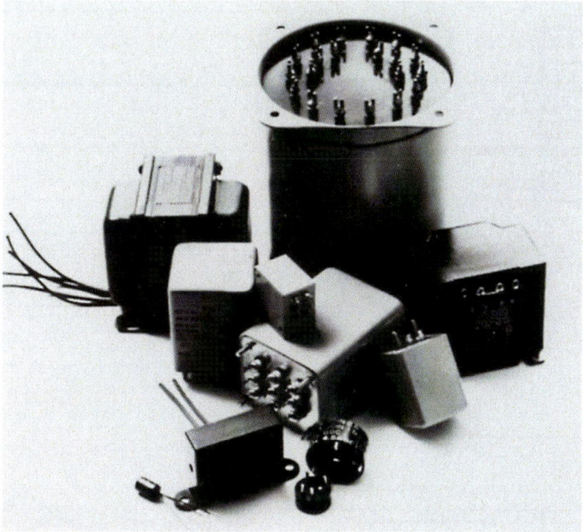

Figure 13-2 Transformers (courtesy of TRW/UTC Transformers).

coils are magnetically linked together by the core material. Several types of transformers are shown in **Figure 13-2**.

Transformer Core

A **core** serves two purposes. First, a core holds the coils of wire in a firm position. Second, a core maximizes the magnetic coupling between the primary and secondary windings.

The **windings** of a transformer can be wound on either open or closed cores. When the magnetic field is completely enclosed within the core, it is called a **closed core**; otherwise, it is called an **open core**. **Figure 13-3** shows selected types of core styles.

Open core construction is frequently made of a stiff cardboard. The windings are sometimes wound on top of each other. Different wiring patterns can be used. This type of transformer is commonly used in **radio frequency (RF)** applications. Open core transformers can be made tunable by moving a ferrite core inside the cardboard cylinder. This changes the frequency response of the transformer. **Closed core** construction is of two types. This includes the standard closed core construction where the core is placed inside the coils, and shell type construction where the core is on the outside, enclosing the coils. The iron cores shown in **Figure 13-3** are used in applications such as **power transformers**, **audio circuits**, and **isolation transformers**. The iron core allows a large number of windings to be wound

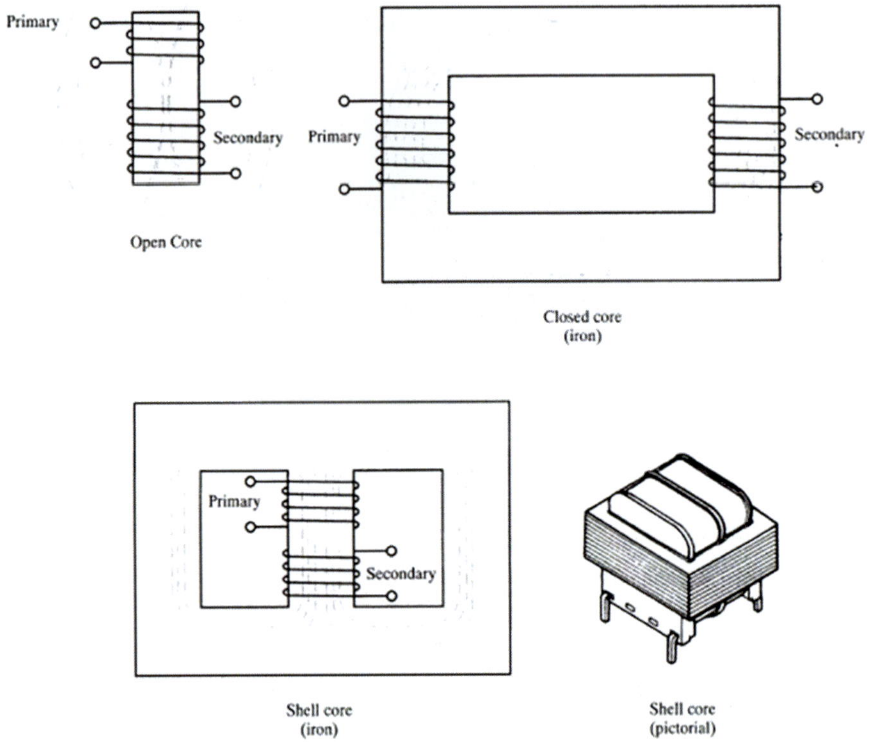

Figure 13-3 In transformer construction, the primary and secondary coils are wound on the same core.

on opposite sides of the core, rather than the windings being placed on top of each other.

Input/Output Phase Relationships

The **primary winding(s)** of a transformer serves as the input and it is where energy is first applied to the device. The **secondary winding(s)** is where the output is derived. The phase relationships between windings can be either 0°, exactly in phase, or 180°, the opposite phase relation. The input signal, applied to the primary, is the reference signal. The signal coming from the secondary windings is compared to it.

Figure 13-4 shows a **transformer** with the windings wrapped around the core in such a way as to cause a 0° phase shift. The drawing allows the use of the left-hand rule to compare the direction of flux to the direction of the current. In this drawing, the polarity of the primary and secondary windings

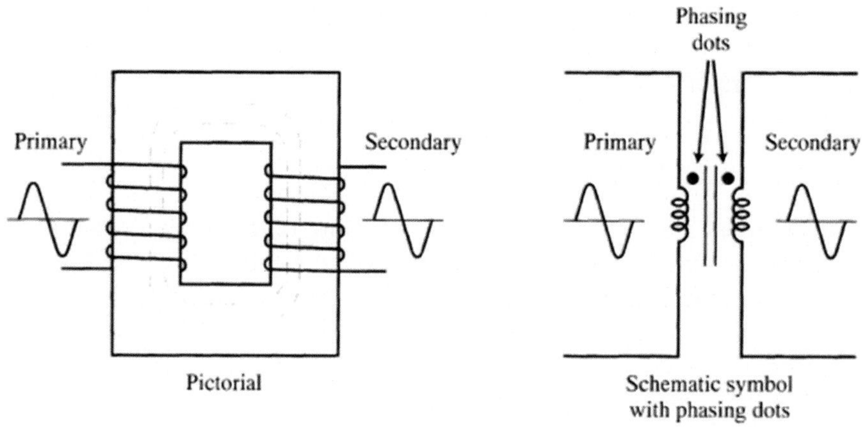

Figure 13-4 Transformer voltages are in phase when the windings have the same relationship as the magnetic field.

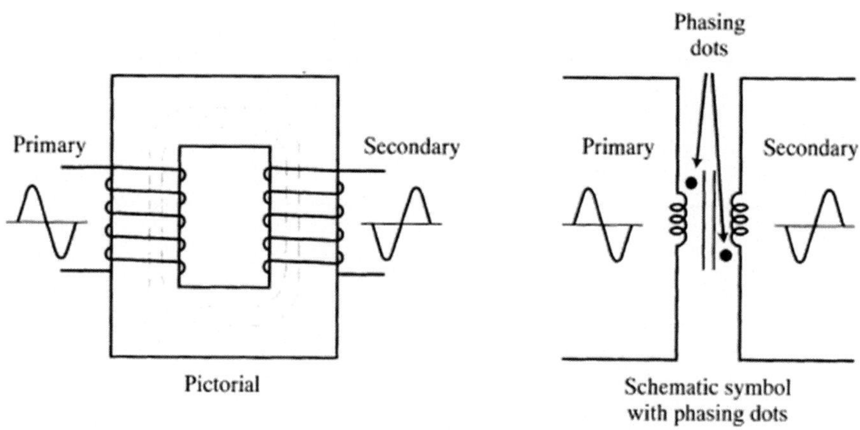

Figure 13-5 Transformer voltages are 1800 out of phase when the windings are wound in opposite directions.

is both the same. This results in the 0° phase shift. The **schematic symbol** to the right shows the use of phasing dots to indicate this phase relationship.

Figure 13-5 shows a transformer with the windings wound to produce a 180° phase shift. Compare direction of windings in this drawing. Note that they are wound in opposite directions. The phasing dots in the schematic symbol show the opposite phase relation.

Do not confuse a 0° or 180° phase shift with series aiding and opposing fields. In the case of a transformer, there is only one magnetic field. Therefore, there is nothing to aid or oppose. Also, the phase relationship does not affect the amount of voltage produced. The need to understand the phase relationship comes in use when working with an oscilloscope, comparing input and output voltages.

13.2 Types of Transformers

Transformers are primarily classified according to the material used in its construction. One important classification of transformers is based on the type of **core material** used. In this regard, an air core construction or metallic core construction may be employed. The schematic symbols for transformers indicate the type of core material used in its construction. **Figure 13-6** shows the schematic symbols for these two transformer types.

Air core transformers are the type used in radio frequency applications. RF circuits cannot use iron cores because the core's response to high frequency sine waves is too slow. Iron core transformers respond best to

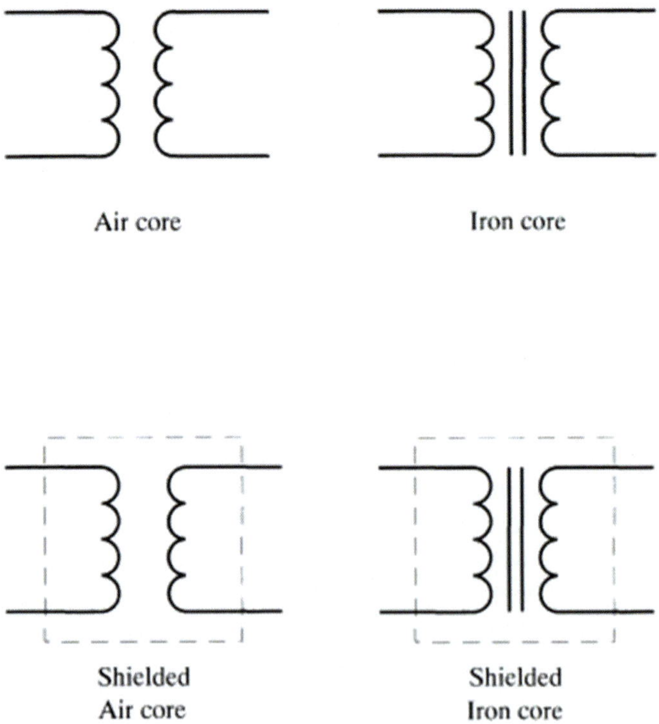

Figure 13-6 Schematic symbols for three types of transformers.

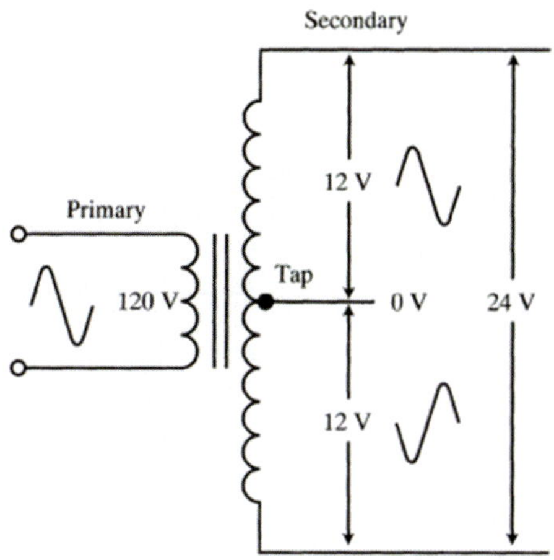

Figure 13-7 This transformer has a center-tapped secondary allowing for three separate outputs.

relatively low frequencies, typically below 20,000 Hz. In schematic symbols, the two parallel lines are used to indicate an iron core.

A **shielded transformer** has a magnetic shield placed around the outside of the windings. The purpose of the shield is to confine the magnetic flux, preventing it from interacting with other nearby circuits. This is shown by a dashed line drawn around the transformer symbol, as shown in **Figure 13-6**. A transformer symbol without a dashed line enclosing it indicates that the transformer is unshielded. Another way of classifying transformer types is based on the connection of its windings. In this regard, a transformer can have **tapped windings**, single or multiple windings, and isolated windings. This is primarily determined by the intended function of the transformer.

Center-tapped Secondary

A **center-tapped secondary** produces two voltages, each equal to half the total secondary voltage. Either half can be used separately, or they can be used together. This is shown in **Figure 13-7**. The center tap can be used as a 0 V or common reference point. When compared to this point, the opposite ends of the secondary winding have reverse polarities. This means that the sine waves produced from the two sides have opposite polarities. This means that the sine waves produced from the two sides of the secondary windings

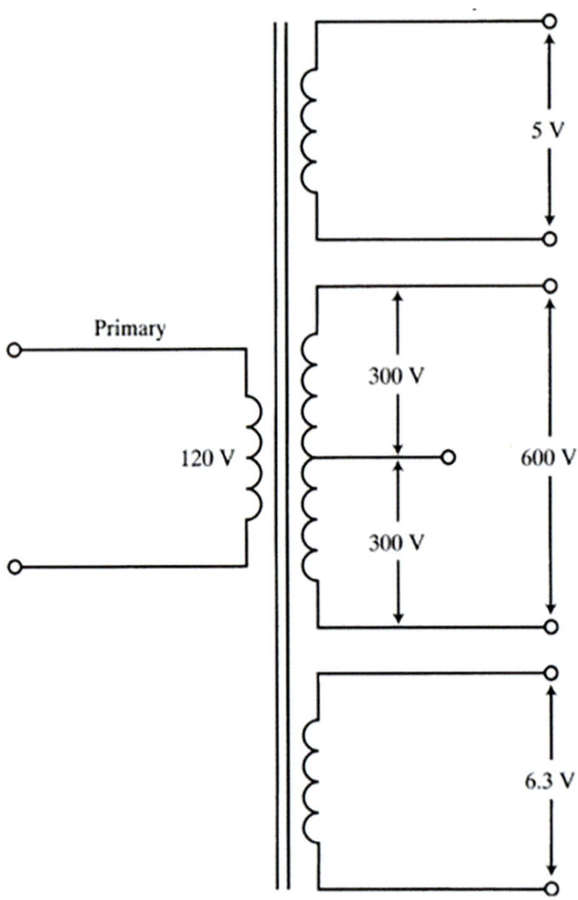

Figure 13-8 Typical power supply transformers have multiple secondary windings.

are 180° out of phase. These dual sine waves are used in applications such as a full-wave rectifier, which is used in a power supply that changes AC to DC.

Multiple Secondary Windings

Multiple secondary windings such as that shown in **Figure 13-8** are used in applications where it is necessary to produce several different voltages from the same power source. This includes power supplies for communications equipment and computer systems.

It is not necessary to use all of the windings that are provided. If the windings are left unconnected, as an open circuit, there is no current. Thus, those windings consume no power. The windings that are not used should be protected. There still will be voltage present in them.

Dual-Primary Transformer

A transformer with two primary windings is used where the electric equipment can be connected to either 120 or 240 V. This is a common situation with large commercial machinery. A higher voltage requires less current to develop the same amount of power.

Figure 13-9 shows a **dual-primary transformer** wired to a 120-V supply. The two windings are connected in parallel. Each coil draws one-half of the total current. If one of the coils is disconnected, the other coil carries

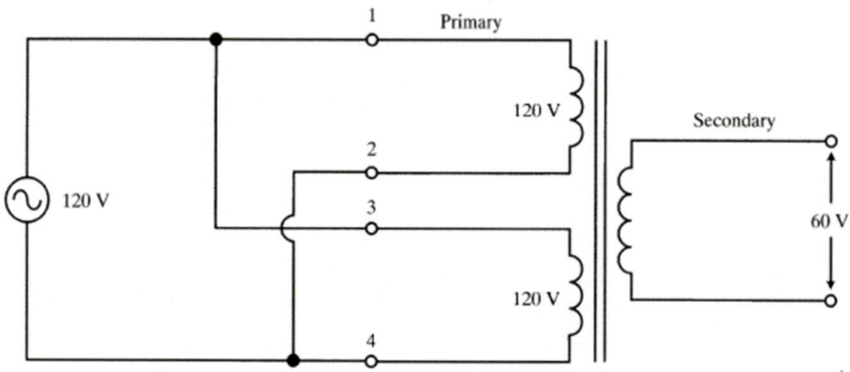

Figure 13-9 Dual-primary transformer wired in parallel for use with the lower primary voltage.

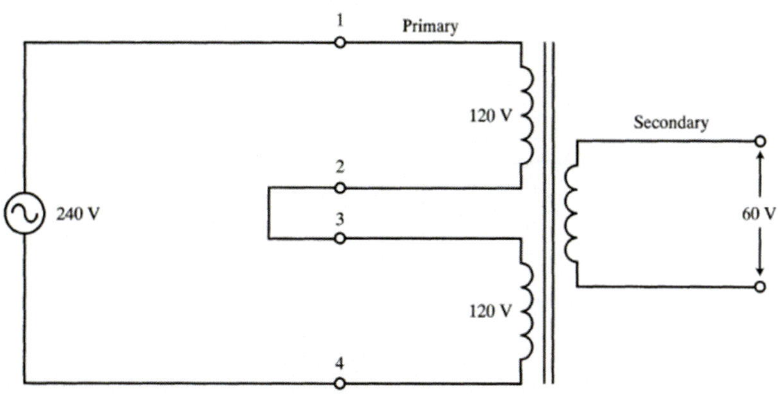

Figure 13-10 Dual-primary transformer wired in series for use with the higher primary voltage.

the full load current. A current this large might be more than the windings are designed to handle and could result in failure.

Figure 13-10 is a dual-primary transformer wired for 240-V operation. In this configuration, the windings are in series. With twice the voltage, the current is one half the total of the 120-V operation. With either the 120-V or the 240-V connection, the individual coils receive the same amount of current. In both circuits, the load voltage is the same.

Autotransformer

An **autotransformer** has only one coil, used for both the primary and secondary. A tap can slide along the coil to select a desired output voltage; see **Figure 13-11**. One application of the autotransformer is on test benches. The autotransformer allows the technician to plug a circuit into an electrical source with a smaller fuse and a more convenient on/off switch. The technician also has the option of using a voltage different from the electrical wall outlet.

Isolation Transformer

An **isolation transformer** has a turns ratio of 1:1. Its function is to isolate electrical equipment from earth ground. Residential wiring uses earth ground as a third wire. When the third wire is properly connected, as in most situations, the ground connection provides protection from electrical shock. Many electrical appliances, including televisions, sound systems, and other consumer products, use only a two-conductor cord. As seen in **Figure 13-12**,

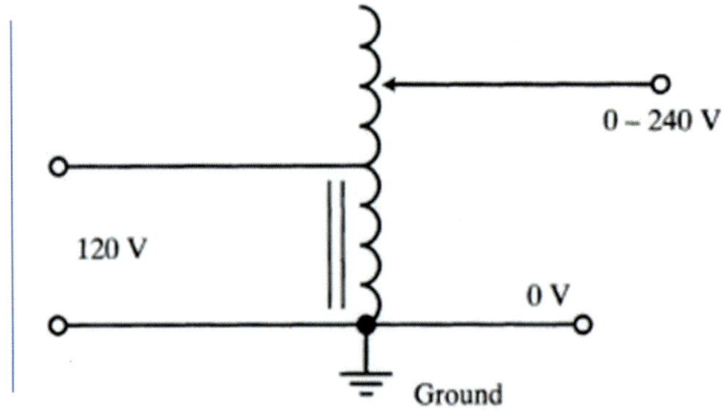

Figure 13-11 Autotransformers use a single coil. A variable sliding tap selects the output voltage.

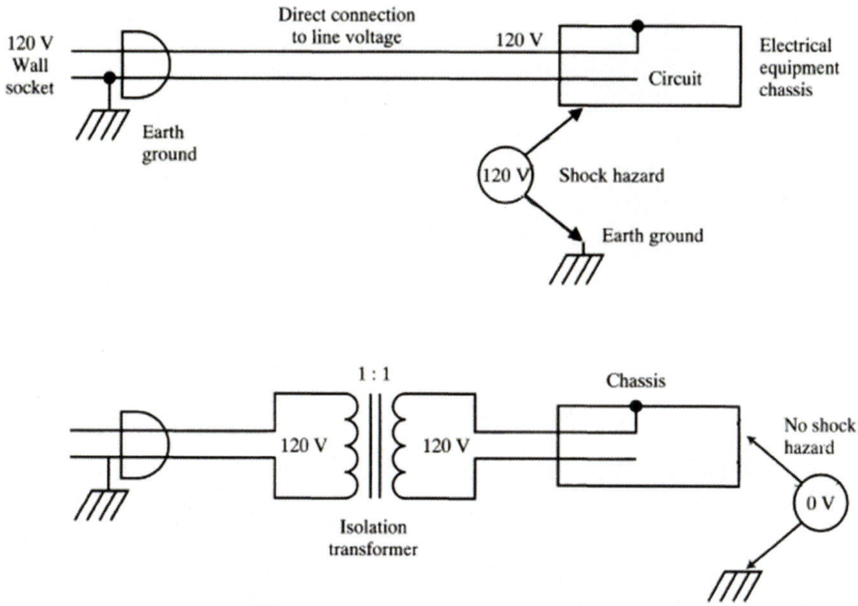

Figure 13-12 When an appliance has one conductor connected to the chassis, there is a significant shock hazard to the technician. An isolation transformer (bottom) eliminates this hazard.

one conductor is usually connected to the equipment chassis. It is inside and safely away from whoever is operating the appliance. A technician, however, must come in contact with the chassis. If the hot wire is connected to the chassis, there is a shock hazard.

The secondary of an isolation transformer does not use ground as a conductor. By plugging an appliance into the isolation transformer, the shock hazard is removed. See the bottom of **Figure 13-12**.

Physical Construction

As with all electronic components, the physical construction is determined by the application. Generally speaking, a larger package is needed for a higher power duty. Smaller sizes, to the point of miniaturizing, are used on circuit boards. **Figure 13-13** shows some of the transformers used in electronic circuits. All the transformers shown here, except the variable power transformer, would be found inside the cabinet of electronic equipment. A variable power transformer is an autotransformer that is used to vary the voltage to a circuit and provide extra protection for a technician.

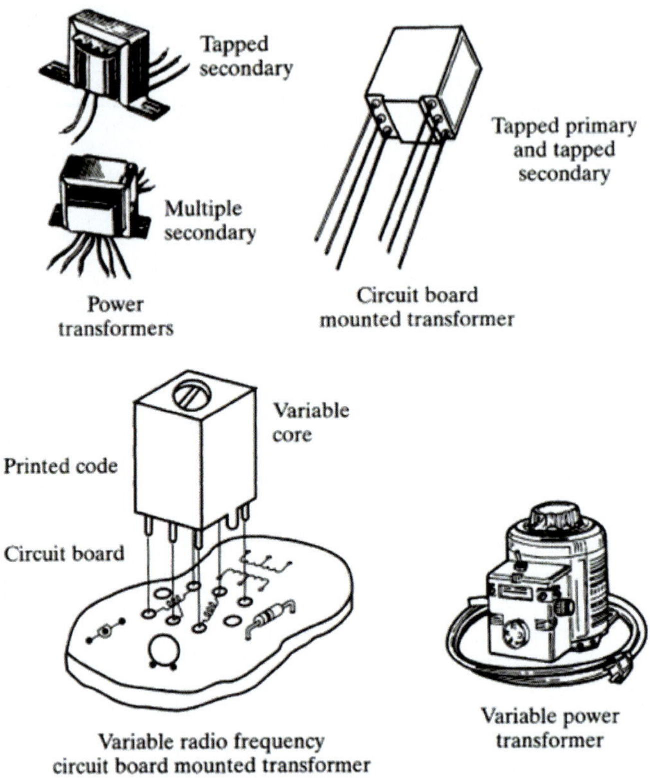

Figure 13-13 Size and shape of a transformer depend on the application.

13.3 Transformer Operation

Transformers are electrical control devices used to either increase or decrease AC voltage. Note in **Figure 13-14** that AC voltage is applied to the primary winding of the transformer. There is no connection of the primary and secondary windings. The transfer of energy from the primary to the secondary winding is due to magnetic coupling or mutual inductance. The transformer relies on electromagnetism to operate.

Although many different types and sizes of transformers exist, the same basic principles of operation apply to all. The operation of a transformer relies on the expanding and collapsing of the magnetic field around the primary winding. When current flows through a conductor, a magnetic field is developed around the conductor. The input of a transformer is called the **primary winding**. Voltage is applied to the primary winding and causes current to flow through the winding. This produces a magnetic field around

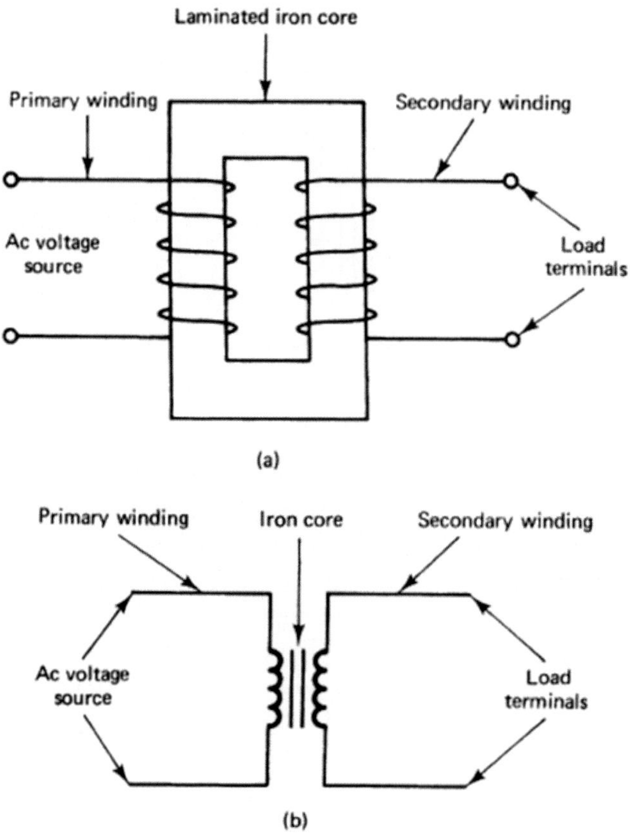

Figure 13-14 Transformer: (a) pictorial; (b) schematic symbol.

the primary winding. The amount of current in the primary winding is determined by a combination of the resistance of the primary winding and the secondary current. When AC voltage is applied to the primary winding, a constantly changing magnetic field is developed by the winding. During times of increasing AC voltage, the magnetic field around the primary winding expands. After the peak value of the AC cycle is reached, the voltage decreases toward zero. When the AC voltage decreases, the magnetic field around the primary winding collapses.

The **secondary winding** is generally used to develop the output of the transformer. The changing magnetic field is transferred to the secondary winding and induces a voltage in it. Thus, the magnetic field developed in the primary windings causes a current to flow in the secondary windings by **electromagnetic induction**. This current is supplied to the load. Generally,

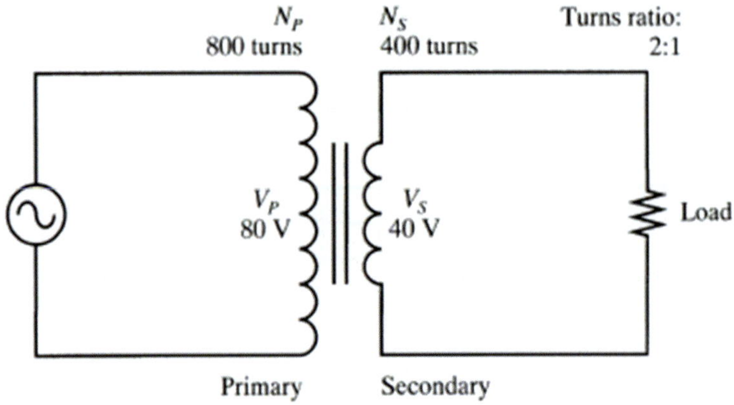

Figure 13-15 Step-down transformers have fewer turns in the secondary than in the primary.

only AC voltages are applied to the primary of a transformer. The AC causes a changing magnetic field to be developed in the primary, which links the secondary windings as well. This, in turn, causes the secondary to develop a voltage. A transformer will not operate with a fixed DC voltage applied to the primary windings because, although a magnetic field is developed by the primary, it does not change (except momentarily when the DC voltage is first applied). In the absence of a changing magnetic field, no voltage can be induced in the secondary of the transformer.

13.4 Transformer Turns Ratio

The voltage developed in the secondary windings of a transformer is determined by the ratio of the number of turns in the primary windings to the number of turns in the secondary windings. The **turns ratio** does not state the actual number of turns. It is a ratio reduced to lowest terms. For example, **Figure 13-15** shows a transformer with 800 turns in the primary and 400 turns in the secondary. The turns ratio would be stated as 2:1. Note, in this figure, that the relationship of the primary and secondary voltages is equal to the turns ratio. The turns ratio, combined with the voltage ratio, can be used as a formula to calculate unknown values.

Transformer Voltage Ratio

A **step-up transformer** has a larger number of turns in the secondary than the primary. In a step-up transformer, the voltage is higher in the secondary.

A **step-down transformer** has a smaller number of turns in the secondary than the primary. In a step-down transformer, the voltage is lower in the secondary. The formula used to calculate the turns ratio and **voltage ratio** is the same for both step-up and step-down transformers.

$$\text{Voltage ratio} = \frac{V_P}{V_S} = \frac{N_P}{N_S}$$

where N_P is the number of turns in the primary, N_S is the number of turns in the secondary, V_P is the primary voltage, and V_S is the secondary voltage.

- With the turns ratio given for the step-down transformer in **Figure 13-15**, determine the secondary voltage with 120 V applied to the primary.

$$\frac{V_P}{V_S} = \frac{N_P}{N_S}$$

$$V_P = \frac{N_P}{N_S} \times V_S$$

$$V_S = \frac{N_S}{N_P} \times V_P = \frac{400}{800} \times 120 = 60\,\text{V}.$$

Transformer Power

A transformer converts energy from one form to another, but it does not create energy. Power is electrical energy. Therefore, power of the primary windings

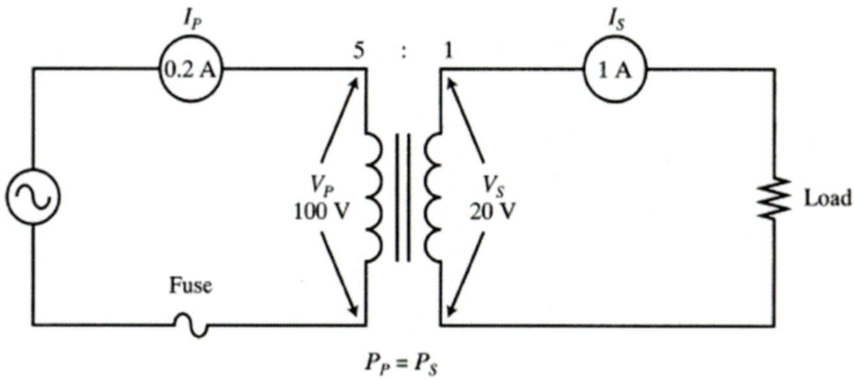

Figure 13-16 Power in the secondary equals power in the primary of an ideal transformer.

is equal to power of the secondary windings, when losses are not considered. If the primary to secondary power was viewed as a ratio, it would be 1:1.

$P_P = P_S$

where P_P is the primary power and P_S is the secondary power.

An **ideal transformer** is one with no losses, such as the example shown in **Figure 13-16**. This example is a step-up transformer with a turns ratio of 1:10. The primary voltage is 10 V, producing a secondary voltage of 100 V with current of 0.1 A.

Calculate the secondary and primary powers of the circuit in **Figure 13-16**.

Secondary power (P_S) = $V_S \times I_S = 100 \times 0.1 = 10$ W

Primary power (P_P), assuming no losses = 10 W.

Transformer Current Ratio

Current in a transformer is the inverse of the voltage ratio. In a **step-up transformer**, the current is stepped down. In a **step-down transformer**, the current is stepped up. This can be proven using the power formula in the voltage ratio, as follows:

Power formula:

$$P = I \times V$$

Power in a transformer:

$$P_P = P_S$$

Substitute into the power formula:

$$I_P \times V_P = I_S \times V_S$$

Transpose to form a ratio:

$$\frac{V_P}{V_S} = \frac{I_S}{I_P}$$

When combined with the voltage/turns ratio:

$$\frac{V_P}{V_S} = \frac{N_P}{N_S} = \frac{I_S}{I_P}$$

This formula relates the current ratio to the turns ratio.

$$\text{Current ratio} = \frac{I_S}{I_P} = \frac{N_P}{N_S}$$

where N_P is the number of turns in the primary, N_S is the number of turns in the secondary, I_P is the primary current, and I_S is the secondary current.

Using the circuit in **Figure 13-17**, determine the primary current if the load current is increased to 2 A.

$$\frac{V_P}{V_S} = \frac{I_S}{I_P}$$

$$\frac{100\text{ V}}{20\text{ V}} = \frac{2\text{ A}}{I_P}$$

$$I_P = \frac{2\text{ A} \times 20\text{ V}}{100\text{ V}}$$

$$I_P = 0.4\text{ A}$$

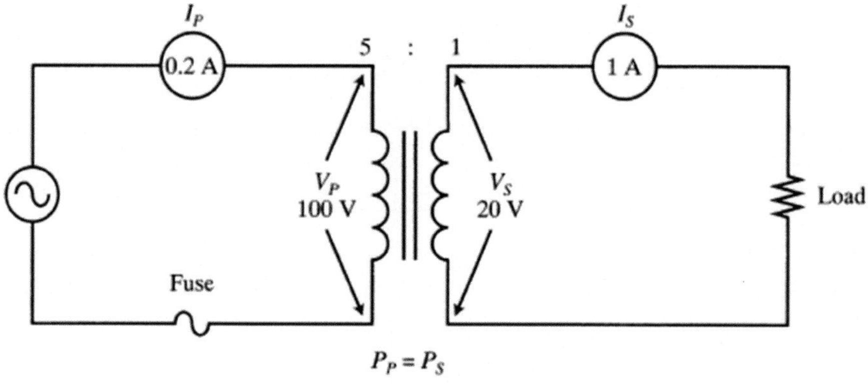

Figure 13-17 Example of an ideal step-down transformer.

Power Company Application

An electric power company must supply voltage and current to a large number of customers. Wire size is determined by the amount of current. Ohm's law showed that voltage drop is determined by resistance. The power company must determine the most cost-effective means of delivering electricity to the consumer. Wire run for long distances has more resistance. Large wires could be used to reduce the resistance, but large wires are very heavy and expensive.

Figure 13-18 is an illustration of the use of transformers. This example covers only 10 residential customers to keep it simple. In this example, 120 kW of power is needed by the customers. The value of 120 kW is the maximum power needed throughout the process. At the power plant, a generator is capable of producing a certain amount of voltage and a maximum current. This generator produces a constant 2400 V. The current varies to match the load demand.

The generator feeds the electricity into a large **step-up transformer**. The voltage is stepped up 50 times. The current is stepped down by the same ratio. This very large voltage is carried over the **long distance transmission lines**. The advantage of using such large voltages is that the amount of current is reduced. Less current means a much smaller wire size can be used, resulting in less cost and a lighter wire. The total amount of power carried in the wire is still the same. **Substation transformers** are located in the general area of the customer, within a few miles. High voltage lines enter the substation. In the substation, the voltage is reduced to much safer levels. In the example shown, the substation transformer uses a 50:1 turns ratio. This medium level voltage is carried at the top of power poles, as seen in **Figure 13-19**. From there, the electricity moves to a transformer located a short distance from the customer.

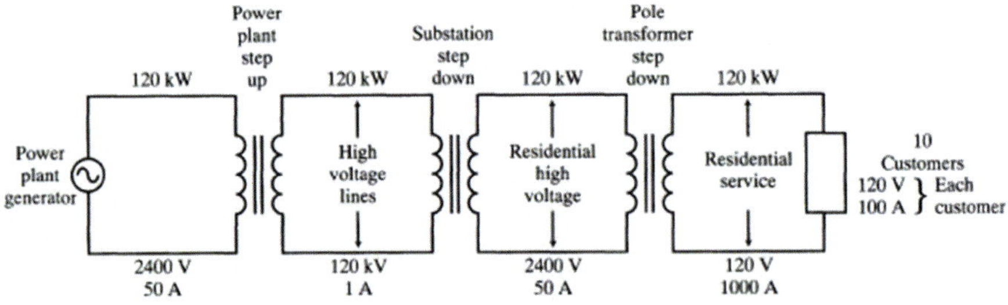

Figure 13-18 Simplified illustration of how transformers are used to deliver electricity to residential customers.

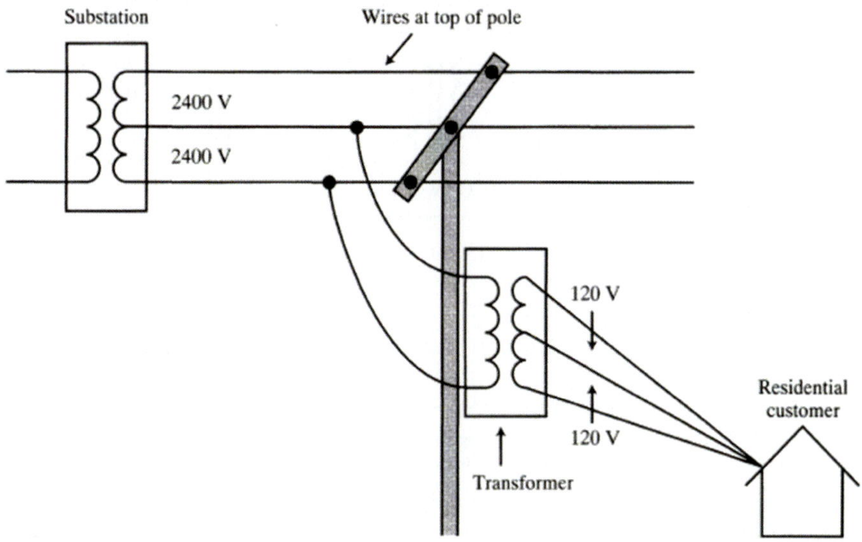

Figure 13-19 Voltage must be stepped down for customers to use.

The figure shows a pole-mounted transformer. Transformers are also located on the ground in tamper-proof boxes in areas where the electrical service is run underground.

13.5 Transformer Core Losses

As with everything else, transformers are not perfect. When discussing theory of operation, it is easiest to start with an understanding of the ideal. However, it is also necessary to understand the performance of non-ideal devices. **Losses** produce heat. In electricity, heat comes from power. Generally speaking, heat is an indication that power is being consumed. When heat is produced from a device that is not intended to produce heat, it is a clear indication of wasted power.

Figure 13-20 shows that power lost in the transformer is not available to the load and must be subtracted from the input power. Transformers have three categories in which power is lost: copper losses, hysteresis loss, and eddy currents. Each of these power losses can be represented as a series resistor, as shown in **Figure 13-21**. Voltage is dropped across these equivalent series resistances that oppose current.

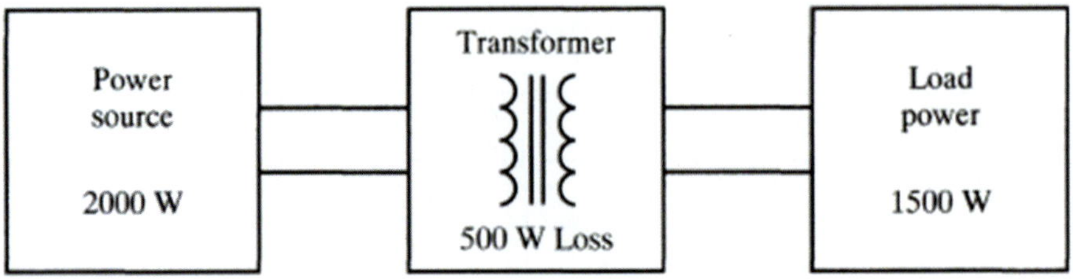

Figure 13-20 Block diagram that represents the power loss that takes place in every transformer.

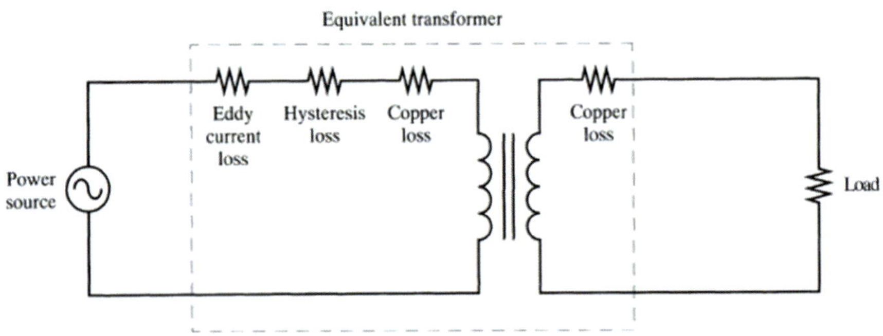

Figure 13-21 Schematic showing transformer losses as equivalent series resistances.

Copper Losses

The wire is used to make the windings of a transformer have some resistance. Even though most wire is thought of as having zero resistance, the coil in a transformer may have several hundred to several thousand turns. This is a very long piece of wire. In addition, the wire that is used in many transformers is very small in diameter. This also gives it a higher resistance.

When current flows in the primary and secondary windings of a transformer, it causes heat to be developed. This represents a loss in power by the transformer during operation. The heating loss is proportional to the square of the current (I) flowing through its conductors which is determined by its resistance (R). In practice, the copper loss is also referred to as the I^2R loss of a transformer.

Hysteresis Loss

Hysteresis is the amount of magnetization or flux density (*B*) that lags the magnetizing force (*H*) because of molecular friction. In the case of

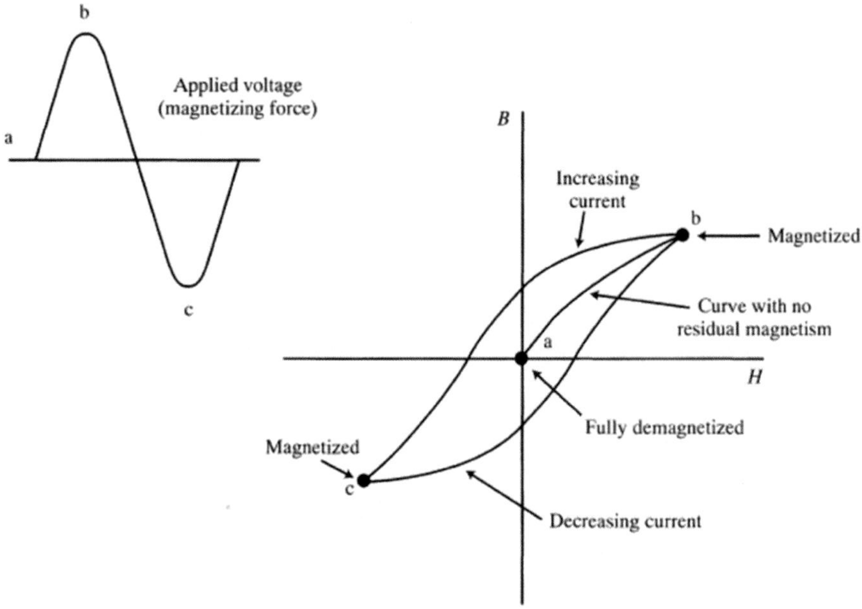

Figure 13-22 Hysteresis loops show flux density (B) vs. magnetizing force (H).

transformers, the greatest cause of hysteresis loss is residual magnetism. The loss of energy comes from overcoming this residual magnetism.

Figure 13-22 shows a **hysteresis loop**. A hysteresis loop represents the amount of energy needed to create a magnetic field in the core of the transformer. The curve starts at the center, the point labeled "a," with no residual magnetism. As the voltage increases in the positive direction, the flux density (B) increases. To continue increasing the strength of the magnetic field, a stronger magnetizing force (H) must be applied. At point "b," the voltage has reached its peak positive and the magnetic field can increase no further. The sine wave decreases and so does the magnetic field but at a somewhat slower pace. As the magnetizing force passes through zero on the *H*-axis, an amount of flux remains. This is **residual magnetism**.

The **magnetizing force** sine wave continues in the negative direction, causing the **magnetic flux** to flow in the opposite direction. At point "c," the sine wave has reached its negative peak and begins to fall back to zero. The magnetic field, again, does not collapse at the same rate as the sine wave. This lagging effect results in a residual magnetism in both the positive and negative half cycles of the magnetizing force. The lower the frequency of the applied voltage, the less significant the **hysteresis loss** will be. Transformers

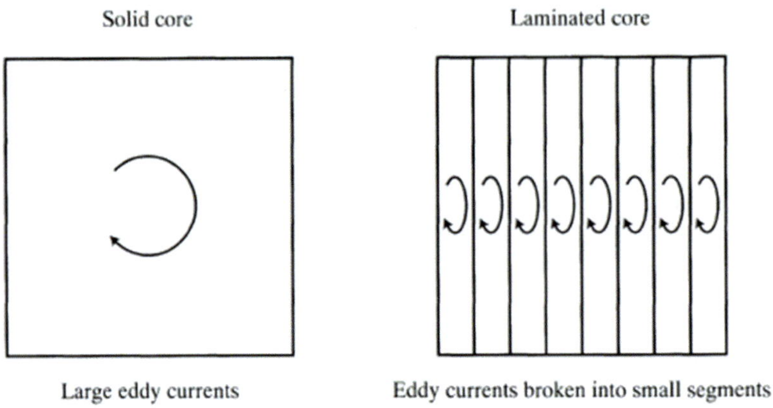

Figure 13-23 Laminated cores reduce the effects of eddy currents. This makes transformers more efficient.

are given a frequency response rating that indicates the best frequency applications for that transformer. Power transformers, with an iron core, used on the 60-Hz line frequency have very small hysteresis losses because of the low frequency. **Radio frequencies**, however, are so high that an iron core is not practical. This is the reason that air core transformers are used with radio frequencies.

Eddy Currents

Eddy currents result from a voltage induced in the iron core of transformers and other electromagnets. These eddy currents oppose the current producing the magnetic field. This results in a loss of power. **Figure 13-23** shows the difference between the eddy currents in a solid iron core and a core made with laminations. In the solid core, the large area allows a large eddy current to circulate. To prevent these large eddy currents, transformer cores are manufactured using thin slices of iron that are glued together. The slices are called laminations.

The laminations break up the path for eddy currents. The result is much smaller losses. A laminated core can be made as large as any solid core.

DC Voltages in Transformers

Any current applied to a coil produces a magnetic field. However, if it is not a changing current, the magnetism produced will not induce a voltage in another coil. When a DC voltage is applied to a transformer, it creates a magnetic field that must be overcome by the AC in order to produce voltage in the secondary. The DC voltage acts like residual magnetism and causes high losses. It is possible for the DC voltage to be strong enough to stop the transformer from producing any further magnetism. When this happens, the transformer is saturated and does not produce secondary current.

13.6 Calculating Transformer Efficiency

When working with transformers, a useful quantity to know is the transformers' efficiency. **Efficiency** is the ratio of power output to power input. Efficiency is expressed in percentage.

$$\text{Efficiency} = \frac{P_{\text{out}}}{P_{\text{in}}} \times 100\%$$

$$= \frac{V_S \times I_S}{V_P \times I_P} \times 100\%$$

where P_P is the primary power, P_S is the secondary power, N_P is the number of turns in the primary, N_S is the number of turns in the secondary, I_P is the primary current, and I_S is the secondary current.

- Calculate the efficiency of the transformer shown in **Figure 13-24**.

$$\text{Efficiency} = \frac{P_{\text{out}}}{P_{\text{in}}} \times 100\%$$

$$= \frac{1500}{2000} \times 100\% = 75\ \%$$

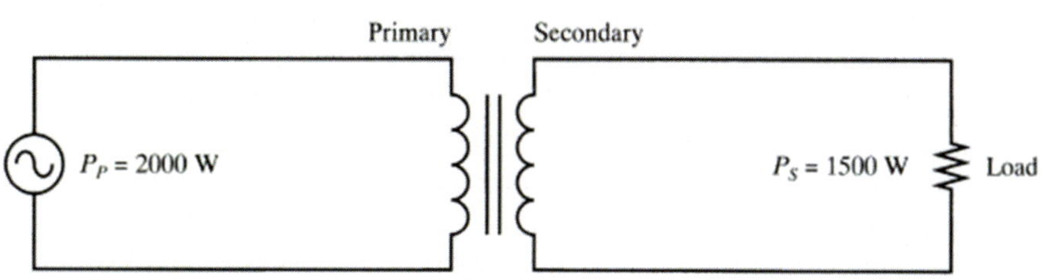

Figure 13-24 Calculate transformer efficiency.

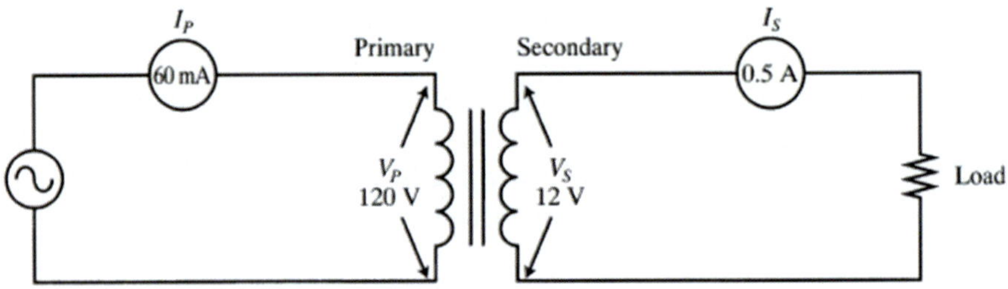

Figure 13-25 Calculate transformer efficiency.

- Use the circuit measurements shown in **Figure 13-25** to determine the efficiency of the transformer.

$$\text{Efficiency} = \frac{P_{\text{out}}}{P_{\text{in}}} \times 100\%$$

$$= \frac{V_S \times I_S}{V_P \times I_P} \times 100\% = \frac{12 \times 0.5}{120 \times 60 \times 10^{-3}} \times 100\% = \frac{6}{7.2} \times 100\% = 83.3\%.$$

13.7 Load Resistance Reflected to the Primary

When the primary winding of a transformer is energized by an AC source, it causes a small current to flow in the circuit. This is needed to establish a magnetic field in the primary winding. With no load connected to the secondary winding, the only power consumed by the transformer is what is developed by the primary. When a load is connected to the secondary winding, the resulting current flow in it is determined by the **impedance** of the load. An increase in secondary current also influences the current flow in the primary winding. This, in turn, causes the primary to develop more current. As a result of this, the magnetic field of the secondary winding will now have some influence on the magnetic field produced by the primary winding.

In order to determine the total power developed by a transformer, the resistance of the load must be taken into account. Since the primary and secondary windings are not electrically connected and are magnetically linked, the effect of the load resistance of the secondary, on the primary winding, is determined indirectly. This is referred to as the **reflected resistance or impedance** of the secondary on the primary. Impedance is a term used to describe AC resistance.

There are two ways to calculate transformer resistance under different load conditions. The first method uses **Ohm's law** to determine the values. The second method uses the **resistance ratio** to determine the values and is used if the load resistance and turns ratio are known. The resistance ratio is also called the impedance ratio. The voltage connected across the load and the current flowing through it is used to determine its power requirement. With a transformer circuit, the load resistance connected to the secondary determines the current drawn in the primary.

Calculating Reflected Resistance Using Ohm's Law

To calculate the resistance of the primary using Ohm's law, it is necessary to know the primary voltage and primary current. The only values given in many circuits are the primary voltage, turns ratio, and load resistance. The turns ratio allows calculating of the secondary voltage. Use Ohm's law to find the secondary current. Then use the turns ratio to find primary current. Finally, use **Ohm's law** to find resistance in the primary circuit. This resistance is a reflection of the load resistance.

- Calculate the primary resistance of the circuit in **Figure 13-26** using Ohm's law and turns ratios.

1. Use turns/voltage ratio to calculate secondary voltage:

$$\frac{N_P}{N_S} = \frac{V_P}{V_S}$$

$$\frac{4}{1} = \frac{120\ \text{V}}{V_S}$$

$$V_S = \frac{120\ \text{V} \times 1}{4}$$

$$V_S = 30\ \text{V}$$

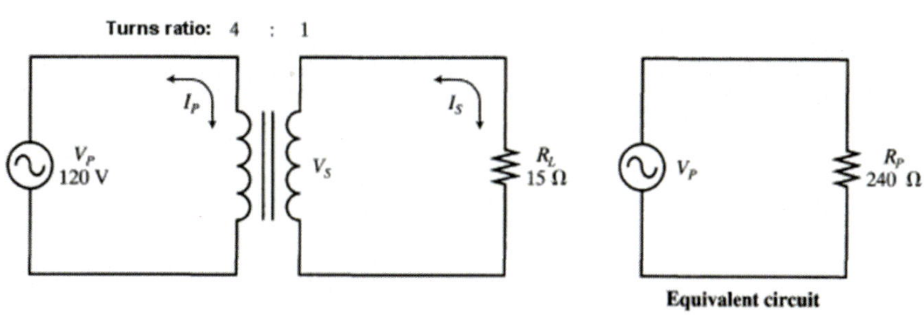

Figure 13-26 Load resistance is reflected to the source voltage through the turns ratio of the transformer.

2. Use Ohm's law to calculate secondary current:

$$I_S = \frac{V_S}{R_L}$$

$$I_S = \frac{30\ \text{V}}{15\ \Omega}$$

$$I_S = 2\ \text{A}$$

3. Use turns/current ratio to calculate primary current:

$$\frac{N_P}{N_S} = \frac{I_S}{I_P}$$

$$\frac{4}{1} = \frac{2\ \text{A}}{I_P}$$

$$I_P = \frac{2\ \text{A} \times 1}{4}$$

$$I_P = 0.5\ \text{A}$$

4. Use Ohm's law to calculate primary resistance:

$$R_P = \frac{V_P}{I_P}$$

$$R_P = \frac{120\ \text{V}}{0.5\ \text{A}}$$

$$R_P = 240\ \Omega$$

- Determine the reflected resistance of the circuit in **Figure 13-26**, using the resistance ratio.

$$\frac{R_P}{R_S} = \left(\frac{N_P}{N_S}\right)^2$$

$$\frac{R_P}{15\ \Omega} = \left(\frac{4}{1}\right)^2$$

$$R_P = \left(\frac{4}{1}\right)^2 \times 15\ \Omega$$

$$R_P = 16 \times 15\ \Omega$$

$$R_P = 240\ \Omega$$

Impedance Matching Transformer

To achieve maximum transfer of power, it is necessary for the load resistance to be equal to the resistance of the voltage source. In **Figure 13-27**, a speaker is connected to an audio amplifier. An impedance matching transformer is

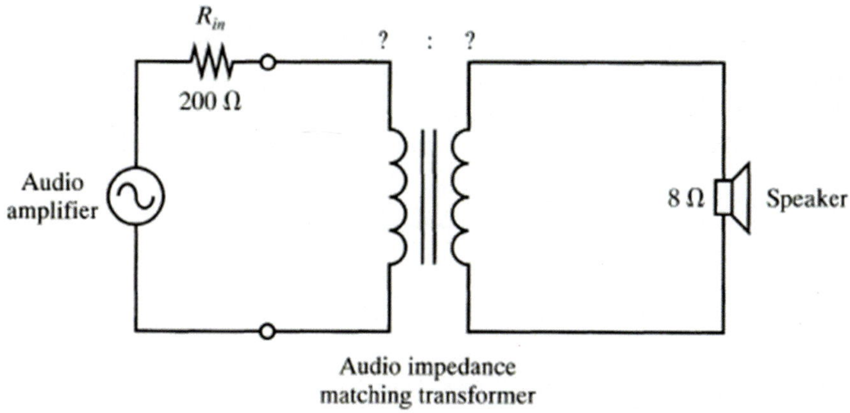

Figure 13-27 An impedance matching transformer is used in audio circuits to achieve maximum transfer of power.

used to reflect the 8-Ω speaker resistance to the amplifier as 200 Ω, which is the amplifier's resistance.

- Determine the turns ratio of the transformer needed in **Figure 13-27** to reflect the 8-Ω speaker resistance as a load equal to the amplifier's internal resistance (200 Ω).

$$\frac{R_P}{R_S} = \left(\frac{N_P}{N_S}\right)^2$$

$$\frac{200\ \Omega}{8\ \Omega} = \left(\frac{N_P}{N_S}\right)^2$$

$$25 = \left(\frac{N_P}{N_S}\right)^2$$

$$\frac{N_P}{N_S} = \sqrt{25} = 5$$

$$\frac{N_P}{N_S} = \frac{5}{1}$$

Impedance matching transformers with resistance ratios of 200:8 are a standard catalog item for use with audio amplifiers.

13.8 Loading a Transformer

When a sine wave flows through an inductor, the current lags behind the applied voltage by 90°. This is represented by the phasors in **Figure 13-28.** In phasor diagrams, a pure resistance has both current and voltage phasors drawn on the horizontal. An ideal transformer is purely inductive (has

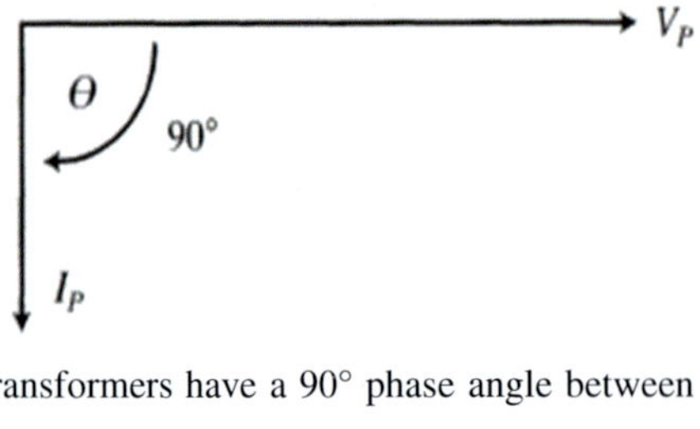

Figure 13-28 Ideal transformers have a 90° phase angle between the primary voltage and current.

no resistance). It is drawn as a vertical phasor. Phasors are used to give a graphical representation of the relative size and angle of two quantities.

Energizing Current

If a transformer were ideal, with no coil resistance, there would be no loss of power to energize the coil. With no load connected across the secondary, there would be an extremely small primary current. **Figure 13-29** shows that an actual transformer requires a slight amount of energizing current to develop the magnetic field, even with no load connected. The phasor diagram shows a shift from the ideal 90° to 78°. The **phase shift** represents losses due to coil resistance.

Power Transformer with a Light Load

The transformer in **Figure 13-30(a)** has a load of approximately 10% of the full load. The primary current is only slightly greater than the current needed to energize. Efficiency is increased to 33% with the increased secondary current. A further shift in the **phasor diagram** at the right of **Figure 13-30(a)** shows that the secondary current causes the transformer to act more like a resistor. The phasor diagram of current in pure resistance, as mentioned previously, is in phase with the voltage at 0°. By becoming more like a resistive current, the transformer becomes more efficient.

Medium and Heavy Loads

Figure 13-30(b) shows a transformer with a load equal to approximately 50% of full secondary power. The transformer's efficiency increases to 56% with the increased load. The **phasor diagram** shows that the current becomes more resistive as it increases.

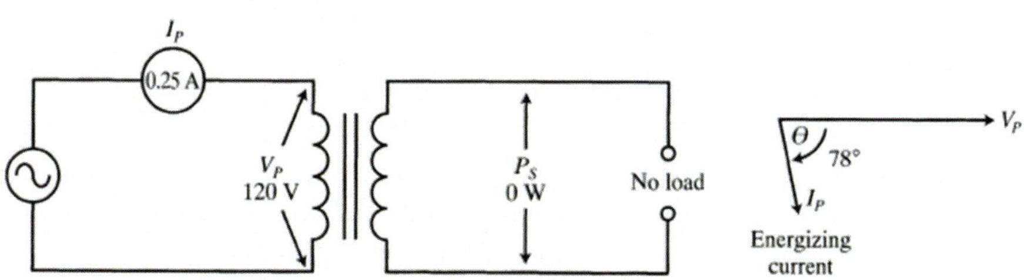

Figure 13-29 Transformers require an energizing current to develop a magnetic field due to the slight coil resistance.

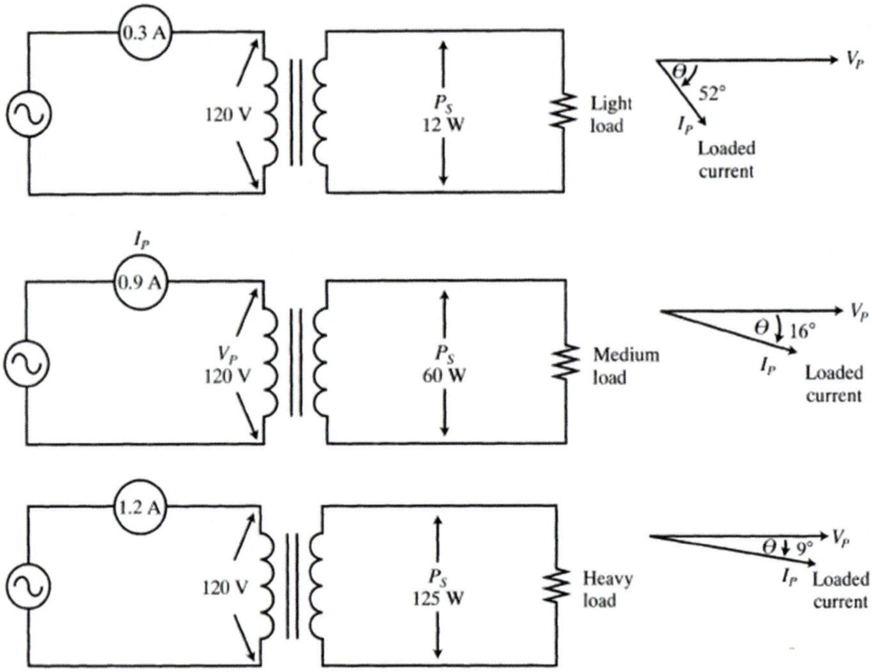

Figure 13-30 Power transformer with various levels of electrical load: (a) a light load. The primary current here is only slightly above the energizing current. (b) Power transformer with a medium load (c) Power transformer with a heavy load. The transformer efficiency has improved to 87%.

Figure 13-30(c) shows the transformer with its heaviest load. **Efficiency** is 87%. The phasor diagram shows that the primary current and voltage have almost the same **phase angle**. When the current has a 0° phase angle, the power produced has no inductive losses. However, it is not possible to achieve 100% efficiency. There will always be some power losses.

13.9 Transformer Ratings

All electronic components have ratings of maximum operating values. In addition, physical measurements are important. They are used to determine if the device fits into a certain size package or if it can be mounted on a circuit board. **Transformers** also have an additional consideration. They can be very heavy with their large number of turns of copper wire.

The turns ratio of a transformer is usually not given. Instead, the primary voltage is given with the nominal secondary for a **nominal secondary**

current. The power rating can be stated in either of two forms. The rating comes as a **maximum wattage** or as a **maximum current** at a specified voltage. Resistance ratios, also called **impedance ratios**, are usually not given for power transformers. Impedance ratios, however, are listed for applications where load to circuit matching is necessary. Impedance, as will recall, is **AC resistance**.

Transformer Troubleshooting

A transformer is two coils of wire or one coil if it is an autotransformer. A coil can have an **open** (a broken wire) or a **short** (where wires are touching together). Measurements from a normal working circuit, **Figure 13-31**, are used as a reference to examine a sample troubleshooting circuit. Four measurements are given: the current and voltage for both primary and secondary circuits. The circuit in **Figure 13-31** also contains a fuse, which can help with troubleshooting problems.

Shorted Primary Windings

A **short circuit** results in an excessively high amount of current. A short can occur as a result of the transformer windings getting so hot that some of their insulation melts.

Symptoms shown in **Figure 13-32**:

- The fuse blows, removing voltage from the circuit.
- All measurements are zero with the blown fuse, except the applied voltage.
- The transformer is cold since there is no current.
- Using an ohmmeter (applied voltage removed).

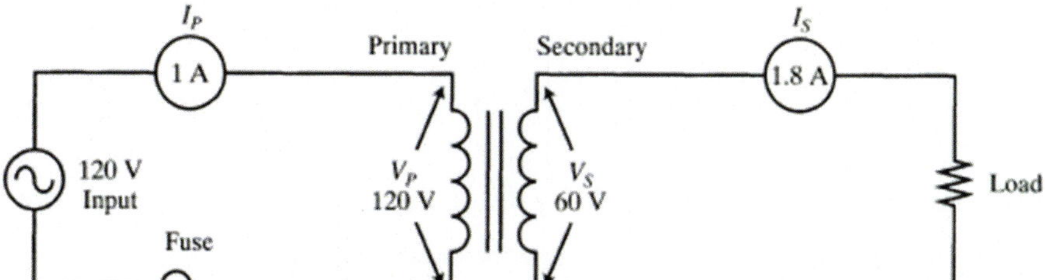

Figure 13-31 Measurements in a normal circuit are used to compare to a defective circuit.

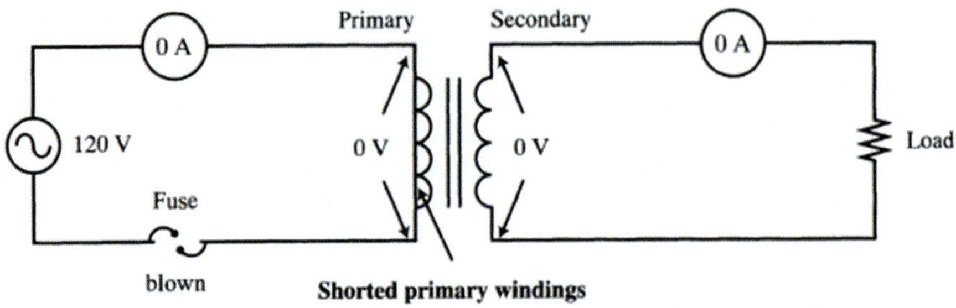

Figure 13-32 Shorted primary windings will blow the fuse.

- Check the fuse with the meter to verify it is blown (infinite resistance).
- Placing the ohmmeter across the primary should show zero resistance. Be sure to use the lowest resistance range.
- Secondary should show a normal resistance measurement. The resistance is determined by the actual number of turns. A reading of 100 Ω to a few thousand is a good approximation.

Shorted Secondary Windings

Shorted secondary windings can be a result of a defect in the load or a defect in the transformer. It is possible for the load to be defective and draw so much current that it damages the transformer. If a **short** is suspected in the secondary, it is best to remove the load to isolate the problem. Symptoms shown in **Figure 13-33**:

- Excessive heat radiates from the transformer to the point of melting insulation. Smoke is possible.

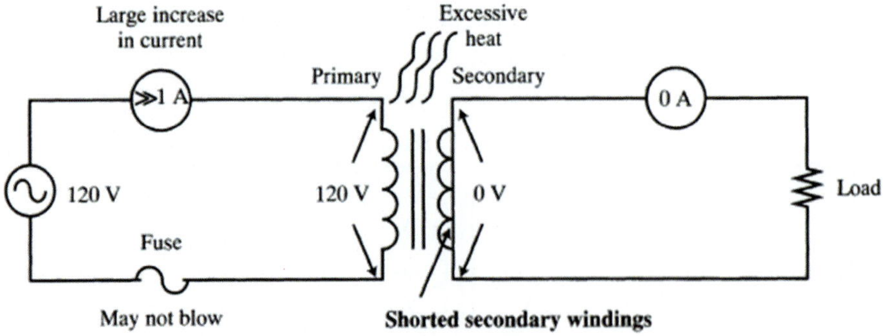

Figure 13-33 Shorted secondary windings do not always blow the fuse. However, it does cause excessive heat buildup in the transformer.

- The primary current is much larger than normal. However, it may not be enough to blow the fuse, especially if the fuse is oversized.
- The primary voltage equals the applied voltage.
- There are zero readings in the secondary.
- Using an ohmmeter (applied voltage removed):
- The primary resistance is normal.
- The secondary resistance is zero. Be sure to use the lowest ohms scale possible.

Partially Shorted Secondary Windings

Excessive current through the load can cause the transformer to get too hot. Too much heat can result in failure of the insulation on the windings. It is possible for only a few of the windings or a section of windings to be shorted together.

Symptoms shown in **Figure 13-34**:

- There is a decrease in secondary voltage and current. Amount of decrease is dependent on the number of turns that are shorted.
- There is a decrease in secondary voltage and current. Amount of decrease is dependent on the number of turns that are shorted.
- There is a decrease in primary current.
- Excessive heat is seen in the transformer.
- Using the ohmmeter (applied voltage removed):
- The reading on the primary is normal.
- The reading on the secondary is lower. An accurate measurement must be taken to see a difference from the normal value.

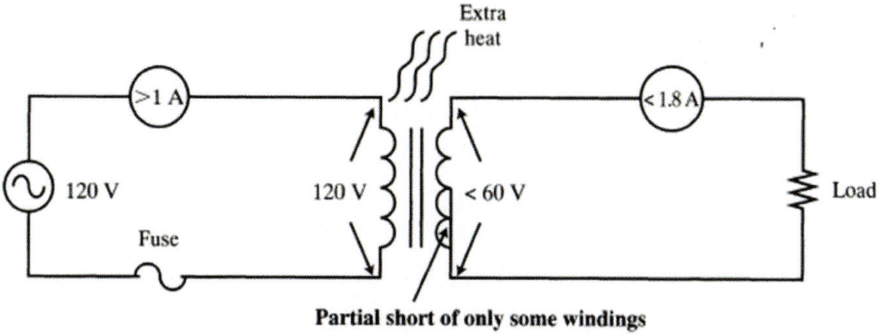

Figure 13-34 A partial short causes extra heat and a decrease in secondary readings.

Open Primary Windings

An **open** means the wire no longer provides a current path. An external open occurs at the point where wires are connected. Other opens can occur internally in the transformer, usually due to excessive heat.

Symptoms shown in **Figure 13-35**:

- There is no primary current.
- There is no secondary voltage or current.
- The transformer is cold. No heat whatsoever is produced.
- The applied voltage is measured across the primary.
- Using an ohmmeter (applied voltage removed):
- An infinite resistance is found in the primary.
- A normal resistance is found in the secondary.

Open Secondary Windings

Opens frequently occur at the connections of the wires to the transformer leads. Check there first whenever an open is suspected. With an open in

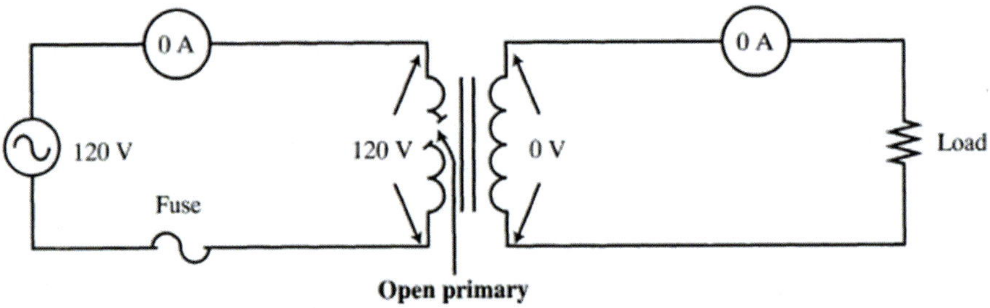

Figure 13-35 Open primary windings stop all current.

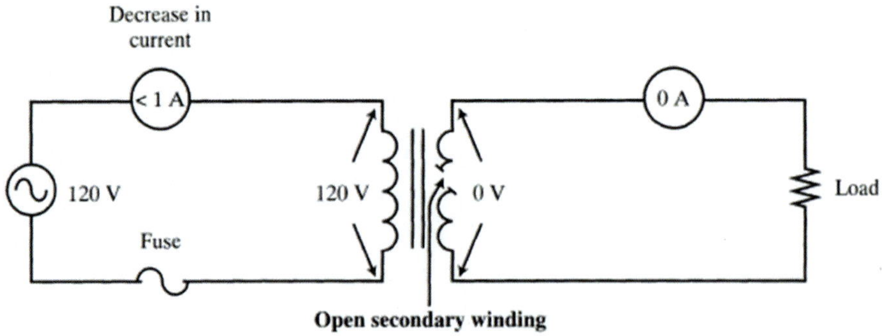

Figure 13-36 Open secondary windings result in a decrease in primary current and zero secondary readings.

the secondary, the primary still develops a magnetic field. However, it does not have to supply secondary current; so only an energizing current will be present. Symptoms shown in **Figure 13-36**:

- The primary current is lower than normal. How much lower depends on the size of the load when the normal measurements were taken.
- The primary voltage is equal to the applied voltage.
- The secondary voltage and current are zero.
- The transformer is cold. Only a very slight amount of heat is developed from the energizing current.
- Using an ohmmeter (after voltage is removed):
- Normal resistance is measured in the primary.
- Infinite resistance is measured in the secondary.

Summary

Transformers use electromagnetic coupling to link the primary and secondary windings.

- The turns ratio of the transformer determines the amount of voltage and current measured red at the secondary as compared to the primary windings.
- A step-up transformer has a larger number of turns on the secondary windings than on the primary windings. It steps up the voltage but steps down the current.
- A step-down transformer has a smaller number of turns on the secondary windings than on the primary windings. It steps down the voltage but steps up the current.
- Power in an ideal transformer is the same in both the primary and secondary windings.
- Transformer losses result in a reduced amount of secondary power.
- Efficiency is the ratio of power out to power in.
- Open transformer windings result in no current, and shorted windings result in too much current.

Formulas

$$Z = \sqrt{R^2 + X^2}$$

$$\text{Turns ratio} = \frac{N_P}{N_S}$$

$$\text{Efficiency} = \frac{P_{\text{out}}}{P_{\text{in}}} \times 100 = \frac{V_S \times I_S}{V_P \times I_P} \times 100$$

$$\text{Resistance ratio} = \frac{R_P}{R_S} = \left(\frac{N_P}{N_S}\right)^2$$

$$\text{Voltage ratio} = \frac{V_P}{V_S} = \frac{N_P}{N_S}$$

Problems

- What is the turns ratio of a transformer with 600 V on the secondary when 50 V are applied to the primary? Is this transformer a step-up or step-down?

  [Turns ratio $= \frac{N_P}{N_S} = \frac{V_P}{V_S} = \frac{50}{600}$

  =1:12, step-up transformer]

  What is the turns ratio of a transformer with 120 V applied to the primary and 12 V on the secondary? Is this transformer a step-up or step-down?

  [Turns ratio $= \frac{N_P}{N_S} = \frac{V_P}{V_S} = \frac{120}{12}$

  =10:1, step-down transformer]

- A transformer has a 2-A fuse in the primary circuit. This primary circuit is connected to 120 V. If the secondary voltage is rated for 20 V, what is the maximum secondary current? (Ignore losses.) Is this transformer a step-up or step-down?

  [$\frac{I_S}{I_P} = \frac{V_P}{V_S}$

  $\frac{I_S}{2} = \frac{120}{20}; I_S = 2 \times 6 = 12\text{ A}$

  12A, step-down transformer]

- Calculate the efficiency of a transformer with 800 W measured at the secondary and a primary power of 1000 W.

  $$[\text{Efficiency} = \frac{P_{\text{out}}}{P_{\text{in}}} \times 100 = \frac{800}{1000} \times 100 = 80\%.]$$

- In a transformer circuit, secondary current is measured at 4 A through a 5-Ω load. If the input voltage is 120 V, calculate the reflected resistance to the primary. Assume an efficiency of 100%.

  [Secondary voltage $V_S = 4 \times 5 = 20$ V

  $$\text{Turns ratio} = \frac{N_P}{N_S} = \frac{V_P}{V_S} = \frac{120}{20} = \frac{6}{1}$$

  $$\text{Resistance ratio} = \frac{R_P}{R_S} = \left(\frac{N_P}{N_S}\right)^2$$

  $$R_P = \left(\frac{N_P}{N_S}\right)^2 \times R_S = \left(\frac{6}{1}\right)^2 \times 5 = 180\,\Omega$$

  180 Ω].

Self-examination/Answers

1. Audio, power, and isolation transformers use ________ cores in their construction.
2. The coils of a transformer are __________linked together.
3. Transformers that have an open core made of paper or cardboard are primarily used in ________ applications.
4. A(n) _____________________ has one coil that is used for the primary and secondary windings.
5. A(n) _____________________ generally has a 1:1 turns ratio.
6. A(n) ____________________ secondary winding produces two voltages that are of equal value and 180° out of phase with each other.
7. The magnetic field of a transformer is first developed by the _____________ winding.
8. The _____________ of a transformer is developed by the secondary windings.
9. (AC/DC) electrical energy must be applied to a transformer in order for it to function.

10. If a transformer primary has 800 turns, and its secondary has 400 turns, the turns ratio is ____________.
11. A transformer with 800 turns in the primary and 400 turns in the secondary will (step-up/step-down) the applied voltage.
12. In a transformer, the power developed by the primary winding is (greater than, equal to, less than) the power developed by the secondary winding.
13. In a step-up transformer, the voltage developed by the secondary is (greater than, equal to, less than) that developed by the primary.
14. ________________ losses in a transformer occur due to heating of its windings.
15. ________________ losses in a transformer are a result of the energy spent in overcoming residual magnetism.
16. Eddy current losses in a transformer can be reduced by using ____________________ of the core material.
17. If the input power of a transformer is 150 W and the output power is 120 W, the transformer efficiency is __________%.
18. The _____________ of a transformer is determined by the ratio of the power output to the power input and is expressed as a percentage.
19. The efficiency of a transformer is generally (greater than, equal to, less than) 100%.
20. When no load is connected to the secondary of a transformer, the power developed by the primary will be (greater than, equal to, less than) the power developed when a load is connected.
21. When a load is connected to the secondary of a transformer, the current developed by the primary will be (greater than, equal to, less than) the current developed when no load is connected.
22. The two ways to calculate the resistance of a transformer are ________________ and __________________.
23. Under (no load , full load) conditions, the primary of the transformer draws reduced current needed only to maintain the magnetic field in the primary winding.
24. When a resistive load is connected to the secondary of a transformer, the phase angle of the energizing current will be (greater than, equal to, less than) 90°.
25. An increase in the resistive load connected to the secondary of a transformer will cause the phase angle of the energizing current to (increase, remain constant, decrease).
26. The transformer ______________ is generally not given and must be derived from the primary and secondary voltage or current ratings.

27. A(n) ______________________________ transformer generally has the primary and secondary impedance values indicated in its ratings.
28. ______________ and ______________ are common fault conditions which may occur in transformer windings.
29. A __________________ in a transformer results in a very heavy current in the transformer.
30. An ohmmeter connected across the windings of a transformer shows a zero resistance. This indicates that the winding is _____________.
31. An ohmmeter connected across the windings of a transformer shows an infinite resistance. This indicates that the winding is _____________.
32. Overheating of a transformer generally causes the insulation to _______________.

Answers

1. metal or iron
2. magnetically or electromagnetically
3. radio frequency or RF
4. autotransformer
5. isolation
6. center-tap
7. primary
8. output
9. AC
10. 2:1
11. step down
12. equal
13. greater
14. Copper or I^2R
15. Hysteresis
16. laminations
17. 80
18. efficiency
19. less than
20. less than
21. greater than
22. Ohm's law, resistance ratio
23. no load
24. less than

25. decrease
26. turns ratio or resistance ratio
27. impedance matching
28. short, open
29. short circuit
30. shorted
31. open
32. melt, break down

Problems Answers

$$\text{Turns ratio} = \frac{N_P}{N_S} = \frac{V_P}{V_S} = \frac{120}{12}$$

=10:1, step-down transformer

$$\frac{I_S}{I_P} = \frac{V_P}{V_S}$$

$$\frac{I_S}{2} = \frac{120}{20}; I_S = 2 \times 6 = 12\,\text{A}$$

12 A, step-down transformer

$$\text{Efficiency} = \frac{P_{\text{out}}}{P_{\text{in}}} \times 100 = \frac{800}{1000} \times 100 = 80\%$$

Secondary voltage $V_S = 4 \times 5 = 20$ V

$$\text{Turns ratio} = \frac{N_P}{N_S} = \frac{V_P}{V_S} = \frac{120}{20} = \frac{6}{1}$$

$$\text{Resistance ratio} = \frac{R_P}{R_S} = \left(\frac{N_P}{N_S}\right)^2$$

$$R_P = \left(\frac{N_P}{N_S}\right)^2 \times R_S = \left(\frac{6}{1}\right)^2 \times 5 = 180\,\Omega.$$

Glossary

Autotransformer
A transformer that has only one coil used for both the primary and the secondary.

Center tap
An electrical connection point at the center of a wire coil or transformer winding.

Current ratio
The ratio of the secondary current to the primary current of a transformer.

$$\text{Current ratio} = \frac{I_P}{I_S}.$$

Eddy current
Electrical current flowing within the core of an electromagnet resulting from voltage induced in the iron core. Eddy currents oppose the current producing the magnetic field resulting in a loss of power.

Efficiency
The ratio of power output available to the load to the power input from the source.

$$\text{Efficiency} = \frac{P_{\text{out}}}{P_{\text{in}}} \times 100.$$

Energizing current
The slight amount of current needed to develop the magnetic field in a winding.

Flux (Φ)
Invisible lines of force that extend around a magnetic material.

Impedance (Z)
The total opposition to current flow in an AC circuit which is a combination of resistance (R) and reactance (X) in a circuit; measured in ohms.

$$Z = \sqrt{R^2 + X^2}.$$

Inductive reactance (X_L)
The opposition to current flow in an AC circuit caused by an inductance (L), measured in ohms. $X_L = 2\pi fL$.

Isolation transformer
A transformer in which the primary and secondary windings are not directly connected to each other.

Maximum power transfer
A condition that exists when the resistance of a load (R_L) equals that of the source which supplies it (R_S).

Mutual inductance (M)
When two coils are located close together so that the magnetic flux of the coils affects one another in terms of their inductance properties.

Phase angle (θ)
The angular displacement between applied voltage and current flow in an AC circuit.

Power (P)
The rate of doing work in electrical circuits, found by using the equation $P == I \times V$

Primary winding
The coil of a transformer to which AC source voltage is applied.

Resistance ratio
The ratio of the primary to the secondary resistance of a transformer.

$$\text{Resistance ratio} = \frac{R_P}{R_S}.$$

Secondary winding
The coil of a transformer into which voltage is induced; energy is delivered to the load circuit by the secondary winding.

Shielded transformer
A transformer with a magnetic shield on the outside of the windings. The shield prevents the magnetic flux from interfering with nearby circuits.

Step-down transformer
A transformer that has a secondary voltage lower than its primary voltage.

Step-up transformer
A transformer that has a secondary voltage higher than its primary voltage.

Transformer
An AC power control device that transfers energy from its primary winding to its secondary winding by mutual inductance and is ordinarily used to increase or decrease voltage.

Turns ratio
The ratio of the number of turns of the primary winding (N_P) of a transformer to the number of turns of the secondary winding (N_S).

$$\text{Turns ratio} = \frac{N_P}{N_S}.$$

Voltage ratio
The ratio of the primary voltage to the secondary voltage of a transformer.

$$\text{Voltage ratio} = \frac{V_P}{V_S}.$$

14

Resistor, Inductor, and Capacitor (RLC) Circuits

Some types of AC circuits are designed to respond to AC frequencies. Circuits that are used to pass some frequencies and block others are called **frequency-sensitive circuits**. Two types of frequency-sensitive circuits are **filter circuits** and **resonant frequency circuits**. Each type of circuit uses reactive devices to respond to different AC frequencies. These circuits have **frequency-response curves**. Frequency is graphed on the horizontal axis and voltage output on the vertical axis. Sample frequency-response curves for each type of filter and resonant circuit are shown in the examples that follow. **Decibels**, as they relate to AC circuits, are also discussed.

Objectives

- Identify low-pass, high-pass, and band-pass filter circuits
- Discuss the performance of series and parallel resonant circuits
- Use decibels and logarithms to plot the frequency response of AC circuits
- Calculate low-frequency cutoff (f_{lc}), high-frequency cutoff (f_{hc}), bandwidth (BW), quality factor (Q), and resonant frequency (f_r) values for frequency-sensitive AC circuits
- Calculate voltage and power amplification and attenuation and voltage and power ratio using decibels

Chapter Outline

14.1 Filter Circuits

The three types of filter circuits are shown in **Figure 14-1**. Filter circuits are used to separate one range of frequencies from another.

Low-pass filters pass low AC frequencies and block higher frequencies. **High-pass filters** pass high frequencies and block lower frequencies.

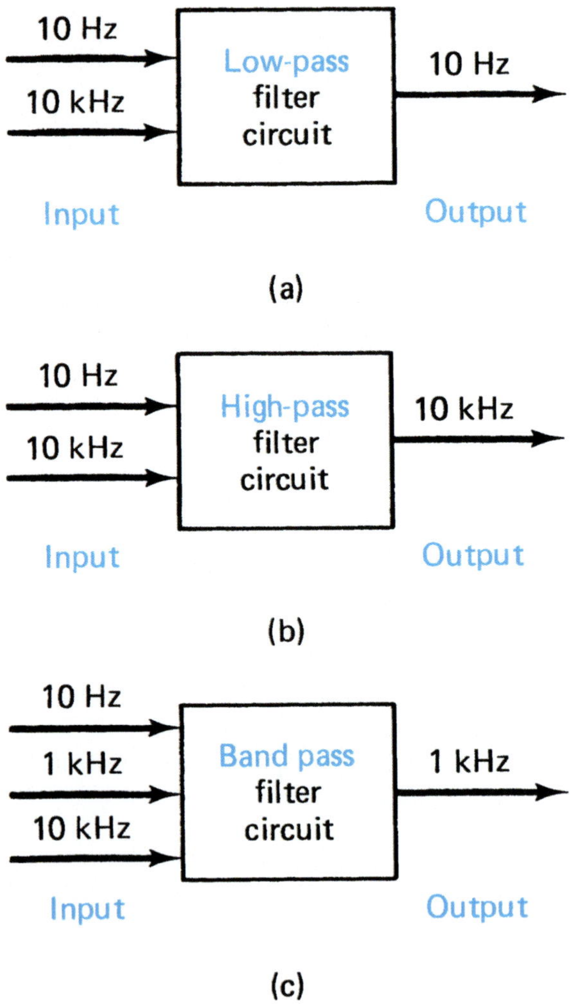

Figure 14-1 Three types of filter circuits: (a) low-pass filter, passes low frequencies and blocks high frequencies; (b) high-pass filter, passes high frequencies and blocks low frequencies; (c) band-pass filter, passes a midrange of frequencies and blocks high and low frequencies.

Band-pass filters pass a midrange of frequencies and block lower and higher frequencies. All filter circuits have resistance and capacitance or inductance. **Figure 14-2** shows the circuits used for low-pass, high-pass, and band-pass filters and their frequency-response curves. Many low-pass filters are series RC circuits, as shown in **Figure 14-2(a)**.

Output voltage (V_{out}) is taken across a capacitor. As frequency increases, capacitive reactance (X_C) decreases, since

$$X_C = \frac{1}{2 \times \pi \times f \times C}.$$

The voltage drop across the output is equal to I times X_C or $V_C = I \times X_C$. So as frequency increases, X_C decreases and voltage output decreases. **Series RL circuits** may also be used as low-pass filters. As frequency **increases**, inductive reactance (X_L) increases since X_L = 2□fL. Any increase in X_L reduces the circuit's current. The voltage output taken across the resistor is equal to I × R. So when I decreases, V_{out} also decreases. As frequency increases, X_L increases, I decreases, and V_{out} decreases.

Figure 14-2(b) shows two types of **high-pass filters**. The series RC circuit is a common type. The voltage output (V_{out}) is taken across the resistor (R). As frequency increases, X_C decreases. A decrease in X_C causes current flow to increase. The voltage output across the resistor (V_{out}) is equal to I × R. So as I increases, V_{out} increases. As frequency increases, X_C decreases, I increases, and V_{out} increases. A **series RL circuit** may also be used as a high-pass filter. The V_{out} is taken across the inductor. As frequency **increases,** X_L increases. V_{out} is equal to $I \times X_L$. So as X_L increases, V_{out} also increases. In this circuit, as frequency increases, X_L increases and V_{out} increases.

The **band-pass filter** of **Figure 14-2(c)** is a combination of low-pass and high-pass filter sections. It is designed to pass a midrange of frequencies and block low and high frequencies. R_1 and C_1 form a low-pass filter and R_2 and C_2 form a high-pass filter. The range of frequencies to be passed is determined by calculating the values of resistance and capacitance.

14.2 Resonant Circuits

Resonant circuits are designed to pass a range of frequencies and block all others. They have resistance, inductance, and capacitance in their construction. **Figure 14-3** shows the two types of resonant circuits – series resonant and parallel resonant circuits – and their frequency-response curves.

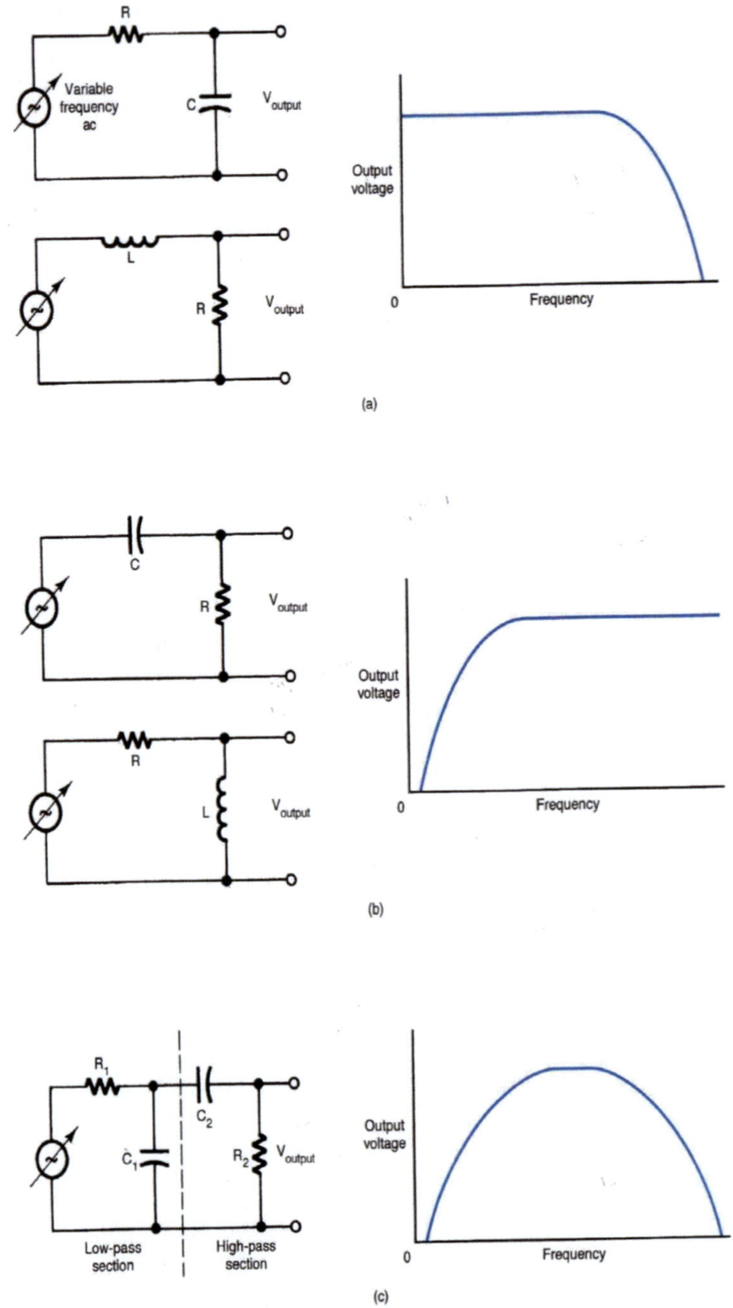

Figure 14-2 Circuits used to filter AC frequencies and their response curves: (a) low-pass filters; (b) high-pass filters; (c) band-pass filter.

Series Resonant Circuits

Series resonant circuits are a series arrangement of inductance, capacitance, and resistance. A series resonant circuit offers a small amount of opposition to some AC frequencies and much more opposition to other frequencies. They are important for selecting or rejecting frequencies. The voltage across inductors and capacitors in AC series circuits are in direct opposition to each other (180° out of phase). They tend to cancel each other out. The frequency applied to a series resonant circuit affects inductive reactance and capacitive reactance. At a specific input frequency, X_L will equal X_C. The voltages across the inductor and capacitor are then equal. The total reactive voltage (V_x) is 0 V at this frequency.

The opposition offered by the inductor and the capacitor cancels each other at this frequency. The total reactance (X_T) of the circuit (X_L minus X_C) is zero. The impedance (Z) of the circuit is then equal to the resistance (R). The frequency at which $X_L = X_C$ is called the **resonant frequency**. To determine the resonant frequency (f_r) of the circuit, use the following formula:

$$f_r = \frac{1}{2\pi\sqrt{LC}}.$$

In the formula, L is in henries, C is in farads, and f_r is in hertz. As either inductance or capacitance increases, resonant frequency decreases. When the resonant frequency is applied to a circuit, a condition called resonance exists. Resonance for a series circuit causes the following:

1. $X_L = X_C$
2. X_T is equal to zero
3. $V_L = V_C$
4. Total reactive voltage (V_x) is equal to zero
5. Z = R
6. Total current (I_T) is maximum
7. Phase angle (θ) is 0°

The ratio of reactance (X_L or X_C) to resistance (R) at resonant frequency is called **quality factor (Q)**. This ratio is used to determine the range of frequencies or bandwidth (BW) that a resonant circuit will pass. A sample resonant circuit problem is shown in **Figure 14-4**. The frequency range that a resonant circuit will pass (BW) is found by using the procedure of steps 5 and 6 in **Figure 14-4(b)**.

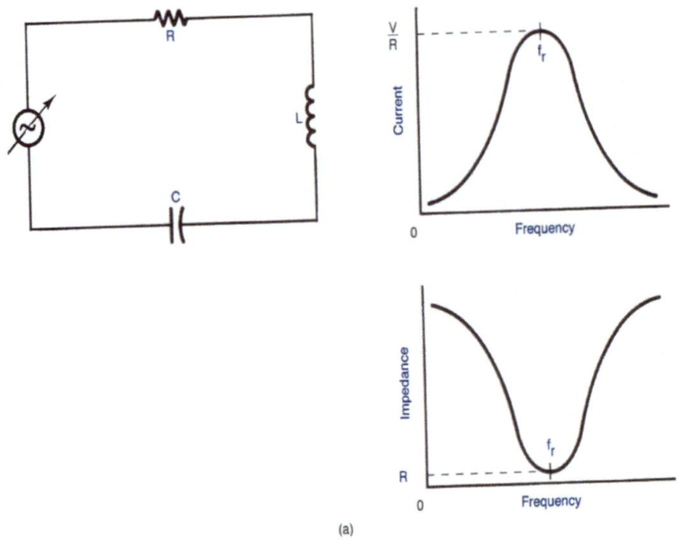

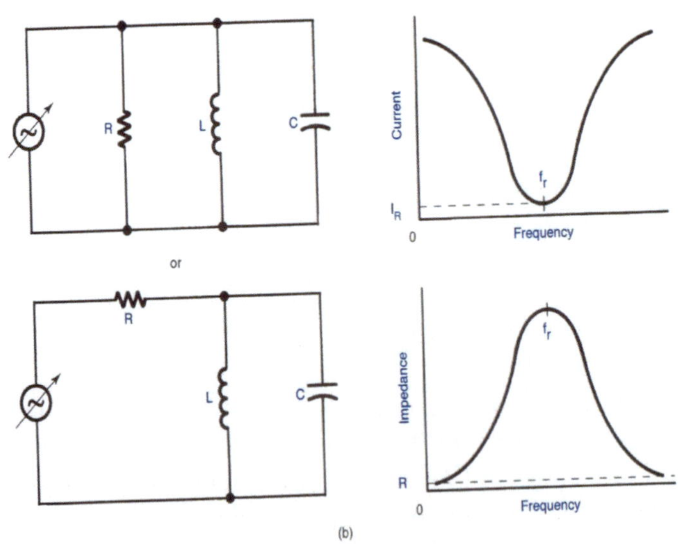

Figure 14-3 Resonant (a) series circuit and frequency-response waveforms and (b) parallel circuits and frequency-response waveforms

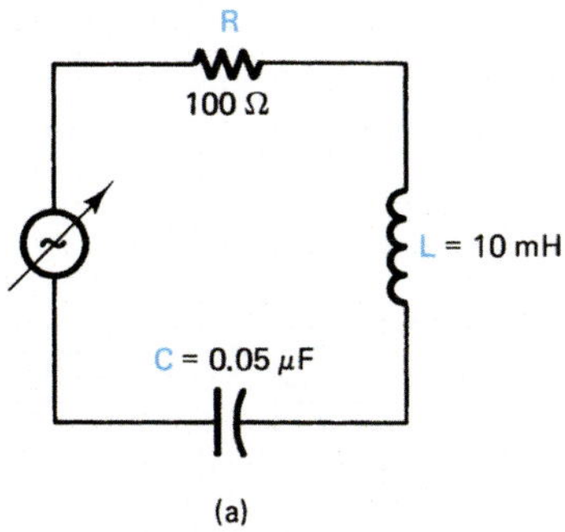

Finding circuit values:

1. Find resonant frequency (f_r):

$$f_r = \frac{1}{2\pi \times \sqrt{L \times C}} = \frac{1}{6.28 \times \sqrt{(10 \times 10^{-3}) \times (0.05 \times 10^{-6})}} =$$

$$= \frac{1}{6.28 \times \sqrt{0.5 \times 10^{-9}}} = \frac{1}{6.28 \times (2.23 \times 10^{-5})} = \frac{1}{1.4 \times 10^{-4}} = 7121 \text{ Hz}$$

2. Find X_L^* and X_C at resonant frequency:

$$X_L = 2\pi \cdot f \cdot L = 6.28 \times 7121 \times (10 \times 10^{-3}) = 447\ \Omega$$

*Easier to calculate.

3. Find quality factor (Q):

$$Q = \frac{X_L}{R} = \frac{447\ \Omega}{100\ \Omega} = 4.47$$

4. Find bandwidth (BW):

$$BW = \frac{f_r}{Q} = \frac{7121 \text{ Hz}}{4.47} = 1593 \text{ Hz}$$

5. Find low-frequency cutoff (f_{lc}):

$$f_{lc} = f_r - \tfrac{1}{2} BW = 7121 \text{ Hz} - 797 \text{ Hz} = 6324 \text{ Hz}$$

6. Find high-frequency cutoff (f_{hc})

$$f_{hc} = f_r + \tfrac{1}{2} BW = 7121 + 797 = 7918 \text{ Hz}$$

(b)

Figure 14-4 Sample resonant circuit problem: (a) circuit; (b) procedure for finding circuit values.

The **cutoff points** are at about 70% of the maximum output voltage, which indicates the location of the half-power point. At this point, the output power changes in value to half of the applied input power. These are called

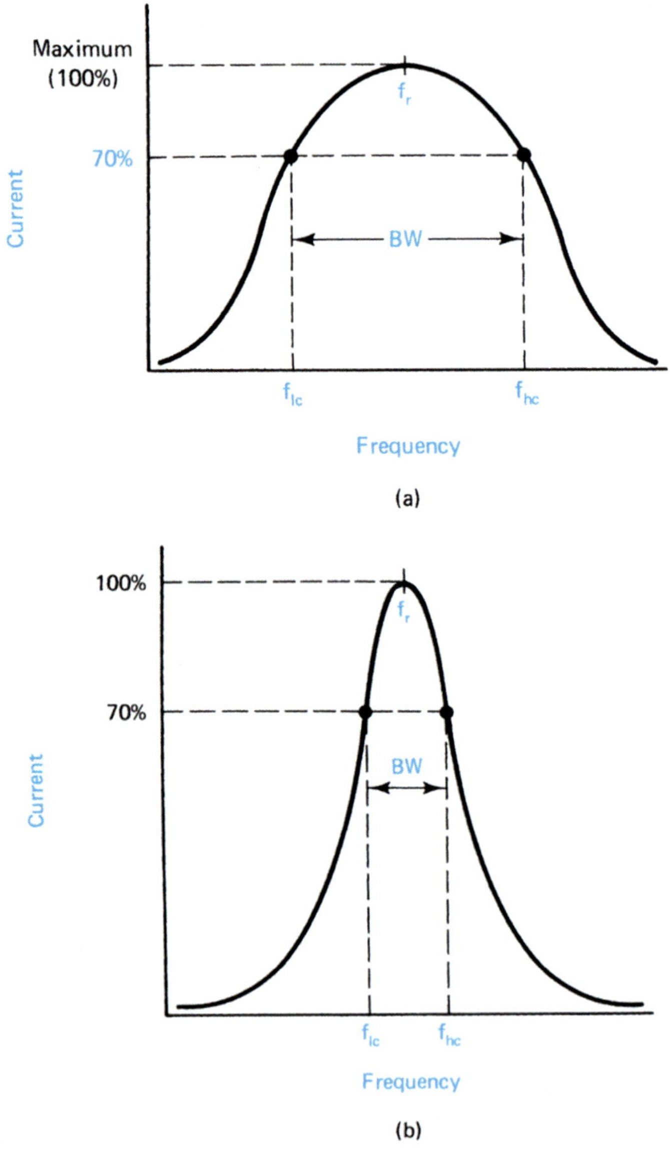

Figure 14-5 Effect of resistance on bandwidth of a series resonant circuit: (a) high resistance, low selectivity; (b) low resistance, high selectivity.

the **low-frequency cutoff (f_{lc}) and high-frequency cutoff (f_{hc})**. Once the resonant frequency (f_r) and the bandwidth (BW) of a specific filter circuit are known, the appropriate cutoff frequencies can be determined as follows:

$$f_{\text{lc}} = f_r - \frac{1}{2}\text{BW} \text{ and } f_{\text{hc}} = f_r + \frac{1}{2}\text{BW}.$$

The **bandwidth** of a resonant circuit is determined by the Q. Q is determined by the ratio of X_L and X_C to R. Resistance mainly determines bandwidth. This effect is summarized as follows:

1. When R is increased, Q decreases, since. $Q = \frac{X_L}{R}$
2. When Q decreases, BW increases, since. $\text{BW} = \frac{f_r}{Q}$
3. When R is increased, BW increases.

Two curves in **Figure 14-5** show the effect of resistance on bandwidth. The curve of **Figure 14-5(b)** has high selectivity. This means that a resonant circuit with this response curve would select a small range of frequencies. This is very important for radio and television tuning circuits.

Parallel Resonant Circuits

Parallel resonant circuits are similar to series resonant circuits. Their electrical characteristics are somewhat different, but they accomplish the same purpose. Another name for parallel resonant circuits is tank circuit. A tank circuit is a parallel combination of L and C. With the resonant frequency applied to a parallel resonant circuit, the following occurs:

1. $X_L = X_C$.
2. $X_T = 0$
3. $I_L = I_C$.
4. I_x = zero; so the circuit current is minimum
5. Z = R and is maximum
6. Phase angle (θ) = 0°

The calculations used for parallel resonant circuits are similar to those for series circuits. There is one exception. The **quality factor (Q)** is found by using this formula for parallel circuits: $Q = \frac{X_L}{R}$. A parallel resonant circuit problem is shown in **Figure 14-6**.

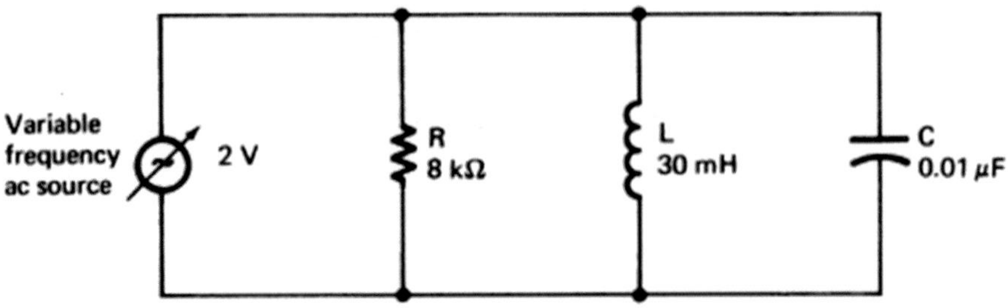

(a)

Finding circuit values:

1. Find resonant frequency (f_r):

$$f_r = \frac{1}{2\pi\sqrt{L \times C}} = \frac{1}{6.28 \times \sqrt{(30 \times 10^{-3}) \times (0.01 \times 10^{-6}}} = \frac{1}{1.0877 \times 10^{-4}} = 9194 \text{ Hz}$$

2. Find X_L at resonant frequency:

$$X_L = 2\pi \cdot f \cdot L = 6.28 \times 9194 \times (30 \times 10^{-3}) = 1732\ \Omega$$

3. Find quality factor (Q):

$$Q = \frac{R}{X_L} = \frac{8000\ \Omega}{1732\ \Omega} = 4.62$$

4. Find bandwidth (BW):

$$BW = \frac{f_r}{Q} = \frac{9194 \text{ Hz}}{4.62} = 1990 \text{ Hz}$$

5. Find low cutoff frequency (f_{lc}):

$$f_{lc} = f_r - 1/2\ BW = 9194 \text{ Hz} - 995 \text{ Hz} = 8199 \text{ Hz}$$

6. Find high cutoff frequency (f_{hc}):

$$f_{hc} = f_r + 1/2\ BW = 9194 \text{ Hz} + 995 \text{ Hz} = 10{,}189 \text{ Hz}$$

7. Find I_R at resonant frequency:

$$I_R = \frac{V_A}{R} = \frac{2 \text{ V}}{8 \text{ k}\Omega} = 0.25 \text{ mA}$$

8. Find I_L and I_C at resonant frequency:

$$I_L = \frac{V_A}{X_L} = \frac{2 \text{ V}}{1732\ \Omega} = 1.15 \text{ mA}$$

Figure 14-6 Frequency response for a low-pass filter circuit: circuit; (b) procedure for finding frequency response.

- Refer to the series resonant circuit of **Figure 14-4**. Use values of R = 1 kΩ, L = 100 mH, V_{in} = 2 V, and C = 0.025 μF. Determine the following: resonant frequency (f_r), X_L and X_C at f_r. [

$$f_r = \frac{1}{2\pi\sqrt{L \times C}} = \frac{1}{2\pi\sqrt{0.1 \times 0.025 \times 10^{-6}}} = \frac{1}{2\pi \times 5 \times 10^{-5}} = 3180.09 \text{ Hz}$$

$$X_L = 2 \times \pi \times f \times L = 2 \times \pi \times 3180.09 \times 0.1 = \mathbf{2000\ \Omega}$$

$$X_C = \frac{1}{2 \times \pi \times f \times C} = \frac{1}{2 \times \pi \times 3180.09 \times 0.025 \times 10^{-6}} = 2001.89\,\Omega.$$

The values of X_L and X_C are approximately the same.]

- Refer to the parallel resonant circuit of **Figure 14-6**. Use values of R = 5 kΩ, L = 20 mH, C = 0.02 μF, and V_{in} = 1 V. Determine the following: resonant frequency (f_r), X_L and X_C at f_r.

$$f_r = \frac{1}{2\pi\sqrt{L \times C}} = \frac{1}{2\pi\sqrt{0.02 \times 0.02 \times 10^{-6}}} = \frac{1}{2\pi \times 2 \times 10^{-5}} = 7957.75 \text{ Hz}$$

$$X_L = 2 \times \pi \times f \times L = 2 \times \pi \times 7957.75 \times 0.02 = 1000\ \Omega$$

$$X_C = \frac{1}{2 \times \pi \times f \times C} = \frac{1}{2 \times \pi \times 7957.75 \times 0.02 \times 10^{-6}} = 1000\,\Omega$$

The values of X_L and X_C are equal.]

14.3 Decibels and Power Calculations in Filter Circuits

The human ear does not respond to sound levels in the same manner as electronic circuits. An electronic amplifier, for example, has a linear rise in signal level. An input signal level of 1 V could produce an output of 10 V. The voltage amplification would be 10:1 or 10. The human ear, however, does not respond in a linear manner. The response is essentially **nonlinear**. As a result of this, sound systems are usually evaluated on a logarithmic scale. This is an indication of how our ears will actually respond to specific signal levels. **Gain** expressed in **logarithms** is much more meaningful than linear gain relationships.

This is illustrated by the following:

$10^4 = 10{,}000$
$10^3 = 1000$
$10^2 = 100$
$10^1 = 10$
$10^0 = 1$

This means that the logarithm of any number between 9999 and 1000 would have a **characteristic** value of 3. The characteristic is an expression of the magnitude range of the number. Numbers between 999 and 100 would have a characteristic of 2. Numbers between 99 and 10 would have a characteristic of 1. Between 9.0 and 1.0, the characteristic would be 0. Number values less than 1.0 would be expressed as a negative characteristic. It is generally not customary to use negative characteristic values.

When a number is not an even multiple of 10, it must have a decimal expression. The decimal part of the logarithm is called the **mantissa**. A number such as 4000 would be expressed as 3.6018. The characteristic is 3 because 4000 is between 9999 and 1000. The mantissa of 4000 is 0.6018. Prior to the widespread use of scientific calculators, the mantissa was found from a table of common logarithms. On a calculator, the log of 4000 is displayed as 3.6021. The characteristic is 3 and the mantissa is 0.6021.

The **mantissa** is always the same for a given sequence of numbers regardless of the location of the decimal point. For example, the mantissa is the same for 1630, 163.0, 16.3, 1.63, 0.163, and so on. The only difference in these values would be the **characteristic**. The mantissa for 1630 is 2122. The log of the five values would be 3.2122, 2.2122, 1.2122, 0.2122, and –1.2122.

Scientific calculators with logarithms make the conversion process very easy. Simply enter the number into the calculator and then press the log button. For example, the log of the number 1590 is 3.2012.

Decibel Applications

The **gain** of a sound system with several stages of amplification can best be expressed as a ratio of two signal levels. Specifically, gain is expressed as the output level divided by the input level. This is determined by the following expression:

$$\text{Power amplification: } A_P = \log \frac{P_{\text{out}}}{P_{\text{in}}} \text{ bels.}$$

For an amplifier with 0.1 W of input and 100 W of output, the A_P would be

$$A_P = \log\frac{100}{0.1} = 3 \text{ bels.}$$

The fundamental unit of sound level gain is the **bel (B)**. As noted in this calculation, the bel represents a rather large ratio in sound level. A **decibel (dB)** is a more practical measure of sound level. A decibel is one-tenth of a bel. The gain of a single stage of amplification within a system can be determined with decibels. A single amplifier stage could have an input of 10 mW and output of 150 mW. The power gain would be determined by the following formula:

$$\text{Power amplification: } A_P = 10\log\frac{P_{out}}{P_{in}} \text{ dB}$$

$$= 10\log\frac{150\,\text{mW}}{10\,\text{mW}} = 10\log 15 = 10\times 1.1761 = 11.761.$$

The **voltage gain** of an amplifier can also be expressed in dB values. To do this, the power-level expression must be adapted to accommodate voltage values. The dB voltage gain formula is

$$\text{Voltage amplification: } A_V = 20\times\log\frac{\text{voltage output}}{\text{voltage input}} = 20\log\frac{V_{out}}{V_{in}} \text{ dB.}$$

Note that the logarithm of V_{out}/V_{in} is multiplied by 20 in this equation. Power is expressed as V^2/R. Power gain using voltage and resistance values would therefore be expressed as follows:

$$A_V = \frac{\dfrac{V_{out}^2}{R_{out}}}{\dfrac{V_{in}^2}{R_{in}}} = \frac{V_{out}^2\times R_{in}}{V_{in}^2\times R_{out}}.$$

If the values of R_{in} and R_{out} are equal, the equation would be simplified to be

$$A_V = \frac{V_{out}^2}{V_{in}^2} = \left(\frac{V_{out}}{V_{in}}\right)^2.$$

The squared voltage values can be expressed as two times the log of the voltage value.

Decibel voltage gain therefore becomes

$$\text{Voltage amplification: } A_V = 10\log\left(\frac{V_{out}}{V_{in}}\right)^2 = 2\times 10\log\frac{V_{out}}{V_{in}} = 20\log\frac{V_{out}}{V_{in}} \text{ dB.}$$

To demonstrate the use of the dB voltage gain equation, let us apply it to a circuit which has an input voltage of 0.25 $V_{p\text{-}p}$ and the output voltage is 1.25 $V_{p\text{-}p}$. The voltage gain in dB is

$$A_V = 20\log\frac{V_{out}}{V_{in}} \text{ dB}$$

$$= 20\log\frac{1.25V_{p-p}}{0.25V_{p-p}} = 20\log 5 = 20\times 0.69897 = 13.9794 \approx 14 \text{ dB.}$$

When the dB value of an amplifier is known, the **power gain** or **voltage gain** may be determined by using **inverse logarithms** or **antilogarithms**. An inverse logarithm is the number from which a logarithm is derived. The process of finding an inverse logarithm is the reverse of finding a logarithm. They may easily be determined by using a scientific calculator.

As an example of using inverse logarithms, assume that an amplifier has a dB value of +3.5. The power gain of the amplifier is found as follows:

$$A_P = 10\log\frac{P_{out}}{P_{in}} \text{ dB}$$

$$3.5 = 10\log\frac{P_{out}}{P_{in}}$$

$$0.35 = \log\frac{P_{out}}{P_{in}} \quad \text{(divide both sides of the equation by 10)}$$

$$\frac{P_{out}}{P_{in}} = \text{inv log } 0.35 \text{ (find inverse logarithm)}$$

$$\frac{P_{out}}{P_{in}} = 2.24.$$

The value of 2.24 obtained in the example is the power ratio. An amplifier with a gain of +3.5 dB thus has a power gain of 2.24 to 1. The inverse logarithm value of the example is found by using a scientific calculator. This will indicate that the **power ratio** is 2.24. This indicates that the value of the output power has increased to approximately twice the value of the input power.

Decibels are also used to express reduction in power or voltage levels. When reduction of input signal level occurs in a circuit, this is called **attenuation**. A circuit that attenuates a signal is compared to an amplifier circuit in **Figure 14-7**. Note that the dB value is marked with a (–) sign when the circuit attenuates the input signal. A common example of attenuation

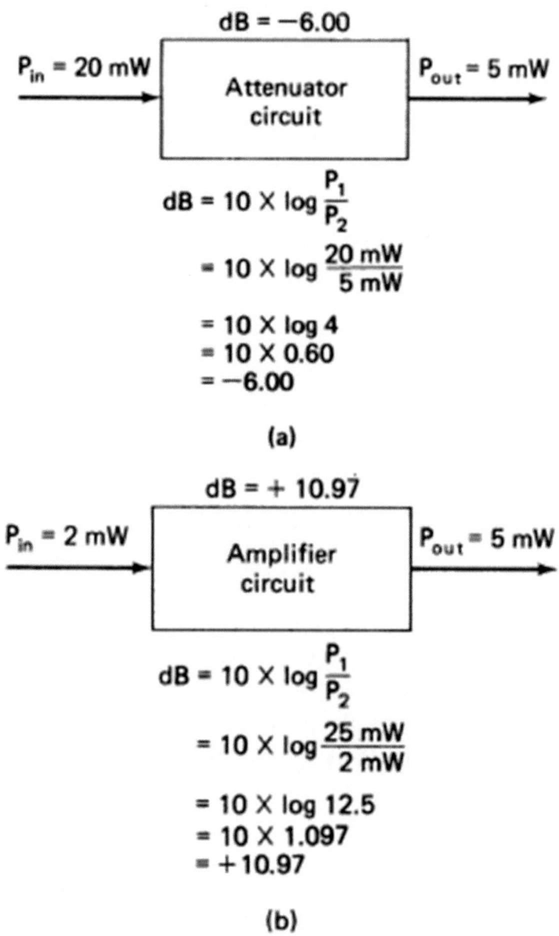

Figure 14-7 Comparison of (a) attenuator and (b) amplifier circuits.

occurs in coaxial cable or other signal transmission cable in which a reduction of signal occurs from input to output.

An example of **attenuation** occurs when a filter reduces the output power by a certain decibel value at a given frequency. This means that the power ratio of a filter is identified by a negative decibel value. Consider a filter circuit which has a –3.5 dB attenuation at 10 kHz. The power ratio of the filter can be found as follows:

$$A_P = 10\log\frac{P_{out}}{P_{in}}\,\text{dB}$$

$$-3.5 = 10\log\frac{P_{out}}{P_{in}}$$

$$-0.35 = \log\frac{P_{out}}{P_{in}} \quad \text{(divide both sides of the equation by 10)}$$

$$\frac{P_{out}}{P_{in}} = \text{inv log } -0.35 \text{ (find inverse logarithm)}$$

$$\frac{P_{out}}{P_{in}} = 0.4467.$$

This indicates that the value of the output power is reduced to approximately half of the input power.

Filter Circuits

Decibels are commonly used to plot **frequency-response curves** for filter circuits using the type of graph paper shown in **Figure 14-8**.

One example of a **low-pass filter** circuit is shown **in Figure 14-9(a)**. The procedure for plotting a frequency-response curve for the low-pass circuit is shown in **Figure 14-9(b)**, and the plot is shown in **Figure 14-9(c)**.

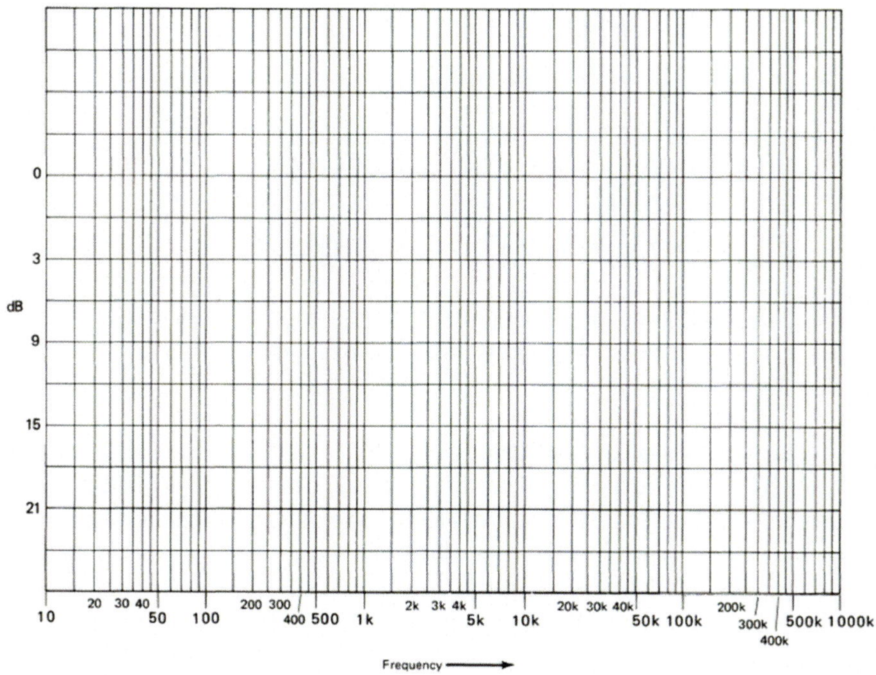

Figure 14-8 Decibel values used to plot frequency response.

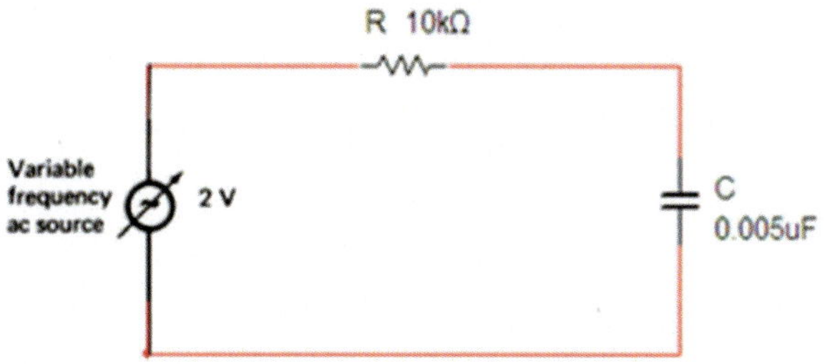

1. Determine the 3-dB frequency using the following formula:

$$f_{3\text{dB}} = \frac{1}{2 \times \pi \times R \times C}$$

$$= \frac{1}{2 \times \pi \times (10 \times 10^{3}) \times (0.005 \times 10^{-6})}$$

$$= \frac{1}{3.1416 \times 10^{-4}} = 3183.1\,\text{Hz}.$$

2. Find the 9-, 15-, and 21-dB frequencies:

$$f_{9\text{dB}} = 2 \times f_{3\text{dB}} = 2 \times 3183.1 = 6366.2\ \text{Hz}.$$

$$f_{12\text{dB}} = 4 \times f_{3\text{dB}} = 4 \times 3183.1 = 12732.4\,\text{Hz}.$$

$$f_{21\text{dB}} = 8 \times f_{3\text{dB}} = 8 \times 3183.1 = 25464.8\,\text{Hz}.$$

3. Label the points on a sheet of frequency-response paper.
4. Connect each of the points to form a low-pass frequency-response curve.

The selection of decibel values of 3, 9, 15, and 21 dB is standard for plotting frequency response in terms of voltage output of a circuit. First, locate the 3-dB line. Note that at **3-dB reduction** of a signal, the power output is approximately 0.5 or 50% of the 0-dB reference level and the voltage is approximately 0.707 or 70.7% of the 0-dB level. Since the power output of a circuit reduces to about 50% of its original value (0 dB), the 3-dB frequency is called the **half-power point**.

(b)

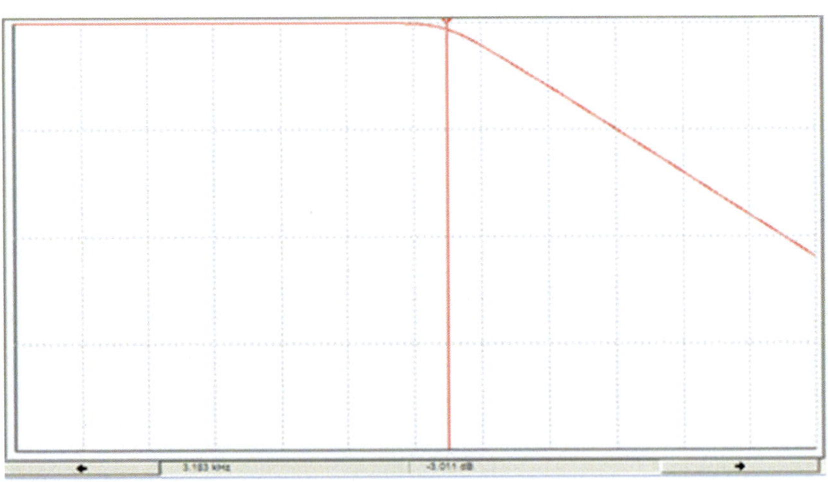

(c)

Figure 14-9 Frequency response for a low-pass filter circuit: (a) circuit; (b) procedure for finding frequency response; (c) frequency-response curve.

The selection of dB points for a **high-pass filter circuit** *is similar to the process used for low-pass filter circuits. A high-pass filter circuit and the procedure* for plotting a frequency-response curve are shown in **Figure 14-10**.

Band-pass filter circuits are a combination of low-pass and high-pass filter circuits. The 3-dB frequency on the low-frequency end of the response curve is called the low-cutoff frequency (f_{lc}). The high-frequency 3-dB point is called the high-cutoff frequency (f_{hc}).

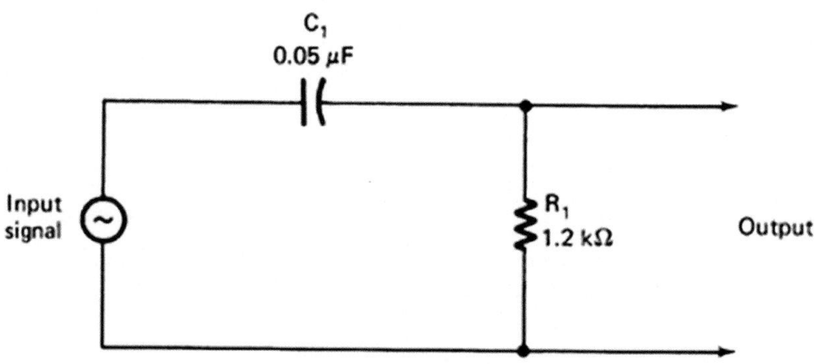

(a)

Finding frequency response:

1. Determine the 3-dB frequency:

$$f_{3dB} = \frac{1}{2\pi \times R \times C} = \frac{1}{6.28 \times (1.2 \times 10^3) \times (0.05 \times 10^{-6})}$$

$$= \frac{1}{3.768 \times 10^{-4}} = 2654 \text{ Hz}$$

2. Find the 9-, 15-, and 21-dB frequencies:

$$f_{9dB} = f_{3dB} \div 2 = 2654 \div 2 = 1327 \text{ Hz}$$

$$f_{15dB} = f_{3dB} \div 4 = 2654 \div 4 = 663.5 \text{ Hz}$$

$$f_{21dB} = f_{3dB} \div 8 = 2654 \div 8 = 332 \text{ Hz}$$

3. Label each of the points on a sheet of frequency-response paper.

4. Connect each of the points to form a high-pass frequency-response curve.

(b)

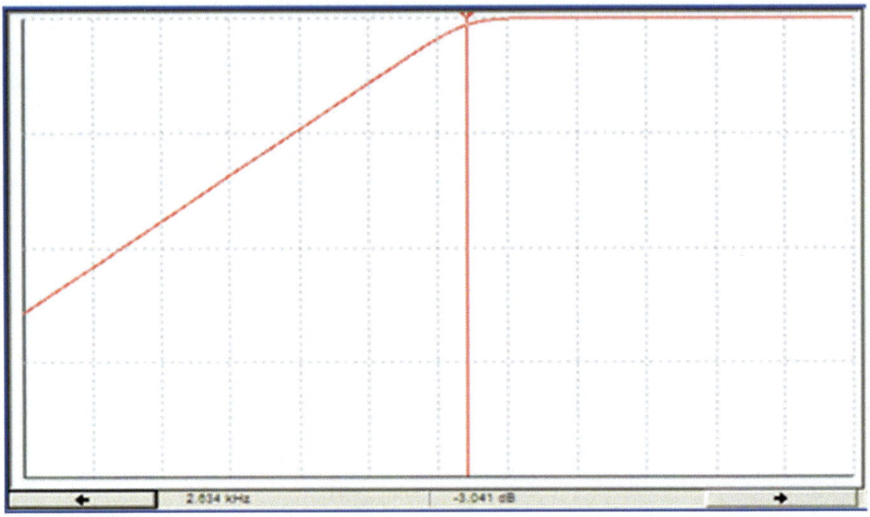

Figure 14-10 Frequency response for a high-pass filter circuit: (a) circuit; (b) procedure for finding frequency response; (c) frequency-response curve.

Review Questions

1. The logarithmic characteristic of the number 5000 is __________.[3]
2. The fractional or decimal part of a logarithm is called the (characteristic, mantissa). [mantissa]
3. An amplifier has input of 0.1 W and an output of 2 W. Determine the power amplification (A_P) in dB.

   $$[\,A_P = 10\log\frac{P_{out}}{P_{in}} = 10\log\frac{2}{0.1} = 10\log 20 = 10\times 1.301 = 13.01\,\text{dB}\,]$$

4. An amplifier has voltage input of 0.2 $V_{P\text{-}P}$ and an output of 1.5 $V_{P\text{-}P}$. Determine the voltage amplification (A_V) in dB.

   $$[\,A_V = 10\log\frac{V_{out}}{V_{in}} = 10\log\frac{1.5}{0.2} = 10\log 7.5 = 10\times 0.8751 = 8.751\,\text{dB}\,]$$

5. A circuit has a loss of –16 dB. What power ratio corresponds to this loss? [

   $$A_P = 10\log\frac{P_{out}}{P_{in}}\,dB$$

$$-16 = 10\log\frac{P_{out}}{P_{in}}$$

$$-1.6 = \log\frac{P_{out}}{P_{in}}$$

$$\frac{P_{out}}{P_{in}} = \text{inv}\ \log -1.6$$

$$\frac{P_{out}}{P_{in}} = 0.025]$$

Troubleshooting Filter and Resonant Circuits

Troubleshooting of RLC or filter circuits can be carried out by energizing the circuit with a **function generator** that will produce different frequencies. An **oscilloscope** can be used to compare the values and shape of the input and output signals. The resonant frequency of a series or parallel RLC circuit can be determined mathematically. When an input signal with the **given resonant frequency** is applied, the output voltage of the circuit should be identical in value and be in phase with the input voltage. This indicates that the RLC circuit is operating satisfactorily and can be used to select a desired resonant frequency. If the output voltage is not the same as the input voltage at the identified resonant frequency, this indicates that the values of either the inductor or the capacitor of the circuit has changed. This means that the circuit may not be able to select or reject a specific frequency. The **bandwidth** of a resonance circuits is an important factor, as it is used in turning circuits for selecting a particular range of frequencies and rejecting others. Variations in the values of resistance in an RLC circuit can affect the bandwidth.

When troubleshooting **filter circuits**, specific low-frequency and high-frequency cutoff values must be determined corresponding to the type of filter being used. A function generator is used to apply input signals to the circuit and an oscilloscope is used to observe the output response. The output can be used to verify whether the circuit is responding properly to these frequencies. For example, a **low-pass filter** will attenuate frequencies above its high cutoff frequency. A **high-pass filter** will attenuate frequencies below its low cutoff frequency. A **band-pass filter** will attenuate frequencies below its low-frequency cutoff and those above its high-frequency cutoff. By observing the frequency below and above the cutoff values, the output response and condition of the filter can be evaluated. Since filters are commonly constructed using resistors, inductors, and capacitors, a change in value of these components will adversely affect the performance of the filter.

Summary

Low-pass, high-pass, and band-pass filter circuits allow certain AC frequencies to pass from input to output and offer high impedance to other frequencies.

- The frequency-sensitive characteristics of filter circuits are determined by the value of inductive reactance (X_L) or capacitive reactance (X_C). As frequency increases, X_L increases and X_C decreases.
- Series and parallel resonant circuits have R, L, and C components and provide the capability of passing or "tuning" a desired frequency range.
- Proper calculations of circuit values are essential for filter and resonant circuits.
- Decibels and logarithms are used to plot the frequency response of AC circuits.
- Frequency-response curves show how a circuit will respond graphically to different frequencies.
- Decibels are also used to indicate the amplification, attenuation, voltage ratio, and power ratio of a circuit.

Formulas

$$X_C = \frac{1}{2 \times \pi \times f \times C}$$

$$f_r = \frac{1}{2\pi\sqrt{LC}}$$

$$Q = \frac{X_L}{R}$$

$$\text{BW} = \frac{f_r}{Q}$$

$$f_{\text{lc}} = f_r - \frac{1}{2}\text{BW}$$

$$f_{\text{hc}} = f_r + \frac{1}{2}\text{BW}$$

Power amplification: $A_P = \log\frac{P_{out}}{P_{in}}$ bels, $A_P = 10\log\frac{P_{out}}{P_{in}}$ dB

Voltage amplification: $A_V = 20\log\frac{V_{out}}{V_{in}}$ dB

Problems

Refer to the series RLC shown in **Figure 14-11**. Determine the resonant frequency, the quality factor, bandwidth, and lower and higher cutoff frequencies. Also determine the current as well as the voltage drops across the resistor, inductor, and capacitor, at the resonant frequency.

An amplifier has a voltage input of 0.7 V_{P-P} and an output of 4.2 V_{P-P}. Determine the voltage amplification (A_V) in dB.

$$[\, A_V = 10\log\frac{V_{out}}{V_{in}} = 10\log\frac{4.2}{0.7} = 10\log 6 = 10\times 0.7781 = 7.781\,\text{dB}\,]$$

What is the voltage ratio of a circuit if there is a gain of 8 dB?

$$[\, A_V = 10\log\frac{V_{out}}{V_{in}}\,\text{dB}$$

$$8 = 10\log\frac{V_{out}}{V_{in}}$$

$$0.8 = \log\frac{V_{out}}{V_{in}}$$

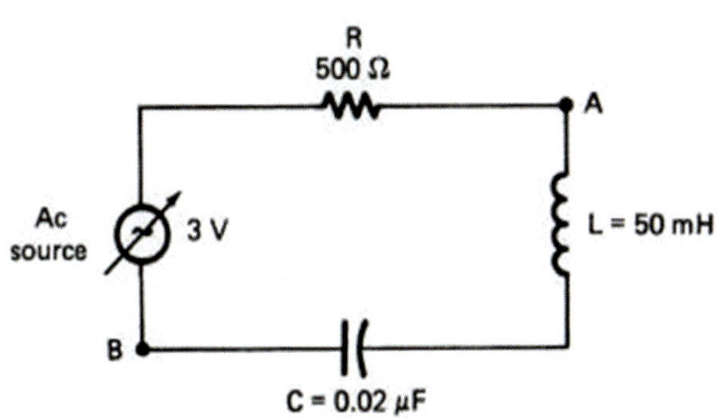

Figure 14-11 Series resonant circuit problem.

$$\frac{V_{out}}{V_{in}} = \text{inv log } 0.8$$

$$\frac{V_{out}}{V_{in}} = 6.31]$$

Refer to the circuit diagram of a low-pass filter shown in **Figure 14-9**. Use the values of R = 20 kΩ and C = 0.004 μF for determining the 3-, 9-, 15-, and 21-dB frequencies. [Determine the 3-dB frequency using the following formula:

$$f_{3dB} = \frac{1}{2\times\pi\times R\times C}$$

$$= \frac{1}{2\times\pi\times(20\times10^{3})\times(0.004\times10^{-6})}$$

$$= \frac{1}{5.0265\times10^{-4}} = 1989.44\,\text{Hz}.$$

Find the 9-, 15-, and 21-dB frequencies:

$$f_{9dB} = 2\times f_{3dB} = 2\times1989.44 = 3978.87\,\text{Hz}$$
$$f_{12dB} = 4\times f_{3dB} = 4\times1989.44 = 7957.75\,\text{Hz}$$
$$f_{21dB} = 8\times f_{3dB} = 8\times1989.44 = 15915.49\,\text{Hz}]$$

Develop a frequency-response curve of the circuit of the low-pass filter shown in **Figure 14-12**.

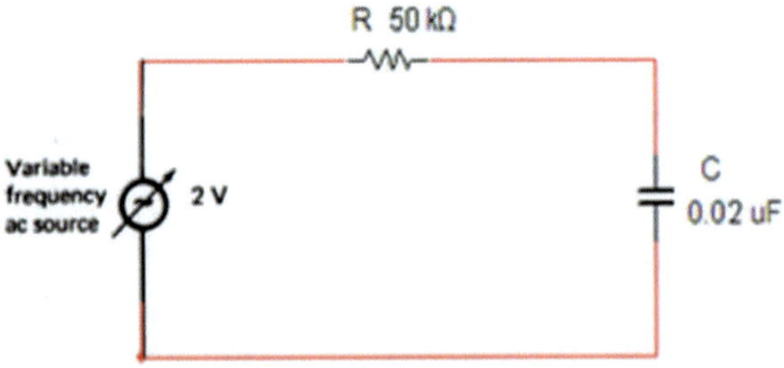

Figure 14-12 Develop a frequency response for a low-pass filter circuit.

[The steps for developing the frequency-response curve of the low-pass filter are shown below.

(a) Determine the 3-dB frequency using the following formula:

$$f_{3\text{dB}} = \frac{1}{2\times\pi\times R\times C}$$

$$= \frac{1}{2\times\pi\times(50\times10^{3})\times(0.02\times10^{-6})}$$

$$= \frac{1}{6.283\times10^{-3}} = 159.1\,\text{Hz}.$$

(b) Find the 9-, 15-, and 21-dB frequencies:

$$f_{9\text{dB}} = 2\times f_{3\text{dB}} = 2\times159.1 = 318.\text{Hz}.$$

$$f_{12\text{dB}} = 4\times f_{3\text{dB}} = 4\times159.1 = 636.6\,\text{Hz}.$$

$$f_{21\text{dB}} = 8\times f_{3\text{dB}} = 8\times159.1 = 1273.2\,\text{Hz}.$$

(c) Label the points on a sheet of frequency-response paper.

(d) Connect each of the points to form a low-pass frequency-response curve, which is shown:

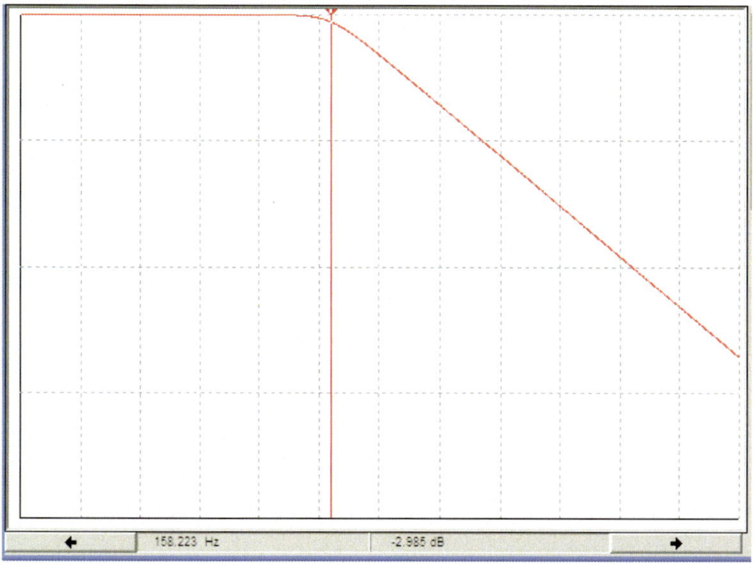

Figure 14-13 Frequency response curve for a low-pass filter circuit.

Self-examination/Answers

1. Low-pass filters pass ___________ frequencies and block ___________frequencies.
2. High-pass filters pass ___________frequencies and block ___________frequencies.
3. Band-passfiltersaredesignedtopass__________________frequencies.
4. Resonant circuits are designed to (pass, block) a range of frequencies, and to (pass, block) frequencies above and below this range.
5. The phase difference between the voltages developed across the inductor and capacitor of a series resonant circuit is _______°.
6. ____________ frequency occurs when the capacitive and inductive reactance are equal.
7. The logarithmic characteristic of the number 5000 is __________.
8. The fractional or decimal part of a logarithm is called the (characteristic, mantissa).
9. An amplifier has input of 0.1 W and an output of 2 W. Determine the power amplification (A_P) in dB.
10. An amplifier has voltage input of 0.2 V_{P-P} and an output of 1.5 V_{P-P}. Determine the voltage amplification (A_V) in dB.
11. A circuit has a loss of –16 dB. What power ratio corresponds to this loss? [$A_P = 10\log\frac{P_{out}}{P_{in}}\, dB$
12. $-16 = 10\log\frac{P_{out}}{P_{in}}$
13. $-1.6 = \log\frac{P_{out}}{P_{in}}$
14. $\frac{P_{out}}{P_{in}} = \text{inv}\log -1.6$
15. $\frac{P_{out}}{P_{in}} = 0.025$]
16. A low-pass filter has a cutoff frequency of 10 kHz. When a 20-kHz signal is applied to the input, the filter does not show any attenuation. This indicates that the filter is (operating properly, faulty).

17. A high-pass filter has a cutoff frequency of 10 kHz. When a 20-kHz signal is applied to the input, the filter does not show any attenuation. This indicates that the filter is (operating properly, faulty).
18. The resonant frequency of a series RLC circuit is 10 kHz. When a 10-kHz signal is applied to the input, the output voltage displayed by an oscilloscope should be (amplified, zero, same as the input, attenuated).

Answers

1. low, high
2. high, low
3. mid-range
4. pass, block
5. 180
6. resonant
7. 3
8. mantissa
9. $A_P = 10\log\frac{P_{out}}{P_{in}} = 10\log\frac{2}{0.1} = 10\log 20 = 10\times 1.301 = 13.01\,\text{dB}$
10. $A_V = 10\log\frac{V_{out}}{V_{in}} = 10\log\frac{1.5}{0.2} = 10\log 7.5 = 10\times 0.8751 = 8.751\,\text{dB}$
11. $A_P = 10\log\frac{P_{out}}{P_{in}}\,\text{dB}$
12. $-16 = 10\log\frac{P_{out}}{P_{in}}$
13. $-1.6 = \log\frac{P_{out}}{P_{in}}$
14. $\frac{P_{out}}{P_{in}} = \text{inv log } -1.6$
15. $\frac{P_{out}}{P_{in}} = 0.025$
16. faulty

17. operating properly
18. same as the input

Problems Answers

Finding circuit values:

1. Find resonant frequency (f_r):

$$f_r = \frac{1}{2\pi \times \sqrt{L \times C}} = \frac{1}{6.28 \times \sqrt{50 \times 10^{-3}) \times (0.02 \times 10^{-6})}}$$

$$= \frac{1}{6.28 \times \sqrt{1.0 \times 10^{-9}}} = \frac{1}{1.9 \times 10^{-4}} = 5035 \text{ Hz}$$

2. Find X_L* and X_C at resonant frequency:

$X_L = 2\pi \cdot f \cdot L = 6.28 \times 5035 \times 0.050 = 1581\ \Omega$

*Easier to calculate

3. Find quality factor (Q):

$$Q = \frac{X_L}{R} = \frac{1581}{500} = 3.162$$

4. Find bandwidth (BW):

$$BW = \frac{f_r}{Q} = \frac{5035 \text{ Hz}}{3.162} = 1592 \text{ Hz}$$

5. Find low-frequency cutoff (f_{LC}):

$$f_{lc} = f_r - \frac{1}{2} BW = 5035 - 796 = 4239 \text{ Hz}$$

6. Find high-frequency cutoff (f_{hc}):

$$f_{hc} = f_r + \frac{1}{2} BW = 5035 + 796 = 5831 \text{ Hz}$$

7. Find I_T at resonant frequency:

$$I_T = \frac{V_A}{R} = \frac{3 \text{ V}}{500\ \Omega} = 0.006 \text{ A} = 6 \text{ mA}$$

8. Find V_R at resonant frequency:

$V_R = I \times R = 0.006 \text{ A} \times 500\ \Omega = 3 \text{ V}$

9. Find V_L and V_C at resonant frequency:

$V_L = V_C = I \cdot X_L = 0.006 \text{ A} \times 1581\ \Omega = 9.49 \text{ V}$*

*Note that this voltage exceeds source voltage—
this is called *"voltage magnification."*

$$A_V = 10\log\frac{V_{out}}{V_{in}} = 10\log\frac{4.2}{0.7} = 10\log 6 = 10\times 0.7781 = 7.781\,\text{dB}$$

$$A_V = 10\log\frac{V_{out}}{V_{in}}\,\text{dB}$$

$$8 = 10\log\frac{V_{out}}{V_{in}}$$

$$0.8 = \log\frac{V_{out}}{V_{in}}$$

$$\frac{V_{out}}{V_{in}} = \text{inv log } 0.8$$

$$\frac{V_{out}}{V_{in}} = 6.31$$

Determine the 3-dB frequency using the following formula:

$$f_{3dB} = \frac{1}{2\times\pi\times R\times C}$$

$$= \frac{1}{2\times\pi\times(20\times10^{3})\times(0.004\times10^{-6})}$$

$$= \frac{1}{5.0265\times10^{-4}} = 1989.44\,\text{Hz}.$$

Find the 9-, 15-, and 21-dB frequencies:

$$f_{9dB} = 2\times f_{3dB} = 2\times 1989.44 = 3978.87\,\text{Hz}$$
$$f_{12dB} = 4\times f_{3dB} = 4\times 1989.44 = 7957.75\,\text{Hz}$$
$$f_{21dB} = 8\times f_{3dB} = 8\times 1989.44 = 15915.49\text{Hz}.$$

The steps for developing the frequency-response curve of the low-pass filter are shown below.

(a) Determine the 3-dB frequency using the following formula:

$$f_{3dB} = \frac{1}{2\times\pi\times R\times C}$$

$$= \frac{1}{2\times\pi\times(50\times10^{3})\times(0.02\times10^{-6})}$$

$$= \frac{1}{6.283\times10^{-3}} = 159.1\,\text{Hz}.$$

(b) Find the 9-, 15-, and 21-dB frequencies:

$$f_{9\text{dB}} = 2\times f_{3\text{dB}} = 2\times159.1 = 318.3\text{Hz}.$$

$$f_{12\text{dB}} = 4\times f_{3\text{dB}} = 4\times159.1 = 636.6\text{Hz}.$$

$$f_{21\text{dB}} = 8\times f_{3\text{dB}} = 8\times159.1 = 1273.2\,\text{Hz}.$$

(c) Label the points on a sheet of frequency-response paper

(d) Connect each of the points to form a low-pass frequency-response curve, which is shown below:

Glossary

Admittance (Y)

The total ability of an AC circuit to conduct current, measured in Siemens, the inverse of impedance.

$$Y = \frac{1}{Z}.$$

Angle of lead or lag

The angle between applied voltage and current flow in an AC circuit, in degrees; in an inductive (L) circuit, voltage (V) leads current (I); in a capacitive (C) circuit, current (I) leads voltage (V).

Apparent power (volt-amperes)

The applied voltage times current delivered to an AC circuit.

Attenuation

A reduction in value.

Band-pass filter
A frequency-sensitive AC circuit that allows incoming frequencies within a certain band to pass through but attenuates frequencies that are below or above this band.

Bandwidth
The band (range) of frequencies that will pass easily through a band-pass filter or resonant circuit.

Bel (B)
A logarithmic unit for comparing the level of intensity of two quantities.

Capacitive reactance (X_C)
An opposition to the flow of AC current caused by a capacitor which is measured in ohms.

$$X_C = \frac{1}{2 \times \pi \times f \times C}.$$

Conductance (G)
The ability of a resistance of a circuit to conduct current, which is measured in Siemens. It is the inverse of resistance.

$$G = \frac{1}{R}.$$

Decibel (dB)
A unit used to express an increase or decrease in power, voltage, or current in a circuit; one tenth of a bel.

Filter
A circuit used to pass certain frequencies and attenuate all other frequencies.

Frequency
The number of AC cycles per second, measured in hertz (Hz).

Frequency response
A circuit's ability to operate over a range of frequencies.

Hertz (Hz)
The international unit of measurement of frequency equal to one cycle per second.

Impedance (Z)
The total opposition to current flow in an AC circuit which is a combination of resistance (R) and reactance (X) in a circuit; measured in ohms.

$$Z = \sqrt{R^2 + X^2}.$$

Inductance (L)
The property of a circuit to oppose changes in current due to energy stored in a magnetic field.

Inductive reactance (X_L)
The opposition to current flow in an AC circuit caused by an inductance (L), measured in ohms. $X_L = 2\pi fL$.

Lagging phase angle
The angle by which current lags voltage (or voltage leads current) in an inductive circuit.

Leading phase angle
The angle by which current leads voltage (or voltage lags current) in a capacitive circuit.

Maximum power transfer
A condition that exists when the resistance of a load (R_L) equals that of the source which supplies it (R_S).

Mho
Ohm spelled backward; the old unit of measurement for conductance, susceptance, and admittance; now replaced by Siemens.

Parallel resonant circuit
A circuit that has an inductor and capacitor connected in parallel to cause response to frequencies applied to the circuit.

Quality factor (Q)
The "figure of merit" or ratio of inductive reactance and resistance in a frequency-sensitive circuit. $Q = \frac{X_L}{R} = \frac{2 \times \pi \times f \times L}{R}$.

Reactance (X_L)
The opposition to ac current flow due to inductance (X_L) or capacitance (X_C).

Reactive circuit
An AC circuit that has the property of inductance or capacitance.

Resonant circuit
See **Parallel resonant circuit and Series resonant circuit.**

Resonant frequency (f_r)
The frequency that passes most easily through a frequency-sensitive circuit when $X_L = X_C$ in the circuit:

$$f_r = \frac{1}{2\pi\sqrt{L \times C}}$$

Selectivity
The ability of a resonant circuit to select a specific frequency and reject all other frequencies.

Series resonant circuit
A circuit that has an inductor and capacitor connected in series to cause response to frequencies applied to the circuit.

Siemens (S)
See **Mho.**

Signal
An electrical waveform of varying value which is applied to a circuit.

Susceptance (B)
The ease with which an inductor or capacitor passes AC current and is expressed as: $B_L = \frac{1}{X_L}$ for inductors and $B_C = \frac{1}{X_C}$ for capacitors. It is measured in Siemens.

15

Time-constant and Wave-shaping Circuits

Time-constant circuits have several applications in electronics. They are used in many industrial processes, computer circuits, and communications systems. Timing functions may be as simple as on–off control or as complex as elaborate sequential operations. The controlled element of a timing circuit could be a motor, a relay, a solenoid, a lamp, or some other circuit or device. Time-constant circuits use the properties of inductance or capacitance to operate as timing circuits to control load devices. Inductance (L) in circuits opposes changes in current and capacitance (C) opposes changes in voltage. The reaction time of inductors and capacitors to oppose changes of current or voltage depends on the resistance (R) in the circuit. The time (in seconds) for a capacitor or an inductor to react is called its time constant.

Objectives

Upon the completion of this chapter, you should be able to:

1. Discuss RL and RC time constants
2. Discuss the charging and discharging characteristics of RL and RC circuits
3. Solve problems dealing with RC and RL time constants for charging and discharging
4. Apply calculator and computer procedures to solve RL and RC time-constant problems
5. Discuss wave-shaping circuits, including differentiators and integrators

Chapter Outline

15.1 Resistor–Inductor (RL) Time Constants

In **Figure 15-1(a)**, a resistor and an inductor are connected in series to a voltage source. Current rises from zero to maximum after a certain time period due to the **counter electromotive force (CEMF)** of the magnetic field surrounding the inductor. The time required for the current to reach maximum is controlled by the values of R and L.

Resistance opposes current flow and **inductance** opposes changes in current. The change in current occurs from its zero to maximum values. The time required for the current to reach about 63% of its maximum value is found by using the formula $t = L/R$. **Time constant (t)** is in seconds, L is in henries, and R is in ohms. After **one time constant**, current through the inductor is about 63% of maximum, as shown in **Figure 15-1(b)**. After **five time constants**, current through the inductor is maximum. Each value of the **time constant curve** is exponentially related to the previous value.

Assume that the **maximum current (I_{max})** in the circuit is 100 mA and the time required for the current to reach 63% of 100 mA is 1.0 s. The time required for the current to reach the maximum (100 mA) is 5 s ($t \times 5 = 1.0 \times 5 = 5$ s). The current flow through the inductor (I_L) varies after the circuit switch is turned on until it reaches its maximum value after five time constants. The values of current are as follows:

- After one time constant: 63% × 100 mA = 63 mA
- After two time constants: 86% × 100 mA = 86 mA
- After three time constants: 95% × 100 mA = 95 mA
- After four time constants: 98% × 100 mA = 98 mA
- After five time constants: 100% × 100 mA = 100 mA

Thus, after **five time constants** (5 s for this example), I_L is approximately equal to its maximum value. The buildup of I_L in an RL circuit is called its charging condition. Charging occurs until I_L stabilizes at its maximum value, as shown in **Figure 15-1(b)**.

RL circuits also have a discharge condition. When the switch of **Figure 15-1(a)** is opened, I_L will reduce at a rate determined by the circuit time constant. Assume that a wire is placed across the points where the power source was connected at the same time the switch is turned off. The current through the inductor (I_L) will discharge, as shown in **Figure 15-1(c)**, until it stabilizes. The approximate values of current during **discharge** are as follows:

- After one time constant: 37% of 100 mA = 37 mA
- After two time constants: 14% of 100 mA = 14 mA

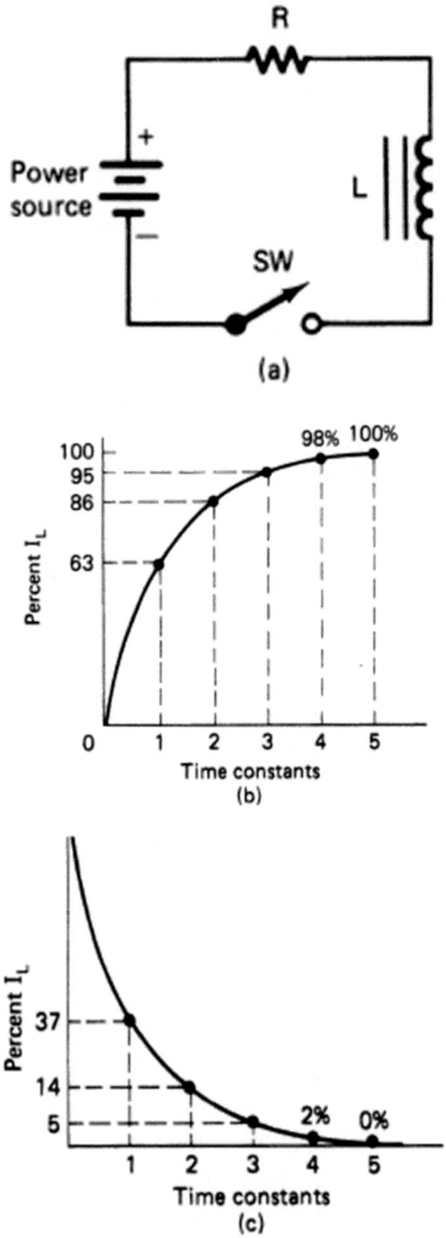

Figure 15-1 RL time-constant circuit: (a) circuit diagram; (b) time versus current curve – charging; (c) time versus current curve – discharging.

- After three time constants: 5% of 100 mA = 5 mA
- After four time constants: 2% of 100 mA = 2 mA
- After five time constants: 0% of 100 mA = 0 mA

15.2 Resistor–Capacitor (RC) Time Constants

Figure 15-2 shows a circuit for charging and discharging a capacitor. In **Figure 15-2(a)**, the switch is in the off position and the capacitor is assumed to be fully discharged. When the switch is moved to the charging position, as shown in **Figure 15-2(b)**, the resistor and capacitor are connected in series with the voltage source.

The 2-μF capacitor begins charging through the 1.5-MΩ resistor. The time for the voltage across the capacitor to reach maximum is controlled by the values of R and C. **Resistance** opposes current flow in the circuit, while **capacitance** opposes changes in voltage. The change in voltage across the capacitor is from zero to maximum. The amount of time for a capacitor to charge to about 63% of the applied voltage is found by using the formula $t = R \times C$, and this is called the **time constant** of the circuit. Time (t) is in seconds, R is in ohms, and C is in farads.

After **one time constant**, the voltage across the capacitor is about 63% of the source voltage. **Figure 15-2(c)** shows a time versus voltage **charging curve**. In **five time constants**, the voltage across the capacitor approximately equals the source voltage. The time for the capacitor of **Figure 15-2(b)** to charge to 63% of the source voltage is **one time constant**, or 3 s. The time for the voltage across the capacitor to equal the source voltage is 15 s ($t \times 5 = 3 \times 5 = 15$ s). Each value of the **time constant curve** is exponentially related to the previous value.

The voltage across the capacitor (V_C) in an RC circuit increases until it reaches the source voltage value. When the switch in **Figure 15-2(a)** is moved to the charging position, V_C will equal approximately 12 V after **five time constants**. The approximate values of voltage across the capacitor are as follows:

- After one time constant: 63% of 12 V = 7.56 V
- After two time constants: 86% of 12 V = 10.32 V
- After three time constants: 95% of 12 V = 11.40 V
- After four time constants: 98% of 12 V = 11.76 V
- After five time constants: 100% of 12 V = 12.00 V

Thus, after **five time constants** (15 s for this example), V_C is approximately equal to the source voltage. The buildup of voltage across the

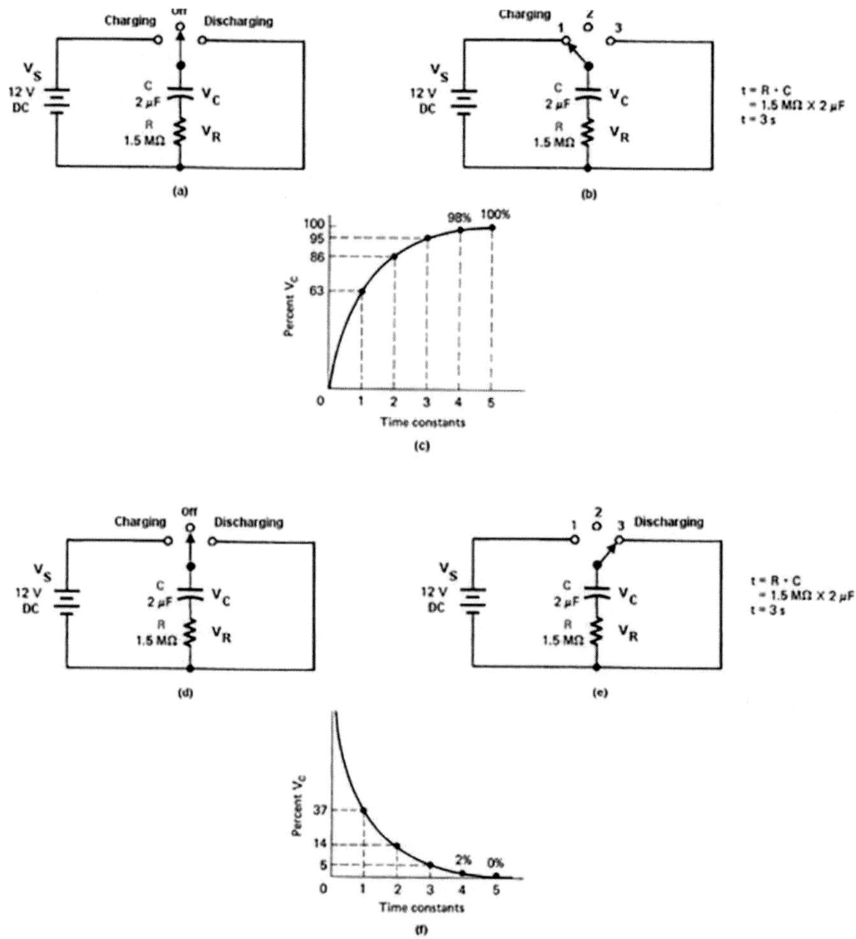

Figure 15-2 *RC* circuit with resulting charging–discharging time constant curves. (a) Switch in off position with no charge on the capacitor. (b) Switch in charging position. (c) RC time-constant charging curve. (d) Switch in off position and the capacitor fully charged. (e) Switch in discharging position. (f) RC time-constant discharging curve.

capacitor is called its **charging** condition. Once the capacitor is fully charged, the switch is then moved to the off position, as shown in **Figure 15-2(d)**. The capacitor will now retain the accumulated charge for a long period of time. In order to discharge the capacitor, it must be connected to a load. When the switch is moved to the discharging position as shown in **Figure 15-2(e)**, the resulting circuit consists of only the capacitor and the resistor connected in series. Note that a wire is placed connecting one terminal of the capacitor directly to the resistor when the switch is in this position.

The **discharging** of the capacitor is similar to its charging action. The capacitor discharges about 63% of the source voltage in **one time constant**. In **five time constants**, a capacitor discharges 100% of its voltage. **Figure 15-2(f)** shows a time versus voltage **discharge curve**. When the switch of **Figure 15-2(d)** is moved to the discharging position, V_C will reduce at a rate determined by the time constant of the circuit. The voltage across the capacitor (V_C) will discharge, as shown in **Figure 15-2(f)** until it reaches minimum value. The approximate values of capacitor voltage during discharge are as follows:

- After one time constant: 37% of 12 V = 4.44 V
- After two time constants: 14% of 12 V = 1.68 V
- After three time constants: 5% of 12 V = 0.60 V
- After four time constants: 2% of 12 V = 0.24 V
- After five time constants: 0% of 12 V = 0.00 V

15.3 Universal Time-constant Curves

The time for current changes in inductive circuits and voltage changes in capacitive circuits are similar. The curves shown in **Figure 15-3** are called universal time-constant curves.

The **vertical axis** is marked as percent of capacitor voltage (V_C) or percent of inductor current (I_L). The **horizontal axis** is marked in time constant values. Sometimes the **charging curve** is called the **rise time** and the **discharging curve** is called the **decay time**. Each value of the **time constant curve** is exponentially related to the previous value.

These curves are useful since they contain the voltage or current values for RC and RL circuits to be determined at any given time. For example, the charge on a capacitor of the series RC circuit is shown in **Figure 15-4**. After 2.5 time constants, the value of the charge on the capacitor will be between 8.6 and 9.5 V. A close estimation of voltage across the capacitor may be found by interpolation of values as follows:

$$V_C = 8.6 + \frac{9.5 - 8.6}{2} = 8.6 + 0.45 = 9.05 \text{ V}.$$

The charge on a capacitor after a given period of time can also be estimated on a **universal time-constant curve**. For example, after 25 s, what is the approximate charge on the capacitor of **Figure 15-4**? This is a difference of 2.3 V. The value of 25 s is 10 s after the first time constant

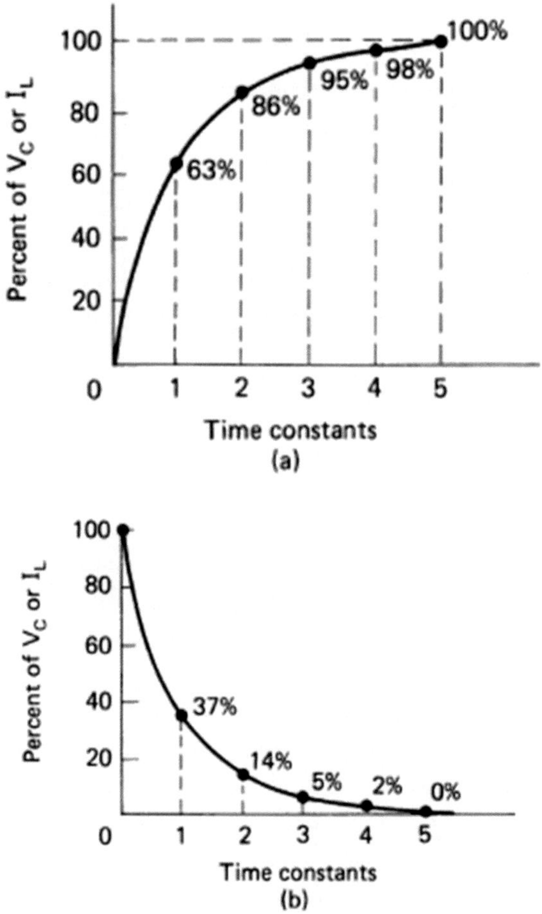

Figure 15-3 Universal time-constant curves: (a) charging or rise time curve; (b) discharging or decay time curve.

(25 s = 15 + 10). The time is then 10/15, or 2/3, of the distance from the first time constant (15 s). The approximation of V_C after 25 s is as follows:

$$V_C = 6.3 + \frac{2}{3} \times 2.3 = 6.3 + 1.53 = 7.83 \text{ V}.$$

It should be pointed out that values along the exponential **universal time-constant curve** using the methods outlined before are approximations. The exact values may be obtained using calculus procedures and natural logarithms. Approximations such as those explained are commonly used for electronic circuit design where extremely precise values are not necessary.

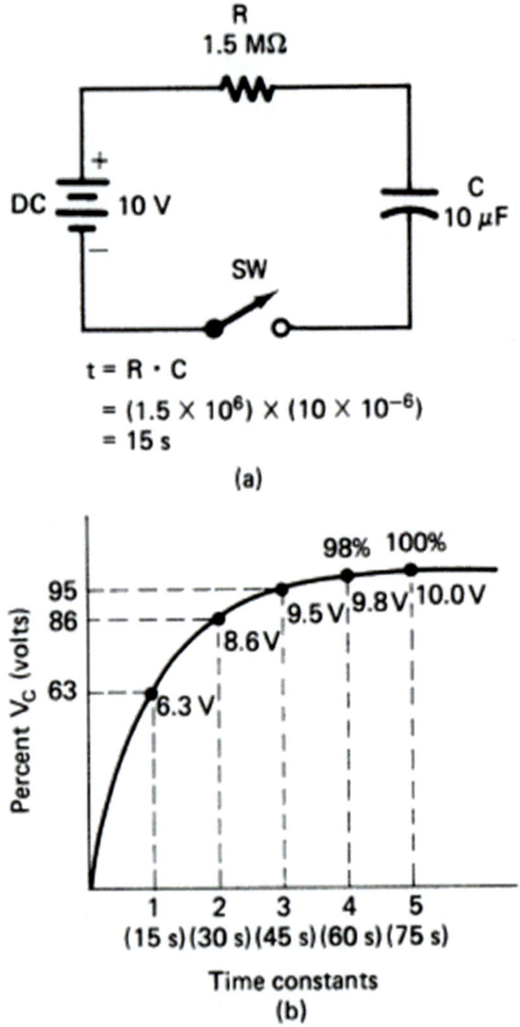

Figure 15-4 (a) Series RC circuit; (b) charging time-constant curve.

15.4 Wave-shaping Control with Time-constant Circuits

A circuit that is designed to change the shape of a waveform is very essential in electronics today. The type of wave shape achieved is based primarily on the shape of the applied signal and the characteristics of the shaping circuit. **Sine**, **square**, and **sawtooth waves** are the basic input waves to be processed. Square and sawtooth waveforms are shown in **Figure 15-5**.

Waves may be of the periodic type or have a pulsing characteristic. A **periodic wave** continually changes value within a given time frame.

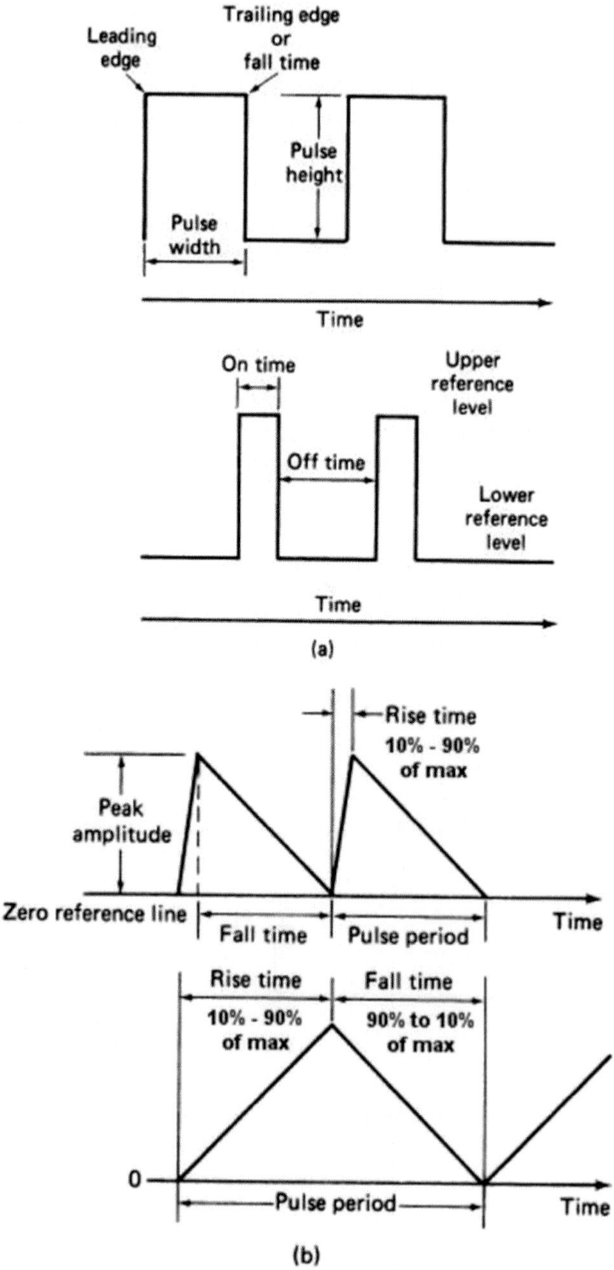

Figure 15-5 (a) Square or rectangular waveforms; (b) sawtooth waveforms.

Pulse waveforms occur momentarily at intervals. Repetitive pulses occur at definite time periods. **Transient pulses** occur randomly and are the result of a switching action or some other electrical change. **Wave-shaping circuits** may be used to minimize transient pulse conditions so that they will not cause irregular control of a circuit.

Numerous devices and components have been developed and are being used to achieve various wave-shaping operations. Wave-shaping is considered to be an altering process that changes the basic form of the wave, its conduction time, or causes a variation in its amplitude. A rather simple method of altering the basic form of a wave is to apply it to an **RC time-constant circuit**. Two distinct form changes can be produced according to the RC component combination selected. With a **square or rectangular wave** applied, a **differentiated** or **integrated wave** shape may be produced. Refer to the wave-shaping circuits of **Figure 15-6**.

Differentiator Circuits

The **differentiator circuit** of **Figure 15-6(a)** has a square-wave input signal applied to a capacitor, and its output is developed across a resistor. Any abrupt change in the input signal will cause a pronounced change or difference in the output voltage developed across R. The term differentiator is used to describe the changing output of this circuit.

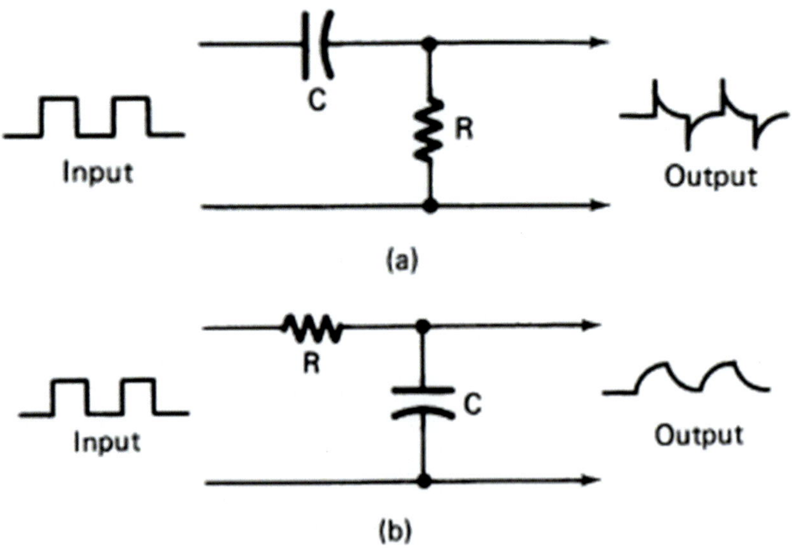

Figure 15-6 RC wave-shaping circuits: (a) differentiator circuit; (b) integrator circuit.

When the **leading edge** of an applied square wave rises quickly, it causes the capacitor of a differentiator to charge immediately. The series-connected resistor of this circuit "sees" a current that is identical to that produced by the capacitor. This current rises sharply to a peak and then drops off very quickly when the capacitor becomes charged. Current passing through R develops an output voltage that conforms to the charging current of the capacitor.

When the **trailing edge** of an applied square wave is reached, it causes an abrupt drop in the applied voltage to the capacitor. The electrostatic charge on the capacitor now begins to discharge back through the resistor. This current is in a direction opposite to that of the original charging current. The resulting current through R causes a corresponding output voltage to be developed. The polarity of this voltage is opposite to that of the charging cycle. The output voltage will then drop below the zero reference line.

When a **differentiator** has a short **time constant** with respect to the applied square wave, the output will have only sharp positive and negative spikes. A longer time constant, however, will cause the trailing edge of the output to be extended somewhat. This indicates that the **discharge** action of the capacitor has been prolonged. The **RC time constant** of a differentiator with respect to **the pulse repetition rate (PRR)** of the square-wave input largely determines the form of the output of this circuit. The **reciprocal** of the time that it takes for a wave of this type to repeat itself is called the pulse repetition rate (PRR). This term compares to the frequency of a sine wave, which is expressed in hertz (Hz).

Integrator Circuits

The **integrator** of **Figure 15-6(b)** is a wave-shaping circuit that has the input signal applied to a resistor and the output developed across a capacitor. The value of the resistor in this case tends to restrict the charging time of the capacitor. The voltage developed across C, therefore, has a gradual rise time and fall time when a square wave is applied.

The output of an **integrator** can be altered by changing the PRR of the input or by changing the **time constant** of the RC network. Time constants longer than the one time constant of a square wave tend to cause it to have a **sawtooth** appearance. A shorter time constant tends to cause a slope on the leading and trailing edge followed by straight-line areas. Integrators are primarily used to sum, or total, signal values that are variable.

Troubleshooting Time-constant and Wave-shaping Circuits

Troubleshooting of time constant and wave-shaping circuits is achieved by energizing the circuit with an energy source that can produce different waveforms with a variable frequency. The resulting **input and output waveforms** can be observed on an **oscilloscope** to determine if the circuit is functioning satisfactorily. The **time-constant** of an RL or RC circuit can be determined mathematically.

When an appropriate **input signal** *of a specific shape is applied to a* **wave**-*shaping circuit, it should cause a corresponding change in the shape of its output. The frequency of the input wave*

should be the same as that of the output. A change in the value of the time constant of the circuit should produce a corresponding change in the shape of the waveform. A **larger** time constant will cause a gradual change in the applied waveform, whereas a **small** time constant will cause a rapid change in the shape of the output.

Summary

- Time-constant circuits use the properties of inductance or capacitance to operate as timing circuits.
- Inductive circuits rely on the delay of time for currents through a series RL circuit to rise and fall.
- Capacitive circuits utilize the time delay for voltage to rise and fall in a series RC circuit.
- Any given value of a universal time-constant curve is exponentially related to its previous and succeeding value.
- The time constant curve graphically shows the rise time or decay time for RL or RC time-constant circuits.
- Wave-shaping may also be accomplished by time-constant circuits.
- Differentiator and integrator circuits function as wave-shaping circuits.
- Troubleshooting a wave-shaping circuit requires energizing the circuit with a waveform of a given shape and frequency and observing the resulting output on an oscilloscope.

Formulas

t = L/R
t = R × C

Problems

Refer to **Figure 15-1(a)** which shows an RL time-constant circuit. When the value of R = 500 Ω and L = 200 mH, calculate the RL time constant (t) of the circuits. [$t=\frac{L}{R}=\frac{200\times10^{-3}}{500}=4\times10^{-4}\,\text{s}=40\,\text{ms}$]

- In a **series RL circuit** with a maximum current of 200 mA, what are the approximate values of current through the circuit (I_L) after two time constants during its charging condition? [After **one time constant**, the value of current increases by 63% of the maximum current value, which is $\frac{63}{100}\times200\times10^{-3}=126\times10^{-3}=126\,\text{mA}$. After **two time constants**, the value of the current increases by an additional 63% of the difference between the present value and the maximum value (200 – 126m = 74 mA), which is $\frac{63}{100}\times74\times10^{-3}=46.62\times10^{-3}=46.62\,\text{mA}$. Thus, the new value of the current will be 126 mA + 46.62 mA= 172.62 mA]
- A **series RL circuit** has a maximum current of 200 mA. What is the approximate value of current through the circuit after two time constants when it is being discharged?
  [After **one time constant**, the value is reduced by 63% of the original value. When this occurs, only 37% of 200 mA will flow, which is $\frac{37}{100}\times200\times10^{-3}=74\times10^{-3}=74\,\text{mA}$.

  After **two time constants**, the value is reduced by an additional 63% of the previous value (74 mA). When this occurs, only 37% of 74 mA will flow, which is: [$\frac{37}{100}\times74\times10^{-3}=27.38\times10^{-3}=27.38\,\text{mA}$]

- Refer to **Figure 15-2** which shows an RC time-constant circuit. When the value of R = 50 kΩ and C = 10 μF, calculate the RC time constant (t) of the circuit.

  [$t = R \times C = 50 \times 10^{3} \times 10 \times 10^{-6} = 0.5\,\text{s}$]

- Assume that the capacitor of a series RC circuit is being energized by 5 V. What is the approximate value of capacitor voltage (V_C) after two time constants?

  [After **one time constant**, the value of charge on the capacitor increases by 63% of the maximum voltage value, which is $\frac{63}{100} \times 5 = 3.15\,\text{V}$. After **two time constants**, the value of the charge on the capacitor increases by an additional 63% of the difference between the present value and the maximum value (5 – 3.15 = 1.85 V), which is $\frac{63}{100} \times 1.85 = 1.1655\,\text{V}$. Thus, the new value of the voltage will be 3.15 + 1.1655 = 4.3155 V]

- Assume that the capacitor of a series RC circuit charged to 5 V. When the capacitor is being discharged, determine the approximate value of the capacitor voltage (V_C) after the three time constants?

  [After **one time constant**, the value is reduced by 63% of the original value. When this occurs, the remaining charge on the capacitor will drop to 37% of 5 V, which is $\frac{37}{100} \times 5\,\text{V} = 1.85\,\text{V}$.

  After **two time constants**, the value is reduced by an additional 63% of the previous value (1.85 V). When this occurs, the remaining charge on the capacitor will drop to 37% of 1.85 V, which is $\frac{37}{100} \times 1.85 = 1.3505\,\text{V}$]

Self-examination/Answers

1. The time required for the current of an RL circuit to reach its maximum value is controlled by the value of __________ and __________.
2. The time required for current of an RL circuit to reach about 63% of its maximum value is called the ___________.

3. In an RL circuit, __________ time constants are needed to cause a maximum current to flow through the inductor.
4. In an RC circuit, the value of the voltage across the capacitor (V_C) will charge to 63% of the applied voltage after (1, 2, 3, 4, 5) time constant(s).
5. In an RC circuit, the value of the voltage across the capacitor (V_C) will equal the applied voltage after (1, 2, 3, 4, 5) time constant(s).
6. Assume that a capacitor is charged to 10 V. When a resistor is connected across this capacitor, the voltage on this capacitor will decrease to _______V in one time constant.
7. Each value of the charging–discharging response curves of an RC circuit is (linearly, exponentially, inversely) related to the previous and successive values.
8. A fully charged capacitor has a resistor connected across it terminals. After five time constants, the resulting voltage of across the capacitor will be (0, 37, 50, 100)% of the initial voltage.
9. A fully discharged capacitor of an RC circuit is connected to a DC voltage supply. After five time constants, the resulting voltage of the capacitor will reach a value corresponding to (0, 37, 50, 100)% of the supply voltage.
10. With a square wave applied to an RC time constant circuit, a ______________or ____________ waveform is produced.
11. In an RC differentiator circuit energized by a square wave, the output is taken across the (capacitor, resistor).
12. In an RC integrator circuit energized by a square wave, the output is taken across the (capacitor, resistor).
13. In an RC wave-shaping circuit energized by an AC waveform, the (shape, frequency) of the output will be different from that of the input.
14. In an RC wave-shaping circuit energized by AC waveform, the (shape, frequency) of the output will be the same as that of the input.
15. If the RC time constant of a circuit is much smaller as compared to the time period of the applied input, the output waveform will have a (gradual, rapid) change in its shape.
16. If the RC time constant of a circuit is comparable to the time period of the applied input, the output waveform will have a (gradual, rapid) change in its shape.

Answers

1. resistance, inductance
2. time constant
3. 5
4. 1
5. 5
6. 3.7 V
7. exponentially
8. 0
9. 100
10. differentiated, integrated
11. resistor
12. capacitor
13. shape
14. frequency
15. rapid
16. gradual

Problems Answers

$$t = \frac{L}{R} = \frac{200 \times 10^{-3}}{500} = 4 \times 10^{-4}\,\text{s} = 40\,\text{ms}$$

After one time constant, the value of current increases by 63% of the maximum current value, which is $\frac{63}{100} \times 200 \times 10^{-3} = 126 \times 10^{-3} = 126\,\text{mA}$. After two constants, the value of the current increases by an additional 63% of the difference between the present value and the maximum value (200 – 126 m = 74 mA), which is $\frac{63}{100} \times 74 \times 10^{-3} = 46.62 \times 10^{-3} = 46.62\,\text{mA}$. Thus, the new value of the current will be 126 mA + 46.62 mA= 172.62 mA.

After one time constant, the value is reduced by 63% of the original value. When this occurs, only 37% of 200 mA will flow, which is $\frac{37}{100} \times 200 \times 10^{-3} = 74 \times 10^{-3} = 74\,\text{mA}$.

After two constants, the value is reduced by an additional 63% of the previous value (74 mA). When this occurs, only 37% of 74 mA will flow, which is $\frac{37}{100}\times 74\times 10^{-3} = 27.38\times 10^{-3} = 27.38\,\text{mA}$

$$t = R\times C = 50\times 10^{3}\times 10\times 10^{-6} = 0.5\,\text{s}.$$

After one time constant, the value of charge on the capacitor increases by 63% of the maximum current value, which is $\frac{63}{100}\times 5 = 3.15\,\text{V}$. After two constants, the value of the charge on the capacitor increases by an additional 63% of the difference between the present value and the maximum value (5 – 3.15 = 1.85 V), which is $\frac{63}{100}\times 1.85 = 1.1655\,\text{V}$. Thus, the new value of the current will be 3.15 + 1.1655 = 4.3155 V.

After one time constant, the value is reduced by 63% of the original value. When this occurs, the remaining charge on the capacitor will drop to 37% of 5 V, which is $\frac{37}{100}\times 5\,\text{V} = 1.85\,\text{V}$.

After two time constants, the value is reduced by an additional 63% of the previous value (1.85 V). When this occurs, the remaining charge on the capacitor will drop to 37% of 1.85 V, which is $\frac{37}{100}\times 1.85 = 1.3505\,\text{V}$.

Glossary

Admittance (Y)
The total ability of an AC circuit to conduct current, measured in Siemens, the inverse of impedance. $Y = \frac{1}{Z}$.

Angle of lead or lag
The angle between applied voltage and current flow in an AC circuit, in degrees; in an inductive (*L*) circuit, voltage (*V*) leads current (*I*); in a capacitive (*C*) circuit, current (*I*) leads voltage (*V*).

Capacitive reactance (X_C)
The opposition to the flow of AC current caused by a capacitive device (measured in ohms).

$$X_C = \frac{1}{2\times\pi\times f\times C}.$$

Conductance (G)
The ability of a resistance of a circuit to conduct current, measured in Siemens, the inverse of resistance.

$$G = \frac{1}{R}.$$

Decay
A term used for a gradual reduction in value of a voltage or current.

Decay time
The time required for a capacitor to discharge to a certain percentage of its original charge or the time required for current through an inductor to reduce to a percentage of its maximum value.

Impedance (Z)
The total opposition to current flow in an AC circuit which is a combination of resistance (R) and reactance (X) in a circuit; measured in ohms.

$$Z = \sqrt{R^2 + X^2}.$$

Mho
Ohm spelled backward; the old unit of measurement for conductance, susceptance, and admittance; now replaced by Siemens.

Quality factor (Q)
The "figure of merit" or ratio of inductive reactance and resistance in a frequency-sensitive circuit.

Waveform
The pattern of an AC frequency derived by looking at instantaneous voltage values that occur over time; on a graph, a waveform is plotted with instantaneous voltages on the vertical axis and time on the horizontal axis.

16

Mathematics for AC Circuits

When AC is applied to an electrical circuit composed of resistors, inductors, and/or capacitors, its response can be studied mathematically and graphically. This chapter provides a review of mathematical concepts that will be needed for analyzing AC component response. This involves graphical representation of trigonometric functions and the component response to AC.

Objectives

1. Understand the trigonometric relationships of AC
2. Show how AC quantities are expressed mathematically
3. Determine the phase relationship of different AC components
4. Graphically explain the behavior of different AC components
5. Perform mathematical operations on AC quantities

Chapter Outline

16.1 Right Triangles and Trigonometry

Right triangles and **trigonometry** are valuable in the study of electricity and electronics. Trigonometry, or "trig," is a form of mathematics that deals with angles and triangles, particularly the right triangle. A right triangle has one angle of 90°. An example of a right triangle is shown in **Figure 16-1**.

This example illustrates how resistance, capacitive reactance, and impedance are related in an AC series RC circuit. Resistance (R) and capacitive reactance (X_c) are 90° apart. Their angle of intersection forms a right angle. The law of right triangles, known as the **Pythagorean theorem**, can be used to solve circuit problems. The theorem states that "in a right triangle, the square of the **hypotenuse** is equal to the sum of the squares of the other two sides." With reference to **Figure 16-1**, the Pythagorean theorem is used as follows:

$$Z^2 = R^2 + X_C{}^2$$

$$Z = \sqrt{R^2 + X_C{}^2}.$$

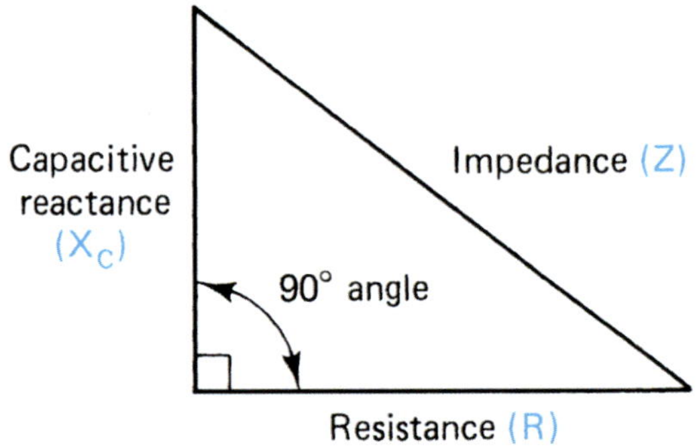

Figure 16-1 Right triangle for an AC series RC circuit.

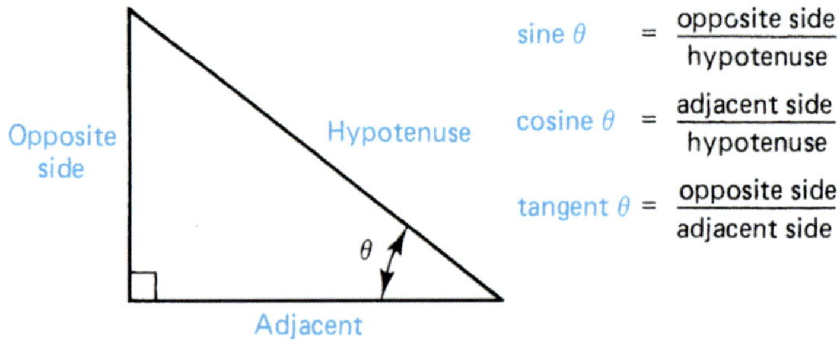

Figure 16-2 Sine, cosine, and tangent ratios.

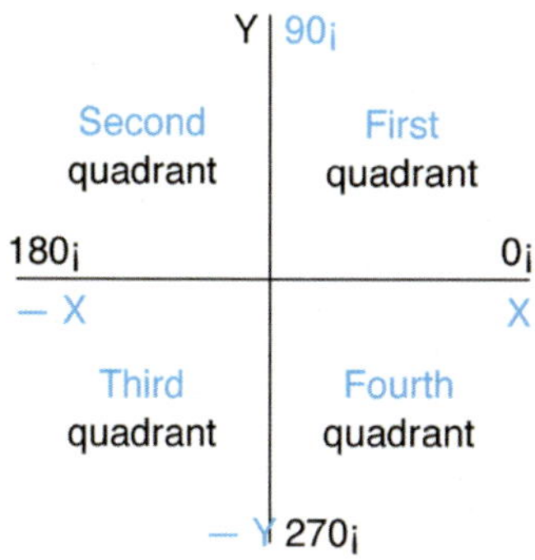

Second quadrant (90¡ to 180¡)

- sin θ□= sin (180¡ – θ)
- cos θ□= –cos (180¡ – θ)
- tan θ□= –tan (180¡ – θ)

Third quadrant (180¡ to 270¡)

- sin θ□= –sin (θ – 180¡)
- cos θ□= –cos (θ – 180¡)
- tan θ□= tan (θ – 180¡)

Fourth quadrant (270¡ to 360¡)

- sin θ□= –sin (360¡ – θ)
- cos θ□= cos (360¡ – θ)
- tan θ□= –tan (360¡ – θ)

Figure 16-3 Standard graph and the method used to find sine, cosine, and tangent values.

By using trigonometry relationships, problems dealing with phase angles, power factor, and reactive power in AC circuits can be solved. The three most-used trig functions are the **sine**, the **cosine**, and the **tangent**. These functions show the ratios of the sides of a triangle. They determine the size of the angle. **Figure 16-2** shows how these ratios are expressed. Their values are found by using a calculator. This process can be reversed to find the size of an angle when the values of the sides are known. The term **inverse** is used for this process.

For example, "inverse sine 0.5 = 30°" means that 30° is the angle whose sine is 0.5. When given "inverse sine 0.866 = 60°," find the angle whose sine is 0.866 which is equal to 60°. A scientific calculator can be used to perform this operation.

Trigonometric ratios hold true for angles of any size. Angles in the **first quadrant** of a standard graph are from 0° to 90°. They are used as a reference. To solve for angles greater than 90° (**second-**, **third-**, and **fourth-quadrant angles**), an angle must be converted to a first-quadrant angle (refer to **Figure 16-3**).

All first-quadrant angles have positive values. Angles in the second, third, and fourth quadrants have two negative values and one positive value.

16.2 Rectangular Coordinates

A two-dimensional relationship can be established by drawing two perpendicular number lines that cross at their zero reference points, as shown

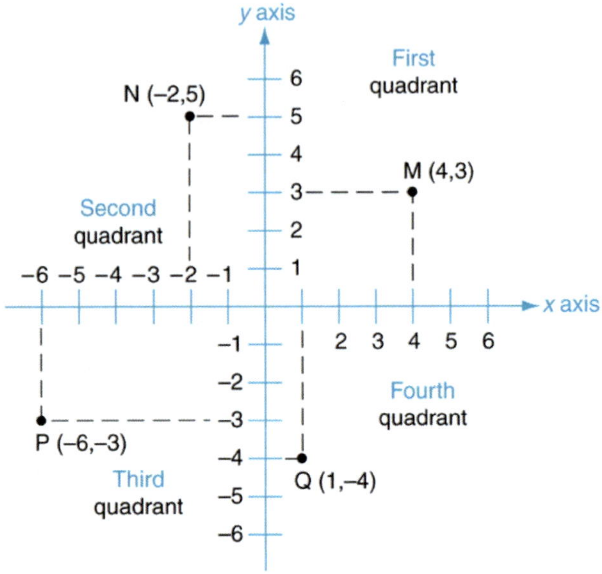

Figure 16-4 Rectangular coordinates.

in **Figure 16-4**. The horizontal line is called the **x-axis** and the vertical line is called the **y-axis**. These coordinates are called **rectangular coordinates** because the axes are at right angles to each other. They are also called **Cartesian coordinates**.

For rectangular coordinates, two axes are placed in a plane along the surface of the paper, and any point on the plane can be located with reference to these axes. That is, any point on the plane must be a certain number of units to the left (negative) or to the right (positive) of the ***y*-axis**. It must lie a certain number of units above (positive) or below (negative) the ***x*-axes**. To locate a point with reference to this set of axes, the *x*-value and the *y*-value of the point must be known. These two values are called the coordinates of the point. The *x*-value is called the **abscissa**, and the *y*-value is called the ordinate. In describing a point in terms of its rectangular coordinates, the abscissa is written first, followed by the **ordinate**. The two numbers are separated by a comma and are usually enclosed in parentheses. Thus, to describe the location of point M in **Figure 16-4**, write **M(4, 3)**. This means that to locate point M, count four divisions to the right from the origin along the *x*-axis, and then count three divisions in the *y*-direction above the *x*-axis. By this method, the other points in the figure are described as N(–2, 5), P(–6, –3), and Q(1, –4).

16.3 Quadrants

The plane containing the *x*- and *y*-axes is called the ***x–y* plane**. The axes divide the plane into four sections, called **quadrants**. **Figure 16-4** shows that in the first quadrant, both coordinates are positive. In the second quadrant, the abscissa is negative, and the ordinate is positive. In the third quadrant, both coordinates are negative. In the fourth quadrant, the abscissa is positive and the ordinate is negative.

16.4 Polar Coordinates

The position of any point on a plane can be determined by using a radius vector and a reference line that meets at a point called the origin. The length of the radius vector is the distance from the origin to the point. The direction of the radius vector is the angle that it makes with the reference line. The length and angle of a vector used to locate a point on a plane are called **polar coordinates**. The term polar refers to the center (or polar) point of

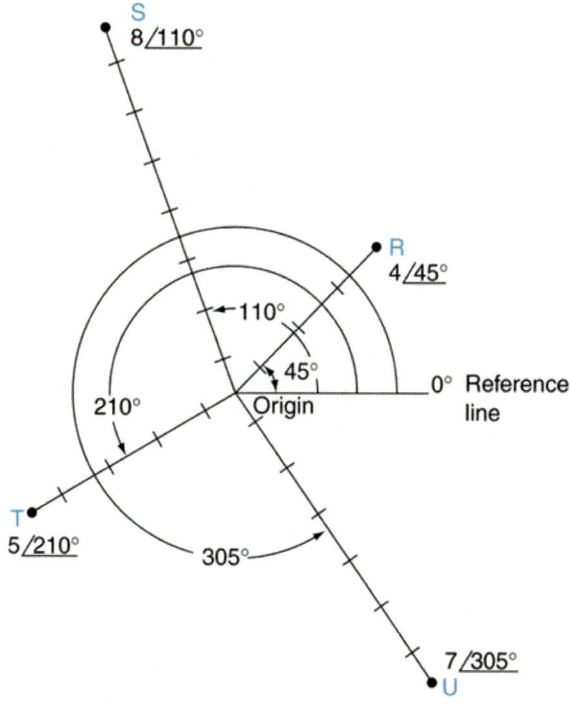

Figure 16-5 Polar coordinates.

the graphed figure. The distance from the origin to the point is called the modulus, and its angular direction is called the argument.

In describing a point in terms of its polar coordinates, the **modulus** is written first and then the **argument**. The general form **r/□** shows the polar coordinates for a point *r* units distance from the origin and six angular units from the reference line. Thus, 4/45° describes the location of point R in **Figure 16-5**. Point R is located by constructing a vector that is four units in length at an angle of 45° to the reference line. By this method, the other points in the figure are found as $S = 8/110°$, $T = 5/210°$, and $U = 7/305°$.

16.5 Angular Velocity

The rate at which an angle is generated by a vector radius is called the **angular velocity**. When the rotation is uniform, the unit of angular velocity is the angle generated per unit of time. Therefore, angular velocity can be measured in degrees per second, radians per second, revolutions per second, and revolutions per minute. For AC waveforms, angular velocity is usually

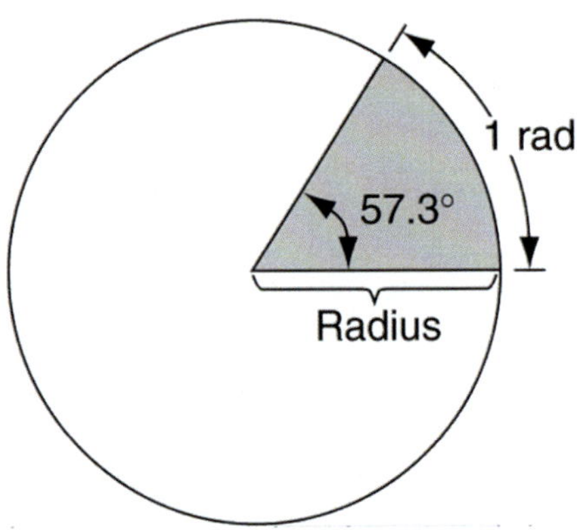

Radian/Degree Conversion

Radians can be converted to degrees using the equation:

$$\text{rad} = \left(\frac{\pi\ \text{rad}}{180°}\right) \times \text{degrees}$$

Similarly, degrees can be converted to radians with the equation:

$$\text{degrees} = \left(\frac{180°}{\pi\ \text{rad}}\right) \times \text{rad}$$

Degrees (°)	*Radians (rad)*
0	0
45	$\pi/4$
90	$\pi/2$
135	$3\pi/4$
180	π
225	$5\pi/4$
270	$3\pi/2$
315	$7\pi/4$
360	2π

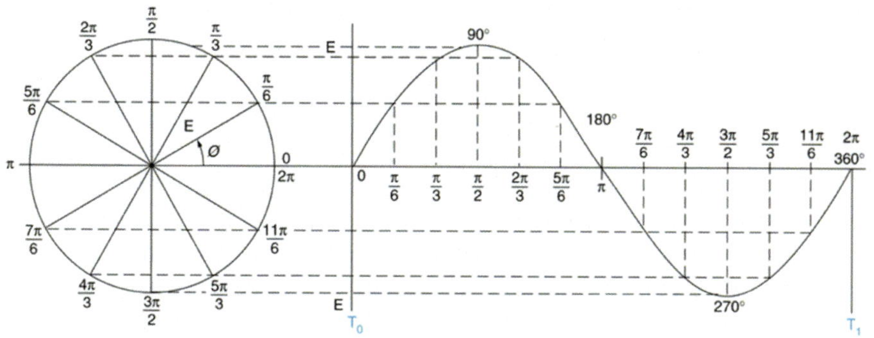

Figure 16-6 Radian values: (a) relation of radians to degrees. (b) radian/degree conversions; (c) table of radian/degree conversions; (d) sine-wave generation.

given in radians per second and is denoted by the Greek lowercase letter omega (ω). The following equation relates ω to revolutions per second:

$$\omega = 2\pi f$$

where f is the number of revolutions per second. Hence, the number of radians generated per second is $2\pi f$. For example, if a motor rotates at 3600 rpm, convert this to radians per second as follows:

$$\begin{aligned}\omega &= 2\pi f \\ &= 2\pi \times 3600/60 \\ &= 120\pi \text{ radians/second.}\end{aligned}$$

Figure 16-6 is helpful in understanding the relationship of sine waves and the rotation of a vector quantity. The angular measurement of a sine wave is based on 360° or 2π radians for a complete cycle. A half-cycle is 180° or π radians; a quarter-cycle is 90° or π/2 radians, and so on. **Figure 16-6** shows angles in degrees over a full cycle of a sine wave.

16.6 Scalars, Vectors, and Phasors

Physical quantities that have magnitude only are called scalar quantities. Scalar quantities may be added or subtracted algebraically. Quantities that have both magnitude and direction are called vector quantities. A vector quantity can be represented by a straight line segment which is drawn in a definite direction as shown in **Figure 16.7**. The direction of the vector is indicated by an arrow, and its direction is specified by angle θ. The magnitude of the vector is represented by the length of the vector.

Vector quantities can be graphically added or subtracted by using the **parallelogram method**. The sum of two vectors can be found by drawing the two vectors with their starting points superimposed as shown in **Figure 16-8(a)**. Next draw a parallelogram with the two vectors as two adjacent sides, as shown in **Figure 16-8(b)**.

A diagonal is then drawn from the common starting point (0) to the opposite vertex of the parallelogram, as shown in **Figure 16-8(c)**. This diagonal represents the sum of the two original vectors and gives the resultant both in length and direction. This method of finding the length and direction from the starting point (0) to the terminating point (T) is called

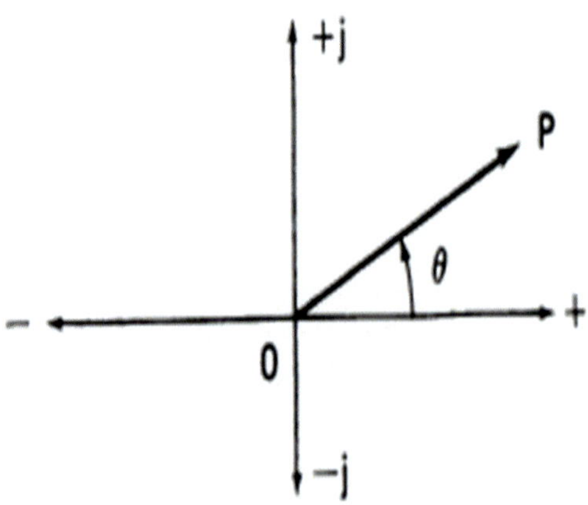

Figure 16.7 Graphical representation of vector quantities.

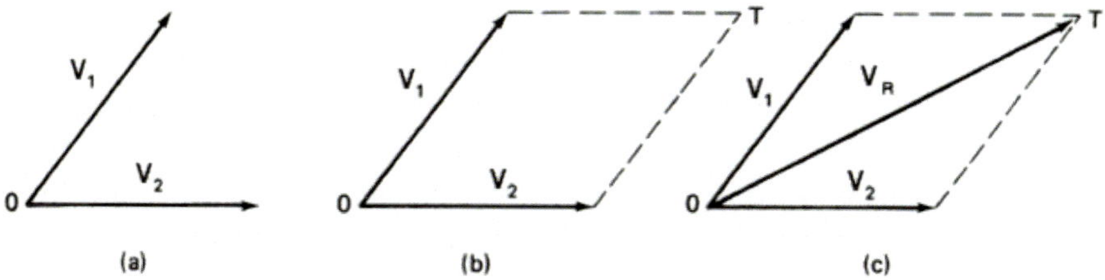

Figure 16-8 Parallelogram method of adding two vectors: (a) superimpose vector starting points; (b) complete the parallelogram; (c) draw resultant.

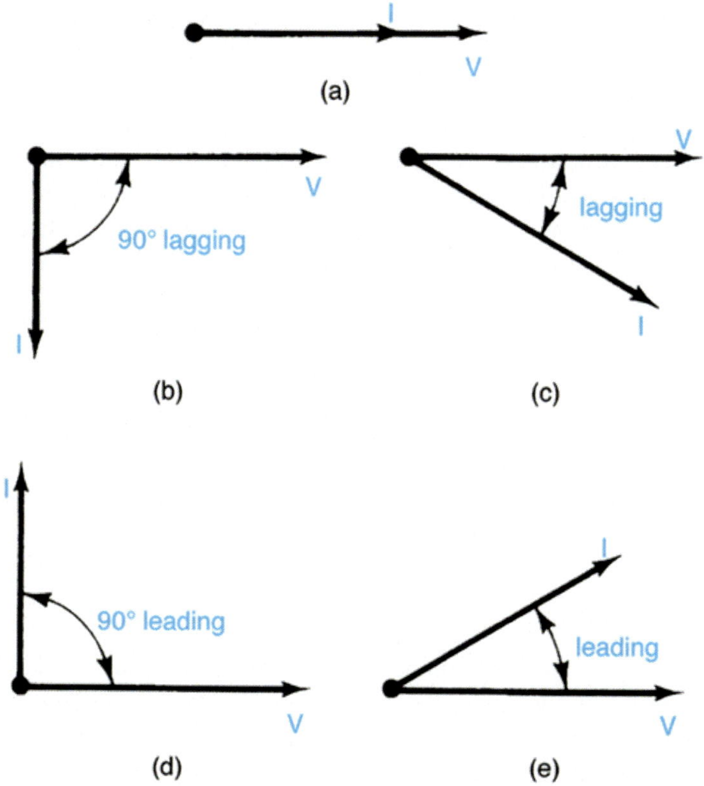

Figure 16-9 Vector diagrams showing voltage and current relationships of AC circuits: (a) resistive (R) circuit; (b) purely inductive (L) circuit; (c) RL circuit; (d) purely capacitive (C) circuit; (e) RC circuit.

the parallelogram method. This method can be used to add any number of vectors, by first finding the resultant of two of them, then adding another vector to this resultant in the same way, and then continuing this process.

A **phasor** is a rotating vector. In an electrical system, the AC input voltage as well as the voltages across and the current through circuit elements such as a resistor, capacitor, or inductor are represented using phasors. For the purpose of circuit analysis, the magnitude of direction of all the phasors in the system can be recorded at any instant. This reduces the analysis of the circuit to just vector manipulation using the parallelogram method. Usually in electrical systems, the term phasor and vector are used interchangeably.

A **vector diagram** for each AC circuit discussed in this chapter is shown in **Figure 16-9**. The understanding of the vector diagram is helpful when working with AC circuits. Rather than using waveforms to show phase relationships, it is easier to use vector diagrams. As previously mentioned, vectors are straight lines that have specific direction and length (magnitude). They are used to represent voltage and current values in AC circuits.

A horizontal line is drawn when beginning a vector diagram. Its left end is the **reference point**. In the diagrams of **Figure 16-9**, the voltage vector line is the reference. For inductive circuits, the current vectors are drawn in a clockwise direction from the voltage vector, which indicates that voltage leads current in an inductive circuit.

Figure 16-10 shows a circuit diagram, the current and voltage waveforms, as well as a phasor diagram for a pure inductor. For the capacitive circuits, the current vectors are drawn in a counter-clockwise direction from the voltage vectors which indicate that current leads voltage in a capacitive circuit. **Figure 16-11** shows the circuit diagram, the current and voltage waveforms, as well as the phasor diagram for a pure capacitor.

Phasor Addition

Circuits often contain many components: resistors, capacitors, and inductors. If two of the components are the same, their effects are combined together by adding their **phasors**. If the components have opposite reactances (X_L and X_C), the component's phasors are subtracted from one another. *RLC* circuits have a least one resistor, one capacitor, and one inductor.

Series circuits have an **impedance triangle** and a **voltage triangle**, as shown in **Figure 16-12**. The impedance triangle has phasors for resistance and reactance. The voltage triangle has phasors representing the voltage drops across the series components.

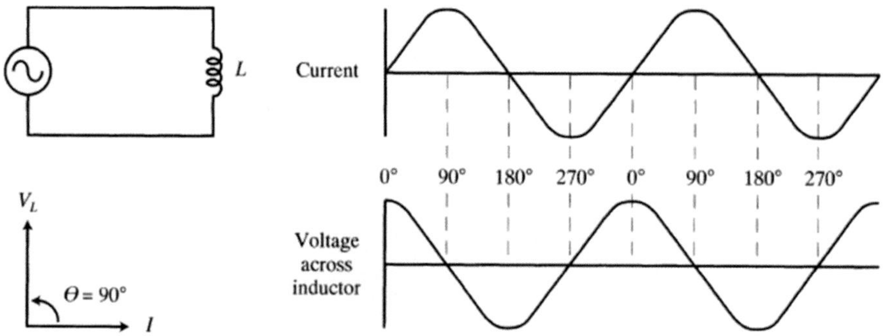

Figure 16-10 Circuit diagram, waveforms, and phasor of a pure inductance.

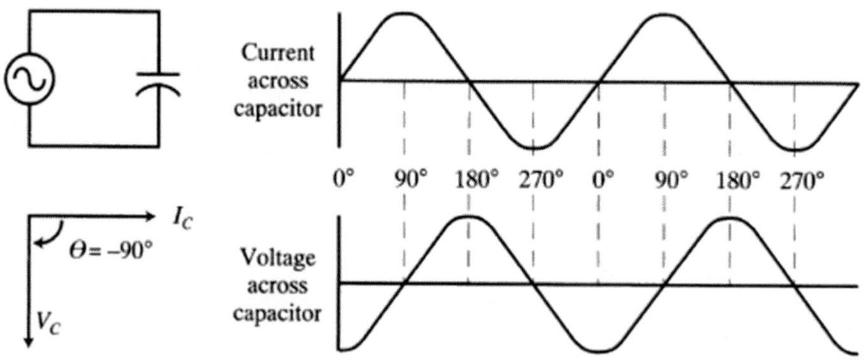

Figure 16-11 Circuit diagram, waveforms, and phasor of a pure capacitance.

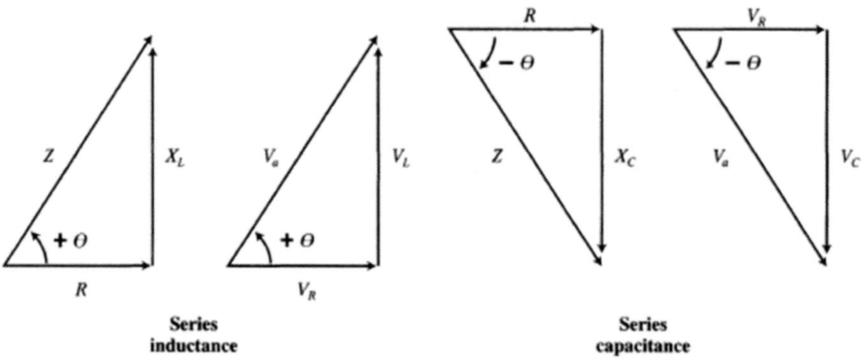

Figure 16-12 Phasors used in series circuits.

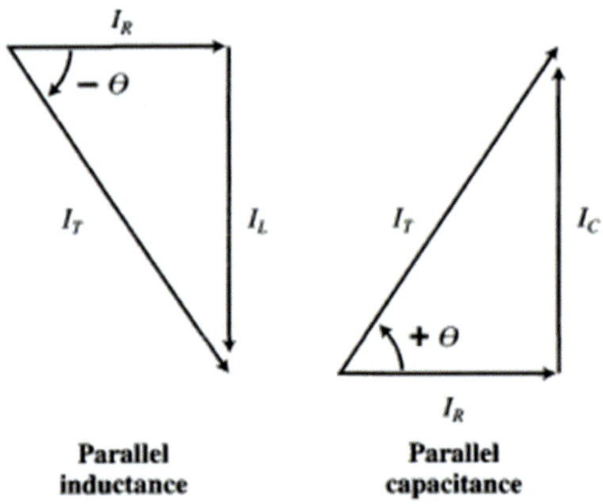

Figure 16-13 Phasors used in parallel circuits.

The voltage drops in an *RLC* series circuit appear to add to a larger value than the applied voltage. A series circuit containing both X_L and X_C has some cancelation due to the opposite phasors created by the inductance and capacitance.

A **parallel circuit** has a **current triangle**, as shown in **Figure 16-13**. The phasors represent the current in each branch of the parallel circuit. Note that the direction of the current phasor in a parallel circuit is opposite to the direction of its corresponding phasor in a series circuit. In a series circuit, inductive reactance is positive, and in a parallel circuit, inductive current is negative.

In-phase Phasors – Addition

Components of the same type have similar **phasors**. These phasors are classified as like phasors. They are plotted end-to-end. The phasors are added together to find the resultant phasor. **Figure 16-14** shows the phasors for three resistors connected in series.

The **phasors** are all drawn at 0° and connected end-to-end. Addition is used to find the resultant phasor. Although this drawing does not show any other components, there can be reactance mixed in the circuit without changing the manner in which the three resistance phasors are plotted.

Figure 16-15 has a resistor and two inductors connected in series. The **inductive reactances** are added by connecting the phasors end-to-end. They are plotted at +90°. The one resistance phasor is plotted at 0°. Phasors for **capacitive reactance** in series are plotted at –90°. They are added by connecting the phasors end-to-end, as shown in **Figure 16-16**.

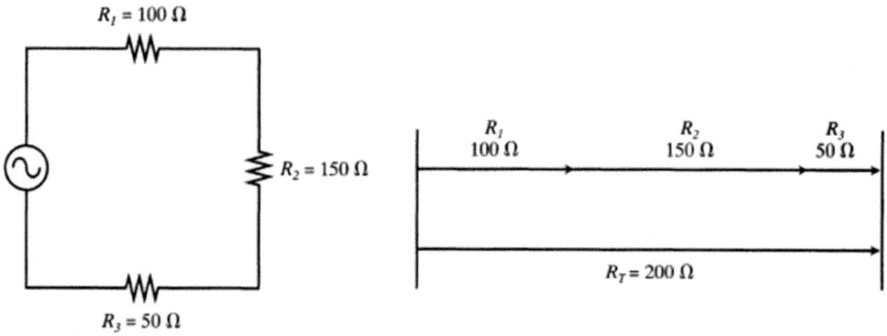

Figure 16-14 Resistor phasors are connected end-to-end. The resultant phasor can be found by simple addition.

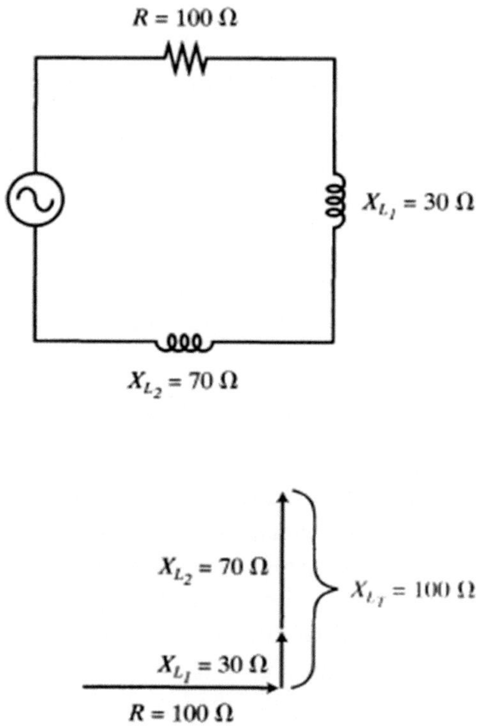

Figure 16-15 The phasors of inductors are added end-to-end.

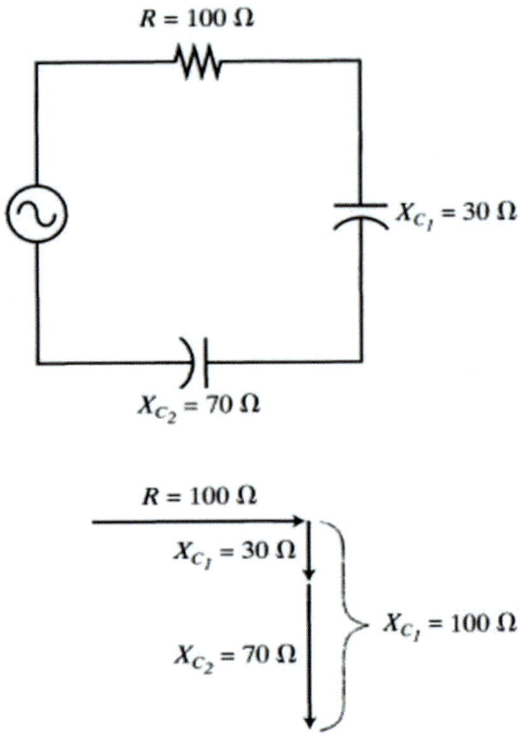

Figure 16-16 The phasors of capacitors are added end-to-end.

Out-of-phase Phasors – Subtraction

Capacitors and inductors have opposite phasors. When these components are contained in the same circuit, they work to cancel each other. Figure 16-17 shows the results of a circuit that has one reactance larger than the other. In these diagrams, resistance is plotted at 0°. It is used as a reference. The reactance phasors have been plotted at the end of the resistor phasor. X_L is positive and X_C is negative. The smaller phasor (X_C in the figure on the left and X_L in the figure on the right) is redrawn as a dotted line starting from the end of the larger phasor. You can see that the opposite phasors subtract, leaving a resultant phasor equal to the difference.

An **alternate method** is to add the phasors, using positive and negative numbers to represent the direction of the phasors. When X_L is larger than X_C, the resultant is positive. When X_L is smaller than X_C, the resultant is negative. The cancelation of reactance phasors has no effect on the resistance phasor.

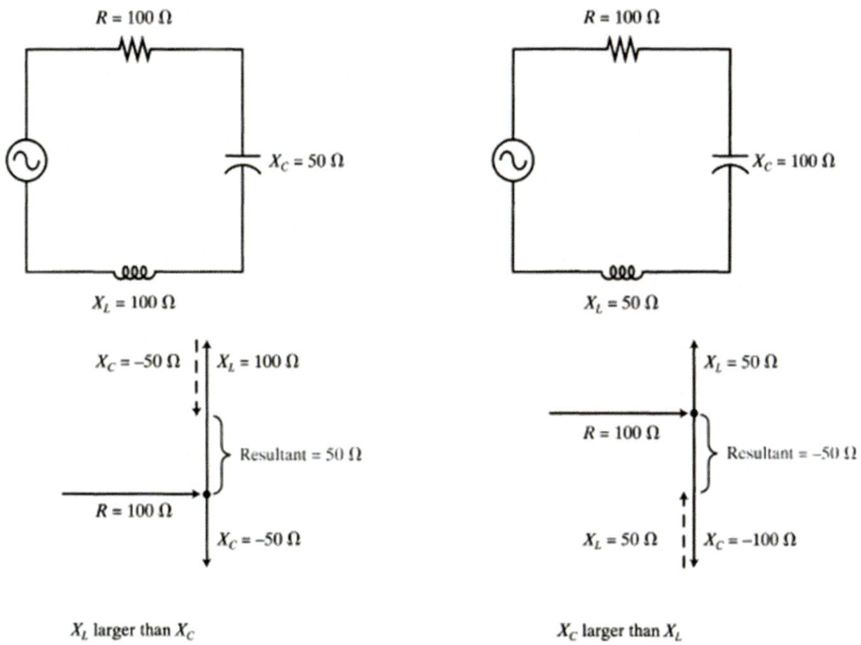

Figure 16-17 Opposite phasors, capacitive and inductive subtract.

16.7 Complex Numbers

Electromotive forces, currents, and impedances are sometimes expressed in the form of **complex numbers**. There are two types of coordinates for locating a point on a plane surface: **rectangular coordinates** and **polar coordinates**. With either type, you use a point and a line for reference; the point is called the **origin** and designated by the letter O, and the line is called the **axis**. With rectangular coordinates, a specific point is located on a plane by moving a measured distance from the origin along the real number axis in either the positive or negative direction, and then moving a measured distance perpendicular to the real number axis. The number written to represent these two distances and their directions is called the **rectangular form** of a complex number. An example of **a complex number** expressed in rectangular for is 3 + j4. The **"j" operator** is used in the rectangular form of complex numbers to represent perpendicular direction or 90°.

With **polar coordinates**, a point on a plane is located by rotating a straight line about the origin as center, until it passes through the given point. You can then indicate the position of the point by the angle that the rotating line makes with the **real number axis**, and the measured distance from the given point to the origin. The number representing these two items of information is called the polar form of a complex number. An example of a complex number expressed in **rectangular form** is: 5/53.13°.

A **trigonometric form** of representing complex numbers is used to convert from the polar to the rectangular form. An example of a complex number expressed in rectangular form is: 5 (cos 53.13°+ j sin 53.13°). These three forms of complex-number representation can be directly related to graphic presentations. A more abstract expression is the exponential form of a complex number, which cannot be shown.

Different mathematical operations can be performed on complex numbers. The conjugate of a complex number is the function in which addition with the complex number itself causes only the **imaginary value** to be canceled, with the **real value** remaining. If the complex number is in rectangular form, the complex conjugate is obtained by changing the sign of the imaginary component. If the complex number is in polar form, the complex conjugate is obtained by reversing the angle.

16.8 Imaginary Numbers

The square root of a negative number is called an **imaginary number**. All other types of numbers, either positive or negative, such as 9, –5, and $\sqrt{3}$, are called **real numbers**. The term $\sqrt{-1}$ can be considered to be an operator that indicates a rotation of 90°. Because the number $\sqrt{-1}$ is clumsy to write and is frequently used, it is represented by a letter. In pure mathematics, $\sqrt{-1}$ is represented by **i** or **j**. However, because "I" is the symbol for instantaneous current, the letter "j" is used to denote $\sqrt{-1}$ in electrical calculations. Thus, the imaginary number $27\sqrt{-1}$ can be written as 27i or 27j, respectively.

16.9 Rectangular Form of Complex Numbers

Combinations of real and imaginary numbers comprise the rectangular form of complex numbers. These can be written as follows: $4 + 3\sqrt{-1}$, –3 + 5i, –6 – j2, and 2 –/3. These numbers are similar to **algebraic binomials**. Such numbers are not restricted to the real or imaginary number axes but may have a magnitude in any direction on a plane surface. A **complex number** is represented by using rectangular coordinates with the horizontal axis for the real numbers and the vertical axis for the imaginary numbers. In **Figure 16-18**,

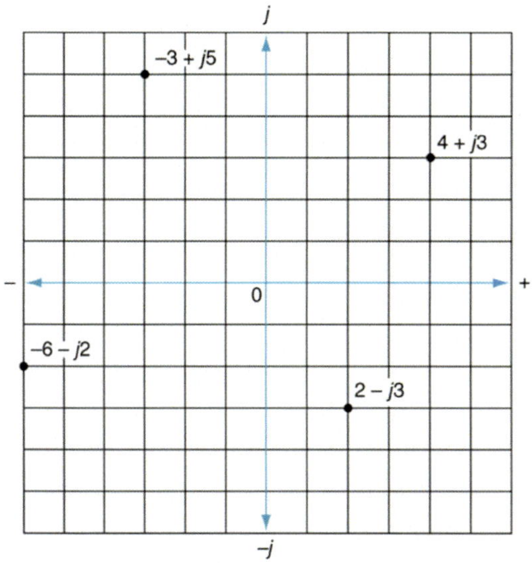

Figure 16-18 Coordinate diagram.

the complex number 4 + j3 consists of a real part (4) and an imaginary part (j3). The point that represents this number is then four units in the positive direction and three units in the +j direction. The absolute value of the number is indicated by the distance from the resulting point to the zero point (origin) of the diagram. By this method, any complex number can be represented.

16.10 Addition and Subtraction of Complex Numbers

To find the **sum** of two complex numbers in **the rectangular form**, algebraically add the real parts and the imaginary parts in the same manner that algebraic binomials are added. The following two examples show the procedure:

$(a + jb) + (c + jd) = (a + c) + j(b + d)$

$= (4 + j3) + (-3 + j5) = (4 - 3) + j(3 + 5) = 1+j8.$

The sum of three or more complex numbers can be obtained by adding two of them, adding the sum to the third of the given numbers, and so on, until all the numbers have been added.

Impedances that are written in the form of complex numbers often have to be added. For example, **Figure 16-19** shows two impedances connected in series. Find the total impedance by adding them, as shown in the following:

$$Z_1 = R_1 - jX_C$$
$$Z_2 = R_2 + jX_L$$
$$Z_1 + Z_2 = (R_1 + R_2) + j(X_L - X_C).$$

Substituting numbers for rcsistances and reactances yields

$$Z_1 = (2 - j750)\Omega$$
$$Z_2 = (280 + j430)\Omega$$
$$Z_1 + Z_2 = (2 + 280) + j(430 - 750)$$
$$= (282 - j320)\Omega.$$

Complex numbers can be **subtracted** as in algebra by changing the signs of the terms in the subtrahend and proceeding as in addition. Examples of subtraction are as follows:

$$(a + jb) - (c + jd) = (a - c) + j(b - d)$$
$$(4 + j3) - (-3 + j5) = 4 - (-3) + j(3 - 5)$$
$$= 7 - j2$$

For the circuit shown in **Figure 16-19**, if Z_{in} = (470 – j35) □, and Z_1 = (4 – j830) □, find Z_2 as follows:

$$Z_2 = Z_{in} - Z_1$$
$$= (470 - j35) - j(4 - j830)$$
$$= (470 - 4) + j(-35 + 830)\ \Omega$$
$$= 466 + j795\ \Omega.$$

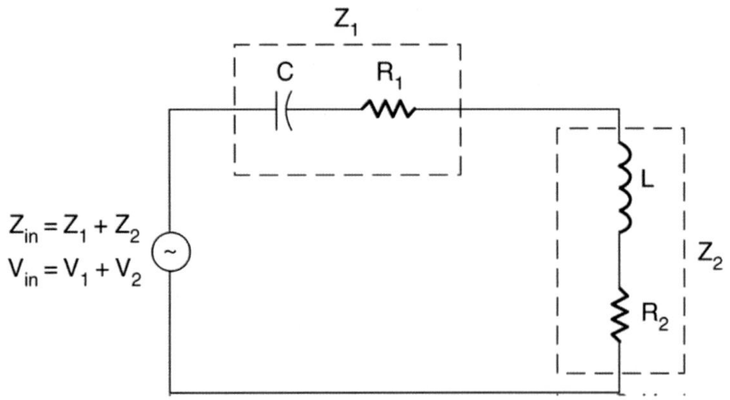

Figure 16-19 Series impedances.

Polar and Trigonometric Forms

Complex numbers in the rectangular form can easily be handled for addition and subtraction operations. Two items of information must be given in both rectangular and polar coordinates to locate a point on a plane. In **polar coordinates**, one item is an **angle** and the other a **distance**. This information is written in the form r/θ, where r is the distance from the origin to the point that represents the complex number, and θ is the angle between the straight line r and the real number axis. The absolute value r is called the modulus, and the angle θ is called the argument. The **imaginary numbers** jr and –jr can then be written as r/90° and r/–90°, respectively. **Figure 16-20** shows the position of complex numbers on a polar diagram. A number expressed in polar form is converted to its rectangular equivalent by using the sine and cosine trigonometric functions.

- Represent 5/53.13° in rectangular form:
  = 5 (cos 53.13° + j sin 53.13°)
  = 5 (0.6 + j 0.8)
  = 3 + j4.
- Represent 3 + j4 in polar form:

  Modulus = $\sqrt{3^2 + 4^2} = \sqrt{9+16} = \sqrt{25} = 5$

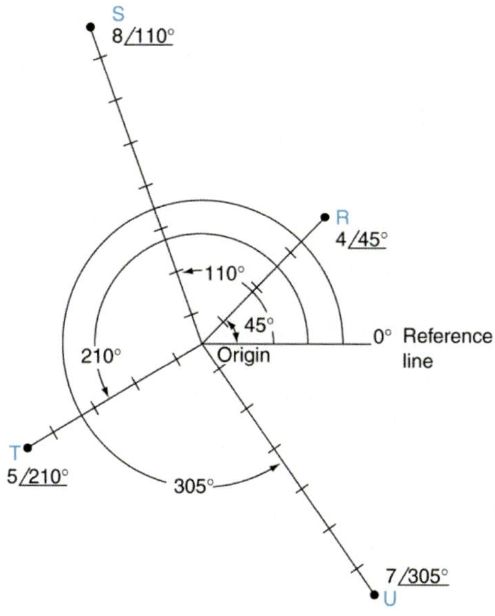

Figure 16-20 Polar diagram.

$$\text{Argument} = \tan^{-1}(\frac{4}{3}) = 53.13\ ^{\circ}.$$

Hence, the complex number in polar form is: 5/53.13°.

Complex numbers in polar form can be multiplied and divided. The magnitudes of the numbers are multiplied and divided during these operations. During multiplication, the angles of the complex numbers are added, and during division, the angles are subtracted.

$$Z_1 = M_1\angle\theta_1$$
$$Z_2 = M_2\angle\theta_2$$
$$Z_1 \times Z_2 = M_1 \times M_2\ \angle(\theta_1 + \theta_2)$$
$$\frac{Z_1}{Z_2} = \frac{M_1}{M_2}\angle(\theta_1 - \theta_2).$$

If Z_1=10/45° and Z_2== 5/15°, determine $Z_1 \times Z_2$ and $\frac{Z_1}{Z_2}$.

$$Z_1 \times Z_2 = 10 \times 5\ \angle(45^{\circ} + 15^{\circ}) = 50\angle 60^{\circ}$$
$$\frac{Z_1}{Z_2} = \frac{10}{5}\angle(45^{\circ} - 15^{\circ}) = 2\angle 30^{\circ}.$$

16.11 Conversions

The purpose of writing circuit values as **complex numbers** is to allow easier manipulation of the arithmetic expressions. Converting between rectangular form and polar form uses three basic functions from trigonometry. These functions are sine (sin), cosine (cos), and tangent (tan).

Rectangular Form to Polar Form

Conversion of rectangular form into polar form involves two steps.

Use the **Pythagorean theorem** to find the hypotenuse.

$$\text{hypotenuse} = \sqrt{\text{adjacent}^2 + \text{opposite}^2}.$$

The angle theta (θ) is calculated using the tangent function. The tangent function is rearranged to solve for the angle:

$$\theta = \tan^{-1}[\frac{\text{opposite}(X)}{\text{adjacent}(R)}]$$

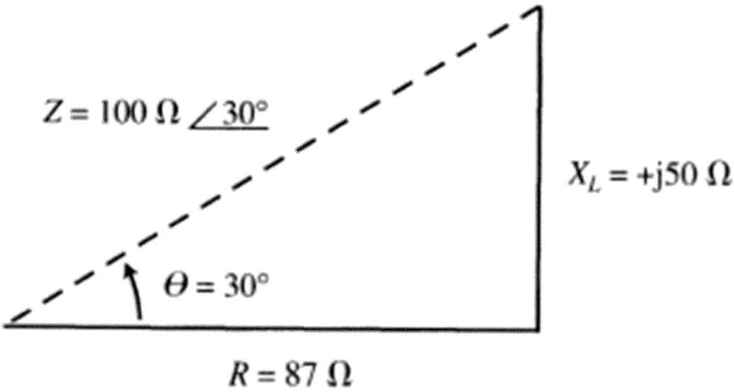

Figure 16-21 Rectangular to polar conversion.

where **θ** *is the angle measured in degrees,*
R *is the resistance measured in ohms,* and
X is the reactance measured in ohms.
The **polar form** of a number at an angle is written as: Z/θ.

- Convert the following circuit values from rectangular form into polar, $R = 87\ \Omega$, $X_L = +j50\ \Omega$. The triangle is shown in **Figure 16-21**.

1. Find the length of the hypotenuse using the Pythagorean theorem.

Formula: $Z = \sqrt{R^2 + X_L^2}$

Substitution: $Z = \sqrt{87^2 + 50^2}$

Answer: $Z = 100\ \Omega$

2. Find the operating angle.

Formula: $\theta = \tan^{-1} \dfrac{X_L}{R}$

Substitution: $\theta = \tan^{-1} \dfrac{50\ \Omega}{87\ \Omega}$

Answer: $\theta = 30°$

3. Express impedance in polar form.

$$Z = 100\ \Omega\ \angle 30°$$

Polar Form to Rectangular Form

A number expressed in polar form is converted to its rectangular equivalent by using the sine and cosine trigonometric functions.

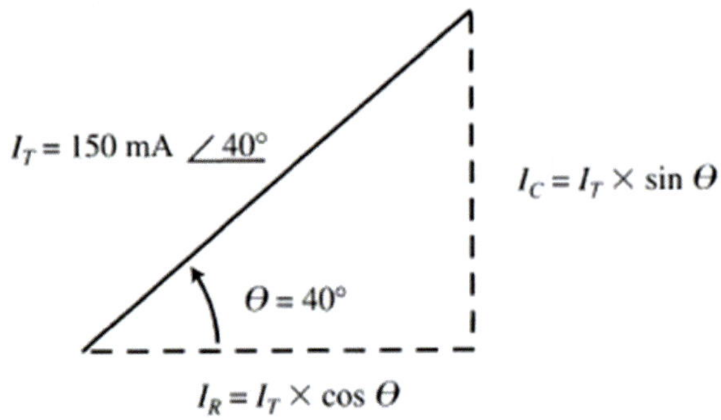

Figure 16-22 Polar to rectangular conversion.

1. Resistive branch current.

Formula: $\cos \theta = \frac{\text{adj}}{\text{hyp}}$

Rearranging: adj = hyp × cos θ

Substitution: $I_R = 150 \text{ mA} \times \cos 40°$

Answer: $I_R = 115 \text{ mA}$

2. Reactive branch current.

Formula: $\sin \theta = \frac{\text{opp}}{\text{hyp}}$

Rearranging: opp = hyp × sin θ

Substitution: $I_C = 150 \text{ mA} \times \sin 40°$

Answer: $I_C = +j96.4 \text{ mA}$

3. Write total current in rectangular form.

$$I_T = 115 + j96.4 \text{ mA}$$

- *Find the reactive and resistive branch currents when given the total current in polar*
- form, $I_T = 150$ mA /40°. The triangle is drawn in **Figure 16-22**.

Writing Complex Numbers from Circuits

All circuit component values can be written in rectangular form and then converted to polar form. **Figure 16-23** lists examples of circuit components and their equivalent rectangular and polar expressions.

Figure 16-23(A) is a resistor with no reactive component. The rectangular form of its impedance is written with the *j*-term having a value

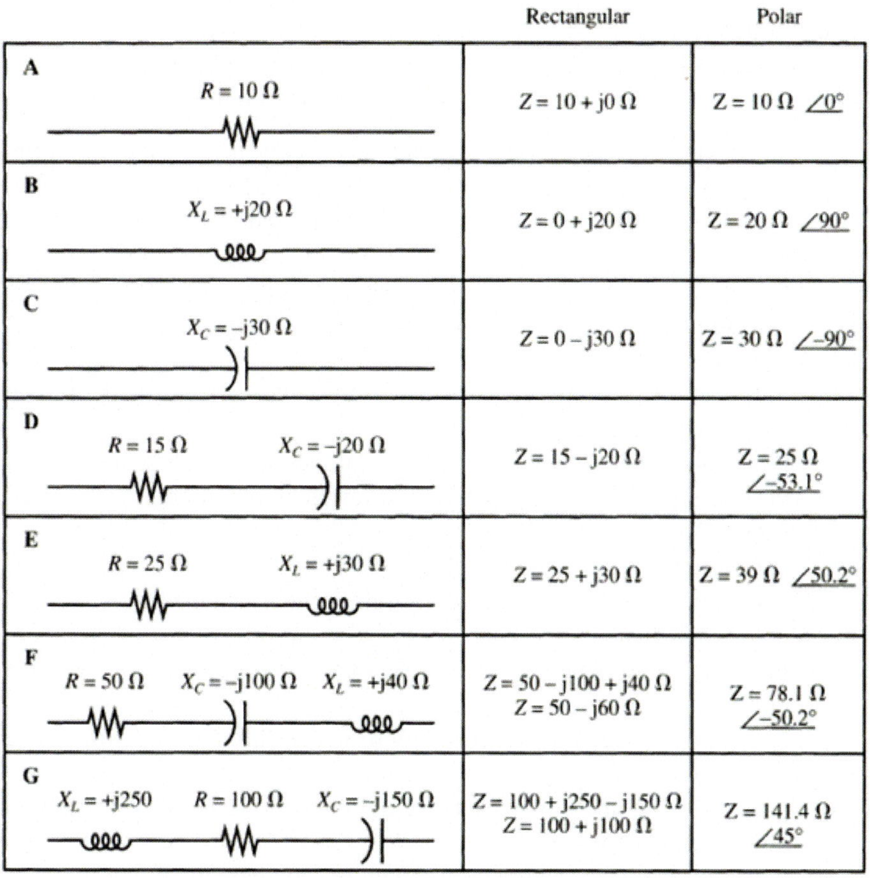

Figure 16-23 Expressing impedances as complex numbers.

of zero. In polar form, this is expressed as the impedance having an angle of zero.

Figures 16-23(B) and **(C)** are reactive components with no resistance. The rectangular form shows a value of zero for resistance. The polar form expresses the impedance at an angle of 900 either + or –.

Figures 16-23(D) and **(E)** show an impedance containing both resistance and reactance. The rectangular form states the values of each component. The polar form is the combined impedance, expressed as a quantity at an angle.

Figures 16-23(F) and **(G)** contain resistance with both inductive and capacitive reactances.

The **impedance** can be written as a rectangular expression showing the value of each component. However, the expression should be simplified to show only the net reactance. The polar form is the combined values of the three components.

SummaryTrigonometry is used for solving problems dealing with phase angle, power factor, and reactive power in AC circuits.

- A rectangular coordinate system is a two-dimensional drawing showing two perpendicular lines that cross at the zero reference point.
- A rectangular coordinate system can be divided into four parts which are called quadrants.
- In the polar coordinate system, the position of any point can be determined by using a radius vector and a reference line that meet at a common point called the origin.
- The angular velocity of a rotating vector is determined by the rate at which the angle changes.
- Physical quantities that have only magnitude are called scalars.
- Physical quantities that have both magnitude and direction are called vectors.
- A rotating vector is called a phasor.
- Complex numbers are commonly represented using rectangular and polar coordinates.
- Imaginary numbers are determined by the square root of a negative number.
- The letter “i” or “j” is used to represent an imaginary number and equals $\sqrt{-1}$.
- A complex number involves real and imaginary numbers when written in rectangular form.
- In circuit analysis, complex numbers are mathematically commonly used to represent the combination of resistance and reactance.

- Algebraic operations such as addition and subtraction can be performed on complex numbers written in rectangular form.
- The real and imaginary parts of the complex numbers can be added or subtracted separately when they are in the rectangular form.
- The polar form of a complex number consists of a magnitude and an angle.
- Multiplication and division of complex numbers is performed using the polar form.
- The magnitudes of complex number in polar form can be multiplied or divided.
- The angles of complex number in polar form can be added when the numbers are being multiplied or subtracted when the numbers are being divided.
- Conversion between different forms of complex numbers is achieved mathematically using trigonometric functions and the Pythagorean theorem.

Formulas

$$Z^2 = R^2 + X_C{}^2$$

$$Z = \sqrt{R^2 + X_C{}^2}$$

$$\omega = 2\pi f$$

$$j = \sqrt{-1}$$

$$Z_1 = R_1 - jX_C$$

$$Z_2 = R_2 + jX_L$$

$$Z_1 + Z_2 = (R_1 + R_2) + j(X_L - X_C)$$

$$Z_1 = M_1 \angle \theta_1$$

$$Z_2 = M_2 \angle \theta_2$$

$$Z_1 \times Z_2 = M_1 \times M_2 \ \angle(\theta_1 + \theta_2)$$

$$\frac{Z_1}{Z_2} = \frac{M_1}{M_2} \angle(\theta_1 - \theta_2)$$

$$\sin\theta = \frac{\text{opposite}}{\text{hypotenuse}}; \cos\theta = \frac{\text{adjacent}}{\text{hypotenuse}}; \sin\theta = \frac{\text{opposite}}{\text{adjacent}}$$

$$\text{hypotenuse} = \sqrt{\text{adjacent}^2 + \text{opposite}^2}$$

$$\theta = \tan^{-1}[\frac{\text{opposite}(X)}{\text{adjacent}(R)}]$$

Problems

Use phasor addition to determine the net reactance of series circuit where $X_C = 40\ \Omega$ and $X_L = 25\ \Omega$. Is the net reactance inductive or capacitive? [$X = X_C - X_L = 40 - 25 = 15\ \Omega$ capacitive]

Find the equivalent impedance of a series circuit in rectangular and polar form, having a 3-Ω resistor in series with an inductive reactance of 4 Ω. [rectangular form: Z = 3 + j4; polar form: 5/53.13°]

Determine the rectangular form of the complex number given as 10/45° in polar form. [$10(\cos 45° + j \sin 45°) = 10(0.707 + j\,0.707) = (7.07 + j7.07)\ \Omega$]

The branch currents of a circuit are given in rectangular form, *I R* = 6 A and *I L* = –j6 A. Find the value of total current in polar form. See **Figure 21-24**.

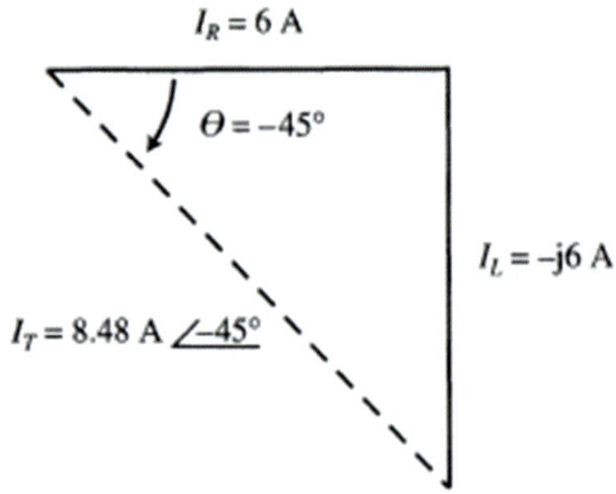

Figure 16-24 Converting rectangular to polar form.

1. Find the length of the hypotenuse using the Pythagorean theorem.

Formula: $I_T = \sqrt{I_R^2 + I_L^2}$

Substitution: $I_T = \sqrt{6^2 + 6^2}$

Answer: $I_T = 8.48\ \text{A}$

2. Find the operating angle.

Formula: $\theta = \tan^{-1} \dfrac{I_L}{I_R}$

Substitution: $\theta = \tan^{-1} \dfrac{-j6\ \text{A}}{6\ \text{A}}$

Answer: $\theta = -45°$

3. Express total current in polar form.

$$I_T = 8.48\ \text{A} \angle -45°$$

Find the voltage drops across the series components with a total voltage, expressed in polar form, of 25 V/–53°. See **Figure 16-25**.

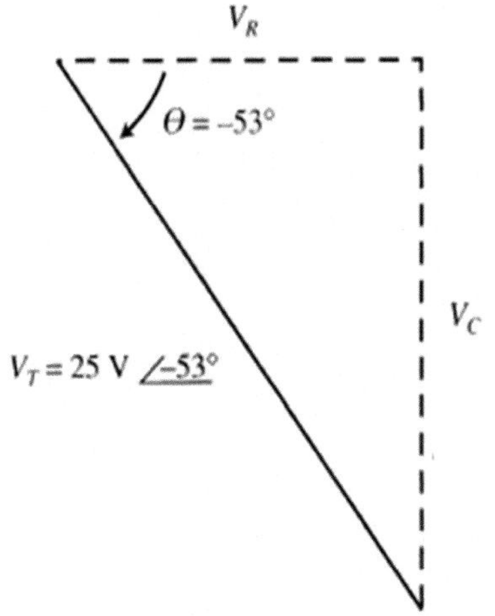

Figure 16-25 Converting polar to rectangular form.

1. Resistive voltage drop.

Formula: $\cos\,\theta = \dfrac{\text{adj}}{\text{hyp}}$

Rearranging: adj = hyp × cos θ

Substitution: $V_R = 25\text{ V} \times \cos -53°$

Answer: $V_R = 15\text{ V}$

2. Reactive voltage drop.

Formula: $\sin\,\theta = \dfrac{\text{opp}}{\text{hyp}}$

Rearranging: opp = hyp × sin θ

Substitution: $V_C = 25\text{ V} \times \sin -53°$

Answer: $V_C = -j20\text{ V}$

3. Write total current in rectangular form.

$$V_T = 15 - j20\text{ V}$$

Self-examination/Answers

1. What is the impedance of a circuit which has a resistance (R) of 10 Ω and an inductive reactance (X_L) of 40 Ω?
2. Calculate the trigonometric function values of sin θ, cos θ, and tan θ for a circuit which has a resistance (R) of 10 Ω and an inductive reactance (X_L) of 40 Ω.
3. The impedance of a circuit is shown as the ____________ of a right-angle triangle.
5. Two perpendicular number lines that cross at a zero reference point constitute the ___________coordinate system.
6. In a rectangular coordinate system, the horizontal line is the ________ axes and the vertical line is the ________axes.
7. In a rectangular coordinate system, the axes cross each other at _______°.
8. In a rectangular coordinate system, there are _____quadrants.

9. In the first quadrant of a rectangular coordinate system, the abscissa of a point is (positive/negative) and the ordinate of the point is __________.
10. In the ____________quadrant of a rectangular coordinate system, both the abscissa and ordinate of a point are negative.
11. In the polar coordinate system, the center point is referred to as the ________. [origin]
12. The distance from the origin in a polar coordinate system is called the __________, and the angular direction is called the ____________.[modulus, argument]
13. A point written in terms of polar coordinates has the ________written first, followed by the ___________. [modulus, argument]
14. A common unit for the measurement of angular velocity of a motor is ____________.
15. The angular measurement of a sine wave is based on _____° for a complete cycle.
16. 360° is equal to _________radians, and ________° is equal to π radians.
17. How does a vector quantity differ from a scalar quantity?
18. The _______________method is used for adding or subtracting vectors.
19. In a pure inductive circuit, the voltage phasor (lags/leads) the current phasor by ______°.
20. In a pure capacitive circuit, the voltage phasor (lags/leads) the current phasor by ______°.
21. In a series *RLC* circuit, the net reactance of the circuit is determined by (adding, subtracting) the phasors of the inductive and capacitive reactances.
22. The _______________form of a complex number represents two distances in perpendicular directions.
23. The _______________form of a complex number which is represented by a modulus and an argument.
24. The trigonometric form of a complex number is used to convert between the _______________and the _____________form.
25. The square root of a negative number is considered to be a(n) ____________number.
26. The square root of a non-negative number is considered to be a(n) ____________number.
27. The letter "j" in electrical calculations represents ___________.
30. A complex number 2 + j3 has a magnitude of ________ on the real axis, and a magnitude of ______on the imaginary axis.

31. Express the rectangular coordinates of a complex number which has a magnitude of 6 on the real axis and 9 on the imaginary axis: _______________.
32. +j and –j are in (same/opposite) directions on the imaginary axis.
33. Add the complex numbers: 2 + j3 and 3 + j4: _______________.
34. Subtract the complex number 2 + j3 from 3 + j4: _______________.
35. Add the complex number 2 + j3 with itself: _______________.
36. In a complex number r/θ, the letter "r" represents the _______________ and θ the ___________.
39. A complex number in polar form $2/0^\circ$ represents a direction along the ___________ axis, while $2/90^\circ$ represents a direction along the __________ axis.
40. The polar form of 4 + j3 is _____________.
41. Determine the rectangular and polar complex number representing the impedance of a series circuit having a10-Ω resistor and an inductor having a reactance (X_L) of 15 Ω.
42. Determine the rectangular and polar complex number representing the impedance of a series circuit having a10-Ω resistor and a capacitor having a reactance (X_C) of 15 Ω.
43. Determine the rectangular and polar complex number representing the impedance of a series circuit having a10-Ω resistor, a inductive reactance (X_L) of 15 Ω, and a capacitive reactance (X_C) of 15 Ω.

Answers

1. $Z^2 = R^2 + X^2 = \sqrt{10^2 + 40^2} = \sqrt{1700} = 41.23$
2. $\sin\theta = \dfrac{40}{41.23} = 0.97$; $\cos\theta = \dfrac{10}{41.23} = 0.24$; $\tan\theta = \dfrac{40}{10} = 4$
3. hypotenuse
4. rectangular
5. *X*, *Y*
6. 90
7. Four
8. positive, positive
9. third
10. origin
11. modulus, argument
12. modulus, argument

13. rpm
14. 360
15. 2π, 180
16. Vector has both direction and magnitude, whereas a scalar has magnitude only Parallelogram
18. leads, 90
19. lags, 90
20. subtracting
21. rectangular
22. polar
23. rectangular, polar
24. imaginary
25. real
26. $\sqrt{-1}$
27. 2, 3
28. 6 + j9
29. Opposite
30. 1 5 + j7
31. 1 + j1
32. 4 + j6
33. modulus, argument
34. real, imaginary
35. 5/36.87°
36. Rectangular form: 10 + j15; Polar form: 18.03/56.31°
37. Rectangular form: 10 – j15; Polar form: 18.03/-56.31°
39. Rectangular form: 10 + j0; Polar form: 10/0°

Problem Answers

$X = X_C - X_L = 40 - 25 = 15$-Ω capacitive
rectangular form: Z = 3 + j4; polar form: 5/53.13°
10 (cos45° + j sin 45°) = 10 (0.707 + j 0.707) = (7.07 + j7.07) Ω
$I_T = 8.48/–45°$

1. Find the length of the hypotenuse using the Pythagorean theorem.

Formula: $I_T = \sqrt{I_R^2 + I_L^2}$

Substitution: $I_T = \sqrt{6^2 + 6^2}$

Answer: $I_T = 8.48$ A

2. Find the operating angle.

Formula: $\theta = \tan^{-1} \dfrac{I_L}{I_R}$

Substitution: $\theta = \tan^{-1} \dfrac{-j6\ \text{A}}{6\ \text{A}}$

Answer: $\theta = -45°$

3. Express total current in polar form.

$$I_T = 8.48\ \text{A}\ \underline{\angle -45°}$$

$V_T = 15 - j20\ \text{V}$

1. Resistive voltage drop.

Formula: $\cos\,\theta = \dfrac{\text{adj}}{\text{hyp}}$

Rearranging: $\text{adj} = \text{hyp} \times \cos\,\theta$

Substitution: $V_R = 25\ \text{V} \times \cos -53°$

Answer: $V_R = 15\ \text{V}$

2. Reactive voltage drop.

Formula: $\sin\,\theta = \dfrac{\text{opp}}{\text{hyp}}$

Rearranging: $\text{opp} = \text{hyp} \times \sin\,\theta$

Substitution: $V_C = 25\ \text{V} \times \sin -53°$

Answer: $V_C = -j20\ \text{V}$

3. Write total current in rectangular form.

$$V_T = 15 - j20\ \text{V}$$

Glossary

Abscissa
The *x*- or horizontal-axis location of a point in a graphical system using Cartesian coordinates.

Amplitude
The vertical height of an AC waveform.

Angle of lead or lag
The angle between applied voltage and current flow in an AC circuit, in degrees; in an inductive (*L*) circuit, voltage (V_L) leads current (I_L); in a capacitive (*C*) circuit, current (I_C) leads voltage (V_C).

Angular velocity
The speed of rotation of a vector around a fixed point, measured in degrees or radians per second.

Argument
The angular position or displacement of points in the polar coordinate system.

Cartesian Coordinates
The location of a point in a two-dimensional graphical system established by drawing two perpendicular number lines that cross at their zero reference points.

Complex numbers
A mathematical method often used in electrical systems for expressing the magnitude and phase relationships of different components.

Conductance (G)
The ability of a resistance of a circuit to conduct current, measured in Siemens; the inverse of resistance: $G = I/R$.

Complex conjugate
A function of a complex number in which addition with the complex number itself causes only the imaginary value to be canceled. If the complex number is in rectangular form, the complex conjugate is obtained by changing the sign of the imaginary component. If the complex number is in polar form, the complex conjugate is obtained by reversing the angle.

Cycle
A sequence of events that causes one complete pattern of alternating current from a zero reference, in a positive direction, back to zero, then in a negative direction, and back to zero.

Effective voltage (V_{eff})
The value of an AC sine-wave voltage which has the same heating effect as an equal value of DC voltage.
$V_{eff} = V_{peak} \times 0.707$

Trigonometric form of complex number
A complex number showing the magnitude and phase angle trigonometrically.

Rectangular form of complex number
A complex number expressed using values that represent the magnitude and direction of a phasor or vector. The operator (j) is used in this form of representation.

Frequency
The number of ac cycles per second, measured in hertz (Hz).

Frequency response
A circuit's ability to operate over a range of frequencies.

Hertz (Hz)
The international unit of measurement of frequency equal to one cycle per second.

In phase
Two waveforms of the same frequency which pass through their minimum and maximum values at the same time and polarity.

Real number
All numbers excluding imaginary numbers.

Imaginary number
The square root of a negative number. It is represented using the product of a real number and 'j' which represents $\sqrt{-1}$.

Instantaneous voltage (V_i)
A value of AC voltage at any instant (time) along a waveform.

j operator
A mathematical tool used to represent reactance values as complex numbers. The "*j*" part of this expression represents the imaginary number or value.

Lagging phase angle
The angle by which current lags voltage (or voltage leads current) in an inductive circuit.

Leading phase angle
The angle by which current leads voltage (or voltage lags current) in a capacitive circuit.

Modulus
The absolute value of a complex quantity regardless of its sign (polarity).

Mho
Ohm spelled backward; the old unit of measurement for conductance, susceptance, and admittance; now replaced by Siemens.

Ordinate
The *y*- or vertical-axis location of a point in a graphical system using Cartesian coordinates.

Peak-to-peak voltage (V_{pp})
The value of AC sine-wave voltage from positive peak to negative peak.

Peak voltage (V_{peak})
The maximum positive or negative value of AC sine-wave voltage.
$V_{peak} = V_{eff} \times 1.41$

Period (time)
The time required to complete one AC cycle; time = 1/frequency.

Phase angle (θ)
The angle between two phasor quantities. In electrical systems, this could be the angular displacement between the applied voltage and current flow, or between any two voltages, currents, powers, or impedances.

Phasor
A rotating vector quantity, which is used to show magnitude and direction.

In-phase
Phasor quantities that have the same direction and may have different magnitudes.

Out-of-phase
Phasor quantities that are not in-phase.

Polar form of complex number
A complex number expressed using a magnitude and phase angle.

Pythagorean theorem
In a right triangle, the square of the hypotenuse is equal to the sum of the squares of the other two sides.

Quadrants
The four regions of a Cartesian coordinate system formed by the intersection of the *x*- and *y*-axes.

Reactance (X)
The opposition to AC current flow due to inductance (X) or capacitance (X_c).

Reactive circuit
An AC circuit that has the property of inductance or capacitance.

Rectangular coordinates
See **Cartesian coordinates.**

Root mean square (RMS) voltage
See **Effective voltage.**

Susceptance (B)
The ability of an inductance (B_L) or a capacitance (B_C) to pass AC current and is measured in Siemens.

$$B_L = \frac{1}{X_L}; B_C = \frac{1}{X_C}$$

Theta (θ)
The Greek letter used to represent the phase angle of an AC circuit.

Vector
A straight line whose length indicates magnitude and position indicates direction.

Wavelength
The distance between two corresponding points that represents one complete wave.

17

AC Problem Solving Methods

When **alternating current** is used in a circuit, it can be solved by using theorems and laws similar to those used for DC circuits. This includes **Kirchhoff's voltage and current laws, superposition theorem** for evaluating circuits with multiple sources, and **Thevenin's theorem** for simplification. The frequency of an AC source influences the values of the inductive and capacitive reactance of the circuit. The problem solving methods of this chapter involve networks having two- or three-terminal configurations. These AC circuit values are represented as complex numbers using rectangular and polar form.

Objectives

1. Determine the impedance of series- and parallel-connected AC circuits
2. Apply Kirchhoff's voltage law to determine AC values in a circuit
3. Apply Kirchhoff's current law to determine AC values in a circuit
4. Use the superposition theorem to determine AC values for circuits containing multiple sources
5. Simplify a complex AC circuit using Thevenin's theorem
6. Identify various three-phase impedance network configurations
7. Convert three-phase networks into different configurations
8. Determine the power relationships of single- and three-phase AC systems

Chapter Outline

17.1 Impedance Network Representations

An **impedance network** consists of a combination of two or more electrical components that have been discussed previously. A basic impedance network is composed of resistors, inductors, and capacitors as shown in **Figure 17-1(a)**.

A network of these components can also be connected to other electrical and electronic components forming a more complex circuit. In a **schematic diagram**, an impedance network is often represented by a **rectangular box** as shown in **Figure 17-1(b)**. Note that the letter Z is placed inside the box to indicate the impedance of the network. The impedance of a network may be represented in either rectangular or polar form as shown by the expression in **Figure 17-1**.

Series Impedance

When the resistor, capacitor, and inductor of an impedance network are connected in **series**, the total impedance or opposition offered by a network

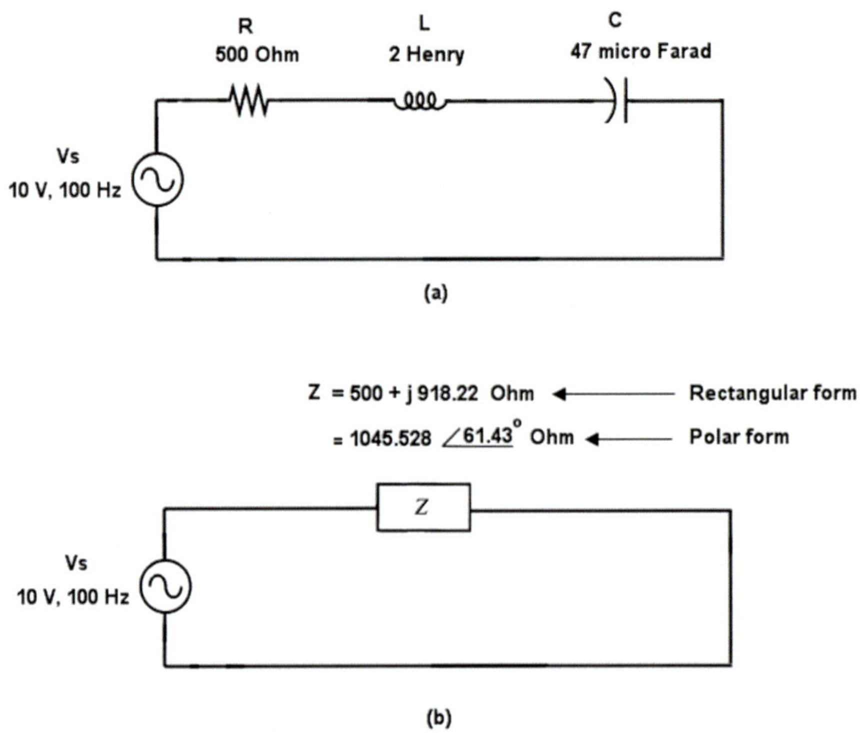

Figure 17-1 Impedance network: (a) conventional representation; (b) single impedance symbol.

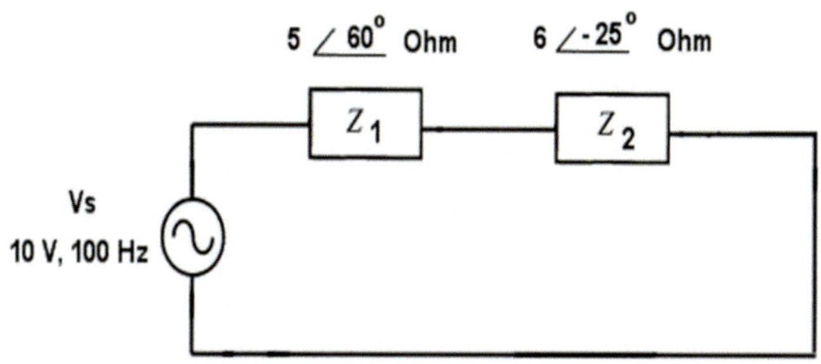

Figure 17-2 Series-connected impedance network.

is determined by the sum of the individual impedance values produced by each component. The impedance of a series network is usually represented in rectangular form.

- Determine the total impedance of the series-connected impedance network shown in **Figure 17-2**.

$$
\begin{aligned}
Z_T &= Z_1 + Z_2 \\
&= 5\angle 60^\circ + 6\angle -25^\circ\ \Omega \\
&= (2.5 + j4.330) + (5.438 - j2.536)\ \Omega \\
&= 7.938 + j1.794\ \Omega \\
&= 8.138\angle 12.738^\circ\ \Omega.
\end{aligned}
$$

Parallel Impedance

When the resistor, capacitor, and inductor of **an impedance network** are connected in **parallel**, the reciprocal of the total impedance is equal to the sum of the reciprocals of the individual impedance values. The **reciprocal** of the total impedance indicates the ease with which current can pass through the network and is referred to as the **admittance**. It is represented as *Y* and is measured in Siemens. The reciprocal of a resistance is called the conductance (G), while the reciprocal of a reactive component such as an inductor or capacitor is called the **susceptance (B)**. The total admittance of a network is the sum of the admittance values produced by each component. The impedance or admittance of a parallel network is usually represented in polar form.

- Determine the total impedance (Z_T or $Z_{1//2}$) of the parallel-connected impedance network shown in **Figure 17-3**.

$$\frac{1}{Z_T} = \frac{1}{Z_1} + \frac{1}{Z_2} = \frac{Z_1 + Z_2}{Z_1 \times Z_2}$$

$$\begin{aligned} Z_T &= \frac{Z_1 \times Z_2}{Z_1 + Z_2} \\ &= \frac{5\angle 60^\circ \times 6\angle -25^\circ}{5\angle 60^\circ + 6\angle -25^\circ} = \frac{30\angle 35^\circ}{(2.5 + j4.330) + (5.438 - j2.536)} \\ &= \frac{30\angle 35^\circ}{7.938 + j1.794\,\Omega} = \frac{30\angle 35^\circ}{8.138\angle 12.738^\circ} \\ &= 3.686\angle 22.262^\circ\,\Omega \quad \text{(polar form)} \\ &= 3.416 + j1.396\,\Omega \quad \text{(rectangular form)}. \end{aligned}$$

17.2 Kirchhoff's Circuit Laws

The values of voltage and current of an AC circuit can be determined using Kirchhoff's laws. The **Kirchhoff voltage law (KVL)** states that the sum of voltages in any close loop is equal to zero. These laws were also used to describe the opposition produced by components in a DC circuit. The **Kirchhoff current law (KCL)** states that the sum of currents entering a junction equals the sum of the currents exiting the junction.

Verifying KVL

When the resistor, capacitor, and inductor of an impedance network are connected in **series** and energized by an AC source, there will be a voltage drop across each component. **Kirchhoff's voltage law** states that the sum of the voltages drop across each component will equal zero in a closed-loop circuit configuration. Alternatively, the sum of the **voltage drops** across the resistor, inductor, and capacitor equals the supply or source voltage.

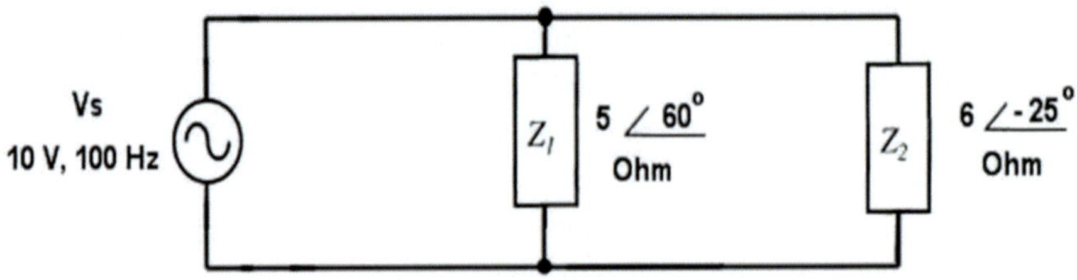

Figure 17-3 Parallel-connected impedance network.

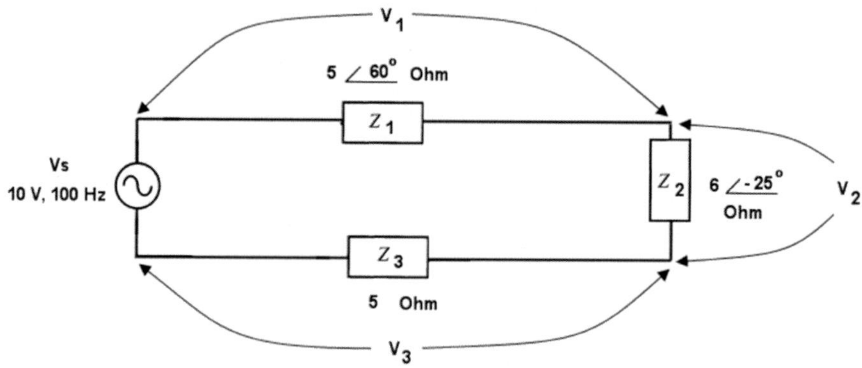

Figure 17-4 Voltage values of a series-connected impedance network.

When using the impedance network of **Figure 17-4**, KVL can be verified by using either:

$$V_S = V_1 + V_2 + V_3 \text{ or } V_S - V_1 - V_2 - V_3 = 0$$

The series-connected impedance network of **Figure 17-5(a)** is used to verify KVL. The **total impedance (Z_T)** of the circuit is first determined by adding the individual impedance values. This converts the impedance network into a single impedance connected to a voltage source, as indicated in **Figure 17-5(b)**.

The resulting current (I_T) can then be determined by using the value of the voltage applied to Z_T. This represents the current flow through the series-connected impedance network. The voltage drop across each impedance can then be determined by using Ohm's law ($V = I \times Z$). The sum of the individual voltage values should then be equal to the applied source voltage (V_s).

For the series-connected impedances Z_1, Z_2 and Z_3, the total impedance is determined using the following equations:

$$\begin{aligned} Z_T &= Z_1 + Z_2 + Z_1 \\ &= 5\angle 60° + 6\angle -25° + 5\,\Omega \\ &= (2.5 + j4.330) + (5.438 - j2.536) + 5\,\Omega \\ &= 12.938 + j1.794\,\Omega \\ &= 13.062\angle 7.894°\,\Omega. \end{aligned}$$

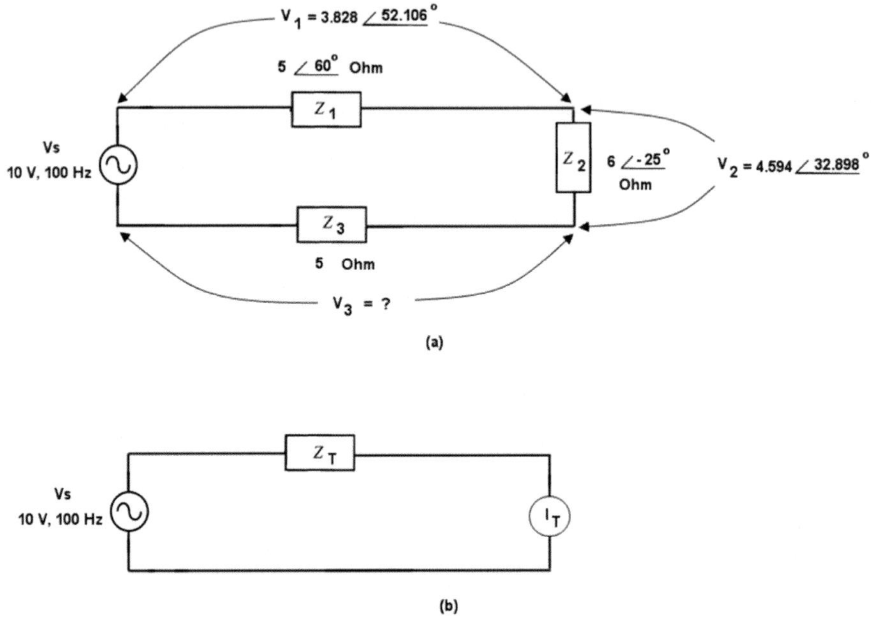

Figure 17-5 Determining an unknown voltage value of a series-connected impedance network.

The current (I_T) in a series impedance network is determined by using Ohm's law:

$$\begin{aligned} I_T &= \frac{V_s}{Z_T} \\ &= \frac{10\,\text{V}}{13.062\angle 7.894^\circ} \\ &= 0.758 - j0.105 \\ &= 0.766\angle -7.894^\circ \text{ A.} \end{aligned}$$

The voltage drop across each of the impedances is shown by the following:

$$V_1 = I \times Z_1 = 0.766\angle -7.894^\circ \times 5\angle 60^\circ = 3.828\angle 52.106^\circ = 2.351 + j3.021 \text{ V}$$
$$V_2 = I \times Z_2 = 0.766\angle -7.894^\circ \times 6\angle -25^\circ = 4.594\angle -32.898^\circ = 3.857 - j2.495 \text{ V}$$
$$V_3 = I \times Z_3 = 0.766\angle -7.894^\circ \times 5 = 3.828\angle -7.894^\circ = 3.792 - j0.526 \text{ V.}$$

Determining the Voltage Drops across a Circuit Component

KVL can be used to determine the voltage drop across an unknown component in an AC circuit. The **sum** of the voltage drops across all the circuit components equals the supply voltage. This means that the unknown voltage can be determined by subtracting the sum of the known voltage values from the supply voltage.

- Using Kirchhoff's voltage law, determine the value of the voltage across impedance Z_3 of the series-connected impedance network shown in **Figure 17-5**. KVL for this circuit can be expressed as follows:

$$V_S = V_1 + V_2 + V_3$$

$$\begin{aligned} V_3 &= V_S - (V_1 + V_2) \\ &= 10 - (3.828\angle 52.106° + 4.594\angle -32.898°)\ \text{V} \\ &= 10 - (2.351 + j3.021 + 3.857 - j2.495)\ \text{V} \\ &= 3.791 + j0.526\ \text{V} \\ &= 3.828\angle -7.894°\ \text{V}. \end{aligned}$$

Kirchhoff's Current Law (KCL)

Kirchhoff's current law states that the current entering a junction equals the current leaving a junction. In a parallel circuit, the total current splits in the individual branches, as shown in **Figure 17-6**. The total current, therefore, is the sum of the branch currents. When the currents are given in polar form, the values should be changed to rectangular form in order to perform the addition operation.

- For the parallel-connected impedance network shown in **Figure 17-6**, the current through impedances Z_1 and Z_2 can be determined using Ohm's law:

$$\begin{aligned} I_1 &= \frac{V_S}{Z_1} \\ &= \frac{10\ \text{V}}{5\angle 60°} \\ &= 1 - j1.732 \\ &= 2\angle -60°\ \text{A} \end{aligned}$$

$$I_2 = \frac{V_S}{Z_2}$$
$$= \frac{10\,\text{V}}{6\angle -25^\circ}$$
$$= 1.511 + j0.704$$
$$= 1.667\angle 25^\circ \text{ A.}$$

Kirchhoff's current law (KCL) states that the **sum** of the individual currents entering a junction is equal to the **sum** of the currents leaving a junction. Alternately, when the individual currents leaving a junction are subtracted from the currents entering the junction, the resulting current will be zero.

In the impedance network of **Figure 17.6**, currents I_{T1} and I_{T2} indicate the total current flow of the network. This current will alternately enter and exit the network at junctions A and B.

The current entering junction A is I_{T1}. This current is divided into two branch currents, which are I_1 and I_2. These currents exit junction A and are recombined at junction B. Applying KCL at junction A:

$$I_{T1} = I_1 + I_2.$$

Similarly, at junction B, the current entering is I_1 and I_2, while the current exiting the junction is I_{T2}. Applying KCL at junction B:

$$I_1 + I_2 = I_{T2}.$$

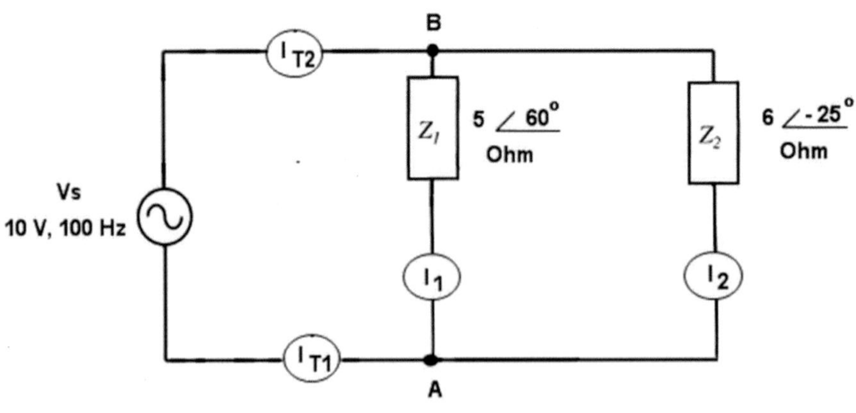

Figure 17-6 Current values of a parallel-connected impedance network.

17.3 Superposition Theorem

The **superposition method** is used to determine the current flowing through any branch of a circuit which has multiple sources of AC voltage applied. The total current flowing in the circuit is the combined current values produced by each voltage source. In AC circuits, the direction (polarity), magnitude, and phase of the applied voltage and resulting current flow in each component must be taken into account.

The number of AC voltage sources in a given circuit determines the procedure that will be followed. For example, a circuit with **two voltage sources** is first reduced to two individual circuits. Each voltage source is considered separately, with other voltage sources short-circuited for making current calculations. The magnitude, direction, and phase of the AC current flowing in each branch of the circuit is then determined independently. When these values are combined, it shows the resulting operation of the composite circuit.

- Determine the current flowing through the load impedance and the voltage across the load for the circuit shown in **Figure 17-7** using the superposition theorem.

Using the **superposition theorem**, individual voltage sources are replaced with short circuits and the current flowing through the load is determined in turn. **Figure 17-8(a)** shows the voltage source V_{S2} replaced with a short circuit, and **Figure 17-8(b)** shows the voltage source V_{S1} replace with a short circuit. The total current flowing through the load is determined by the vector addition of the current with voltage sources V_{S1} and V_{S2} acting

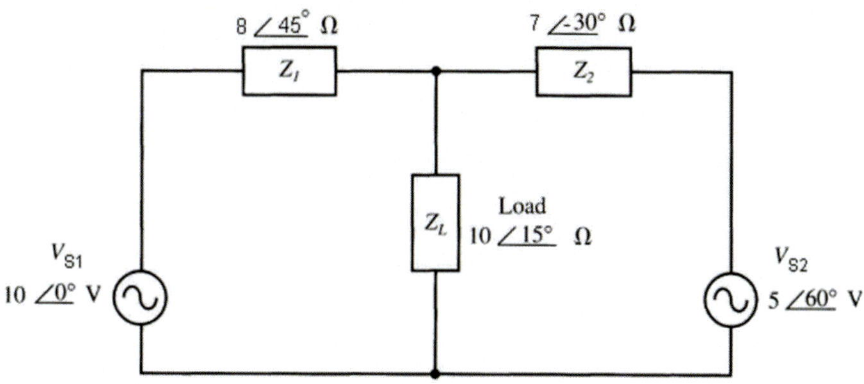

Figure 17-7 Using superposition theorem to find the current through a load with two voltage sources.

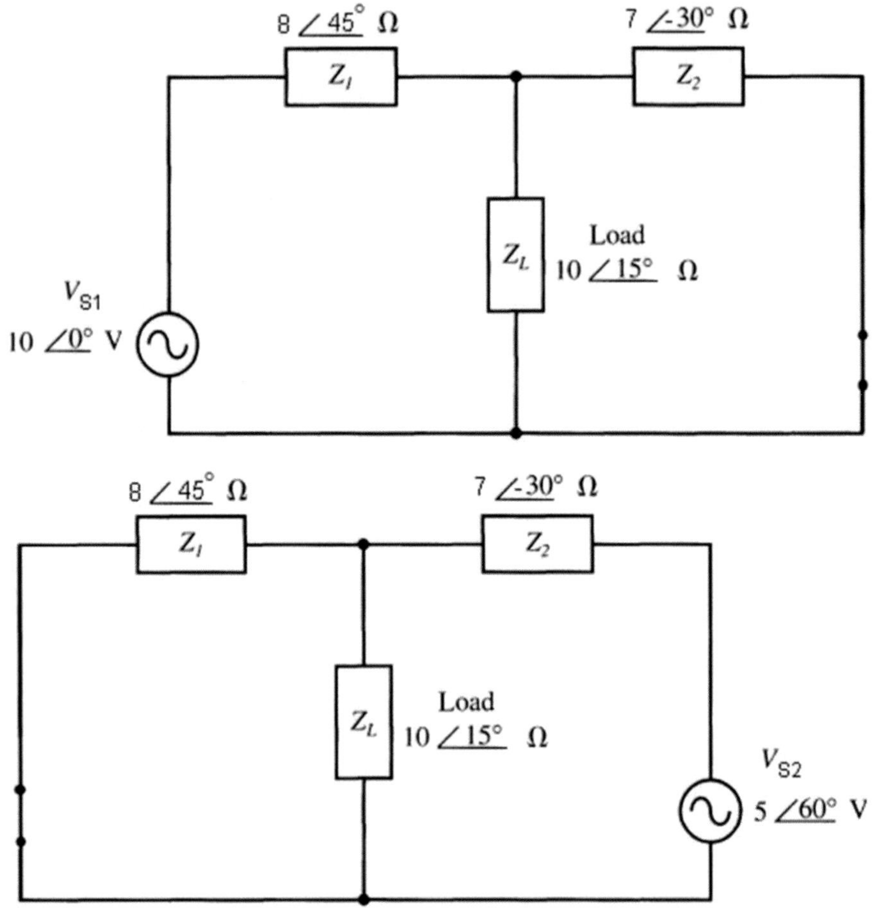

Figure 17-8 Superposition theorem replaces each voltage source (V_{S1} and V_{S2}) in turn with a short circuit (a) V_{S2} replaced with short circuit and (b) V_{S1} replaced with short circuit.

individually. The voltage across the load is then determined by the product of the current and the impedance of the load.

The current flowing through the load with only a single voltage source V_{S1} in the circuit as shown in **Figure 17-8(a)** is determined in steps 1–5:

1. Find the parallel impedance of Z_L and Z_2 as $Z_{L//Z2}$

$$Z_{L//Z2} = \frac{Z_L \times Z_2}{Z_L + Z_2}$$

$$= \frac{10\angle 15° \times 7\angle -30°}{10\angle 15° + 7\angle -30°} = \frac{70\angle -15°}{(9.659 + j2.588) + (6.062 - j3.5)}$$

$$= \frac{70\angle -15°}{15.721 - j0.912\,\Omega} = \frac{70\angle -15°}{15.747\angle -3.32°}$$

$$= 4.353 - j0.899 = 4.445\angle -11.68°\,\Omega.$$

2. Find the series impedance of Z_1 and $Z_{L//Z2}$.
3. $Z_{T1} = Z_1 + Z_{L//Z2}$
4. $= 8\angle 45° + (4.353 + j0.899) = (5.657 + j5.657) + (4.353 + j0.899)$

$= 10 + j4.757 = 11.083\angle 25.418°\,\Omega.$

5. Find the current supplied by source V_{S1}.
6. $I_{T1} = \dfrac{V_{S1}}{Z_{T1}} = \dfrac{10\angle 0°}{11.083\angle 25.418°} = 0.815 - j0.387 = 0.902\angle -25.418°$ A.
7. Find the voltage drop across the parallel impedance ($Z_{L//Z2}$).
8. $V_{Z(L//Z2)} = I_{T1} \times Z_{L//Z2}$.
9. $= 0.902\angle -25.418° \times 4.445\angle -11.68°$

$= 3.227 - j2.381 = 4.01\angle -36.418°$ V.

10. Find the current through load.
11. $I_{L1} = \dfrac{V_{Z(L//Z2)}}{Z_L} = \dfrac{4.01\angle -36.418°}{10\angle 15°} = 0.401\angle -51.418° = 0.2502 - j0.313$ A.
12. The current flowing through the load with only a single voltage source V_{S2} in the circuit as shown in **Figure 17-8(b)** is determined in steps 6–9.

Find the parallel impedance of Z_L and Z_1 as $Z_{L//Z1}$:

$$Z_{L//Z1} = \frac{Z_L \times Z_1}{Z_L + Z_1}$$

$$= \frac{10\angle 15° \times 8\angle 45°}{10\angle 15° + 8\angle 45°} = \frac{80\angle 60°}{(9.659 + j2.588) + (5.657 + j5.657)}$$

$$= \frac{80\angle 60°}{15.316 + j8.245\ \Omega} = \frac{80\angle 60°}{17.394\angle 28.29°}$$

$$= 3.913 + j2.417 = 4.599\angle 31.705°\ \Omega.$$

Find the series impedance of Z_2 and $Z_{L//Z1}$:

$$Z_{T2} = Z_2 + Z_{L//Z1}$$

$$= 7\angle -30° + (4.353 + j0.899) = (6.062 - j3.5) + (3.913 + j2.417)$$

$$= 9.975 - j1.083 = 10.033\angle -6.196°\ \Omega.$$

Find the current supplied by source V_{S2}:

$$I_{T2} = \frac{V_{S2}}{Z_{T2}} = \frac{5\angle 60°}{10.033\angle -6.196°} = 0.201 + j0.456 = 0.498\angle 66.197°\ \text{A}.$$

Find the voltage drop across the parallel impedance ($Z_{L//Z2}$):

$$V_{Z(L//Z1)} = I_{T2} \times Z_{L/\,Z1}$$

$$= 0.498\angle 66.197° \times 4.599\angle 31.705°$$

$$= -0.315 + j2.27 = 2.292\angle 97.902°\ \text{V}.$$

Find the current through load:

$$I_{L2} = \frac{V_{Z(L//Z1)}}{Z_L} = \frac{2.292\angle 97.902°}{10\angle 15°}$$

$$= 0.229\angle 82.90° = 0.028 + j0.227\ \text{A}.$$

Find the total current flowing through the load as the vector sum of currents I_{L1} and I_{L2}:

$$I_L = I_{L1} + I_{L2} = (0.2502 - j0.313) + (0.028 + j0.227)$$

$$= 0.278 - j0.086 = 0.2913\angle -17.28°\ \text{A}.$$

Find the voltage across the load Z_L:

$$V_L = I_L \times Z_L = 0.2913\angle -17.28° \times 10\angle 15° = 2.913\angle -2.28° \text{ V}.$$

17.4 Thevenin's Theorem

The **Thevenin equivalent circuit** method is used to simplify AC electrical circuits. It allows a complex circuit to be reduced to one **equivalent voltage source (V_{Th})**, and a **series impedance (Z_{Th})** connected to a **load (Z_L)**. It is a practical method that is used to calculate load currents and load voltages for any value of load impedance. When the proper circuit values are selected, it will permit the maximum power to a load device.

The **Thevenin equivalent voltage source (V_{Th})** is determined by first **disconnecting** the load from the circuit. The value of the voltage which appears across the open load terminals is then determined. Next, the **series impedance (Z_{Th})** is determined by first shorting out all voltage sources and disconnecting the load from the circuit. Then the value of the impedance which appears across the **open load** terminals is determined.

Maximum power will be transferred from the source to the load device when the impedance of Z_{Th} **matches** Z_L. The magnitude of the impedances should be the same and the phase angle should be opposite. Thus, the reactive portions of the source and the load impedance will be canceled. This means that a large portion of applied voltage and current from the source will be effectively delivered to the load device. For example, if the source has a given capacitive reactance value, then a maximum transfer of power will occur when it is connected to a load that has an **equa***l* value of inductive reactance. The resistance of the source should also approximate the load resistance and produce a maximum transfer of power.

- Determine the Thevenin equivalent of the circuit shown in **Figure 17-9(a)**. In order to determine the Thevenin equivalent circuit, first the Thevenin equivalent impedance is determined by disconnecting the load connected between terminals A and B. The voltage source V_S is replaced with a short circuit and the impedance across terminals A and B is then evaluated. This causes impedance Z_1 to be placed in parallel with Z_3. Impedance Z_2 is in series with their parallel equivalent ($Z_{1//3}$).

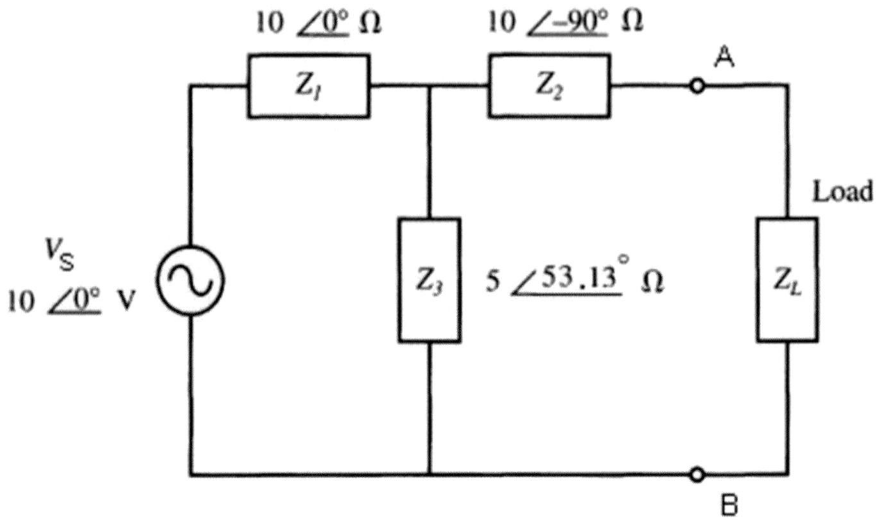

10 ∠0° Ω
10 ∠–90° Ω
Z_1
Z_2
A
Load
V_S
10 ∠0° V
Z_3
5 ∠53.13° Ω
Z_L
B

(a)

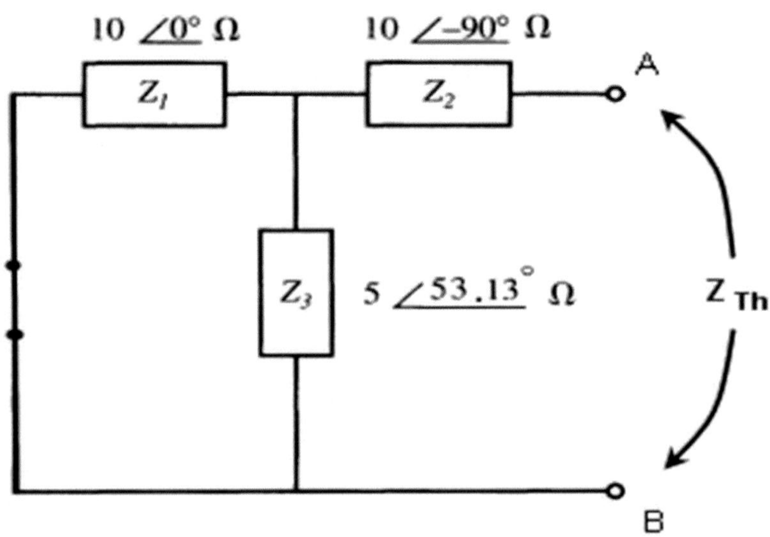

10 ∠0° Ω
10 ∠–90° Ω
Z_1
Z_2
A
Z_3
5 ∠53.13° Ω
Z Th
B

(b)

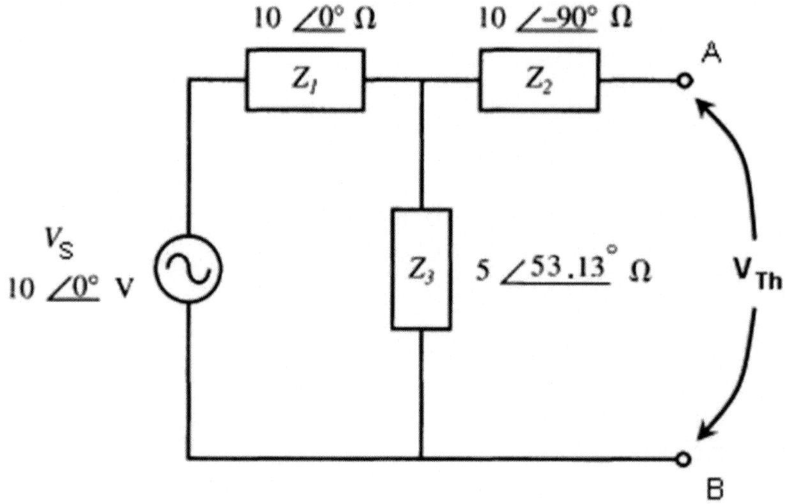

(c)

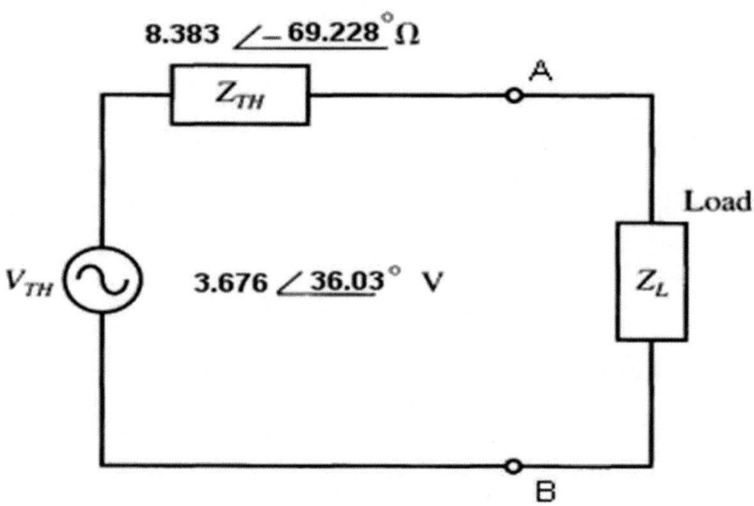

(d)

Figure 17-9 Thevenin's theorem is used to find the voltage and its internal resistance at the load terminals: (a) original AC circuit; (b) Thevenin impedance; (c) Thevenin voltage; (d) Thevenin equivalent AC circuit.

- Find the equivalent parallel impedance of Z_1 and Z_3 as $Z_{1//3}$ as shown in **Figure 17-9(b)**:

$$Z_{1//3} = \frac{Z_1 \times Z_3}{Z_1 + Z_3}$$

$$= \frac{10\angle 0° \times 5\angle 53.13°}{10\angle 0° + 5\angle 53.13°} = \frac{50\angle 53.13°}{(10 + j0) + (3 + j4)}$$

$$= \frac{50\angle 53.13°}{13 + j4} = \frac{50\angle 53.13°}{13.601\angle 17.103°}$$

$$= 2.973 + j2.162 = 3.676\angle 36.027°\ \Omega.$$

- Find the Thevenin equivalent impedance (Z_{Th}) as the series arrangement of impedances Z_2 and $Z_{1//3}$:

$$Z_{Th} = Z_2 + Z_{1//3}$$

$$= 10\angle -90° + 3.676\angle 36.027°\ \Omega$$

$$= (-j10) + (2.973 + j2.162) + 5\ \Omega$$

$$= 2.973 - j7.838\ \Omega$$

$$= 8.383\angle -69.228°\ \Omega.$$

To find the Thevenin equivalent voltage (V_{Th}) and determine the voltage that appears across the impedance Z_3, as shown in **Figure 17-9(c)**:

$$V_{Th} = V_S \times \frac{Z_3}{Z_1 + Z_3} = 10\angle 0° \times \frac{5\angle 53.13°}{10 + (3 + j4)}$$

$$= \frac{50\angle 53.13°}{13 + j4} = \frac{50\angle 53.13°}{13.601\angle 17.103°}$$

$$= 3.676\angle 36.03° = 2.973 + j2.162\ \Omega.$$

The Thevenin equivalent circuit is shown in **Figure 17-9(d)**.

17.5 Three-phase Impedance Networks

A **three-phase impedance network** is a combination of multiple impedances, connected in different configurations. The names of various

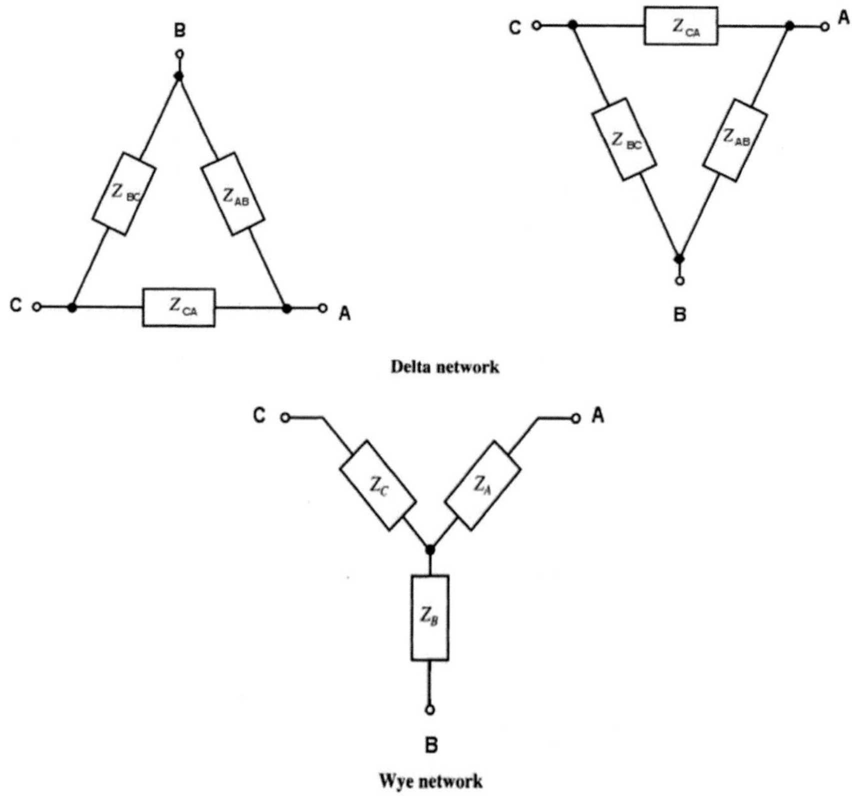

Figure 17-10 Three-terminal impedance networks are named by their shape.

impedance configurations are derived from the manner in which individual impedances are connected together. In this regard, the two common network configurations are used in three-phase circuitry. These are **the Delta and Wye configurations** as shown in **Figure 17-10**.

In a three-phase **Delta configuration**, three impedances are interconnected. Each one of the impedances has two leads or terminals. One lead of each impedance is connected to a lead of an adjoining impedance forming a common connection point. In a delta network, three such connection points are established. A three-phase input or output signal can be derived from or applied to these connection points. A delta configuration is shown in **Figure 17-10(a)**. Note that the shape of the resulting impedance network is triangular and resembles the Greek letter Delta (Δ).

In a three-phase **Wye network**, three impedances are interconnected. One lead from each impedance is connected to a common point. The remaining three terminals serve as the input or output of the configuration as shown in

Figure 17-10(b). A three-phase input or output signal can be derived from these connection points. Note that the resulting impedance network is shaped like the capital letter "Y."

When working with three-phase AC networks, there is often a need for converting between different circuit configurations. In this regard, conversion would be from **Wye to Delta** or from **Delta to Wye**. The following formulas are used for converting between network configurations:

- Wye to Delta:

$$Z_{AB} = \frac{Z_A \times Z_B + Z_B \times Z_C + Z_C \times Z_A}{Z_C}$$

$$Z_{BC} = \frac{Z_A \times Z_B + Z_B \times Z_C + Z_C \times Z_A}{Z_A}$$

$$Z_{CA} = \frac{Z_A \times Z_B + Z_B \times Z_C + Z_C \times Z_A}{Z_B}.$$

- Convert the Wye network shown in **Figure 17-11** to a delta network.

$$Z_{AB} = \frac{Z_A \times Z_B + Z_B \times Z_C + Z_C \times Z_A}{Z_C}$$

$$= \frac{(20 \times 30\angle -20°) + (30\angle -20° \times 10\angle 45°) + (10\angle 45° \times 20)}{10\angle 45°}$$

$$= \frac{600\angle -20° + 300\angle 25° + 200\angle 45°}{10\angle 45°}$$

$$= \frac{979.158\angle 3.689°}{10\angle 45°} = 73.548 - j64.639 = 97.916\angle -41.311°\ \Omega$$

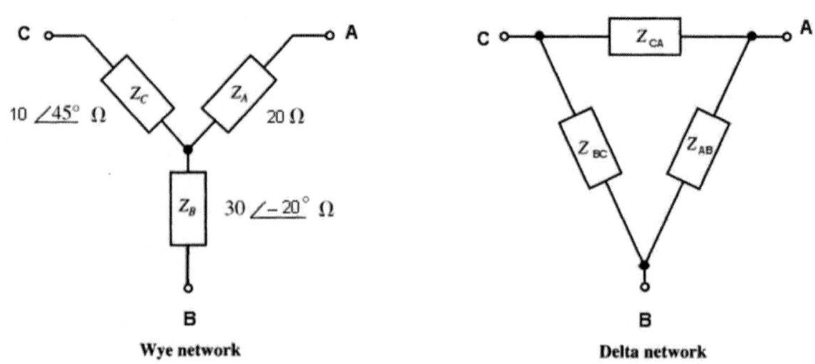

Figure 17-11 Transform a Wye network into a Delta network.

$$Z_{BC} = \frac{Z_A \times Z_B + Z_B \times Z_C + Z_C \times Z_A}{Z_A}$$
$$= \frac{979.158\angle 3.689^\circ}{20} = 48.859 + j3.1497$$
$$= 48.958\angle 3.689^\circ\ \Omega$$

$$Z_{CA} = \frac{Z_A \times Z_B + Z_B \times Z_C + Z_C \times Z_A}{Z_B}$$
$$= \frac{979.158\angle 3.689^\circ}{30\angle -20^\circ} = 29.889 + j13.113$$
$$= 32.639\angle 23.689^\circ\ \Omega.$$

- Delta to Wye:

$$Z_A = \frac{Z_{AB} \times Z_{CA}}{Z_{AB} + Z_{BC} + Z_{CA}}$$

$$Z_B = \frac{Z_{AB} \times Z_{BC}}{Z_{AB} + Z_{BC} + Z_{CA}}$$

$$Z_C = \frac{Z_{BC} \times Z_{CA}}{Z_{AB} + Z_{BC} + Z_{CA}}.$$

- Convert the delta network shown in **Figure 17-12** to a Wye network.

$$Z_A = \frac{Z_{AB} \times Z_{CA}}{Z_{AB} + Z_{BC} + Z_{CA}} = \frac{97.916\angle -41.311^\circ \times 32.639\angle 23.689^\circ}{97.916\angle -41.311^\circ + 48.958\angle 3.689^\circ + 32.639\angle 23.689^\circ}$$
$$= \frac{3195.880\angle -17.622^\circ}{73.548 - j64.639 + 48.859 + j3.1497 + 29.889 + j13.113} = \frac{3195.880\angle -17.622^\circ}{152.293 - j48.376}$$
$$= \frac{3195.880\angle -17.622^\circ}{159.792\angle -17.623^\circ} \approx 20\ \Omega$$

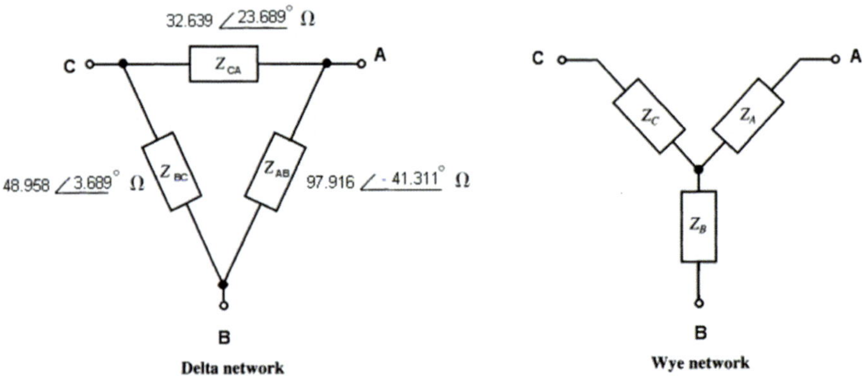

Figure 17-12 Transform a Delta network into a Wye network.

$$Z_B = \frac{Z_{AB} \times Z_{BC}}{Z_{AB} + Z_{BC} + Z_{CA}}$$

$$= \frac{97.916\angle -41.311^\circ \times 48.958\angle 3.689^\circ}{97.916\angle -41.311^\circ + 48.958\angle 3.689^\circ + 32.639\angle 23.689^\circ}$$

$$= \frac{4793.7715\angle -37.622^\circ}{159.792\angle -17.623^\circ} \approx 30\angle 20^\circ\ \Omega$$

$$Z_C = \frac{Z_{BC} \times Z_{CA}}{Z_{AB} + Z_{BC} + Z_{CA}}$$

$$= \frac{48.958\angle 3.689^\circ \times 32.639\angle 23.689^\circ}{97.916\angle -41.311^\circ + 48.958\angle 3.689^\circ + 32.639\angle 23.689^\circ}$$

$$= \frac{1597.94\angle 27.378^\circ}{159.792\angle -17.623^\circ} \approx 10\angle 45^\circ\ \Omega.$$

17.6 Power in AC Circuits

Power values in AC circuits are found by using **power triangles**, as shown in **Figure 17-13**. The volt-amperes (VA) delivered to an AC circuit are called the **apparent power**. The unit of measurement is the **volt-ampere (VA)** or **kilo-volt-ampere (kVA)**. Meters are used to measure the apparent power in AC circuits. The apparent power is the voltage applied to a circuit multiplied by the total current flow. The power converted to another form of energy by the load is measured with a wattmeter. This actual power converted is called **true power** and is equal to VI × PF. The ratio of true power converted in a circuit to apparent power delivered to an AC circuit is called the **power factor (PF)**. The power factor is found by using the following formula:

$$\text{PF} = \frac{\text{true power (W)}}{\text{apparent power (VA)}}$$

or

$$\%\,\text{PF} = \frac{\text{true power (W)}}{\text{apparent power (VA)}} \times 100\%.$$

- Given: a 240-V, 60-Hz, 30-A electric motor is rated at 6000 W. Find the power factor at which the motor operates.

$$\%\,\text{PF} = \frac{W}{\text{VA}} \times 100 = \frac{6000}{240 \times 30} \times 100 = 83.3\%.$$

The term W is the **true power** in watts and VA is the **apparent power** in volt-amperes. The highest power factor of an AC circuit is 1.0 or 100%. This value is called **unity power factor**.

The **phase angle (θ)** of an AC circuit determines the power factor and is equal to PF = cos **θ**. Remember that the phase angle is the angular separation between voltage applied to an AC circuit and current flow through the circuit.

- Given: a circuit has the following values: applied voltage = 240 V, current = 12 A, and power factor = 0.83. Find the true power of the circuit.
  W = VA × cosine (θ)
  = 240 × 12A × 0.83
  = 2390 W.

The expression $\frac{\text{true power } (W)}{\text{apparent power (VA)}}$ is the power factor of a circuit; therefore, the power factor is equal to the cosine of the phase angle (PF = cosine θ).

Right triangle relationships can be expressed as equations that determine the value of any of the sides of the **power triangle** when the other two values are known. These expressions are as follows:

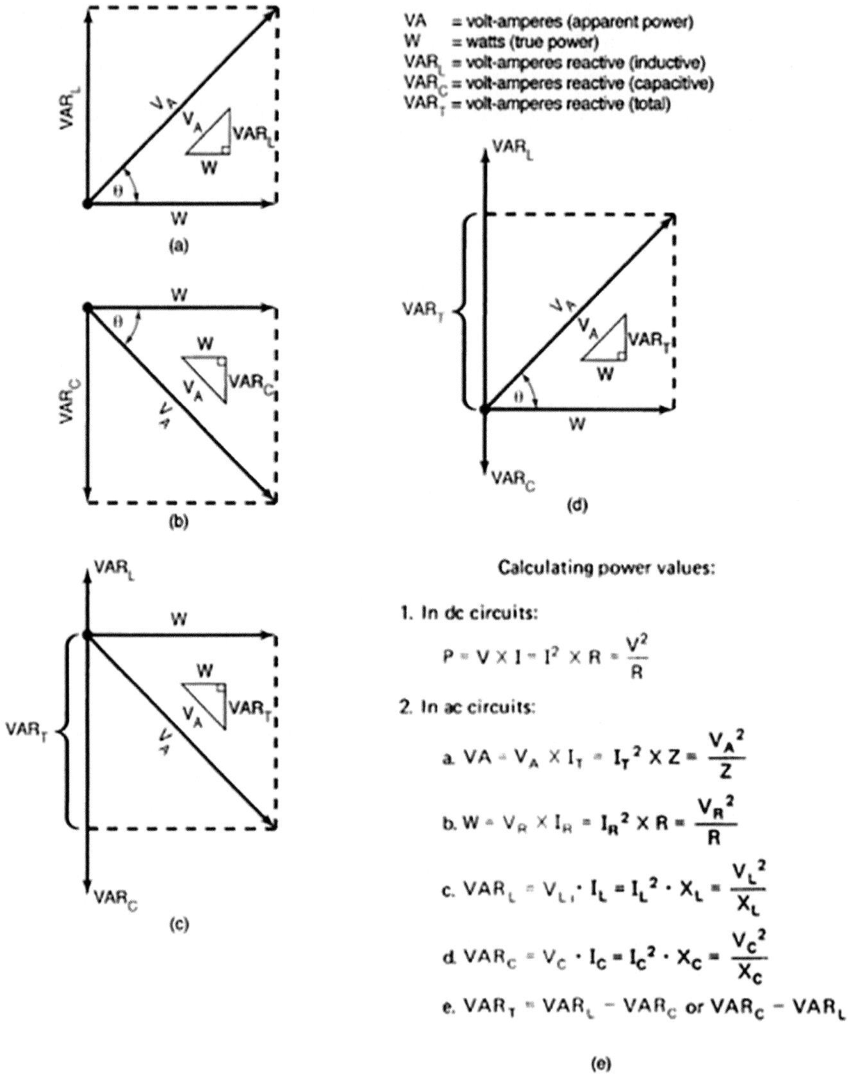

Figure 17-13 Power triangles for AC circuits: (a) *RL* circuit; (b) *RC* circuit; (c) *RLC* circuit with X_L larger than X_C; (d) RLC circuit with X_C larger than X_L; (e) formulas for calculating power values.

$$W = \sqrt{\mathrm{VA}^2 - \mathrm{VAR}^2}$$

$$\mathrm{VA} = \sqrt{W^2 - \mathrm{VAR}^2}$$

$$\mathrm{VAR} = \sqrt{\mathrm{VA}^2 - W^2}.$$

- Given: total reactive power = 54 VAR; applied voltage = 120 V; current = 0.5 A. Find the true power of the circuit.

$$W = \sqrt{\text{VA}^2 - \text{VAR}^2}$$
$$= \sqrt{(120 \times 0.5)^2 - 54^2}$$
$$\text{VAR} = \sqrt{3600 - 2916} = 26.15 \text{ W}.$$

More inductive or capacitive reactance causes a **larger** phase angle. In a purely inductive or capacitive circuit, a 90° phase angle causes a **power factor of zero**. The power factor varies according to the values of resistance and reactance in a circuit. There are two types of power that affect power conversion in AC circuits. Power conversion in the **resistive** part of the circuit is called **active power or true power**. True power is measured in **watts**. The other type of power is caused by **inductive or capacitive** loads. It is 90° out of phase with the true power and is called **reactive VA**. It is a type of "unused" power. Reactive VA is measured in **volt-amperes reactive (VAR)**.

The **power triangle** of **Figure 17-13** has **true power (watts)** marked as a horizontal line. **Reactive power (VAR)** is drawn at a 90° angle from the true power. **Volt-amperes or apparent power (VA)** is the longest side (**hypotenuse**) of the right triangle. This right triangle is similar to the impedance triangle and the voltage triangle for series AC circuits and to the current triangle and admittance triangle for parallel AC circuits. Each type of right triangle has a horizontal line used to mark the resistive part of the circuit. The vertical lines are used to mark the reactive part of the circuit. The longest side of the triangle (hypotenuse) is used to mark total values of the circuit. The size of the hypotenuse depends on the amount of resistance and reactance in the circuit. **Vector diagrams** and right triangles are extremely important for understanding AC circuits.

Power in Three-phase Circuits

The formulas for three-phase power relationships are as follows:
Apparent power (volt-amperes) per phase:

$$\text{VA}_p = V_p \times I_p.$$

Total apparent power (volt-amperes):

$$\text{VA}_T = 3 \times V_p \times I_p = 1.73 \times V_L \times I_L.$$

Power factor:

$$\text{PF} = \frac{\text{true power } (W)}{1.73 \times V_L \times I_L} = \frac{W}{3 \times V_P \times I_P}.$$

True power per phase:

$$P_p = V_p \times I_p \times \text{PF}.$$

Total true power:

$$P_T = 3 \times V_p \times I_p \times \text{PF} = 1.73 \times V_L \times I_L \times \text{PF}.$$

Note that the symbols used previously are as follows: V_L is the line voltage in volts, I_L is the line current in amperes, V_p is the phase voltage in volts, and I_p is the phase current in amperes.

- Determine the line voltage of a three-phase wye-connected system which has a phase voltage of 120 V.

  $V_L = V_p \times 1.73$

  $= 120 \text{ V} \times 1.73$

  $= 208 \text{ V}.$

- Determine the power per phase of a three-phase delta system that has a phase voltage of 240 V, a phase current of 20 A, and a power factor of 0.75.

  $P_p = V_p \times I_p \times \text{PF}$

  $= 240 \text{ V} \times 20 \text{ A} \times 0.75$

  $= 3600 \text{ W}$

  where PF is the power factor of the load.

  The total power (P_T) developed by all three phases of a three-phase system is expressed as follows:

  $P_T = 3 \times P_p$

  $= 3 \times V_p \times I_p \times \text{PF}$

  $= 1.73 \times V_L \times I_L \times \text{PF}.$

- Determine the total three-phase power of a wye-connected system that has a phase voltage of 277 V, a phase current of 10 A, and a power factor of 0.85.
  $P_T = 3 \times 277\ V \times 10\ A \times 0.85$
  $= 7063.5\ W.$
  To summarize three-phase power relationships:
  Volt-amperes per phase (VA_p) = V_pI_p
  Total volt-amperes (VA_T) = 3 V_pI_p
  = 1.73 V_LI_L.
- Determine the total three-phase volt-amperes of a delta-connected system that has a line voltage of 208 V and a line current of 4.86 A.
  VA = 1.73 × 208 V × 4.86 A
  = 1748.8 VA.

$$\text{power factor (PF)} = \frac{\text{true power } (W)}{1.73\ V_L I_L} = \frac{W}{3V_P I_P}.$$

Power per phase (P_P) = $3 \times V_P \times I_P \times PF = 1.73 \times V_L \times I_L \times PF$
where V_L *is the line voltage in volts,* I_L *is the line current in amperes,* V_p *is the phase voltage in volts*, and I_p is the phase current in amperes.

- Determine the total three-phase power of a wye-connected system that has a phase voltage of 120 V, a phase current of 18.5 A, and a power factor of 0.95.
  $P_T = 3 \times V_p \times I_p \times PF$
  $= 3 \times 120\ V \times 18.5\ A \times 0.95$
  $= P_T = 6.327\ W = 6.327\ kW.$
  Calculations involving three-phase power are somewhat more complex than for single-phase power.

Summary

- An impedance network consists of two or more electrical components connected together.
- Complex numbers are used to represent AC values
- The total impedance of components connected in series is determined by adding the reactance values of the individual components.
- Admittance is the reciprocal of the impedance.

- The total admittance of components connected in parallel is determined by adding the reciprocal values of the reactance of individual components.
- Kirchhoff's voltage law states that the vector sum of voltages in any loop equals zero.
- Kirchhoff's current law states that vector sum of currents entering a node or junction equals the sum of the currents exiting the junction.
- Superposition theorem is method for determining the current in a circuit containing multiple voltage sources. The current flowing in any branch of the circuit can be determined by considering the current produced when each source is acting individually with the other voltage sources short circuited.
- Thevenin's theorem provides a method of simplifying a complex electrical circuit which consists of multiple sources and components. The simplified series circuit consists of a voltage source V_{Th} and impedance Z_{Th} which is then connected to a load (Z_L).
- A three-phase impedance network consists of electrical components which are connected between three terminals. This includes the delta and wye three-phase network configurations.
- In a delta network, the beginning of each impedance is connected in series with the ending of the adjacent impedance, thus forming a closed loop.
- In a wye network, one end of each impedance is connected to a common point and the other three ends are connected to the power source and/or a load.
- Power factor of an electrical circuit or system is the relationship of true power (measured in watts) to apparent power (measured in volt-amperes).
- The amount of power consumed by a system is dependent on the power factor.
- In a balanced three-phase AC electrical system, the total power consumed is three times the power consumed by each phase.

Formulas

Equivalent impedance (Z_T) for series impedances Z_1 and Z_2:

$$Z_T = Z_1 + Z_2.$$

Equivalent impedance (Z_T) for parallel impedances Z_1 and Z_2:

$$Z_T = \frac{Z_1 \times Z_2}{Z_1 + Z_2} \text{ or } \frac{1}{Z_T} = \frac{1}{Z_1} + \frac{1}{Z_2}.$$

Determining the resistance and reactance of a component from the polar form Z:

$R = Z \times \cos\theta$

$X = Z \times \sin\theta.$

Determining the magnitude and phase angle of a component in rectangular form:

$$Z = \sqrt{R^2 + X^2}$$

$$\theta = \tan^{-1}\left(\frac{X}{R}\right).$$

Ohm's law:

$$V = I \times Z$$

KVL:

In any loop: Algebraic sum of voltage drops = Algebraic sum of source voltages.

KCL:

At any node: Algebraic sum of incoming currents = Algebraic sum of exiting currents.

Delta network to Wye network conversion:

$$Z_A = \frac{Z_{AB} \times Z_{CA}}{Z_{AB} + Z_{BC} + Z_{CA}}$$

$$Z_B = \frac{Z_{AB} \times Z_{BC}}{Z_{AB} + Z_{BC} + Z_{CA}}$$

$$Z_C = \frac{Z_{BC} \times Z_{CA}}{Z_{AB} + Z_{BC} + Z_{CA}}.$$

Wye network to Delta network conversion:

$$Z_{AB} = \frac{Z_A \times Z_B + Z_B \times Z_C + Z_C \times Z_A}{Z_C}$$

$$Z_{BC} = \frac{Z_A \times Z_B + Z_B \times Z_C + Z_C \times Z_A}{Z_A}$$

$$Z_{CA} = \frac{Z_A \times Z_B + Z_B \times Z_C + Z_C \times Z_A}{Z_B}$$

$$\text{PF} = \frac{\text{true power } (W)}{\text{apparent power (VA)}}$$

$$\%\ \text{PF} = \frac{\text{true power } (W)}{\text{apparent power (VA)}} \times 100\%$$

$$W = \text{VA} \times \text{cosine } (\theta)$$

$$W = \sqrt{\text{VA}^2 - \text{var}^2}$$
$$\text{VA} = \sqrt{W^2 - \text{VAR}^2}$$
$$\text{VAR} = \sqrt{\text{VA}^2 - W^2}.$$

Apparent power (volt-amperes) per phase: $VA_p = V_p \times I_p$.

Total apparent power (volt-amperes): $VA_T = 3 \times V_p \times I_p = 1.73 \times V_L \times I_L$

$$\text{PF} = \frac{\text{true power } (W)}{1.73 \times V_L \times I_L} = \frac{W}{3 \times V_P \times I_P}.$$

True power per phase: $P_p = V_p \times I_p \times \text{PF}$

Total true power: $P_T = 3 \times V_p \times I_p \times \text{PF} = 1.73 \times V_L \times I_L \times \text{PF}$

$$\text{power factor (PF)} = \frac{\text{true power } (W)}{1.73\ V_L I_L} = \frac{W}{3 V_P I_P}.$$

Problems

Determine the total impedance of the network part of the circuit shown in **Figure 17-14**, in both rectangular and polar forms. [(10 + j 25) Ω; 26.93 Ω/68.2°]

Determine the total impedance of the circuit shown in **Figure 17-14**. [$Z_T = 9.27\ \angle 19.51^\circ$]

The incoming currents at a node are $I_1 = (2 + j4)$ A and $I_2 = (4 - j5)$ A. Determine the value of the exiting current in rectangular and polar forms. [(6 – j1) A; 6.083A/–9.46°]

If the internal impedance of a circuit is 25/30°, determine the value of the load which should be connected, so that a maximum transfer of power can be obtained. [25/–30°]

The three impedances connected to a wye network are $Z_A = Z_B = Z_C =$ 3/30°. Determine the value of equivalent delta network impedances Z_{AB}, Z_{BC}, and Z_{CA}. [9/30°]

Problem Answers

$Z_T = Z_1 + Z_2 = 10 - j50 + j75 = 10 + j25$ (rectangular form);

$$Z = \sqrt{10^2 + 25^2}\ \tan^{-1}(\frac{25}{10}) = 26.93/68.2°]\ \text{(polar form)}$$

$$\frac{1}{Z_T} = \frac{1}{Z_1} + \frac{1}{Z_2};\ \frac{1}{Z_T} = \frac{1}{25\angle -20^\circ} + \frac{1}{10\angle 50^\circ};$$

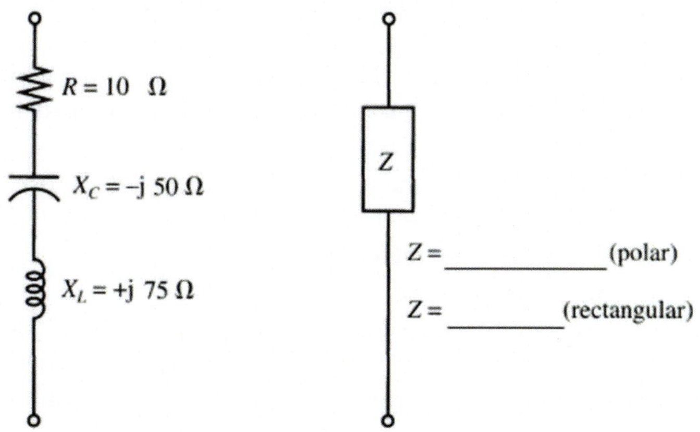

Figure 17-14 Determine the total impedance of series-connected components.

$$\frac{1}{Z_T} = 0.1016 - j0.063 = 0.1078\angle -19.51^\circ$$

$$Z_T = 9.27\ \angle 19.51^\circ$$

$I_T = I_1 + I_2 = (2 + 4) + j(4 - 5) = 6 - j1$ (rectangular form), $6.083/-9.46^\circ$ (polar form).

Maximum power transfer occurs when the load is a complex conjugate of the source impedance. Hence, it should be $25/-30^\circ$.

Assume $Z_A = Z_B = Z_C = Z$;

$$Z_{AB} = \frac{Z_A Z_B + Z_B Z_C + Z_A Z_C}{Z_A} = \frac{Z^2 + Z^2 + Z^2}{Z} = \frac{3Z^2}{Z} = 3Z.$$

Hence, $Z_{AB} = 3 \times 3/30^\circ = 9/30^\circ$.

Z_{BC} and Z_{CA} can be determined similarly and will equal $Z_{AB} = 9/30^\circ$.

Self-examination/Answers

1. List at least three components which may be used in the construction of an impedance network.
2. Impedances being added or subtracted must use the (polar/rectangular) form of a complex number.
3. Impedances being multiplied or divided must use the (polar/rectangular) form of a complex number.
4. According to KVL, the sum of AC voltages lost across each component in a loop is (greater than/equal to/smaller than) the sum of the source voltages.
5. According to KCL, the sum of AC currents entering any node in the circuit is (greater than/equal to/smaller than) the currents leaving the node.
6. The (polar/rectangular) form of complex numbers is used while applying KCL at a node in a circuit.
7. While using superposition theorem, all but one of the voltages sources in the circuit are replaced by a(n) (open/short) circuit, in order to determine circuit values caused by each source acting independently.
8. The current in any given branch of a circuit using superposition is obtained by (adding/multiplying) the individual branch current values caused by each source acting independently.

9. The superposition method is most commonly used when the number of voltage sources in the circuit is (exactly one/small/large).
10. Thevenin's equivalent circuit consists of ________________ and ________________.
11. The main use of Thevenin's equivalent circuit is to ________________.
12. The Thevenin equivalent impedance is obtained by first (shorting/removing) the load from the circuit.
13. ____________ is the type of impedance network formed when three impedances are connected to a common point.
14. ____________ is the type of impedance network formed when three impedances are connected such that the beginning and end of each component are in series forming a closed loop.
15. Impedance networks of a three-phase system are connected in either a ____________ or a ____________circuit configuration.
16. The three-phase __________-connected impedance network has a single common connection point.
17. Power factor is the ratio of (true, apparent) power and (true, apparent) power.
18. True power is measured in ________ and apparent power is measured in ________.
19. In a balanced three-phase power distribution system, the power converted by each phase is (, , same) that of the total power.

Answers

1. resistors, inductors, capacitors
2. rectangular
3. polar
4. equal
5. equal
6. rectangular
7. short
8. adding
9. small
10. Thevenin equivalent voltage, V_{Th}, Thevenin equivalent impedance, Z_{Th}
11. match the impedance of the external load with the internal impedance of the circuit

12. removing
13. wye
14. delta
15. delta, wye
16. wye
17. true, apparent
18. W or watts, VA or volt-ampere
19. 1/3

Glossary

Admittance
The reciprocal of the impedance, which is represented as $Y = G \pm jB$, where G is the conductance (reciprocal of a resistance) and B is the susceptance (reciprocal of the reactance).

Apparent power (volt-amperes)
The applied voltage times current delivered to an AC circuit.

Delta impedance network
A three-terminal/phase network where one lead of an impedance is connected to the lead of an adjoining impedance forming a common connection point. In a delta network, three such connection points are established. A three-phase input or output signal can be derived from or applied to these connection points.

Junction
A common point where three or more electrical components are connected.

Kirchhoff's voltage law
The algebraic sum of voltages in any loop equals zero. The sum of voltage drops in a loop equals the sum of the voltages applied.

Kirchhoff's current law
The vector sum of current(s) entering and leaving a node or junction equals zero. Thus, the vector sum of current entering a node or junction equals the current leaving it.

Loop
Any closed path in an electrical circuit. A circuit may contain multiple loops.

Maximum power transfer
A condition that exists when the impedance of a load (Z_L) matches the impedance of the source (Z_S).

Network
A combination of electrical components connected together.

Node
A common point where two or more electrical components are connected.

Power factor
The ratio of true power in an AC circuit and apparent power and is expressed as follows:

$$\frac{\text{true power } (W)}{\text{apparent power } (VA)}.$$

Power (P)
The rate of doing work in electrical circuits, found by using the equation $P = I \times V$

Reactive power (VAR)
The "unused" power of an AC circuit has inductance or capacitance, which is absorbed by the magnetic or electrostatic field of a reactive circuit.

Susceptance (B)
The ease with which an inductor or capacitor passes AC current and is expressed as follows: $B_L = \frac{1}{X_L}$ for inductors and $B_C = \frac{1}{X_C}$ for capacitors. It is measured in Siemens.

Superposition theorem
A method for determining the current in a circuit containing multiple voltage sources. The current flowing in any branch of the circuit can be determined by considering the current produced when each source is acting individually with the other voltage sources short circuited. The contribution of each voltage source to the current in any branch of the circuit can thus be determined.

Thevenin's theorem

A method of simplifying a complex electrical circuit which consists of multiple sources and components. The simplified series circuit consists of a voltage source V_{Th} and an impedance Z_{Th} which is then connected to a load (Z_L).

Thevenin equivalent impedance

The impedance that would be measured at the load terminals if the load was removed and the voltage source was removed and replaced with a short circuit.

Three-terminal network

A network of electrical components that are connected between three terminals. This includes Delta, Wye, Tee, and Pi network configurations.

True power (W)

The power actually converted by an AC circuit, as measured with a wattmeter.

Volt-ampere (VA)

The unit of measurement of apparent power.

Volt-amperes reactive (VAR)

The unit of measurement of reactive power.

Wye impedance network

A three-terminal/phase network where the impedances have one end connected to a common point and the other ends are connected to the power source and/or a load.

18

Electrical Energy Conversion

Electrical energy conversion takes place in all electrical systems. The part of the system where energy is converted to another form is called the **load**. There are many different types of loads that are part of electrical circuits. A load can be a **resistance**, **inductance**, or **capacitance** as discussed earlier. In this chapter, the function of the load is considered. Electrical energy is commonly converted to **light energy**, **heat energy**, **and mechanical energy**. Many types of electrical **lights** are used today. Electrical **heating** has also become very popular in homes and other buildings. Electrical **motors** are loads that convert electrical energy into mechanical energy. Several different types of electrical energy conversion will also be studied in this chapter.

Objectives

Upon the completion of this chapter, you should be able to:

1. Describe the ways in which electrical energy conversion takes place
2. Describe incandescent, fluorescent, and vapor lighting systems
3. Explain resistive, inductive, and capacitive heating effects
4. Describe the basic principles of electric motor operation
5. Identify and describe the following types of electric motors: (1) direct current motors, (2) single-phase AC motors, and (3) three-phase AC motors
6. Explain the operation of synchro systems, servo systems, and DC stepping motors
7. Describe basic motor control methods

Chapter Outline

18.1 Lighting Systems

Lighting systems convert electrical energy to light energy. Several types of lighting systems are used today, including incandescent, fluorescent, mercury vapor, and several other special-purpose lighting systems.

Incandescent Lighting

For many years, incandescent lighting was by far the most widely used lighting method. Today, fluorescent lamps are used extensively; however, incandescent lamps are still used for several applications. The construction of an **incandescent lamp** is shown in **Figure 18-1**.

This lamp is simple to install and maintain. The bulbs cost less to buy than most other types of light bulbs. They have lower efficiency and a shorter life span than many other lighting systems. Incandescent lamps have thin tungsten **filaments** in the shape of coiled wires. The filament is connected directly to a voltage source (usually 120-V AC) through the lamp base.

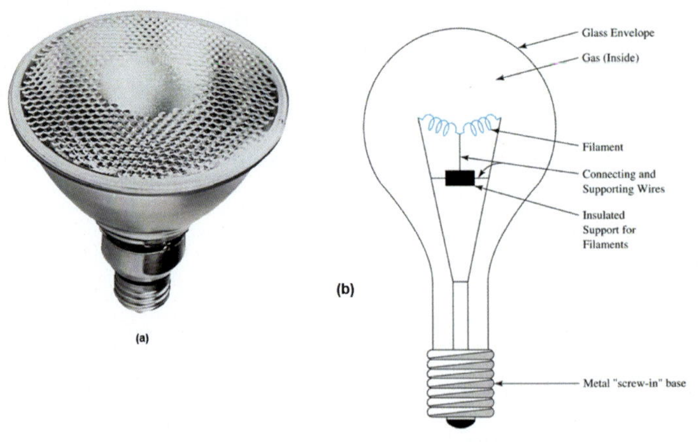

Figure 18-1 Incandescent lamps: (a) high-wattage lamp; (b) parts of the lamp [(a) courtesy of Litetronics].

When electrical current is passed through the filament, the temperature of the filament rises from 3000 to 5000°F. At this temperature range, the tungsten filament produces a high-intensity white light. When incandescent lamps are made, the air is removed from inside the bulb and replaced with an inert gas. This prevents the filament from burning out instantly.

Fluorescent Lighting

Fluorescent lighting is also used in many places today. Fluorescent lamps are tube-shaped bulbs with a filament on each end, with no electrical connection between the two filaments.

The operating principle of the fluorescent lamp is shown in **Figure 18-2**. The tube is filled with **mercury vapor**. High-speed electrons passing between the filaments collide with the mercury atoms. This collision produces ultraviolet radiation. The inside of the lamp tube has a phosphor coating, which reacts with the ultraviolet radiation to produce visible light. Several types of phosphors may be used to produce different colors of light.

The circuit of a type of fluorescent lamp is shown in **Figure 18-3**. A thermal starter is used – it is a bimetallic strip and a heater connected in series with the filaments. The bimetallic strip remains closed long enough for the filaments to heat. Heat causes the mercury in the tube to become a vapor and conduct electrical current. The bimetallic strip then bends due to the heat produced by current 1low through the heater. The contact points then separate, which opens the filament circuit. A capacitor is connected across the switch to reduce contact sparking. When the contacts of the starter open, a high voltage is placed across the lamp filaments. The high voltage is caused

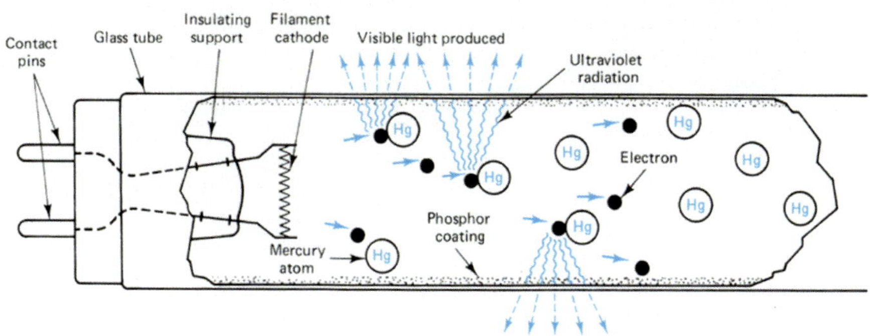

Figure 18-2 Operating principle of a fluorescent lamp: (I) filament glows when AC current flows through it; (2) when the starter opens, the ballast discharges to ionize the mercury (g) atoms; (3) ultraviolet radiation is given off by Hg atoms; (4) visible light produced.

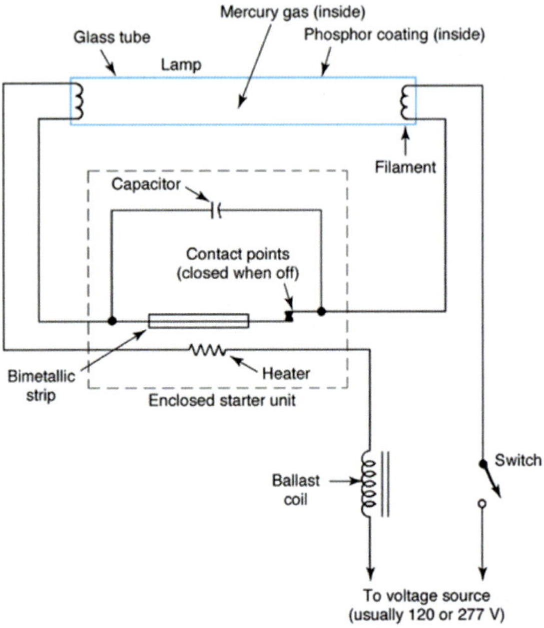

Figure 18-3 Circuit for a fluorescent lamp.

Figure 18-4 Fluorescent lighting fixture.

by action of the ballast coil. The **ballast coil** has many turns of fine wire and produces a strong magnetic field. When the contacts of the starter separate, the magnetic field collapses. The high voltage across the filaments causes the mercury inside the lamp to ionize. A current flow then starts through the lamp. The ballast also limits the current flow. Other types of starter circuits are used.

Fluorescent lamps produce more light than incandescent lamps of the same wattage. They are cheaper to operate because they use less energy. The light produced by fluorescent tubes also has less glare and is very similar to natural light. Various sizes and shapes of fluorescent lamps are available. Fluorescent lamps have the advantage over incandescent lamps of lower energy consumption. A fluorescent lighting fixture is shown in **Figure 18-4**.

Vapor Lighting

Another type of electrical lighting is the vapor type. The **mercury vapor lamp** is one of the most common types. Another type is the sodium vapor lamp. These lamps are filled with a gas that produces a color. For instance, **mercury** produces a greenish-blue light and **argon** produces a bluish-white light. Gases can be mixed to produce other color combinations.

Mercury vapor lamps are shown in **Figure 18-5**. They are made of a glass arc tube placed inside a bulb. The inner tube contains mercury. A voltage is applied between the starting probe and an electrode, which causes an arc to start between them. The arc strength is limited by a resistor. The arc causes the mercury in the inner tube to ionize. When the mercury has ionized, greenish-blue light is produced.

Mercury vapor lamps are long lasting and easy to maintain. They are used in industry and commercial buildings to provide intense light outputs. Mercury vapor lamps may be used for outdoor lighting. The **sodium vapor lamp** shown in **Figure 18-6** is popular for outdoor industrial and commercial lighting and for highway lighting. This lamp contains low-pressure neon gas and sodium. When electrical current passes through a heater, electrons are given off. An ionizing circuit causes positive charges to be placed on the positive electrodes. Electrons pass from the heater to the positive electrodes. This causes the neon gas to ionize. The neon produces enough heat to cause the sodium to ionize. A yellowish light is then produced by the sodium vapor. Sodium vapor lamps can produce about three times as much light as incandescent lamps of the same wattage.

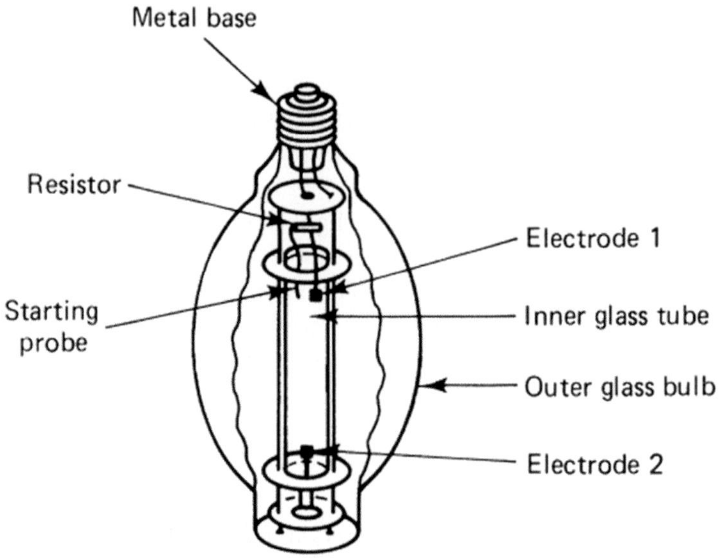

(a)

Figure 18-5 Mercury vapor lamps: (a) cutaway view of a mercury vapor lamp; (b) commercial mercury vapor lamp [(b) courtesy of Philips Lighting Co.].

The types of lighting loads discussed in this section are three of the most common – incandescent, fluorescent, and vapor lamps – used for many lighting applications. A few other specialized types of lighting are also used.

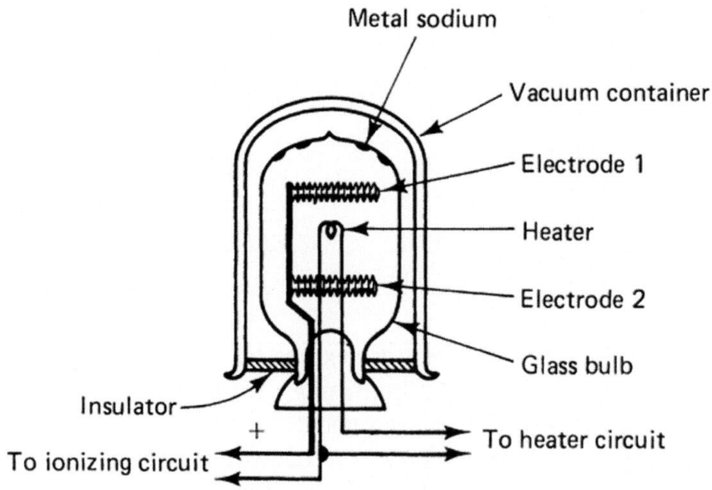

Figure 18-6 Sodium vapor lamp.

Table 18-1 Lighting terms.

Lighting quantity	**Definition**	**Symbol SI**	**Unit English**	**Definition of unit**
Luminous intensity (candlepower)	Ability of source to produce light in a given direction	I	Candela (cd)	Approximately equal to the luminous intensity produced by a standard candle
Luminous flux	Total amount of light		Lumen (lm)	Luminous flux emitted by a 1-candela uniform point source
Illumination	Amount of light received in a unit area of surface (density)	*E* Lux (lx)	Footcandle (fc)	One lumen equally distributed over one unit area of surface
Luminance (brightness)	Intensity of light per unit of area reflected or transmitted from	*L* ed/ m^2 (foot lambert)	ed/ $in.^2$	A surface reflecting or emitting light at the rate of 1 candela per unit of projected area

Table 18-2 Comparison of some different types of light sources.

Lamp type	Efficacy (Lumens/watt)
200-W incandescent lamp	20
400-W mercury lamp	50
40-W fluorescent lamp	70
400-W metal halide lamp	75
400-W high-pressure sodium lamp	110

Units of light measurement are expressed in both English and metric units, which can be somewhat confusing. **Table 18-1** summarizes some common terms used for light measurement.

The purpose of a light source is to convert electrical energy into light energy. A measure of how well this is done is called efficacy. Efficacy is the lumens of light produced per watt of electrical power converted. A comparison of some different types of light sources is shown in **Table 18-2**.

Solid State Lighting

A type of lighting system that has steadily been gaining attention today is generated by solid state/semiconductor devices such as **light emitting diodes (LEDs)**. These are much smaller in size as compared to traditional lighting devices and are generally grouped together in a single unit. LED lighting units are significantly more efficient than incandescent lamps and last much longer for up to 50,000 hours as compared to the 1000 hours for an incandescent lamp. Over time, their output gradually decreases. It is possible to manufacture LEDs in numerous colors without the use of color filters. LEDs have a rugged construction and are well-suited for applications where there is vibration or frequent switching. Unlike incandescent or fluorescent lamps, LEDs produce a directional light output. LED lighting can be to illuminate specific objects or areas. The efficacy of a residential type of LED application is compared with other lighting devices in **Table 23-2**. LEDs generally consume very low power.

18.2 Heating Systems

Most **electrical loads** produce a certain amount of heat. **Heating** is caused when current flows through a resistive device. In some equipment, heat causes a power loss. Lamps, for example, produce heat energy and light energy. Several types of loads are mainly heating loads – their main purpose is to convert electrical energy into heat energy. These load devices include resistance heating, inductive heating, and dielectric (capacitive) heating.

Resistance Heating

Heat energy is produced when electrical current flows through a resistive material. In some cases, the heat energy produced is undesirable. **Resistance heating** produces useful heat that can be transferred from a resistive material to where it is used. The heating element has coiled resistance wire designed as a heat-conducting material. The element is contained in a metal sheath. This principle can be used to heat water, oil, or a room. Resistive heaters may be used in open air or immersed in a material that is to be heated. The heat energy produced depends on the amount of current flow and the resistance of the element. A typical application is shown in **Figure 18-7**, where resistance heating coils are used as supplemental elements for a heat pump unit.

Induction Heating

The principle of **induction heating** is shown in **Figure 18-8**. Heat is produced in magnetic materials when alternating current is applied. In the example, current is induced into the material to be heated.

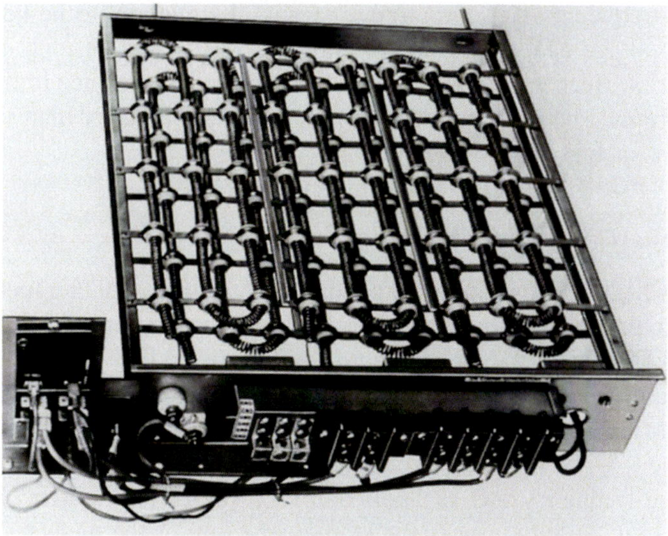

Figure 18-7 Resistance heating element (courtesy of Williamson Co.)

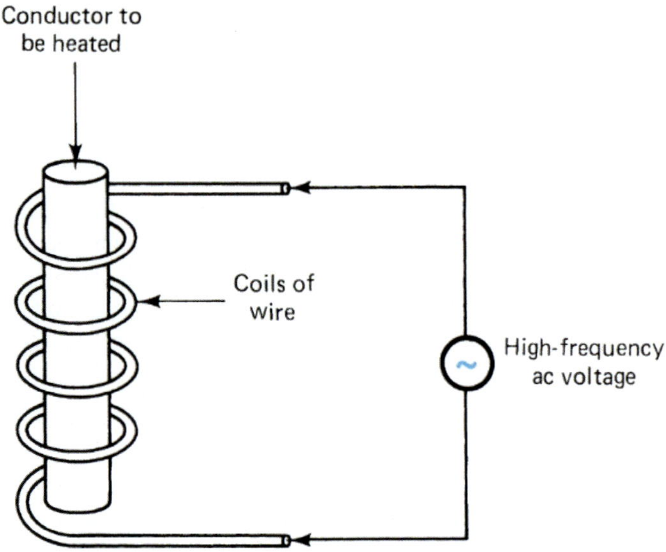

Figure 18-8 Principle of induction heating.

Electromagnetic induction is caused by the application of alternating current to the heating coil. The material to be heated must be a conductor. Current is induced into the conductor. A high-frequency AC source is used to produce higher heat output. The high-speed magnetic field created by the high-frequency AC source moves across the material to be heated. The induced voltage causes circulating currents called eddy currents to flow in the material. Heat is produced due to the material's resistance to the flow of eddy currents. The heat produced by this method is used for many industrial processes, such as heating metals.

Dielectric (Capacitive) Heating

Induction heating is used only for conductive materials. Another method must be used to heat nonconductive materials. **Dielectric or capacitive heating** is shown in **Figure 18-9**. Nonconductors may be heated by placing them in an electrostatic field between two metal electrodes. The electrodes are supplied by a high-frequency AC source. The material to be heated is like the dielectric of a capacitor. The metal electrodes are like two plates of a capacitor.

High-frequency AC is applied to the **dielectric heater**. The high-frequency AC voltage causes the atomic structure of the dielectric material

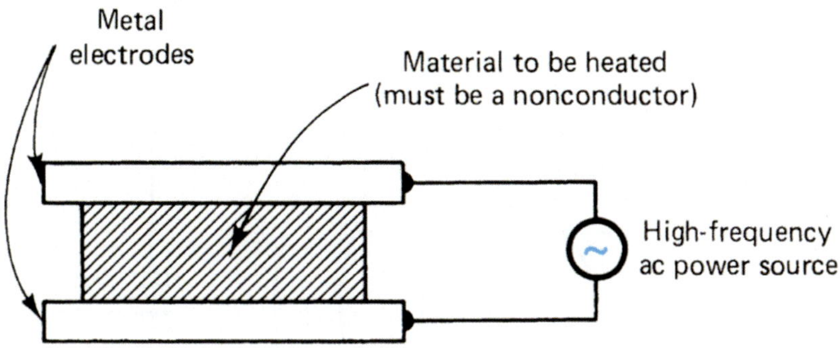

Figure 18-9 Dielectric (capacitive) heating.

to become distorted. As frequency increases, the amount of distortion also increases. Internal friction is caused which produces a large amount of heat in the nonconductive material. Dielectric heating produces rapid heat that spreads throughout the heated material. Common applications of dielectric heating are for making plywood and bonding plastic sheets.

18.3 Mechanical Loads (Motors)

Equipment that converts electrical energy into mechanical energy is another type of electrical load. **Electric motors** are the main type of **mechanical load**. There are many types of motors used today. Motors are our major energy-consuming load. Motors of various sizes are used for many applications. All motors have several basic parts in common, including (1) a **stator**, which is the frame and other stationary parts; (2) a **rotor**, which is a rotating shaft and other parts that rotate; and (3) other equipment, such as a **brush commutator assembly** for DC motors and a **starting circuit** for certain AC motors.

The basic parts of a **DC motor** are shown in **Figure 18-10**. DC motors are constructed in the same way as DC generators. Their basic parts are the same. The function of a motor is to convert electrical energy into **mechanical energy**. Motors produce **rotary motion**. A motor has an electrical input and a mechanical output. Mechanical energy is produced by a motor due to the interaction of a magnetic field and a set of conductors.

The **motor principle** is shown in **Figure 18-11**. In **Figure 18-11(a)**, no current is flowing through the conductors due to the position of the brushes against the commutator, and no motion is produced at this time. When current flows through the conductors, a circular magnetic field is developed around the conductors. The direction of current flow determines the direction of the

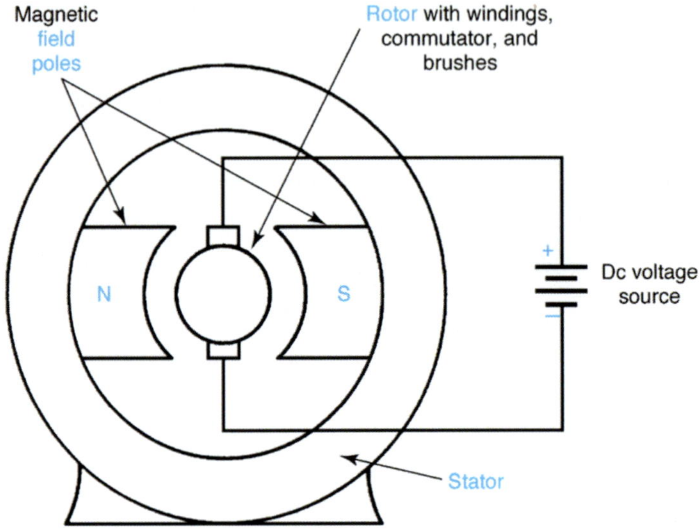

Figure 18-10 Basic parts of a DC motor.

circular magnetic field, as shown in **Figure 18-11(b)**. The **magnetic field** around the conductors interacts with the main magnetic field. The interaction of these two magnetic fields results in rotary motion. The circular magnetic field around the conductors causes a compression of the main magnetic field at points A and B in **Figure 18-11(c)**. This compression causes the magnetic field to react in the opposite direction. Motion is produced away from points A and B. **Rotary motion** is produced in a clockwise direction. To change the direction of rotary motion, the direction of current flow through the conductors is reversed. This can be observed by using the right-hand rule shown in **Figure 18-12**.

The rotary force produced by the interaction of two magnetic fields is called **torque** or motor action. The amount of torque produced by a motor depends on the strength of the main magnetic field and the amount of current flowing through the rotor conductors. If the magnetic field or the current through the conductors **increases**, the amount of torque will increase.

18.4 Direct Current Motors

Motors that operate from direct current power sources are often used. **Direct current motors** are classified as **series**, **shunt**, or **compound machines**. The method of connecting the armature and field determines the type. Also, permanent magnet DC motors are used.

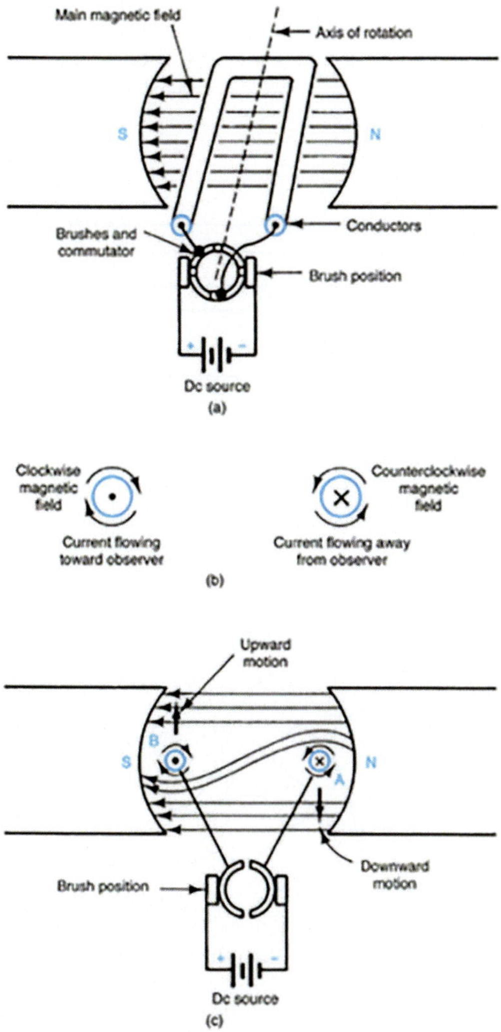

Figure 18-11 Motor principle: (a) no current flows through rotor conductor; (b) direction of circular magnetic fields around the motor conductors when current flows; (c) compression points produced at A and B when current flows through rotor conductors-causing rotation.

The operation of DC motors can be explained by referring to **Figure 18-13**. Some important terms are used here. These terms are **load**, **speed**, **counter electromotive force (CEMF)**, **current**, **and torque**. The amount of **mechanical load** applied to the shaft of a motor depends on how it is used. Load applied to the motor affects its operation. As the mechanical

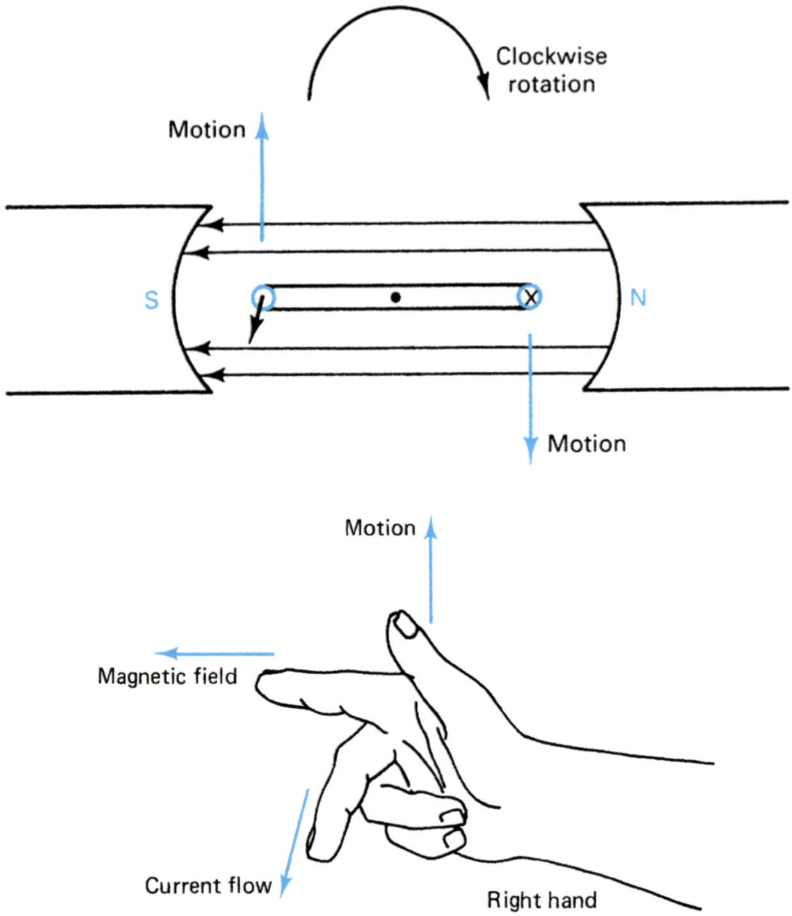

Figure 18-12 Right-hand motor rule.

load **increases**, the **speed** of a motor tends to decrease. As speed **decreases**, **CEMF** induced into the rotor conductors decreases. This generated voltage (**counter electromotive force**) depends on the number of rotating conductors and the speed of rotation, just like a generator. The counter electromotive force produced by a motor opposes the applied voltage. The actual voltage applied to a motor **increases** as the CEMF decreases. When the working voltage increases, more current flows from the power source. The **torque** of a motor depends on current flow. Torque will increase as current increases.The opposite situation occurs when the mechanical load connected to the shaft of a motor **decreases**. The speed of the motor tends to **increase**. Increased speed

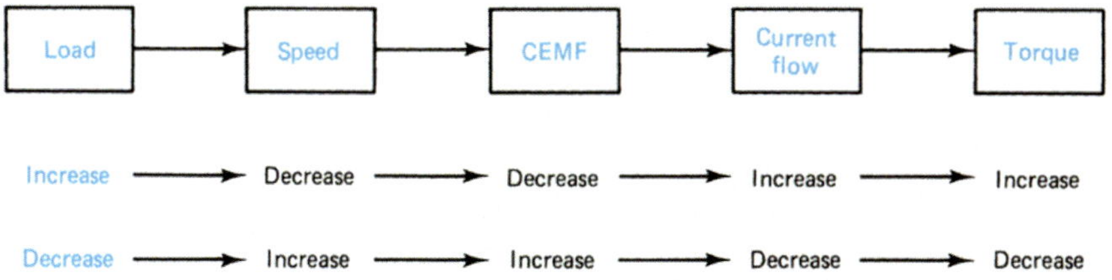

Figure 18-13 DC motor operating characteristics.

causes increased CEMF. The **CEMF** is in opposition to the applied voltage; so the CEMF **increases**. Current decreases as CEMF increases. Decreased current causes a **decrease** in torque. Torque varies with changes in load. As the load on a motor is **increased**, its torque also increases. The current drawn from the power source also increases when the load is increased. CEMF is important in motor operation. When a motor is started, a large starting current flows. The starting current is much larger than the current when full speed is reached. Current flow at full speed is called running current. Maximum current flows when there is no CEMF (when the rotor is not turning). As the CEMF increases, the current flow decreases. A resistance in series with the armature circuit is often used to reduce the starting current of a motor. After the motor has reached full speed, this resistance is bypassed by automatic or manual switching systems. This allows the motor to produce maximum torque.

The **torque** of a motor is measured in **foot-pounds**. Motors are rated in terms of their horsepower. The **horsepower rating** of a motor is based on the amount of torque produced. Horsepower can be found by using the following formula:

$$\text{hp} = \frac{2\pi \times \text{speed} \times \text{torque}}{33{,}000} = \frac{\text{speed} \times \text{torque}}{5252}.$$

The speed is in revolutions per minute (rpm) and the torque is in foot-pounds.

DC motors, like generators, have **armature reaction**. This effect decreases and distorts the main magnetic field in DC motors. The **brushes** of a DC motor can be moved to counteract the effect of armature reaction. Coils called interpoles are used to control armature reaction in DC motors. The speed of DC motors can be easily controlled. By adjusting the applied DC voltage with a rheostat, motor speed can be varied from zero to maximum rpm. Some DC motors have better speed control than others and can be

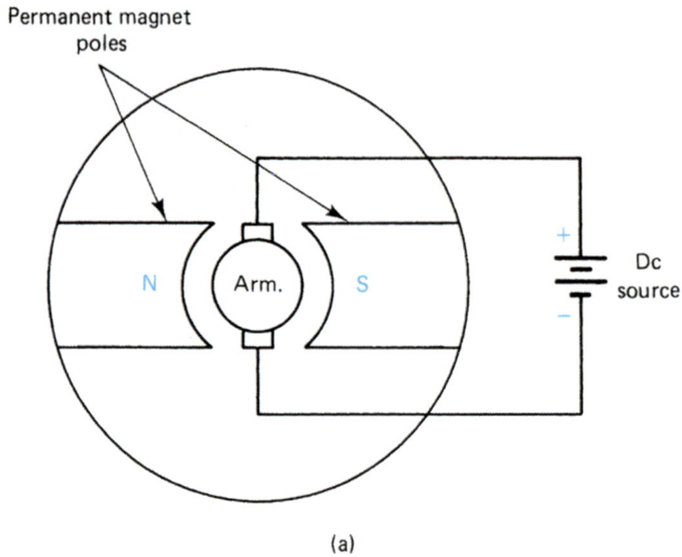

Figure 18-14 Permanent magnet DC motor: (a) pictorial; (b) schematic.

understood by studying the different types of motors. The types of DC motors are (1) permanent magnet DC motors, (2) series-wound DC motors, (3) shunt-wound DC motors, and (4) compound-wound DC motors. Each has different characteristics.

Permanent Magnet DC Motors

The **permanent magnet DC motor** shown in **Figure 18-14** is constructed like a permanent magnet DC generator. The permanent magnet motor is used only when a low amount of torque is needed. The DC power supply is connected directly to the armature conductors through the brush/commutator assembly. The main magnetic field is produced by permanent magnets mounted on the stator. Its speed can be changed by adjusting the supply voltage.

Series-wound DC Motors

The way in which the armature and field circuits of a DC motor are connected affects its operation. Each type of DC motor is similar in construction to that of a corresponding type of DC generator. The only difference is that a generator is a voltage source and the motor is a load.

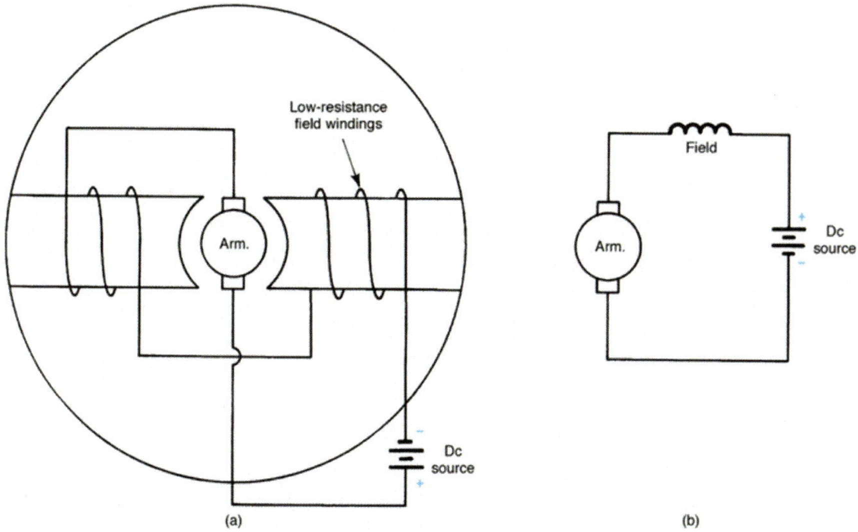

Figure 18-15 Series-wound DC motor: (a) pictorial; (b) schematic.

The **series-wound motor** shown in **Figure 18-15** has its armature and field circuits connected in series. There is only one path for current flow. The field coils are wound with a few turns of a large-diameter wire, which gives the field circuit a low resistance. Changes in mechanical load applied to the motor shaft cause charges in the current flow through the field. If the load **increases**, the current flow also increases. Increased current causes a stronger magnetic field. The speed of a series-wound motor varies from very fast with no load to very slow with a heavy load applied. A high current flows through the low-resistance field. The series-wound motor produces a high starting torque due to the large current flow. Series-wound motors are used when heavy loads must be moved.

Shunt-wound DC Motors

Shunt-wound DC motors are used more than any other type of DC motor. As shown in **Figure 18-16**, the **shunt-wound DC motor** has its field coils connected in parallel with the armature. The field coils are wound with many turns of small-diameter wire, which gives the field a high resistance. The field coils are a high-resistance parallel path, which causes a small amount of current flow through the field. A strong electromagnetic field is developed due to the many turns of wire that the field windings have. Remember that

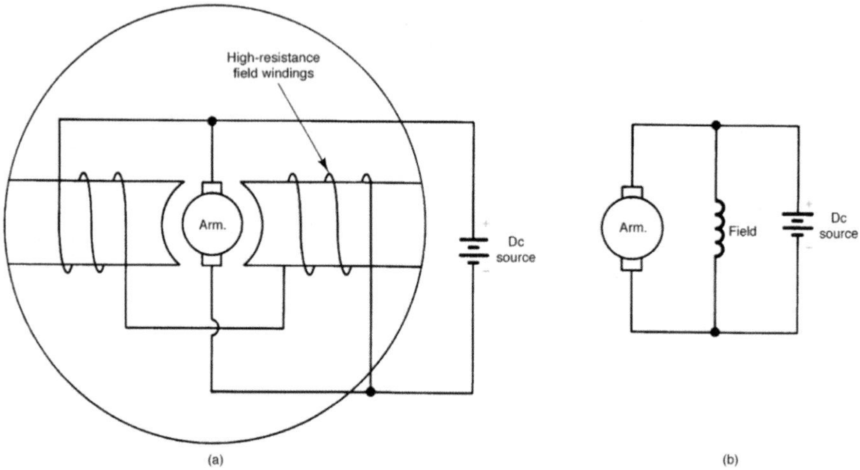

Figure 18-16 Shunt-wound DC motor: (a) pictorial; (b) schematic.

the amount of current flow and the number of turns affect the strength of an electromagnet.

Most of the current drawn by a shunt motor flows in the **armature circuit** path. The armature current does not affect the strength of the field. Motor speed is not affected by changes in load. The **field current** can be varied by placing a variable resistance in series with the field windings. The current in the field is low, which allows a low-wattage rheostat to be used to vary the speed of the motor. As the field resistance is increased, the field current will decrease. A decrease in field current reduces the strength of the electromagnetic field around the field coils. When the field strength is decreased, the motor will rotate faster due to the weaker magnetic lines of force that the rotor moves through. The speed of DC shunt-wound motors can be varied quite easily by using a field rheostat, as shown in **Figure 18-17**.

As rheostat resistance increases, field current decreases, causing magnetic field strength to decrease and motor speed to increase.

Shunt-wound DC motors have fairly **constant speed** with changes in load. The speed decreases slightly when the mechanical load increases due to increased voltage drop across the armature circuit. Because of its constant speed under load and its ease of speed control, the shunt-wound DC motor is used for many industrial applications. Variable-speed machines and tools are driven by shunt-wound DC motors.

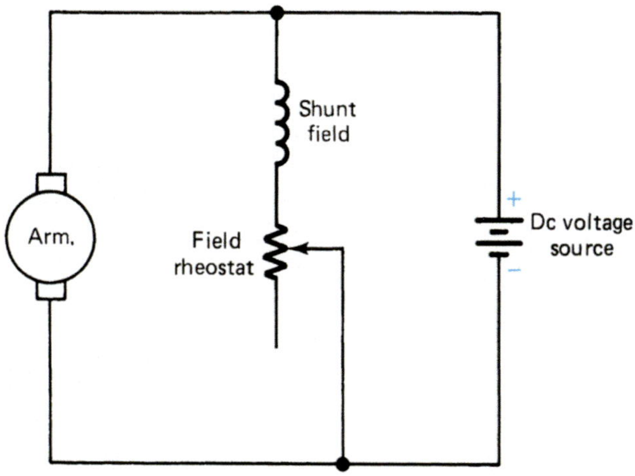

Figure 18-17 Field rheostat used to control DC motor speed.

Compound-wound DC Motors

Compound-wound DC motors have two sets of field windings – one connected in series with the armature and one in parallel. These motors combine the desirable aspects of the series-wound and shunt-wound motors. They have **high starting torque** similar to a series-wound motor. They also have **constant speed** similar to a shunt-wound motor. When high torque and constant speed are needed, the compound-wound DC motor can be used. Compound-wound motors are more expensive than series-wound or shunt-wound DC motors. DC motors are designed so that their shaft will rotate in either direction. It is simple to reverse the direction of rotation of a DC motor. By reversing the connections of the armature windings or field windings, reversal of direction is achieved. This is done by changing the terminal connections. Four terminals are ordinarily used. They are labeled A–A2 for the armature connections and F–F2 for the field connections. If either the armature connections or the field connections are reversed, the rotation of the motor will reverse (see **Figure 18-18**). If both the armature and field connections are reversed, the motor shaft will rotate in the same direction. The relative direction of current flow through the armature and field circuits would still be the same.

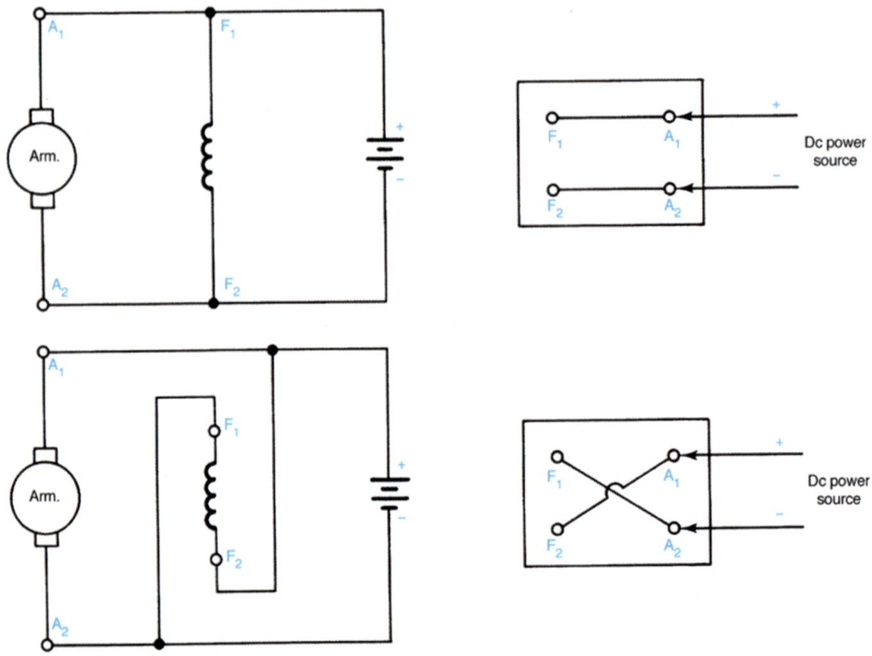

Figure 18-18 Reversing the direction of rotation of DC motors.

18.5 Three-phase AC Motors

Three-phase AC motors are used in industry and commercial buildings. Three-phase AC voltage is not supplied to homes. Most three-phase motors are integral horsepower sizes, which are ratings of 1 or more horsepower. They may be as large as 10,000 hp! Three-phase AC motors are simpler in construction and operation than single-phase AC motors. Two types of three-phase AC motors are **three-phase AC induction motors** and **three-phase AC synchronous motors**.

Three-phase AC Induction Motors

A **three-phase AC induction motor** is shown in **Figure 18-19**. The construction of the motor is quite simple. It has wound stator coils that are connected in either a wye or delta circuit. A squirrel-cage rotor is used and three-phase AC voltage is applied to the stator. A 120° separation makes the motor "self-starting." No starting winding is needed. Three-phase AC induction motors are made in many integral horsepower sizes. They have

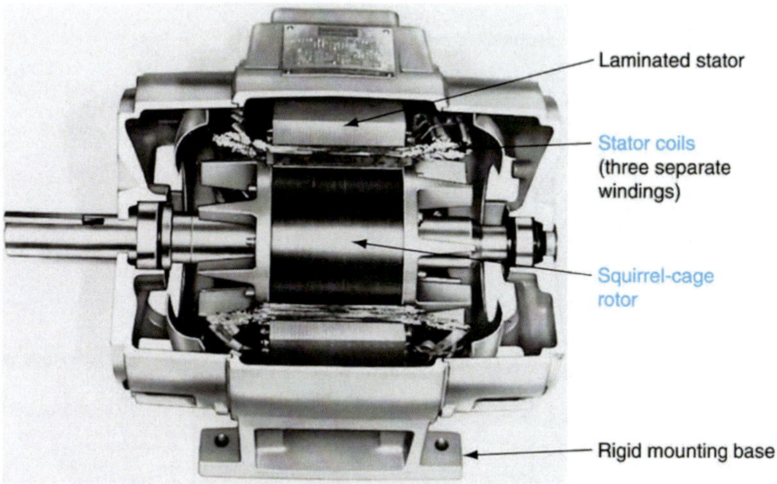

Figure 18-19 Three-phase AC induction motor (courtesy of Westinghouse Electric Corp.).

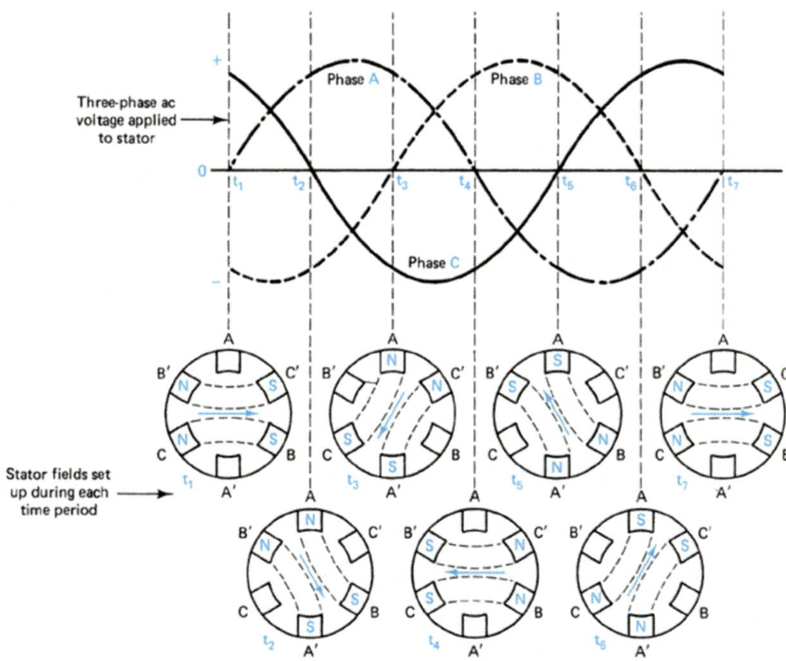

Figure 18-20 Operational principle of a three-phase ac induction motor.

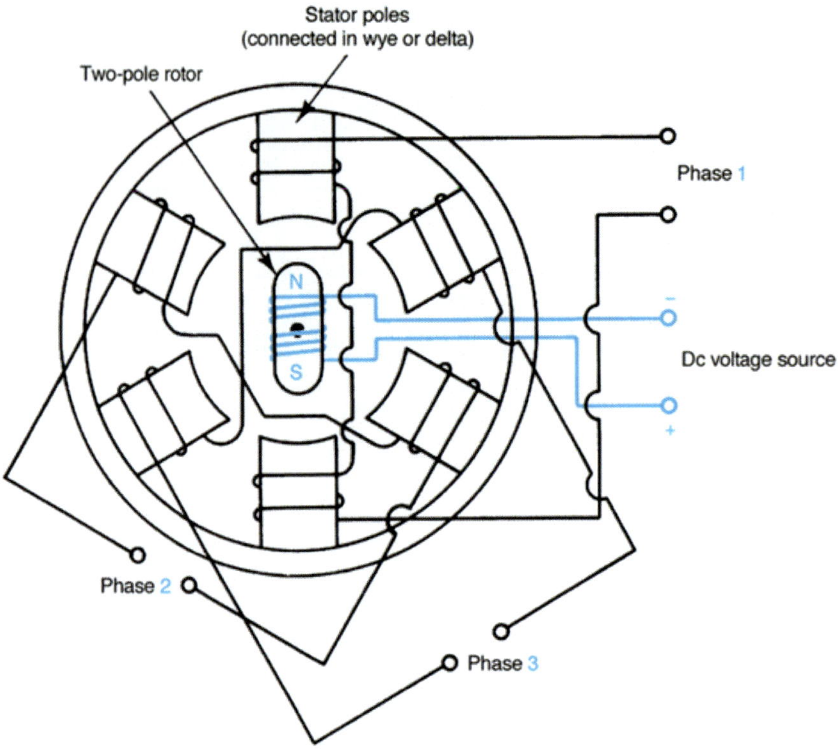

Figure 18-21 Diagram of a three-phase AC synchronous motor.

good starting and running torque. The operational principle is shown in **Figure 18-20**.

The direction of rotation of a three-phase AC induction motor can be changed easily. Two power lines coming into the stator windings can be **reversed**. The direction of rotation of the shaft will then change. Three-phase AC induction motors are used for many industrial applications. They are used to drive machine tools, pumps, elevators, hoists, conveyors, and many other machines.

Three-phase AC Synchronous Motors

Three-phase AC synchronous motors are highly specialized motors. They are constant-speed motors which can be used to "correct" power factor of three-phase systems. A diagram of a three-phase AC synchronous motor is shown in **Figure 18-21**. Three-phase AC synchronous motors are constructed like three-phase alternators. DC voltage produces an electromagnetic field.

Note the north and south polarities. The stator windings are connected in either a wye or delta circuit. Three-phase voltage is applied to the stator.

The three-phase AC synchronous motor is different from the three-phase AC induction motor. The rotor has wire windings. It is connected by **slip rings and brushes** to a DC power source. Three-phase AC synchronous

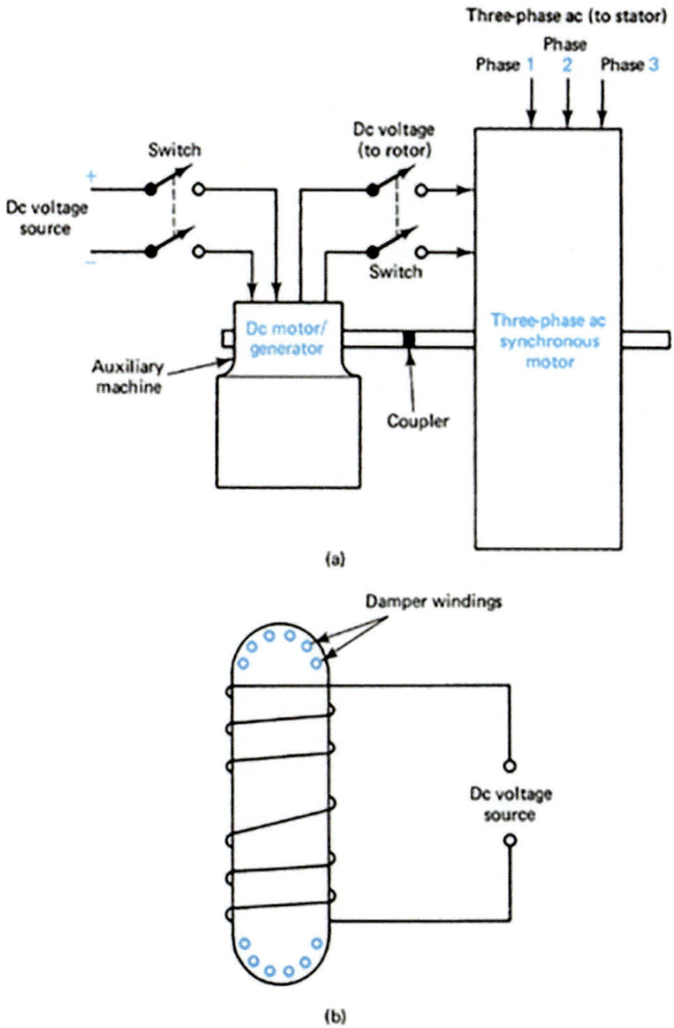

Figure 18-22 Methods of starting three-phase as synchronous motors: (a) auxiliary DC motor/generator used to accelerate the motor; (b) squirrel-cage or damper windings placed in the rotor.

motors are not self-starting – some method must be used to start the motor. Synchronous motors are designed so that they will rotate at one speed regardless of the load. This speed is called synchronous speed. The speed of an AC synchronous motor may be found by using the following formula:

$$\text{speed}(\text{rpm}) = \frac{\text{frequency}(\text{Hz}) \times 120}{\text{number of stator pole pairs per phase}}.$$

- The number of poles must be divided by 3 (the number of phases). For example, if a three-phase AC motor with 12 stator poles has 4 poles per phase (12 = 3 = 4), its speed is 1800 rpm, as follows:

$$\text{speed}(\text{rpm}) = \frac{60 \times 120}{12/3} = \frac{7200}{4} = 1800\,\text{rpm}.$$

Synchronous AC motors have speeds based on the number of stator poles and the AC frequency. One method of starting a three-phase AC synchronous motor is to use a small DC machine connected to its shaft, as shown in **Figure 18-22(a)**. It is one of the most complex types of motors in use.

Listed below are the steps for starting this motor.

1. DC power is applied to the auxiliary motor. This causes it to increase in speed.
2. Three-phase AC power is applied to the stator of the synchronous motor.
3. When motor speed nears the synchronous speed of the motor, the DC power circuit is opened. At the same time, the line terminals of the auxiliary machine are connected across the slip ring/brush assembly of the synchronous motor.
4. The auxiliary machine now converts to generator operation. It supplies DC current to the rotor of the synchronous motor. The synchronous motor now acts as a prime mover to turn the auxiliary machine.
5. When the rotor is magnetized, it will "lock in step" or "synchronize" with the revolving stator field.
10. The speed of the motor will remain the same even with changes in load as long as the rotor is magnetized.

Another starting method is shown in **Figure 18-22(b)**. This method uses "**damper**" **windings**. They are similar to the conductors of a squirrel-cage rotor. A damper winding is placed inside the laminated iron of the rotor. No auxiliary machine is required to start a synchronous motor when damper windings are used.

The steps of operation are now simplified as follows:

1. Three-phase AC power is applied to the stator.
2. The motor operates like a three-phase induction motor due to the damper windings.
3. The rotor speed builds up to slightly less than the motor's synchronous speed. Remember that the speed of an induction motor is always slower than its synchronous speed.
5. DC power is applied to the slip ring/brush assembly.
6. The rotor becomes magnetized. The rotor then increases speed until it reaches synchronous speed.
7. The speed of the motor remains the same regardless of the load placed on the shaft of the motor.

An **advantage** of the three-phase AC synchronous motor is that it can be connected to a three-phase power system to increase **power factor** of AC circuits. Three-phase AC synchronous motors are sometimes used only to increase power factor. If no load is connected to the shaft of a three-phase AC synchronous motor, it is called a synchronous capacitor. In many cases, this motor is used for its constant-speed capability.

As **load increases** on a three-phase synchronous motor, the stator current will increase. The motor will remain "synchronized" (same speed) unless the load causes "**pull out**" to take place. The motor would then stop rotating due to excessive load. Many three-phase synchronous motors are larger than 100 hp. They are used mainly for industrial applications that require **constant speed**.

18.6 Single-phase AC Motors

Another type of mechanical load is the **single-phase AC motor**. This type of motor is common for industrial, commercial, and home use. They operate from a single-phase AC power source. Many are plugged into 120-V power outlets. The three types of single-phase AC motors are (1) **universal motors**, (2) **induction motors**, and (3) **synchronous motors**.

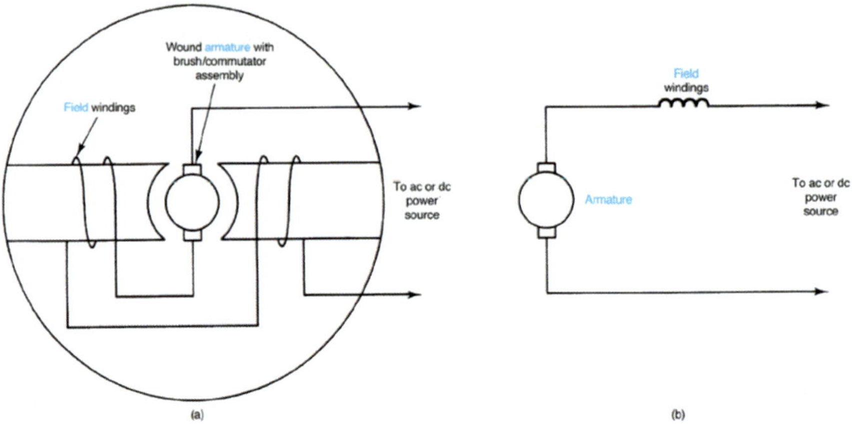

Figure 18-23 Universal motor: (a) pictorial; (b) schematic.

Universal Motors

Universal motors can operate with either **AC** or **DC** voltage applied. A universal motor is shown in **Figure 18-23** is constructed in the same way as a series-wound DC motor. The series-wound arrangement is the only type of motor that will operate with AC voltage applied. Series motors have windings with low inductance (few turns of large-diameter wire). They offer low opposition to the flow of alternating current. The universal motor has concentrated field windings which are similar to those of DC motors.

The operating principle of the **universal motor** with AC applied involves the change of field and armature polarities at the same time. A reversal of field and armature polarity is brought about by the changing nature of alternating current. The current changes direction 120 times per second! Reversal of current direction through the armature conductors at the proper time causes the armature to rotate. The principle is the same as that for DC motors. In fact, the universal motor operates in the same manner as a **series-wound DC motor** except that the field polarity and the direction of armature current change at a rate of 120 times per second with 60-Hz AC voltage applied. The speed and torque of universal motors are similar to DC series-wound motors. Applications of universal motors are for many portable tools and small motor-driven equipment. Appliances such as mixers and blenders and tools such as drills and saws use universal motors.

Induction Motors

The operation of a single-phase AC induction motor is shown in **Figure 18-24**. The coil symbols around the stator represent the field coils. The coils are connected to an AC voltage source. The polarity of the coils will change 120 times per second when 60-Hz AC is applied.

Induction motors have a solid rotor, called a **squirrel-cage rotor**. Remember that DC motors have wound rotors. The squirrel-cage rotor shown in **Figure 18-25** has large copper conductors soldered at each end to a round connecting plate. The plates actually short-circuit the copper conductors. When current flows through the stator windings, a current flow is induced into the rotor conductors. This current is caused by "**transformer action**" between the stator and rotor. The stator has AC voltage and acts as a transformer primary winding. The rotor is similar to a transformer secondary winding as current is induced into it.

The **stator** polarity changes with the applied AC voltage. It develops a changing or "rotating" magnetic field. The rotor conductors change polarity due to the **induced current** flow through the copper conductors. The rotor rotates in step with the changing magnetic field of the stator. Some method of starting the rotor to move must be used. Due to inertia, a rotor will not move until it is put into motion. Most AC motors have some method for starting rotation.

The **operating speed** of an AC induction motor is based on the speed of the rotating magnetic field. The number of stator poles also affects speed. The speed of the rotor will never be as high as the speed of the changing stator field. If the two speeds were equal, there would be no relative motion

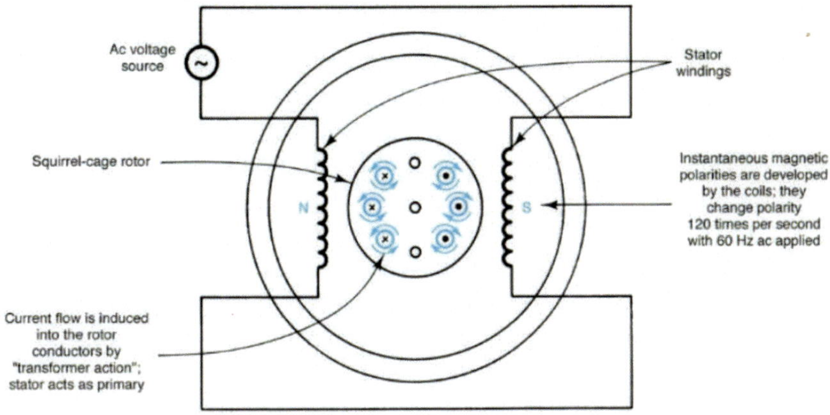

Figure 18-24 Operation of a single-phase AC induction motor.

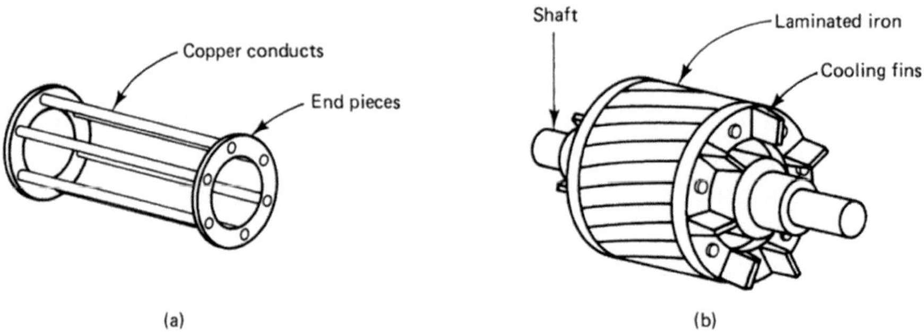

Figure 18-25 Squirrel-cage rotor: (a) internal construction; (b) assembled rotor.

between the rotor and stator conductors. No induced rotor current could flow, and no torque would be developed. The **operating speed (rotor speed)** of an AC induction motor is always less than the speed of the stator field.

The speed of the rotating field may be found by using the following formula:

$$\text{speed (rpm)} = \frac{\text{frequency (Hz)} \times 120}{\text{number of stator pole pairs}}.$$

- A motor with two stator poles operating on a 60-Hz source would have a rotating field speed of 3600 rpm:

$$\text{speed (rpm)} = \frac{60 \times 120}{2} = \frac{7200}{2} = 3600\,\text{rpm}.$$

The rotating field speed is also called the **synchronous speed** *of a motor. The difference* between the speed of the stator field of an AC induction motor and the rotor speed is called **slip**. The rotor speed must lag behind the rotating field. Slip is found by using the following formula:

$$\%\,\text{slip} = \frac{\text{field speed} - \text{rotor speed}}{\text{field speed}} \times 100.$$

- If rotor speed is close to field speed, the percentage of slip is small. For example, a two-pole motor has a field speed of 3600 rpm and a rotor speed of 3450 rpm. The percentage of slip is as follows:

$$\%\,\text{slip} = \frac{3600 - 3450}{3600} \times 100 = \frac{150}{3600} \times 100 = 4.17\%.$$

Figure 18-26 Spilt-phase AC induction motor (courtesy of Dayton Manufacturing Co.).

Single-phase AC induction motors are classified according to the method used for starting. Common types include **split-phase motors**, **capacitor motors**, and **shaded-pole motors**.

Split-phase Motors

A split-phase AC induction motor is shown in **Figure 18-26**. It has two sets of stator windings. One set is called a **run winding**. The run winding is connected across the AC power source. The other winding is called the **start winding**. The start winding is also connected across the AC line. A **centrifugal switch** mounted on the shaft of the motor is connected in series with the start winding. The centrifugal switch is in the closed position when the motor is turned off.

To understand the operation of a **split-phase AC motor**, refer to **Figure 18-27(a)**. **Figure 18-27** shows a two-pole motor with single-phase AC applied. A permanent magnet is placed inside the stator for simplicity. Remember that a squirrel-cage rotor is used in AC induction motors. At time t_0 of the AC sine wave, no stator magnetic field is produced. At time t_1, a north magnetic polarity is produced on the right pole of the stator and a south polarity on the left pole. These **polarities** will cause the rotor to align horizontally. Remember the fact that unlike magnetic poles attract. At time t_2, the stator poles are demagnetized. They then magnetize in the opposite direction at time t_3. The rotor will now align horizontally in the opposite direction. This effect will continue at a rate of 120 polarity changes per second, with 60-Hz

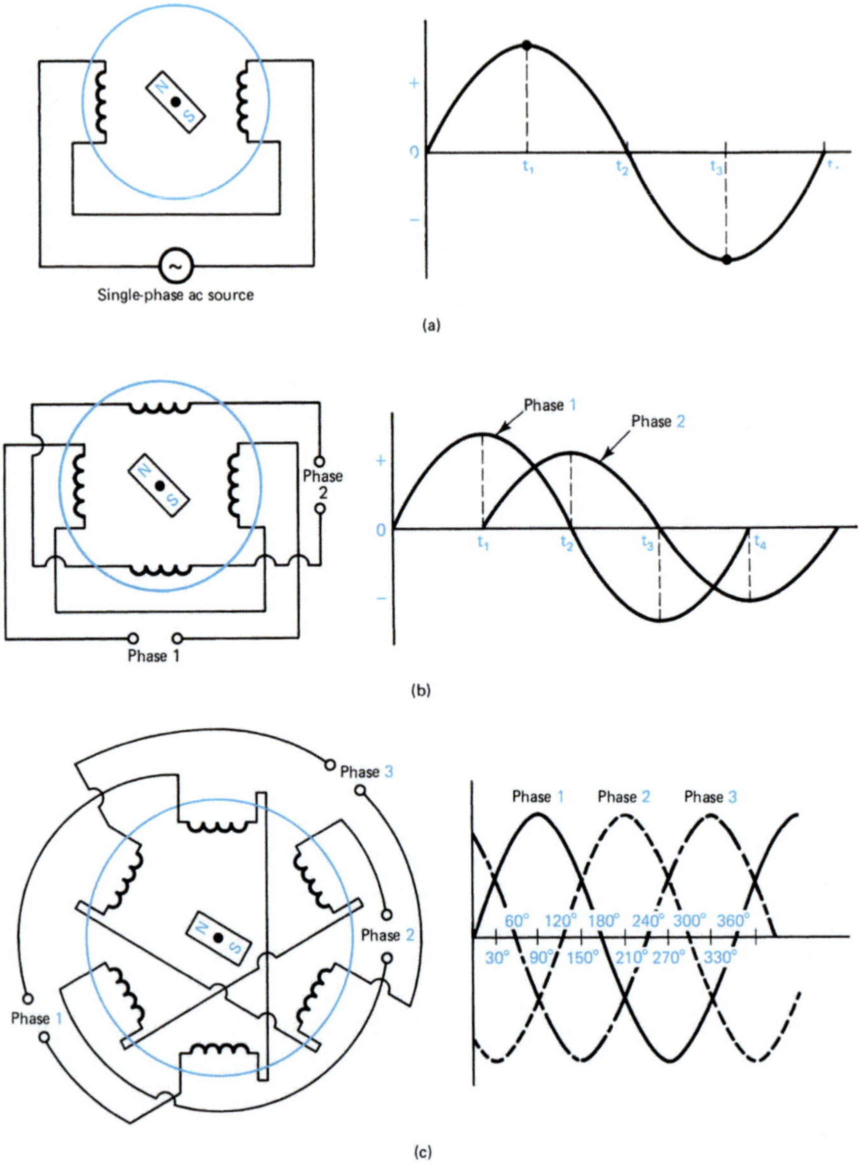

Figure 18-27 Operation of a split-phase AC motor: (a) two-pole motor with single-phase as applied; (b) motor with two-phase AC applied; (c) motor with three-phase AC applied.

AC applied to the stator. The rotor would not move unless it is positioned so that it will be drawn toward one of the poles. Some starting method must be used for single-phase AC motors. They are not self-starting.

A two-phase AC voltage applied to a motor is shown in **Figure 18-27(b)**. Two sets of **stator windings** are used. One phase is connected to each set. Two-phase voltage is not produced by power companies, but this example shows the operating principle of a split-phase motor. Two-phase voltage has one phase at minimum value while the other phase is at maximum value.

At time t_1, phase 1 is maximum and phase 2 is minimum. The right stator pole is now a north polarity, and the left stator pole is a south polarity. The rotor aligns itself horizontally. No polarities are developed by the vertical poles at this time. At time t_2, phase 1 is minimum and phase 2 is maximum. The top **stator pole** is a south polarity, and the bottom stator pole is now a north polarity. The rotor will now align itself vertically, by moving 90°. At time t_3, phase 1 is maximum in the opposite direction and phase 2 is minimum. A north pole is now developed on the left and a south pole on the right. The rotor moves 90° in a clockwise direction. This effect will continue as two-phase voltage is applied to the stator poles. The direction of rotation is established by the relation of phase 1 to phase 2. An AC motor with two-phase voltage applied would be **self-starting**.

The same is true with three-phase voltage applied to a motor as shown in **Figure 18-27(c)**. Movement of the rotor results due to the 120° phase separation of three-phase AC voltage. The voltage is applied to six stator poles. Three-phase induction motors are self-starting. No starting method needs to be built into three-phase motors.

Look at the **split-phase AC motor** diagram in **Figure 18-28(a)**. The purpose of the two sets of windings is to produce a type of two-phase AC voltage to start the motor. The single-phase AC voltage applied to the motor is said to be "split" into a two-phase current. The magnetic field of the stator is created by **phase splitting**. The start winding of the split-phase motor is made of a few turns of small-diameter wire, which gives the winding a **high resistance** (small-diameter wire) and a **low inductance** (few turns of wire). The run winding is wound with many turns of large-diameter wire. It has a lower resistance (large-diameter wire) and a higher inductance (many turns of wire). Inductance in an AC circuit causes the current flow to lag behind the applied voltage; with more inductance, there is a greater lag in current flow.

The single-phase AC voltage waveform of a **split-phase AC induction motor** is shown in **Figure 18-28(b)**. The current flow in the start winding lags the applied AC voltage due to the winding's inductance. The current flow in the run winding lags behind by a greater amount due to its higher inductance. The phase separation of the currents in the start and run windings creates a two-phase current. The phase separation of the currents is usually around 30°

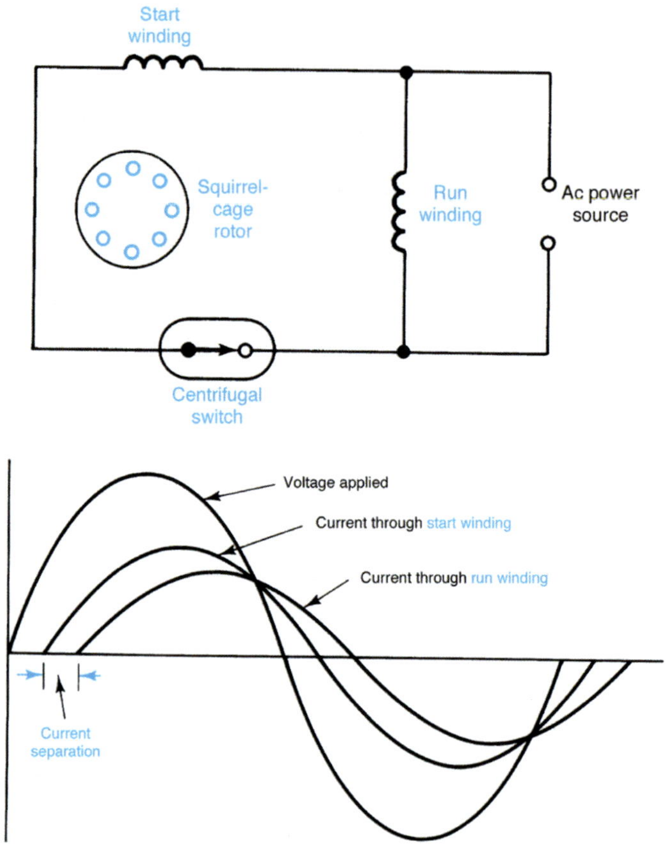

Figure 18-28 Split-phase AC motor: (a) schematic; (b) waveforms.

or less, which causes the motor to have a low starting torque. Higher torque would be produced if phase separation were closer to 90°.

When a split-phase AC induction motor reaches about 75% of its operating speed, the centrifugal switch will open the **start-winding** circuit. The start winding is no longer needed. Removing the start winding reduces power losses and prevents the motor from overheating. When the motor is turned off, the centrifugal switch closes to connect the start winding across the AC voltage source again. Split-phase motors are used when low torque is required to drive mechanical loads such as small machinery and appliances in the home.

Capacitor Motors

Capacitor motors are similar to split-phase AC motors. A **capacitor-start, single-phase AC induction motor** and its wiring diagram are shown in **Figure 18-29**.

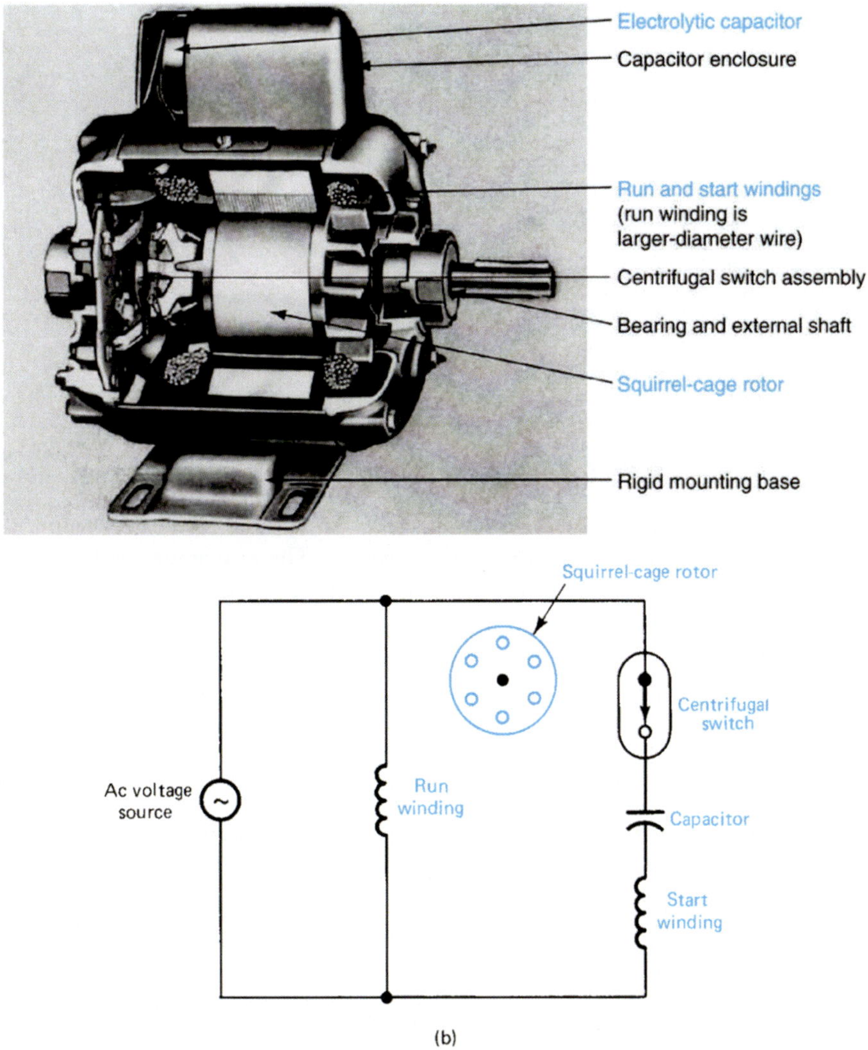

Figure 18-29 Capacitor-start AC motor: (a) cutaway view; (b) schematic diagram [(a) courtesy of Dayton Electric Mfg. Co.].

The **wiring diagram** is the same as the split-phase motor except that a capacitor is placed in series with the start winding. The capacitor causes the current in the start winding to lead (rather than lag) the applied voltage. Remember that current leads voltage in a capacitive circuit. **Figure 18-30** shows the voltage and current waveforms in a capacitor-start motor circuit.

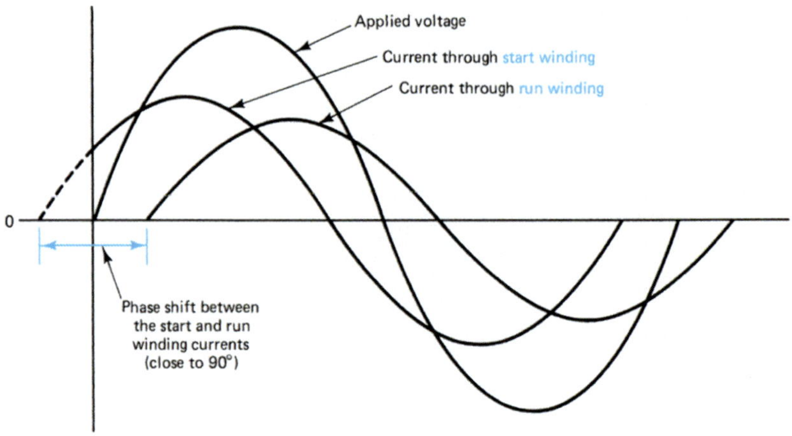

Figure 18-30 Waveforms of a capacitor-start AC motor.

The current in the **start winding** leads the applied AC voltage due to the high capacitance in the start-winding circuit. The **run winding** has only inductance; so the current in the run winding lags the applied voltage. The amount of **phase separation** of the current is close to 90°. It is almost like a two-phase AC voltage. The starting torque of a capacitor-start induction motor is greater than that of a split-phase motor. This type of motor can be used where greater torque is needed. Most capacitor motors and split-phase motors are used in fractional-horsepower sizes (less than 1 hp). Common **fractional-horsepower** sizes are ¼, ½, and ¾ hp.

Another type of capacitor motor is called a **capacitor-start**, **capacitor-run**, or two-value capacitor motor. Its circuit is shown in **Figure 18-31(a)**.

This type of motor has **two** capacitors. One capacitor of higher value is in series with the start winding and centrifugal switch. The other capacitor is lower in value. It is connected in series with the start winding. The smaller capacitor remains in the circuit during operation. The larger capacitor is used to increase starting torque. It is removed from the circuit by the centrifugal switch when operating speed reaches about 75%. The smaller capacitor and the start winding are part of the motor circuit during operation. The smaller capacitor helps to produce a constant running torque. It also reduces vibration.

Another type of capacitor motor is called a **permanent capacitor motor**. Its circuit is shown in **Figure 18-31(b)**. This motor has no centrifugal switch. The capacitor is connected into the circuit at all times. These motors are used when very low torque is needed. They are made in small, fractional-horsepower sizes.

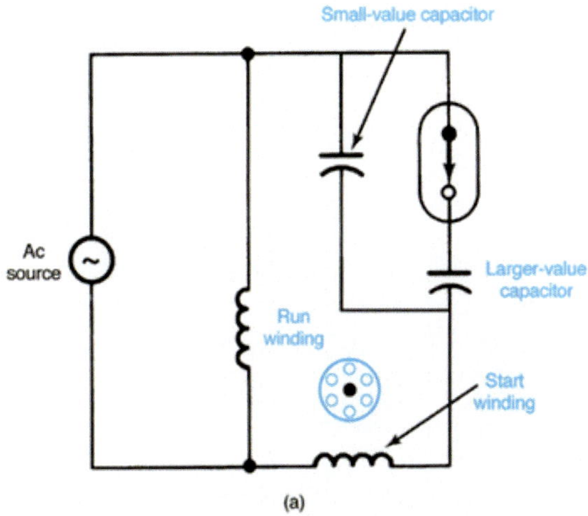

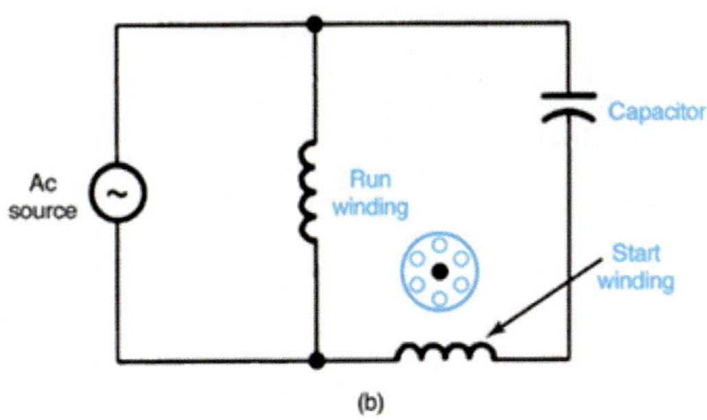

Figure 18-31 Other types of capacitor motors: (a) capacitor-start, capacitor-run motor; (b) permanent capacitor motor.

Split-phase motors and capacitor motors may have their direction of rotation reversed easily. The connection of the start winding or run winding must be changed. By reversing either the start-winding connections or the run-winding connections (but not both), the direction of rotation is reversed.

Shaded-pole Motors Another method of producing a two-phase current is called pole shading. A **shaded-pole AC induction motor** is shown in **Figure 18-32**.

Shaded-pole motors are used for very low torque applications such as fans and small blower units. They are low-cost, rugged, and reliable motors. They are only made in very low-horsepower ratings. The operating principle of a shaded-pole motor is shown in **Figure 18-33**. A single-phase AC alternation is shown to represent the applied AC voltage. The dashed lines show voltage induced into the shaded section of the field poles. The shading

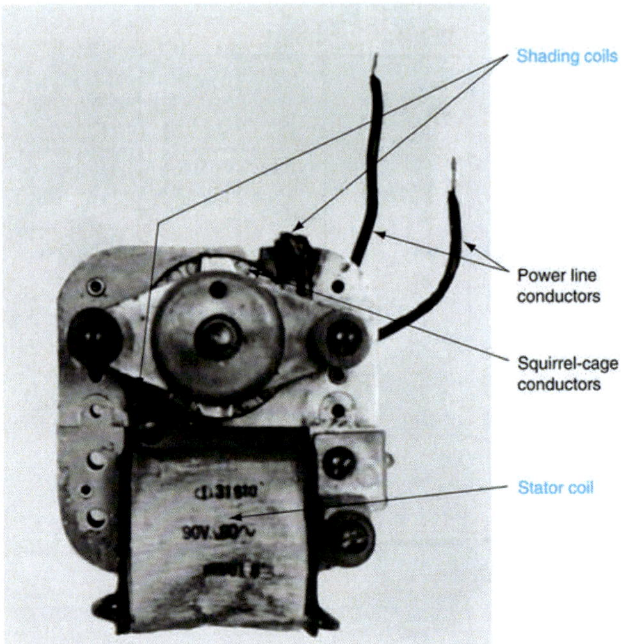

Figure 18-32 Shaded-pole AC induction motor.

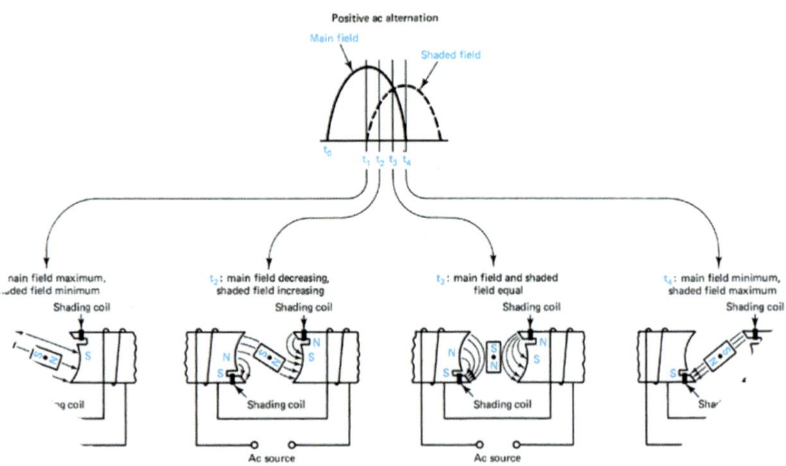

Figure 18-33 Operating principle of a shaded-pole AC motor.

coils are shown in the upper right and lower left of the field poles. The shaded pole is surrounded by a heavy copper conductor, which is actually part of the main field pole. This closed-looped copper conductor has current induced into it when AC voltage is applied to the main field poles.

Single-phase AC voltage is applied to the **stator windings**. The magnetic field of the main field poles induces a current flow into the shaded sections of the main poles. The shaded sections act like transformer secondary windings. The voltage induced into a shaded section is out of phase with the main field voltage. Four time periods are shown in **Figure 18-33**. The voltage induced into the shaded pole from the main field pole causes movement of the rotor. The movement of the rotor at each time period is shown in the **Figure 18-33**. Shaded-pole AC induction motors are inexpensive. They use a squirrel-cage rotor and have no start winding or centrifugal switch. Applications are limited mainly to small fans and blowers.

Single-phase Synchronous Motors

In timing or clock applications, a **constant-speed** motor must be used. A constant-speed motor is called a **synchronous motor**. A **single-phase AC synchronous motor** has stator windings that are connected across the AC voltage source and a rotor made of permanent magnet material. Some method of starting must be used to start the rotor's movement. When the rotor is started, it will rotate in "synchronism" with the changing AC voltage applied to the stator. This motor does not operate on the induction principle. The speed of a synchronous motor is based on the following speed formula:

$$\text{speed}(\text{rpm}) = \frac{\text{frequency}(\text{Hz}) \times 120}{\text{number of stator pole pairs per phase}}.$$

At 60-Hz frequency, these are the synchronous speeds:

Two-pole = 3600 rpm

Four-pole = 1800 rpm

Six-pole = 1200 rpm

Eight-pole = 900 rpm

Ten-pole = 720 rpm

Twelve-pole = 600 rpm

Single-phase AC synchronous motors are used for low-torque applications, which include clocks, phonographs, and timing systems that require constant speed.

18.7 Synchro Systems and Servo Systems

Synchro systems cause angular motion to remotely control angular position. Rotation of one shaft causes positioning of the second device as shown in **Figure 18-34**. Synchro systems are used to control movement. It is possible to transfer rotary motion from one point to another without using mechanical linkage.

A synchro system is two or more motor/generator units connected together. When an operator turns the "generator" shaft of the unit to a different position, it automatically rotates the "motor" shaft to a corresponding position.

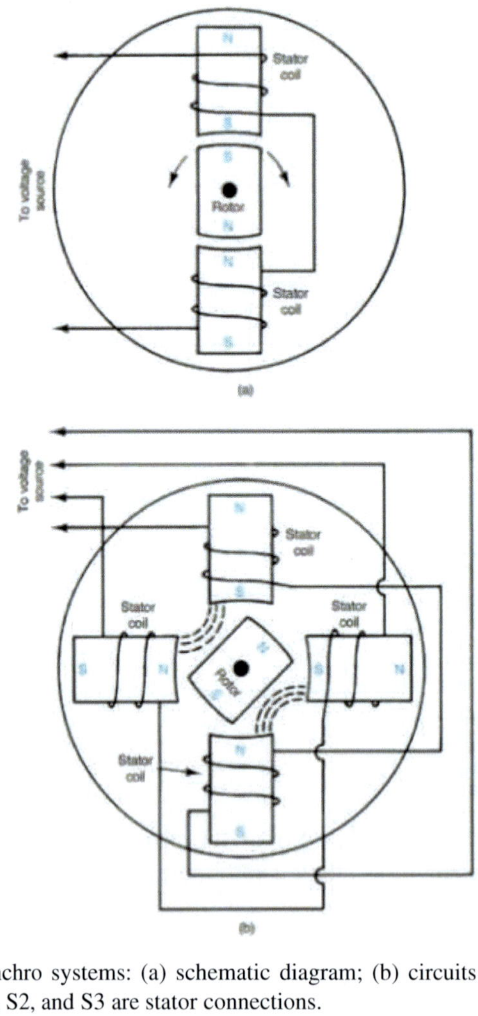

Figure 18-34 Synchro systems: (a) schematic diagram; (b) circuits R1 and R2 are rotor connections and S1, S2, and S3 are stator connections.

The motor may be at a remote location. It is possible to have accurate control over long distances. An antenna rotor for a TV set is an example of a synchro system. In synchro systems that require high torque, a servo system is used. A servo system is a special type of motor that controls movement of equipment. Synchro systems have two or more electromagnetic devices that look like small electric motors. They are connected so that the position of the generator shaft determines the position of the motor shaft. **Figure 18-34(a)** shows a **schematic diagram** of a synchro unit. The generator and motor units look identical. The motor unit has a metal flywheel attached to its shaft to prevent shaft vibration. The letters G and M inside the electrical symbol are used to identify the generator and motor units. **Figure 18-34(b)** shows the circuit of a synchro system. Single-phase AC voltage is used to power the system. The AC voltage is applied to the rotors of both the generator and motor. The stator coils are connected together in a wye connection.

When power is applied, the motor will lock into position according to the location of the generator shaft. Both units remain in the same position until the generator shaft is turned. Turning the generator shaft 30° in a clockwise direction will cause the motor unit to move 30° clockwise. If calibrated dials are attached to the shaft of each unit, they would show the same movement. The stator coils of a synchro unit are wound inside a metal housing. The coils act like the secondary windings of a transformer to transfer energy. The rotor of a synchro unit is wound on a laminated core. **Slip rings and brushes** on the shaft are used to supply AC power to the rotor. The rotor acts like the primary winding of a transformer.

Figure 18-35 Servomotor (courtesy of Superior Electric Co.).

When AC voltage is applied to the rotor of a **synchro unit**, an alternating magnetic field is produced. By transformer action, this field cuts across the stator coils, which induces a voltage in the stator windings. The position of the rotor determines the amount of voltage induced in the stator coils. The stator coils of the generator and motor are connected together as shown in **Figure 18-34(a)**. Voltage induced in the stator coils of the generator causes the same current flow in the stator coils of the motor, which causes the same magnetic field to develop in the motor stator. The rotor of the motor unit will align itself with the magnetic field of the stator coils.

A change in rotor position of the generator unit affects the induced voltage of the stator coils. In this way, movement can be transferred to the motor through three small wires which connect the stators. Synchro and **servo systems** are used to change the position or speed of an object. Position changes are needed for numerically controlled machinery and process control equipment in industry. Speed control is used for conveyor belt units, operating speed of machine tools, and disk or magnetic-tape drives for computers. A servomotor is shown in **Figure 18-35**.

DC Stepping Motors

DC stepping motors are **servomotors** used to drive many types of computer-controlled machines. They convert electrical pulses into rotary motion. The shaft of a DC stepping motor rotates a specific number of degrees when a pulse is applied to it. The amount of movement caused by each pulse can be repeated. The shaft of a DC stepping motor is used to position industrial machinery accurately. The speed, distance, and direction of a machine can be controlled by a **DC stepping motor**. The stator construction and coil placement are shown in **Figure 18-36**.

The rotor is made of permanent magnet material. The **permanent magnet rotor** is placed between two series-connected stator coils as shown in **Figure 18-37**. With power applied to the stator, the rotor can be repelled in either direction. With the stator polarities shown, the rotor aligns itself midway between two pairs of stator coils. The direction of rotation is determined by the polarities of the stator coils.

Adding more stator coils to this motor makes positioning the rotor extremely accurate. Operation of a **DC stepping motor** is achieved by electrical pulses from a computerized power supply. The stepping motor rotor shown in **Figure 18-38** has 50 teeth. It takes 200 pulses or “steps” for each complete revolution. The amount of movement is determined by the number

Figure 18-36 DC stepping motor construction (courtesy of Superior Electric Co.).

of teeth on the rotor and the electrical pulse sequence. A DC stepping motor that takes 200 steps for one revolution will move 1.8° per step (360° divided by 200). DC stepping motors are important for use in industry.

18.8 Motor Performance

The major consumer of electrical power is the electric motor. It is estimated that electric motors account for 50% of the electrical power consumed in industrial usage and that 35% of all electrical power is used by electrical motors. For these reasons, consider the efficient operation of motors to be a major part of **energy conservation** efforts. Both **efficiency** and the **power factor** must be considered to determine the effect of a motor, in terms of efficient power conversion. Remember the following relationships. First,

$$\text{efficiency}\,(\%) = \frac{P_{\text{out}}}{P_{\text{in}}} \times 100$$

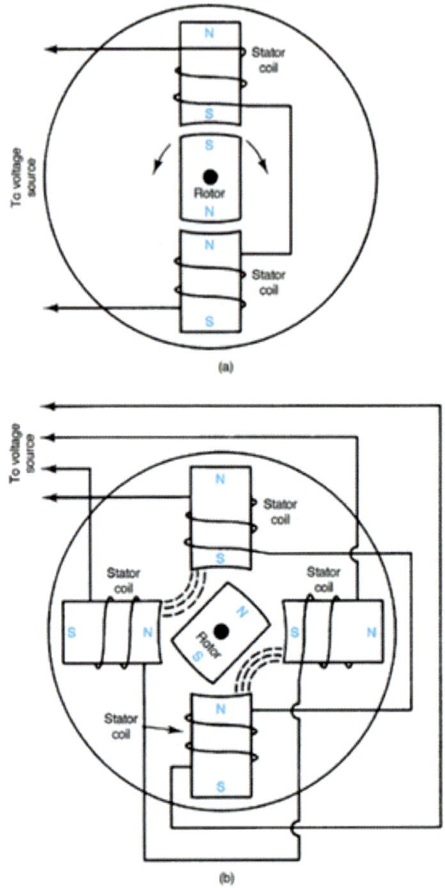

Figure 18-37 Stator coil placement of stepping motor: (a) one set of stator coils rotor can move in either direction; (b) two sets of stator coils rotor will move in a clockwise direction.

where $\mathbf{P}_{in}$ = *the power input in horsepower* and $\mathbf{P}_{out}$ = *the power output in watts* (to convert horsepower to watts, remember that 1 hp = 746 W). Then:

$$PF = \frac{P}{VA}$$

where:

PF = the power factor of the circuit
P = the true power in watts
VA = the apparent power in volt-amperes

Figure 18-38 DC stepping motor rotor construction (courtesy of Superior Electric Co.).

The maximum PF value is 1.0, or 100%, which would be obtained in a purely resistive circuit. This is referred to as unity power factor.

Effect of Load

Because electrical power will probably become more expensive and less abundant, the efficiency and power factor of electric motors will become increasingly important. The **efficiency** of a motor shows mathematically just how well a motor converts electrical energy into mechanical energy. A mechanical load placed on a motor affects its efficiency. Thus, it is particularly important for industrial users to load motors so that their maximum efficiency is maintained.

Power factor is also affected by the mechanical load placed on a motor. A higher power factor means that a motor requires less current to produce a given amount of torque or mechanical energy. Lower current levels mean that less energy is being wasted (converted to heat) in the equipment and circuits connected to the motor. Penalties are assessed on industrial users by the electrical utility companies for having low system power factors (usually less than 0.8 or 0.85 values). By operating at higher power factors, industrial users can save money on penalties and can help, on a larger scale, with more efficient utilization of electrical power. Motor load affects the power factor to a much greater extent than it does the efficiency. Therefore, motor applications should be carefully studied to ensure that motors (particularly very large ones) are not overloaded or under-loaded, so that the available electrical power will be used more effectively.

Effect of Voltage Variations

Voltage variation also has an effect on the power factor and efficiency. Even slight changes in voltage produce a distinct effect on the power factor; however, a less distinct effect results when the voltage causes a variation in the efficiency. Because proper power utilization is becoming more and more important, motor users should ensure that their motors do not operate at **under-voltage** or **overvoltage** conditions.

Considerations for Mechanical (Motor) Loads

The three basic types of **mechanical (motor) loads** connected to electrical power systems are DC, single-phase AC, and three-phase AC systems. **DC motors** are ordinarily used for special applications, because they are more expensive than other types and require a DC power source. Typically, they are used for small, portable applications and powered by batteries or for industrial or commercial applications with AC converted to DC by rectification systems. A major advantage of DC motors is the ease of speed control. The **shunt-wound DC motor** can be used for accurate speed control and good speed regulation. A disadvantage is the increased maintenance caused by the brushes and commutator of the machines. DC shunt-wound motors are used for variable-speed drives on printing presses, rolling mills, elevators, hoists, and automated industrial machine tools.

Series-wound DC motors have a high starting torque. Their speed regulation is not as good as that of shunt-wound DC motors. The series-wound motor also requires periodic maintenance because of the **brush/commutator assembly**. Typical applications of series-wound DC motors are automobile starters, traction motors for trains and electric buses, and mobile equipment operated by batteries. **Compound-wound DC motors** have few applications today.

Single-phase AC motors are relatively inexpensive. Most types have good starting torque and easily provide 120- and 240-V electrical power. Disadvantages include maintenance problems due to centrifugal switches, pulsating torque, and rather noisy operation. They are used in **fractional-horsepower** sizes (less than 1 hp) for residential, commercial, and industrial applications. Some integral horsepower sizes are available in capacitor-start types. Uses include machine cooling system blowers and clothes dryer motors.

Specialized applications tor single-phase motors include:

- Shaded-pole motors used for portable fans, record players, dishwasher pumps, and electric typewriters. They are low cost and small but inefficient.
- Single-phase synchronous motors used for clocks, appliance timers, and recording instruments (compact disk players). They operate at a constant speed.
- Universal (AC/DC) motors used for many types of portable tools and appliances such as electric drills, saws, office machines, mixers, blenders, sewing machines, and vacuum cleaners. They operate at speeds up to 20,000 rpm and have easy speed control.
- Remember that AC induction motors do not have speed control capability without the addition of expensive auxiliary equipment.

Three-phase AC motors of the induction type are simple in construction, rugged, and reliable in operation. They are less expensive (per horsepower) than other motors. Applications of **three-phase induction motors** include industrial and commercial equipment and machine tools. **Three-phase AC synchronous motors** run at constant speeds and may be used for power factor correction of electrical power systems. However, they are expensive, require maintenance of brushes/slip rings, and need a separate DC power supply.

18.9 Motor Control Basics

A block diagram of a **motor control** circuit is shown in **Figure 18-39**. The components used to control motors include circuit breakers, switches, magnetic contactors, protective overloads, limit switches, and many other devices.

Motor Starting Control

On many motor control circuits, **pushbutton switches** such as those shown in **Figure 18-40** are used. A pushbutton may be pressed to cause a motor to start. Another pushbutton may be used to cause it to stop. Pushbuttons either apply power from a source or open the power circuit to stop a motor.

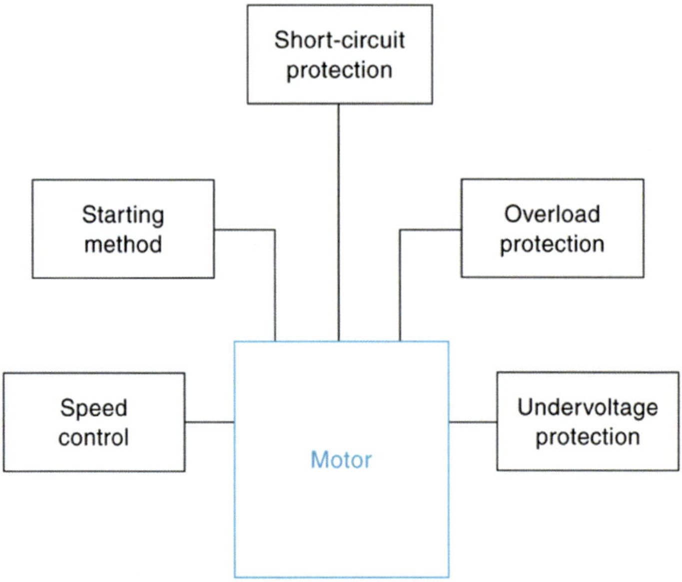

Figure 18-39 Block diagram of a motor control circuit.

Figure 18-40 Pushbutton switches [(a) courtesy of Cutler-Hammer Co.; (b) courtesy of Furnas Electric Co.].

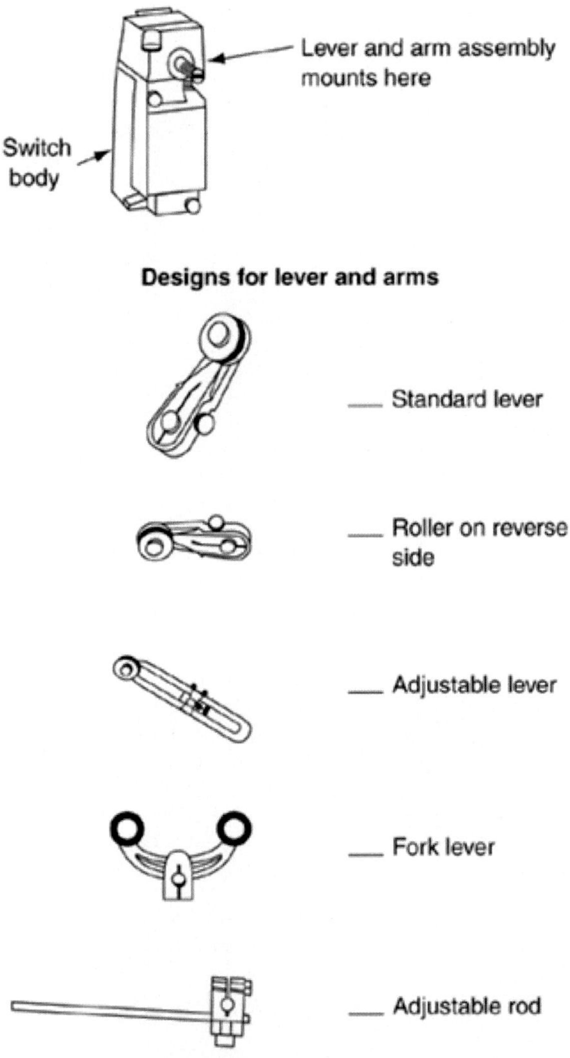

Figure 18-41 Limit switches.

Limit switches may also be used to control a motor's on or off operation. **Limit switches** are used to limit the movement or travel of a piece of equipment; for example, they could be used in an elevator to cause it to stop at the desired floor. Some types of limit switches are shown in **Figure 18-41**.

Fuses or circuit breakers (see **Figure 18-42**) are also part of a motor control circuit. **Fuses** are made so that they will open a circuit by melting. If the current flowing through a circuit exceeds a safe limit, the fuse element

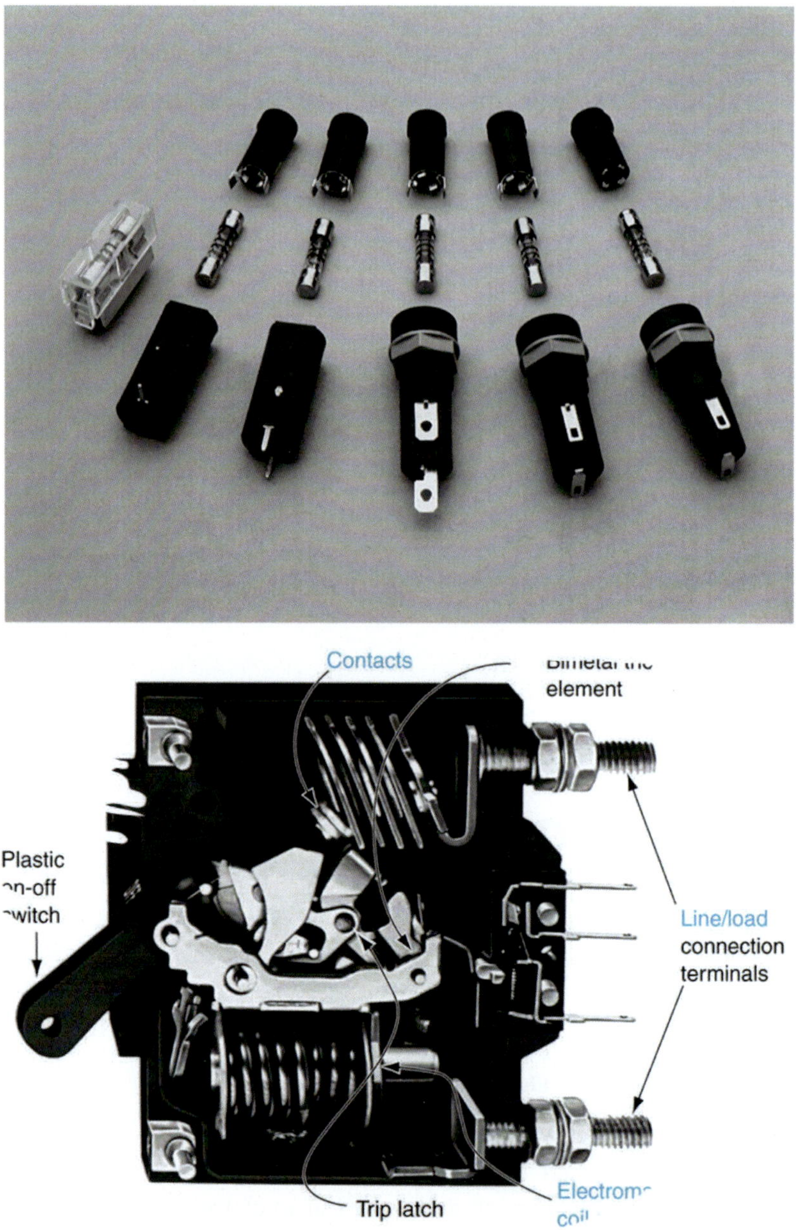

Figure 18-42 Fuses and circuit breakers: (a) fuses with holders; (b) internal construction of a circuit breaker [(a) courtesy of Cooper Industries; (b) courtesy of Heinemann Electric].

will melt. A fuse is used to prevent excessive current from flowing through a motor. It is a form of protective control. **Circuit breakers** have the same function. A circuit breaker must be reset after it is opened. If a circuit breaker is rated at 20 A, a current greater than that amount will disconnect the power source. By automatically disconnecting the power source, circuit breakers act as a protective control. They are usually preferred over fuses because they do not have to be replaced.

There are several types of electromechanical equipment used for the control of electric motors. The selection of control equipment will affect the efficiency of operation and the performance of the machinery. It is important to use the proper type of equipment for each control application. A **motor-starting device** is a type of power control used to accelerate a motor from a stopped condition to its normal operating speed. There are many variations in motor-starter design, the simplest being a manually operated on–off switch connected in series with one or more power lines. This type of starter is used only for smaller motors that do not draw an excessive amount of current (see **Figure 18-43**).

Another type of motor starter is the magnetic starter (see **Figure 18-44**), which relies on an electromagnetic effect to open or close the power source circuit of the motor. Often, motor starters are grouped together for the control

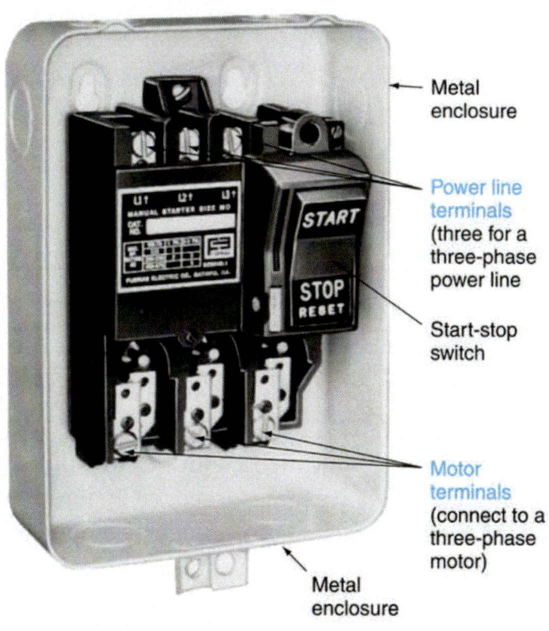

Figure 18-43 Manual motor starter (courtesy of Furnas Electric Co.).

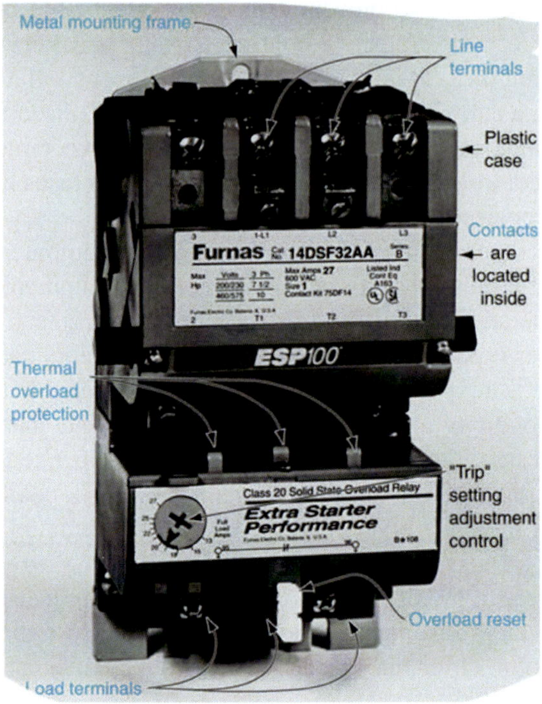

Figure 18-44 Magnetic motor starter (courtesy of Furnas Electric Co.).

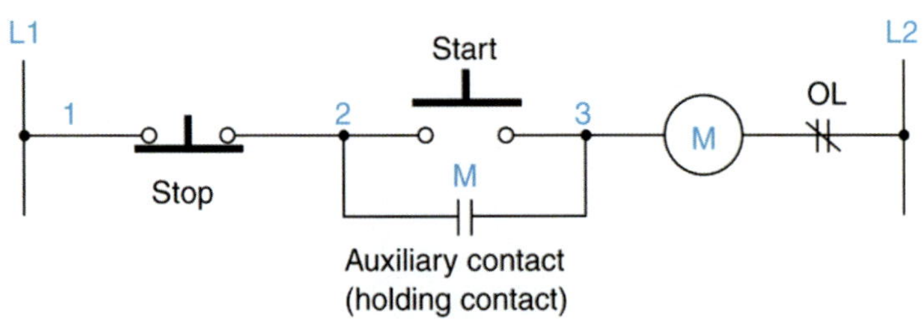

Figure 18-45 A start–stop pushbutton control circuit with overload protection (one-line diagram).

of equipment in an industrial plant. Such groupings of motor starters and associated control equipment are called control centers. Control centers (see **Figure 18-45**) provide a relatively easy access to the power distribution system, because they are compact, and the control equipment is not scattered throughout a large area.

Various types of motor starters are used for control of motors. The functions of a starter vary in complexity; however, motor starters usually perform one or more of the following functions:

1. On and off control
2. Acceleration
3. Overload protection
4. Reversing the direction of rotation

Some starters control a motor by being connected directly across the power input lines. Other starters reduce the level of input voltage that is applied to the motor when it is started, so as to reduce the value of the starting current. Ordinarily, motor overload protection is contained in the same enclosure as the starter. A popular type of motor starter used in control applications is the **combination starter**. These starters incorporate protective devices such as fused-disconnect switches, circuit breakers, or a system of fuses and circuit breakers mounted in a common enclosure.

Electronic controllers are capable of replacing electromagnetic **circuit breakers** and **relays** used for electrical control. The introduction of computerized control for equipment has brought about a new technology of machine control. Many industrial machines, such as automated manufacturing and robotic equipment, are now controlled by computerized circuits. An understanding of the basic principles of electrical control is, however, still important for technicians.

Figure 18-45 is a **start–stop pushbutton control circuit** (**one-line diagram**) with **overload protection (OL)**. Note that the **start pushbutton**

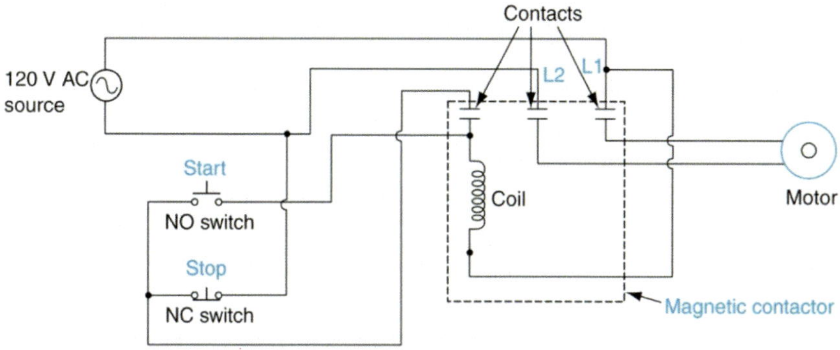

Figure 18-46 Motor-control circuit.

is **normally open (NO)**, and the **stop pushbutton** is **normally closed (NC)**. Single-phase lines L_1 and L_2 are connected across the control circuit. When the start pushbutton is pushed, a momentary contact is made between points 2 and 3. This causes the NO contact (M) to close. This results in a closed circuit between L_1 and L_2, which causes the electromagnetic coil M to be energized. When the NC stop pushbutton is pressed, the circuit between L_1 and L_2 will open. This causes contact M to open and turn the circuit off. A motor is typically connected to a control circuit as shown in **Figure 18-46**.

Criteria for Selecting Motor Controllers

Several important criteria should be considered when selecting electric motor controllers, including:

- The type of motor – AC or DC, induction, or wound rotor
- The motor ratings – voltage, current, duty cycle, and service factor
- Motor operating conditions – ambient temperature and type of atmosphere
- Utility company regulations – power factor, demand factor, load requirements, and the local codes
- Type of mechanical load connected to motor-torque requirement

To become more familiar with the criteria listed, you must be able to interpret the data on a **motor nameplate**. The information contained on a typical nameplate is summarized as follows:

- *Manufacturing Co.* – The company that built the **motor type** – A specific type of motor such as split-phase AC, universal, three-phase induction
- *Identification number* – Number assigned by the manufacturer
- *Model number* – Number assigned by the manufacturer
- *Frame type* – Frame size defined by NEMA
- *Number of phases (AC)* – Single phase or three phase
- *Horsepower* – The amount produced at rated speed
- *Cycles (AC)* – Frequency the motor should be used with (usually 60 Hz)
- *Speed (rpm)* – The amount at rated horsepower, voltage, and frequency
- *Voltage rating* – Operating voltage of motor
- *Current rating (A)* – Current drawn at rated load, voltage, and frequency

- *Thermal protection* – The type of overload protection used
- *Temperature rating (°C)* – Amount of temperature that the motor will rise over ambient temperature, when operated
- *Time rating* – Time the motor can be operated without overheating (usually continuous)
- *Amps* – Current drawn at rated load, voltage, and frequency

Summary

- Electrical energy can be converted into different forms of energy such as lighting, heating, or electromechanical.
- Common types of lighting systems are incandescent, fluorescent, vapor, and solid state.
- Common types of heating systems are resistive, inductive, and capacitive (dielectric).
- Equipment that converts electrical energy into mechanical energy is considered to be mechanical loads.
- Motors are a very common type of mechanical load.
- Motors are classified by the type of input power needed for its operation.
- Common types of DC motors are series, shunt, and compound wound.
- Common types of AC motors are single-phase, three-phase, and universal.
- Common types of single-phase AC motors are identified by their method of starting.
- Synchro systems have two or more motor-generator units connected together.
- Synchro systems are used to control the angular position of a rotor from a remote location.
- A DC stepping motor converts electrical pulses into rotary motion.
- Motor performance is determined by the amount of useful work produced by the amount of electrical energy supplied.
- Common types of motor control are speed control, starting method, short circuit protection, overload protection, and under-voltage protection.
- The nameplate of a motor identifies operational information.

Formulas

$$\text{Horsepower (hp)} = \frac{2\partial \times \text{speed} \times \text{torque}}{33{,}000}$$

$$\text{Speed (rpm)} = \frac{\text{frequency} \times 120}{\text{number of poles per phase}}$$

$$\%\ \text{slip} = \frac{\text{stator field speed} - \text{rotor speed}}{\text{stator field speed}} \times 100$$

$$\%\ \text{efficiency} = \frac{\text{Ouput power}}{\text{Input power}} \times 100$$

$$\text{Power Factor (pf)} = \frac{\text{Power (W)}}{\text{Votlage (V)} \times \text{Amperes (A)}}$$

Problems

1. Calculate the horse power of a motor rotating at 1800 rpm, and developing a torque of 10 ft-lb.

   [$\text{Horsepower (hp)} = \frac{2\pi \times \text{speed} \times \text{torque}}{33{,}000} = 0.54$ hp]

2. Determine the speed of a three-phase motor which has four poles per phase and is supplied with 220-V AC of 60 Hz.

   [$\text{Speed (rpm)} = \frac{\text{frequency} \times 120}{\text{number of poles per phase}} = 1800$ rpm].

3. Determine the slip of a three-phase AC induction motor whose stator magnetic field is rotating at 1800 rpm while the rotor is rotating at 1500 rpm.

   [$\%\ \text{slip} = \frac{\text{stator field speed} - \text{rotor speed}}{\text{stator field speed}} = 16.67\%$].

4. Determine the efficiency of a 1-ph motor which develops an output power of 600 W.

$$[\% \text{ efficiency} = \frac{\text{Ouput power}}{\text{Input power}} \times 100 = 80.4\%].$$

5. When a 500-W motor is connected to a load, the 6 A of current and 120 V. Determine the power factor of a 500-W motor connected to a load which draws 5 A of current when 120-V AC is applied.

$$[\text{Power Factor (pf)} = \frac{\text{Power (W)}}{\text{Votlage (V)} \times \text{Amperes (A)}} = 0.833].$$

Analysis and Troubleshooting

Troubleshooting lighting, heating, and motors should be a systematic process. One should have a basic understanding of the system and how it should respond under normal operating conditions. Each system has a number of unique characteristics which must be taken into account during the troubleshooting process.

Lighting systems convert electrical energy into some form of light. Damage of the lamp will cause the system to be inoperative. In most cases, a faulty lamp cannot be repaired and must be replaced. If the device is not operated within its specifications, it is more likely to become faulty.

Heating systems convert electrical energy into heat. The heat producing element of this type of system is the component that is more likely to require maintenance.

Motors convert electrical energy into a mechanical form. For effective troubleshooting of motors, one should have a good understanding of the type of motor being evaluated. With this knowledge of operation and some test equipment, motor repair can be accomplished. Some of the following instruments are used to make tests on motors: tachometers for speed tests; meggers for insulation tests; and VOMs for resistance, current, and voltage tests.

Problem Answers

1. $$\text{Horsepower (hp)} = \frac{2\pi \times \text{speed} \times \text{torque}}{33{,}000} = 0.54 \text{ hp}$$

2. $$\text{Speed (rpm)} = \frac{\text{frequency} \times 120}{\text{number of poles per phase}} = 1800 \text{ rpm}$$

3. $\% \text{ slip} = \dfrac{\text{stator field speed} - \text{rotor speed}}{\text{stator field speed}} = 16.67\%$

4. $\% \text{ efficiency} = \dfrac{\text{Ouput power}}{\text{Input power}} \times 100 = 80.4\%$

5. $\text{Power Factor (pf)} = \dfrac{\text{Power (W)}}{\text{Votlage (V)} \times \text{Amperes (A)}} = 0.833$

Self-examination/Answers

1. Four common types of lighting systems are: ______________, ______________, ______________, and ______________.
2. Three fundamental parts of a fluorescent light are ______________, ______________, and ______________.
3. Two types of vapor lamps are ______________ and ______________.
4. Three common types of heating systems are: ______________, ______________, and ______________.
5. ______________ is produced when current flows through a resistive material.Induction heating uses (alternating/direct) current.
6. Heat is produced in a (conductive/nonconductive) material using dielectric or capacitive heating.
7. Electric motors are considered to be a ____________ load device.
8. Electrical energy is converted into mechanical energy by a ______________.
9. The turning force of a motor describes its ______________.
10. Four common types of DC motors are ____________, ____________, ____________, and ______________.
11. If the DC supply voltage applied to a permanent magnet motor increases, the speed ____________.
12. The armature and field coils of a series-wound DC motor generally have a (low/high) resistance value.
13. The armature and field coils of a shunt-wound DC motor generally have a (low/high) resistance value.

14. The armature coils of a compound-wound DC motor generally have a (low/high) resistance, and the field coils have a (low/high) resistance value.
15. 18. A motor application requiring high torque is best achieved by a (series/shunt) wound motor.
16. A motor application requiring constant speed is best achieved by a (series/shunt) wound motor.
17. In a (series/shunt) wound motor, the armature and field coils are connected in parallel.
18. If the polarity of the DC voltage applied to a motor is reversed, the direction of rotation will __________________.
19. Two common types of three-phase AC motors are ___________ and ___________.
20. The windings of a three-phase motor are arranged _________° apart for generating a naturally rotating magnetic field.
21. If two of the three power lines of a three-phase AC motor are switched, the direction of rotation will ________________.
22. The stationary coil of an AC motor is generally referred to as the _____________ winding, and the moving coil is referred to as the ___________ winding.
23. A three-phase _______________ motor has a wire wound rotor that is energized by DC.
24. Three common types of single-phase AC motors are: _______________, _______________, and _________________. A _________________ motor is designed to operate on either AC or DC energy. A universal motor has the field and armature windings connected in (series/parallel).
25. When current flows through the stator windings of an AC induction motor, it causes current to be induced into the _____________ winding.
26. The stator of an AC induction motor can be regarded as the (primary/secondary) winding of a transformer and the rotor as the (primary/secondary) winding.The two stator windings of a split-phase AC motor are the __________________ and _______________.

27. A ______________ motor has a shorted copper coil on its stator.
28. The rotor of a single-phase synchronous motor is a ______________.
29. The two units connected together in a synchro system are ______________ and ______________.
30. A position change of the (generator/motor) shaft of a synchro system will cause a corresponding change in the (generator/motor) shaft at a remote location.
31. A servo motor is regarded as a DC ___________ motor. The ___________ of a DC stepping motor changes a number of degrees when a pulse is applied to it.
32. The rotor of a DC stepping motor is as ________________.
33. A purely resistive load has a power factor of ___________.
34. Three common types of motor loads connected to an electrical power system are ___________, ________________, and ________________.
35. A ____________ motor is less expensive per horsepower than other motors.
36. A ______________ is a common type of overload protection device.
37. Three common types of motor control are ____________, ______________, and ______________.
38. The criteria for achieving motor control can be found on the motor ______________ provided by the manufacturer.

Answers

1. fluorescent, incandescent, vapor, solid state
2. ballast, tube, starter
3. sodium, neon
4. resistance, induction, dielectric
5. Heat
6. Alternating
7. nonconductive
8. mechanical

9. motor
10. torque
11. permanent magnet, series, shunt, compound
12. increases
13. low
14. high
15. low, high
16. series
17. shunt
18. shunt
19. reverse
20. induction, synchronous
21. 120
22. Reverse
23. stator, rotor
24. synchronous
25. universal, induction, single-phase synchronous
26. universal
27. series
28. rotor
29. primary, secondary
30. start, run
31. shaded pole
32. permanent magnet
33. generator, motor
34. generator, motor
35. stepping
36. shaft
37. permanent magnet
38. 1
39. DC, single-phase AC, three-phase AC
40. three-phase AC
41. circuit breaker or fuse
42. speed control, starting method, short circuit, overload, under-voltage (any three)
43. nameplate

Glossary

Armature
The rotating coils of a motor that has a wound rotor.

Back EMF
See **Counter electromotive force.**

Ballast
An inductive coil placed in a fluorescent light circuit to develop a high voltage and limit the current through the tube.

Brush
A sliding contact made of carbon and graphite, connected between the commutator and the power source of a DC motor.

Capacitive heating
A method of heating nonconductive materials by placing them between two metal plates and applying a high-frequency AC voltage.

Capacitor-start motor
A single-phase AC induction motor that uses a capacitor in one winding for starting.

Commutation
The process of applying direct current of the proper polarity at the proper time to the rotor windings of a DC motor through a commutator and brushes.

Copper losses
The power losses that occur when heat is produced as electrical current flows through the copper windings of an electric machine; equal to $I^2 \times R$.

Core
Magnetic iron or steel materials used to form the laminated parts of electrical motors, generators, and other electromagnetic equipment.

Counter electromotive force (CEMF)
"Generator action" that takes place in motors when voltage is induced into the rotor conductors which opposes the source voltage (EMF).

Counter EMF
A shortened term that means **counter electromotive force** (CEMF).

Delta connection
A method of connecting three-phase alternator stator windings in which the beginning of one phase winding is connected to the end of the adjacent phase winding; power lines extend from the three beginning–end connections.

Efficiency
The ratio of output power to input power.

Field coils
Electromagnetic coils that develop the magnetic fields of electrical machines.

Field pole
Laminated metal that serves as the core material for field coils.

Fluorescent lighting
A lighting method that uses phosphor-coated lamps which are tubes filled with mercury vapor to produce light.

Horsepower
One horsepower (hp) = 33,000 ft-lb of work per minute or 550 ft-lb per second; it is the unit of measuring mechanical power; 1 hp = 746 W.

Incandescent lighting
A method of lighting that uses bulbs that have tungsten filaments which, when heated by electrical current, produce light.

Induction motor
An AC motor that has a squirrel-cage rotor and operates on the induction (transformer action) principle.

Inductive heating
A method of heating conductors in which the material to be heated is placed inside a coil of wire and high-frequency AC voltage is applied.

Laminations
Thin pieces of sheet metal used to construct the metal parts of motors and generators.

Light emitting diode (LED)
A semiconductor electronic device that emits light.

Motor
A rotating machine that converts electrical energy into mechanical energy.

Repulsion motor
A single-phase AC motor that has a wound rotor and brushes that are shorted together.

Resistive heating
A method of heating that relies on the heat produced when electrical current moves through a conductor.

Right-hand motor rule
The rule applied to find the direction of motion of the rotor conductors of a motor.

Rotor
The rotating part of an electrical generator or motor.

Self-starting
The ability of a motor to begin rotation when electrical power is applied to it.

Shaded-pole motor
A single-phase AC induction motor that uses copper shading coils around its poles to produce rotation.

Single-phase motor
Any motor that operates with single-phase AC voltage applied to it.

Slip rings
Copper rings mounted on the end of a rotor shaft and connected to the brushes and the rotor windings.

Speed regulation
The ability of a motor to maintain a steady speed with changes in load.

Split-phase motor
A single-phase AC induction motor that has start windings and run windings.

Split-ring commutator
A group of copper bars mounted on the end of a rotor shaft of a DC motor and connected to the brushes and the rotor windings.

Squirrel-cage rotor
A solid rotor used in AC induction motors, which has round metal conductors joined together at the ends and placed inside the laminated metal motor core.

Starter
A resistive network used to limit starting current in motors.

Stator
The stationary part of an electrical generator or motor.

Synchronous motor
An AC motor that operates at a constant speed regardless of the load applied.

Torque
The turning force delivered by a motor.

Universal motor
A motor that operates with either AC or DC voltage applied.

Vapor lighting
A method of lighting that uses lamps that are filled with certain gases that produce light when electrical current is applied.

Wye connection
A method of connecting three-phase alternator stator windings in which the beginnings **or** ends of each phase winding are connected together to form a common or neutral point; power lines extend from the remaining beginnings or ends.

Appendices

Appendix A: Electrical and Electronic Safety

Appendix B: Electricity/Electronic Symbols

Appendix C: Units of Measurement/Conversions

Appendix D: Tools

Appendix E: Circuit Boards and Experimentation

Appendix F: Calculator Examples

Appendix A: Electrical and Electronic Safety

Safety First!

Safety can never be stressed enough while working with electrical and electronic equipment. Since the voltages in use with electronic equipment are typically 5 V or lesser, similar to what one may experience while handling a flashlight, it is easy to disregard standard safety practices. Ensuring the safety of people and equipment is very important in a laboratory or workplace. Many dangers are not easy to see. For this reason, safety should be based on understanding basic electrical principles. **Common sense** is also important. The physical arrangement of equipment in the electrical lab or work area should be done in a safe manner. Well-designed electrical equipment should always be used. Electrical equipment is often improvised for the sake of economy. It is important that all equipment be made as safe as possible. This is especially true for equipment and circuits that are designed and built in the lab or shop. The following points should be noted with regard to the use of safety.

For People

- **Electrical Shock**: Humans are, in general, good conductors of electricity. Electricity can cause burns, and when it flows through heart cavity, it can cause the natural electrical rhythm to be disrupted.
- Be aware of what is happening around you, and, if possible, do not work alone.
- Know the location of the main breaker or fuse box in the room.
- Physically disconnect the power cable while working on high power equipment.
- Do not wear metal objects such as bangles, necklaces, or chains while working with electrical systems, as contact though them could cause a severe electrical shock.

For Measuring Equipment

- Examine the power cabling of any equipment. Use grounded power cables for equipment, especially those which require 120-V AC or higher.
- Ensure that all power is switched off while making connections to the circuit. The best way to ensure this is by physically removing the power cable to the circuit.
- Discharge any capacitors prior to use in a circuit. For this, touch their terminals a couple of times, as this will cause the potential difference between the two capacitor terminals to be neutralized. You may place a low value resistor across the leads to regulate the current flow rather than shorting the terminals.

Voltmeters

Place the meter in the proper range of the expected voltage. If needed, start at the highest range and then change the range to successively lower ones if the reading needs to be accurately determined. If the voltage to be measured is too high, and the meter is set in a much lower range of voltage measurement, the high voltage may damage the instrument.

Connect the voltmeter across the element whose voltage is to be measured.

Ammeters

Place the meter in the proper range of the expected current. If needed, start at the highest range and then change the range to successively lower ones if the reading needs to be accurately determined. If the current flow to be measured is too high, and the meter is set in a much lower range of current measurement, the high current may damage the instrument.

Even though most meters have high current protection fuses, it is best to first estimate the approximate range in which the measured value of current may lie, prior to use.

Connect the voltmeter in series (inline) with the circuit branch in which the current is to be measured.

Digital Multi-meters

- Since these instruments can read voltage, current, resistance, etc., it is important that they be set to the proper parameter, prior to use.
- Many digital multi-meters are auto-ranging, meaning thereby that they can sense the proper range of values they are measuring.
- When using a multi-meter to measure voltage, it should be used to measure across the element; and when used to measure current, it should be placed in series with the branch in which the current flow is to be measured.
- When using the multi-meter to measure resistance, ensure that the element is preferably not connected to any other circuit element. If the resistance must be measured without disconnecting the circuit, ensure that there are no voltage sources (capacitors are discharged as well) and that at least one terminal of the resistor is not electrically connected to anything. This will ensure that only the resistance of the element will be measured by the multi-meter.

Logic Probes

A logic probe is a hand-held test instrument which detects the presence or absence of voltage at key test points. The operational status of a logic circuit or device can be evaluated with this device. As a rule, LEDs are used to identify the status of signals applied to the probe. In some probes, an illuminated green LED indicates the on or 1 status of a test point. An off LED or the illumination of an alternate red LED identifies the off or 0 state. Illumination of both LEDs indicates a bad level or open condition that is somewhere between logic 0 and 1.

Manufactured logic probes resemble an oversized pen with a pointed tip. This type of probe is generally energized by the circuit under test. Clips are attached to the $+V_{cc}$ and ground of the circuit to energize the probe. The probe tip is then touched to a selected test point. If a voltage of approximately +2.4 V is present, the 1 LED will light. If the voltage at the tip is 0.6 V or less, the 0 LED will light. Some logic probes have only one LED in their construction. This type of probe illuminates the LED with +2.4 V and has the LED off when the voltage is less than +2.4 V. Manufacturers of the probe generally provide instructions that identify these features.

Logic Pulser

The logic pulser is an ideal companion for the logic probe. Pulsers generate high or low logic levels or trains of pulses. This signal can be injected into a gate while observing the response of the output with a logic probe.

Logic pulsers receive operating energy from the circuit under test. When the probe tip is touched to an input point, a pulse is injected by depressing a pushbutton. An LED generally shows when the tip is 1 or 0. The LED is on with a 1 and off with a 0. A logic pulser is also called a digital signal injector. Pulsers usually have single-pulse or continuous-train generating capabilities.

Miscellaneous Safety

Use the proper tools meant for specific purpose. Do not use a knife to function as a screwdriver.

Make sure that all meters are calibrated.

Cables

- Use insulation tape over any electrical cables which are not insulated.
- Use cables rated for the proper voltage and current. For electronics use, breadboard typically require AWG (American Wire Gauge) #22–24, solid wire.
- Do not use multi-strand wire with a breadboard as errant strands could find their way into neighboring slots and cause the circuit to malfunction.

Static Electricity

- **ESD (Electro Static Discharge)** occurs when one material which is charged comes in contact with one which has a low charge or is grounded.
- Electronic components can be adversely affected by as little as 25 V even though humans will not sense this voltage.
- No polyester.
- Use anti-static bags, anti-static straps, and anti-static mats, especially while dealing with CMOS elements.
- Best environment to prevent static buildup is hot and somewhat humid.

- Worst environment for static buildup is cool and dry.
- Keep away from strong magnetic or electro-magnetic fields.

Laboratory

All activities should be done with low voltages whenever possible. Instructions should be written, with clear directions, for performing lab activities. All lab or shop work should emphasize safety. Experimental circuits should always be checked before they are plugged into a power source. Electrical lab projects should be constructed to provide maximum safety when used.

Disconnect electrical equipment from the source of power before working on it. When testing electronic equipment, such as TV sets or other 120-V devices, an isolation transformer should be used. This isolates the chassis ground from the ground of the equipment and eliminates the shock hazard when working with 120-V equipment.

A good first-aid kit should be in every electrical shop or lab. The phone number of an ambulance service or other available medical services should be in the lab or work area in case of emergency. Any accident should be reported immediately to the proper school officials. Teachers should be proficient in the treatment of minor cuts and bruises. They should also be able to apply artificial respiration. In case of electrical shock, when breathing stops, artificial respiration must be started immediately. Extreme care should be used in moving a shock victim from the circuit that caused the shock. An insulated material should be used so that someone else does not come in contact with the same voltage. It is not likely that a high-voltage shock will occur. However, students should know what to do in case of emergency.

Normally, the human body is not a good conductor of electricity. When wet skin comes in contact with an electrical conductor, the body is a better conductor. A slight shock from an electrical circuit should be a warning that something is wrong. Equipment that causes a shock should be checked immediately and repaired or replaced. Proper grounding is important in preventing shock.

Safety devices called **ground-fault circuit interrupters (GFCIs)** are now used for bathroom and outdoor power receptacles. They have the potential of saving many lives by preventing shock. GFIs immediately cut off power if a shock occurs. The **National Electrical Code®** specifies where GFIs should be used.

Work surfaces in the shop or lab should be covered with a material that is non-conducting, and the floor of the lab or shop should also be non-conducting. Concrete floors should be covered with rubber tile or linoleum. A fire extinguisher that has a non-conducting agent should be placed in a convenient location. Extinguishers should be used with caution. Their use should be explained by the teacher.

Electrical circuits and equipment in the lab or shop should be plainly marked. Voltages at outlets require special plugs for each voltage. Several voltage values are ordinarily used with electrical lab work. Storage facilities for electrical supplies and equipment should be neatly kept. Neatness encourages safety and helps keep equipment in good condition. Tools and small equipment should be maintained in good condition and stored in a tool panel or marked storage area. Tools that have insulated handles should be used. Tools and equipment plugged into convenience outlets should be wired with three-wire cords and plugs. The purpose of the third wire is to prevent electrical shocks by grounding all metal parts connected to the outlet.

Soldering irons are often used in the electrical shop or lab. They can be a fire hazard. They should have a metal storage rack. Irons should be unplugged while not in use. Soldering irons can also cause burns if not used properly. Rosin-core solder should always be used in the electrical lab or shop.

Electricity causes many fires each year. Electrical wiring with too many appliances connected to a circuit overheats wires. Overheating may set fire to nearby combustible materials. Defective and worn equipment can allow electrical conductors to touch one another and cause a short circuit, which causes a blown fuse. It could also cause a spark or arc which might ignite insulation or other combustible materials or burn electrical wires.

Electrical Fires

Fire extinguisher of type C is approved for electrical fires. Type A is used for wood and paper; while Type B is used for flammable liquids such as petroleum-based products.

When using a fire extinguisher, aim at the base of the fire. Always face the fire and, if necessary, back away from it even while facing it. This will ensure that the user is aware if it suddenly spreads to a different location.

Breakers and Fuses

- Fuses should be of the proper rating for handling the rated load. Do not replace fuses with ones of higher rating as these might damage the equipment when rating is exceeded.
- Fuses and circuit breakers are important safety devices. When a fuse blows, it means that something is wrong in the circuit. Causes of blown fuses could be:
    - A short circuit caused by two wires touching
    - Too much equipment on the same circuit
    - Worn insulation allowing bare wires to touch grounded metal objects such as heat radiators or water pipes
- After correcting the problem, a new fuse of proper size should be installed. Power should be turned off to replace a fuse. Never use a makeshift device in place of a new fuse of the correct size. This destroys the purpose of the fuse. Fuses are used to cut off the power and prevent overheated wires.
- Circuit breakers are now very common. Circuit breakers operate on spring tension. They can be turned on or off like wall switches. If a circuit breaker opens, something is wrong in the circuit. Locate and correct the cause and then reset the breaker.
- Always remember to use common sense whenever working with electrical equipment or circuits. Safe practices should be followed in the electrical lab or shop as well as in the home. Detailed safety information is available from the National Safety Council and other organizations. It is always wise to be safe.

Power Conditions

Being aware of the kinds of power problems which can occur can be useful in identifying what can be done to solve the problem.

Blackout

This occurs when there is a complete power failure, and no power is available.

Brownout

This occurs when there is an under-voltage in the system which lasts for a short duration (few seconds).

Sag

This occurs when there is an under-voltage in the system which lasts for a very short duration (milliseconds).

Spike

This occurs when there is an over-voltage in the system which lasts for a very short duration (milliseconds).

Surge

This occurs when there is an over-voltage in the system which lasts for a short duration (seconds).

Maintaining Power

Surge Suppressers/Protectors

These are useful against short-term over-voltages and offer some protection for a limited number of such incidents. These devices often include an indicator which indicates whether it is functioning properly.**Power Conditioners.** These are useful for monitoring the quality of power being received.

UPS (Uninterruptible Power Supply)

These supply power to the equipment they are connected to by first converting the available AC power from the room wall output into DC and then recreating the AC waveform. Owing to this conversion, line conditioning circuits are often incorporated into the UPS, to remove any electrical surges and to smoothen out the resulting AC. These devices have a VA (Voltage × Ampere) rating which should not be exceeded. A 500-VA rating, for example, means that equipment running on 110 V should not consume more than 4.75 A. The UPS keeps functioning for a short while even after there is a blackout, giving users enough time to perform shutdown operations on their equipment.

Lighting

Adequate lighting is needed while performing all electronic wiring and testing.

Adequate laboratory space is needed to reduce the possibility of accidents.

Proper ventilation, heat, and light also provide a safe working environment.

Circuit Connections

- Always use a circuit diagram while connecting up a circuit. Label all component values and mark voltage and ground connections.
- Keep the length of all wires in the circuit to a minimum.
- Use components well within their rated range. For example, a low resistance, low wattage resistor (say 10 Ω, 1/4 W) should not be used directly across with a 120-V AC supply.
- Discharge capacitors prior to connecting them in a circuit.
- Take static precautions.
- Wiring in the electrical lab or shop should conform to specifications of the National Electrical Code® (NEC®). The NEC® governs all electrical wiring in buildings.

Humidity, Dust, and Smoke

- Dust may clog equipment and cause heat buildup.
- Avoid direct sunlight for overheating purpose.
- Not too humid; else, there will be condensation on the equipment which could cause shorting owing to conductive path provided by the water.
- Not too dry; else, there will be a possibility of static electricity.
- Humidity between 40%–60% is recommended for electrical equipment.

Hazardous Material

Handling and disposal of any potentially hazardous equipment, cleaning agents, solvents, chemicals, etc., should be follower as per their MSDS (Material Safety Data Sheets).

Appendix B: Electrical/Electronic Symbols

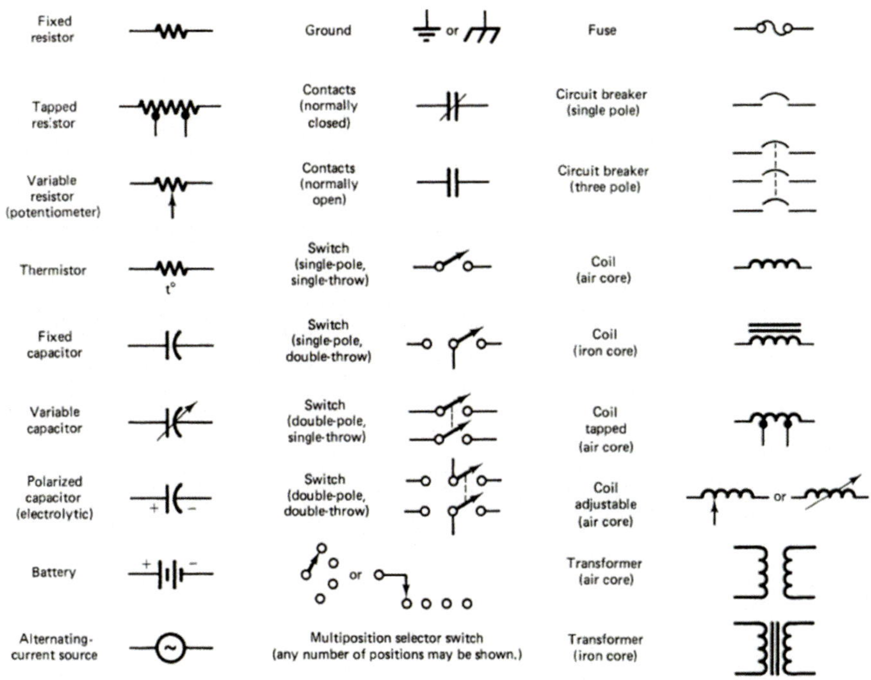

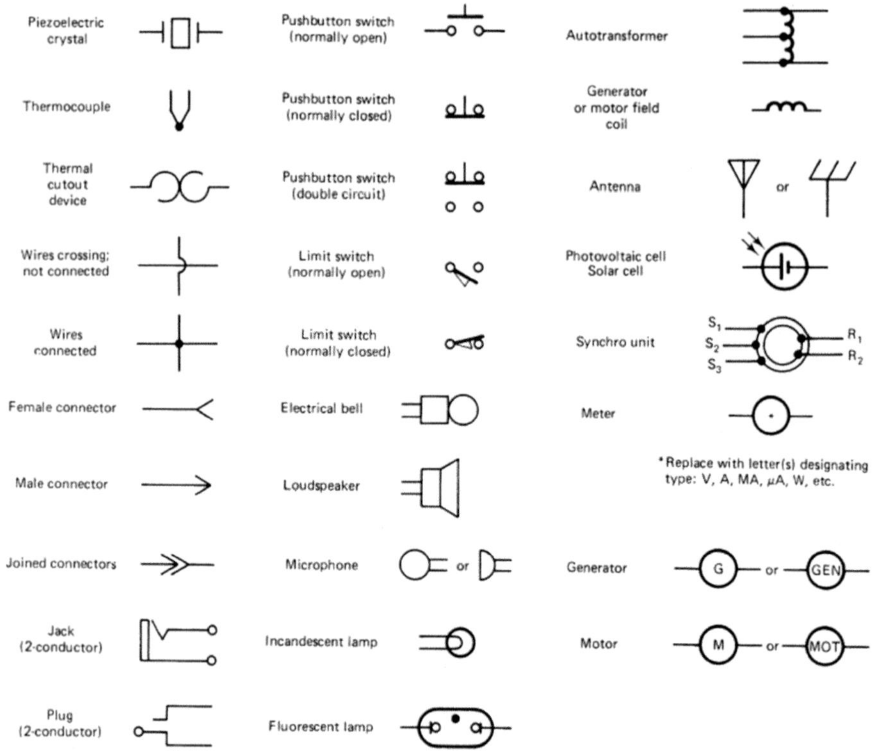
Piezoelectric crystal
Thermocouple
Thermal cutout device
Wires crossing; not connected
Wires connected
Female connector
Male connector
Joined connectors
Jack (2-conductor)
Plug (2-conductor)
Pushbutton switch (normally open)
Pushbutton switch (normally closed)
Pushbutton switch (double circuit)
Limit switch (normally open)
Limit switch (normally closed)
Electrical bell
Loudspeaker
Microphone
or
Incandescent lamp
Fluorescent lamp
Autotransformer
Generator or motor field coil
Antenna
or
Photovoltaic cell Solar cell
Synchro unit
S1
S2
S3
R1
R2
Meter
*Replace with letter(s) designating type: V, A, MA, μA, W, etc.
Generator
G
or
GEN
Motor
M
or
MOT

Appendix C: Units of Measurement/ Conversions

This appendix lists some typical means of converting quantities in the English system into values in the metric system (and vice versa). **Tables C-1** through **C-6** list the typical multiplying factor for various units. However, temperature conversions require specific formulas. These formulas follow:

Temperature Conversions

1. To convert degrees Fahrenheit to degrees Celsius, use one of the following formulas:

$$°C = \frac{5}{9}(°F - 32°)$$

or

$$°C = \frac{°F - 32°}{1.8}.$$

2. To change degrees Celsius to degrees Fahrenheit, use one of the following formulas:

$$°F = \frac{9}{5}(°C)+32°)$$

or

3. Degrees Celsius may be converted to Kelvins by using $°C = K - 273.16°$.
4. Kelvins can be converted to degrees Celsius by simply changing the formula in Step 3 to: $K = °C + 273.16°$.
5. If you wish to change degrees Fahrenheit to degrees Rankine, use the following formula: $°F = °R + 495.7°$.
6. Then, by changing the preceding formula, degrees Rankine can be converted to degrees Fahrenheit:

°R = °F + 459.7°.

By using a combination of the above-listed formulas, temperature values given for any scale can be changed to a temperature on one of the three other scales. **Figure C-1** gives a comparison of the various temperature scales.

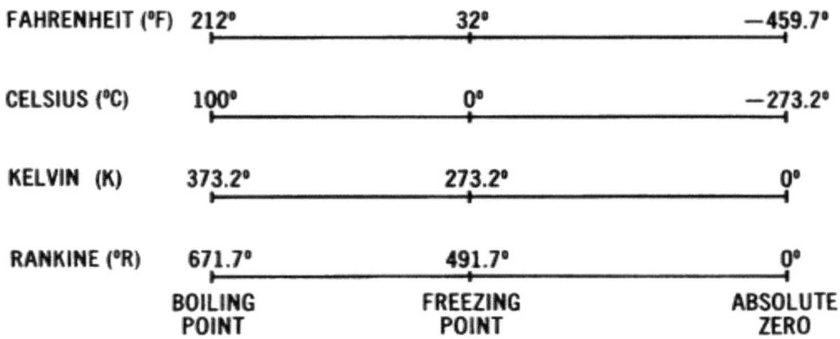

Figure C-1 Comparison of temperature scales.

Table C-1 Length.

Known Quantity	Multiply by	Quantity To Find
inches (in)	2.54	centimeters (cm)
feet (ft)	30	centimeters (cm)
yards (yd)	0.9	meters (m)
miles (mi)	1.6	kilometers (km)
millimeters (mm)	0.04	inches (in)
centimeters (cm)	0.4	inches (in)
meters (m)	3.3	feet (ft)
meters (m)	1.1	yards (yd)
kilometers (km)	0.6	miles (mi)
centimeters (cm)	10	millimeters (mm)
decimeters (dm)	10	centimeters (cm)
decimeters (dm)	100	millimeters (mm)
meters (m)	10	decimeters (dm)
meters (m)	1000	millimeters (mm)
dekameters (dam)	10	meters (m)
hectometers (hm)	10	dekameters (dam)
hectometers (hm)	100	meters (m)
kilometers (km)	10	hectometers (hm)
kilometers (km)	1000	meters (m)

Table C-2 Area.

Known Quantity	Multiply by	Quantity To Find
square inches (in^2)	6.5	square centimeters (cm^2)
square feet (ft^2)	0.09	square meters (m^2)
square yards (yd^2)	0.8	square meters (m^2)
square miles (mi^2)	2.6	square kilometers (km^2)
acres	0.4	hectares (ha)
square centimeters (cm^2)	0.16	square inches (in^2)
square meters (m^2)	1.2	square yards (yd^2)
square kilometers (km^2)	0.4	square miles (mi^2)
hectares (ha)	2.5	acres
square centimeters (cm^2)	100	square millimeters (mm^2)
square meters (m^2)	10,000	square centimeters (cm^2)
square meters (m^2)	1,000,000	square millimeters (mm^2)
ares (a)	100	square meters (m^2)
hectares (ha)	100	ares (a)
hectares (ha)	10,000	square meters (m^2)
square kilometers (km^2)	100	hectares (ha)
square kilometers (km^2)	1,000,000	square meters (m^2)

Table C-3 Cubic measure.

Known Quantity	Multiply by	Quantity To Find
cubic meters (m^3)	35	cubic feet (ft^3)
cubic meters (m^3)	1.3	cubic yards (yd^3)
cubic yards (yd^3)	0.76	cubic meters (m^3)
cubic feet (ft^3)	0.028	cubic meters (m^3)
cubic centimeters (cm^3)	1000	cubic millimeters (mm^3)
cubic decimeters (dm^3)	1000	cubic centimeters (cm^3)
cubic decimeters (dm^3)	1,000,000	cubic millimeters (mm^3)
cubic meters (m^3)	1000	cubic decimeters (dm^3)
cubic meters (m^3)	1	steres
cubic feet (ft^3)	1728	cubic inches (in^3)
cubic feet (ft^3)	28.32	liters (L)
cubic inches (in^3)	16.39	cubic centimeters (cm^3)
cubic meters (m^3)	264	gallons (gal)
cubic yards (yd^3)	27	cubic feet (ft^3)
cubic yards (yd^3)	202	gallons (gal)
gallons (gal)	231	cubic inches (in^3)

Table C-4 Volume.

Known Quantity	Multiply by	Quantity To Find
milliliters (mL)	0.03	fluid ounces (fl oz)
liters (L)	2.1	pints (pt)
liters (L)	1.06	quarts (qt)
liters (L)	0.26	gallons (gal)
gallons (gal)	3.8	liters (L)
quarts (qt)	0.95	liters (L)
pints (pt)	0.47	liters (L)
cups (c)	0.24	liters (L)
fluid ounces (fl oz)	30	milliliters (mL)
teaspoons (tsp)	5	milliliters (mL)
tablespoons (tbsp)	15	milliliters (mL)
liters (L)	1000	milliliters (mL)

Table C-5 Electric.

Known Quantity	Multiply by	Quantity To Find
Btu per minute	0.024	horsepower (hp)
Btu per minute	17.57	watts (W)
horsepower (hp)	33,000	foot-pounds per min (ft-lb/min)
horsepower (hp)	746	watts (W)
kilowatts (kW)	57	Btu per minute
kilowatts (kW)	1.34	horsepower (hp)
watts (W)	44.3	foot-pounds per min (ft-lb/min)

Appendix D: Tools

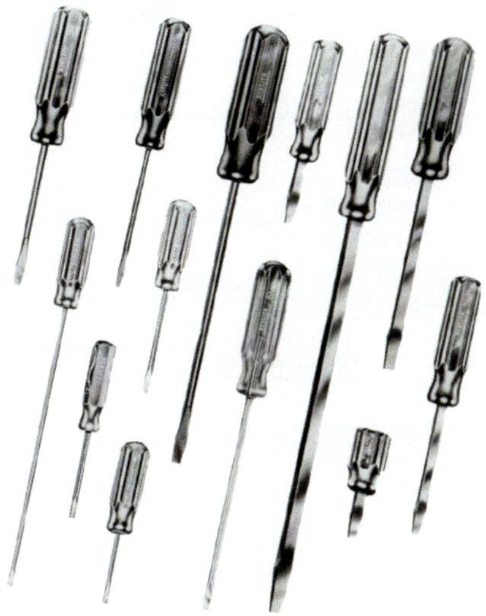

Figure D-1 Various sizes of standard screwdrivers (courtesy of The Cooper Group).

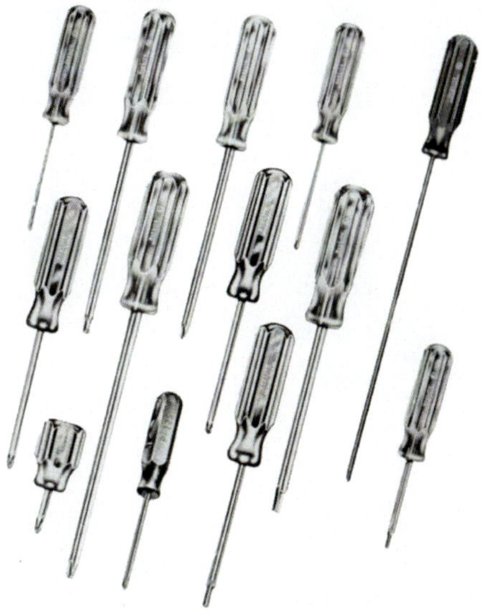

Figure D-2 Various sizes of Phillips screwdrivers (courtesy of The Cooper Group).

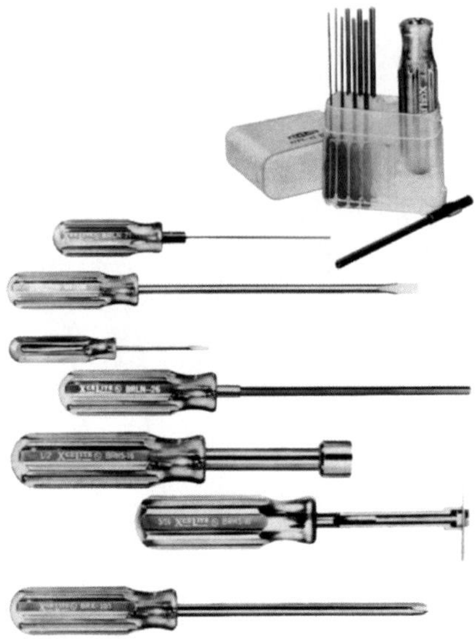

Figure D-3 Miscellaneous: Allen setscrew drivers (top), standard screwdrivers; nut drivers; Phillips screwdriver (bottom) (courtesy of The Cooper Group).

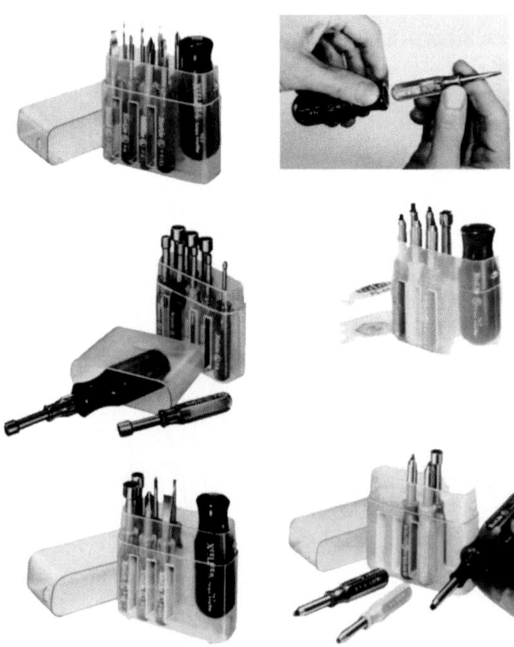

Figure D-4 Driver combination kits (courtesy of The Cooper Group).

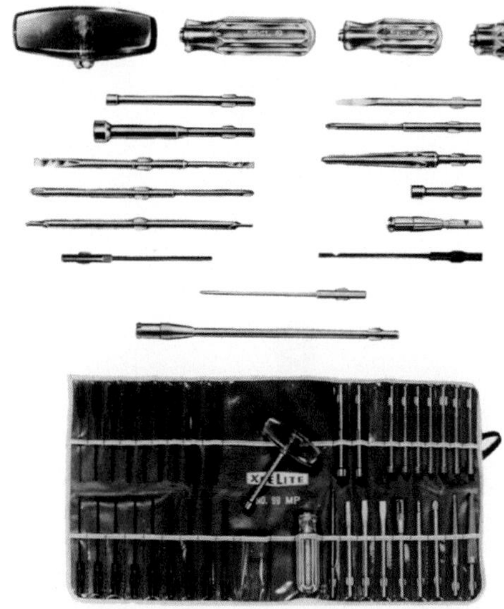

Figure D-5 Screwdriver kit (courtesy of The Cooper Group).

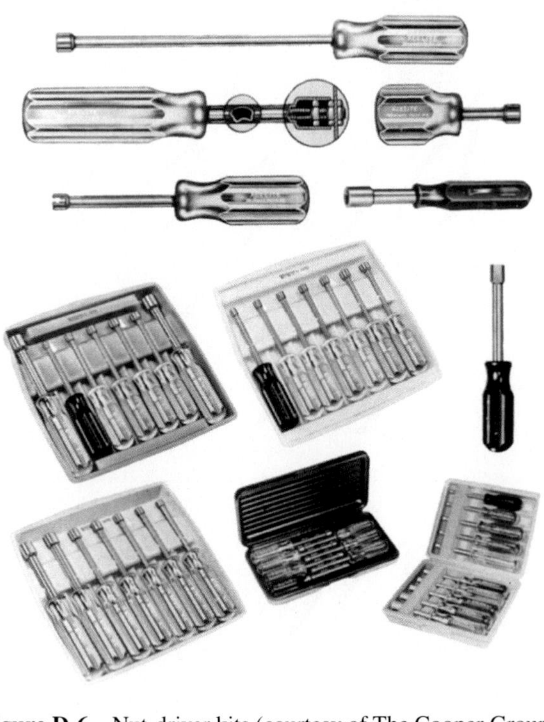

Figure D-6 Nut-driver kits (courtesy of The Cooper Group).

Figure D-7 (top) Round-nose diagonal wire cutters; (bottom) wire strippers (courtesy of Hunter Tools, Division of K-D Manufacturing Co.).

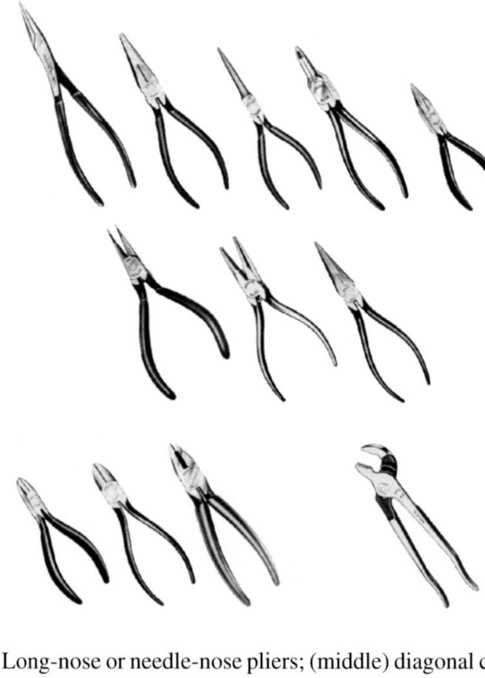

Figure D-8 (top) Long-nose or needle-nose pliers; (middle) diagonal cutting pliers; (bottom) channel lock pliers (courtesy of The Cooper Group).

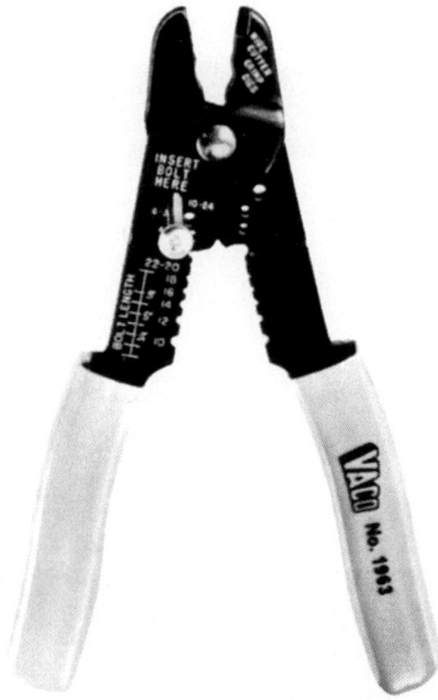

Figure D-9 Wire crimping tool with stripping and bolt-cutting capability (courtesy of Vaco Products Co.).

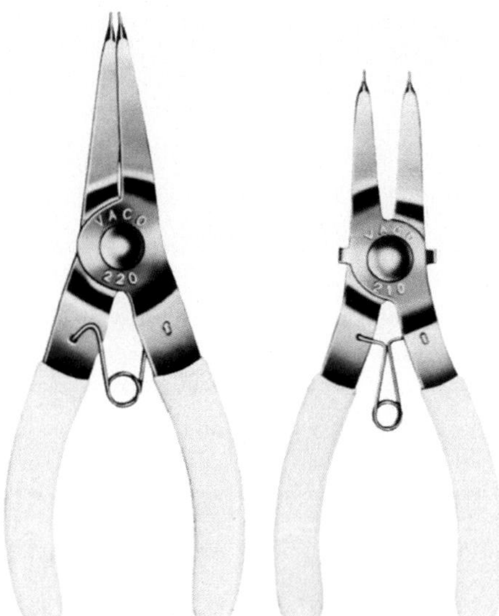

Figure D-10 C-ring pliers (Courtesy of Vaco Products Co.).

Figure D-11 Cheater cord for electric equipment repair (courtesy of GC Electronics).

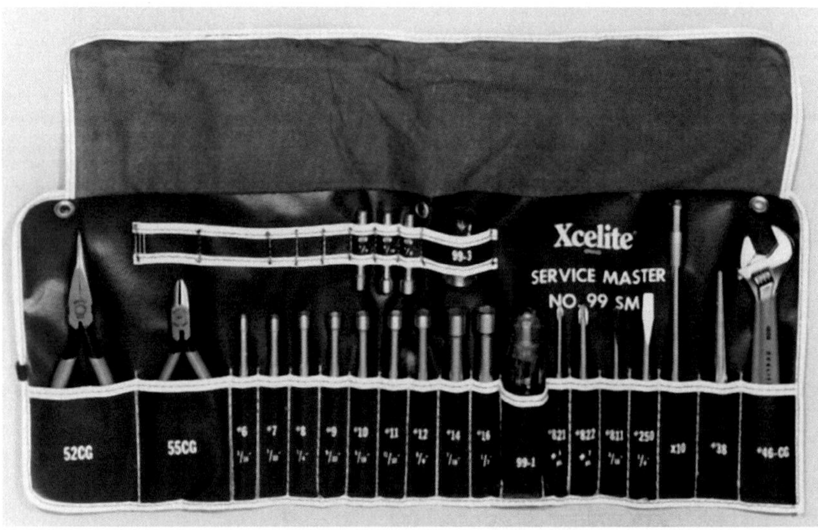

Figure D-12 Roll of tools for servicing (courtesy of The Cooper Group).

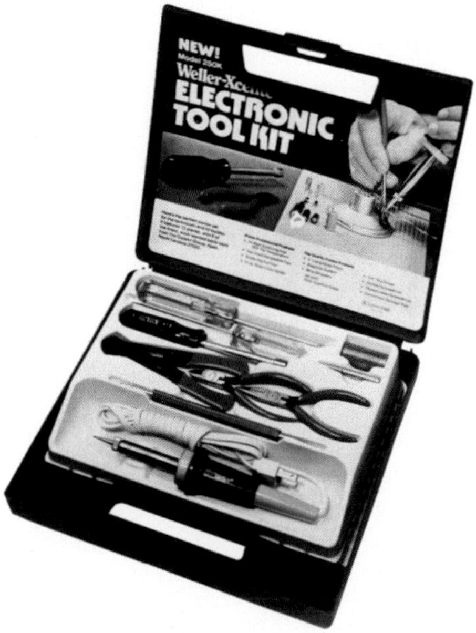

Figure D-13 Electronic tool kit (courtesy of The Cooper Group).

Figure D-14 Portable multipurpose tool (courtesy of The Cooper Group).

Appendix E: Circuit Boards and Experimentation

Sometimes referred to as solder-less breadboards, prototyping boards allow electronics designers, engineers, or students the ability to easily build, test, and evaluate circuits. By design, conductors can be pushed in to the board, rather than a permanent attachment. As a result of spacing, holes for the connectors at 0.1 inches, these boards are compatible with most components with axial or radial leads such as resistors, capacitors, diodes, and transistors, as well as dual inline packages such as integrated circuits. Leads of the components may be pushed through the opening in the board, making contact with a conductive strip below. The conductive strip is somewhat "V" shaped, allowing the component to be held in place, and is suitable for thousands of uses. Additional connections to the electrical components or external equipment may be made by hook-up wire. This wire is typically a 20–28 gauge solid wire, with 22 gauge solid wire being the most popular. This hook-up or jumper wire can be purchased, stripped in precut lengths, or prepared by the user. **Figure E-1** represents a typical solder-less prototyping board.

. Features may also include binding posts, boards that connect to each other to make larger boards, special plating, and conductive backings. This design of solder-less prototyping has become so popular that some manufacturers are including this type of board in the design of various pieces of laboratory and training equipment. The board shown in **Figure E-1** is an example of how a typical board is laid out. Other layouts may vary in the number of rows or columns of connectors. Before beginning a prototyping project on an unfamiliar board, it is good practice to consult the documentation supplied with the board or make a few continuity checks before wiring any circuits.

Figures E-2 through **E-4** show some typical connections using resistors. The remaining figures show tools and equipment used for soldering and electrical and electronic experimentation in a laboratory or the home.

Figures E-5 through **E-11** show illustrations of soldering tools and equipment for experimentation with electrical and electronic circuits.

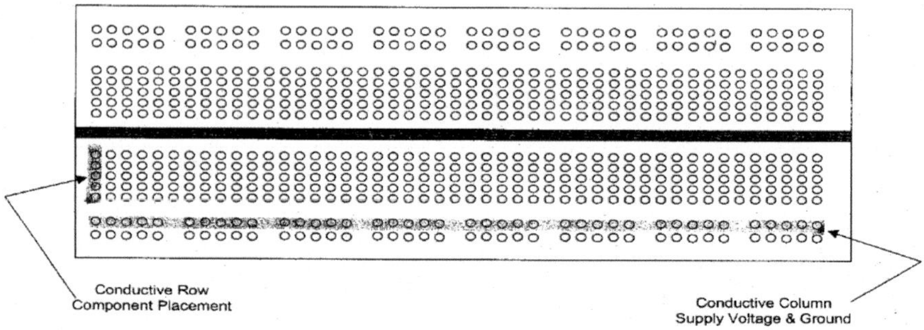

Figure E-1 Experimental protoboard for constructing electrical circuits.

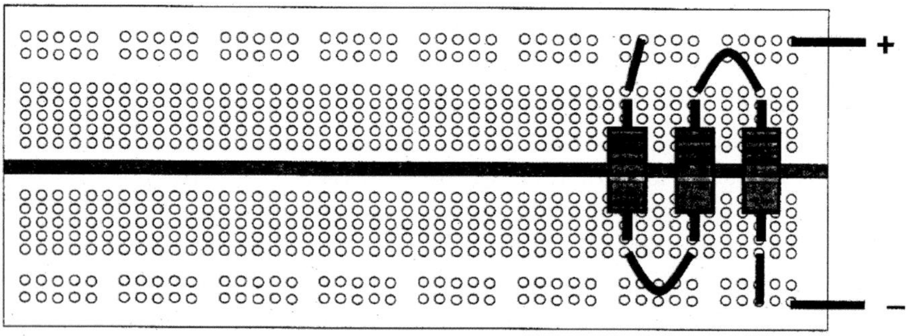

Figure E-2 Experimental protoboard with series-connected electrical circuit.

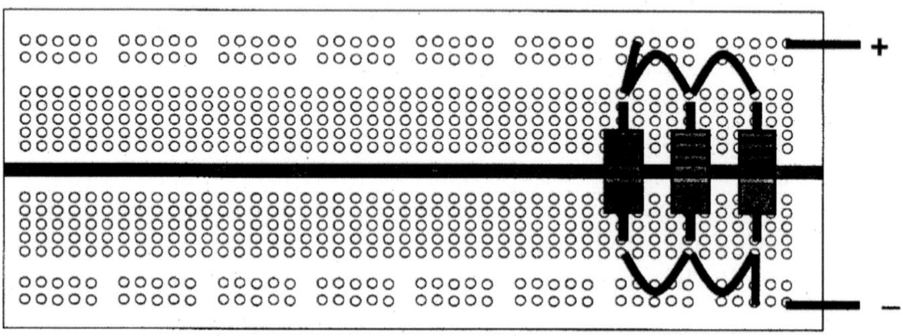

Figure E-3 Experimental protoboard with parallel-connected electrical circuit.

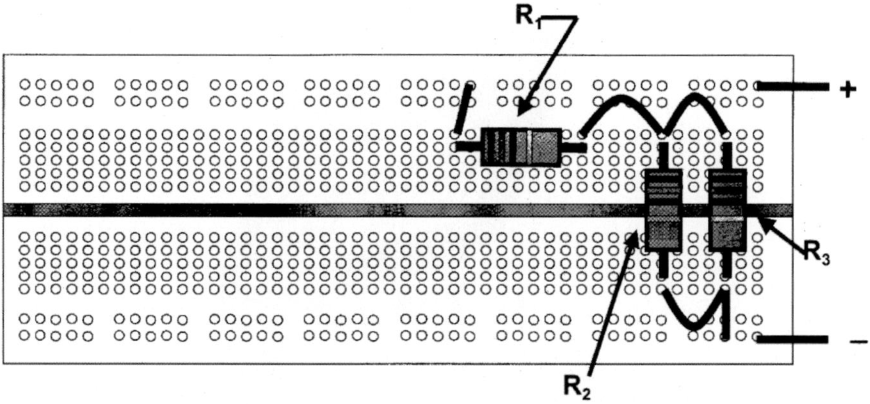

Figure E-4 Experimental protoboard with combination (series–parallel) electrical circuit.

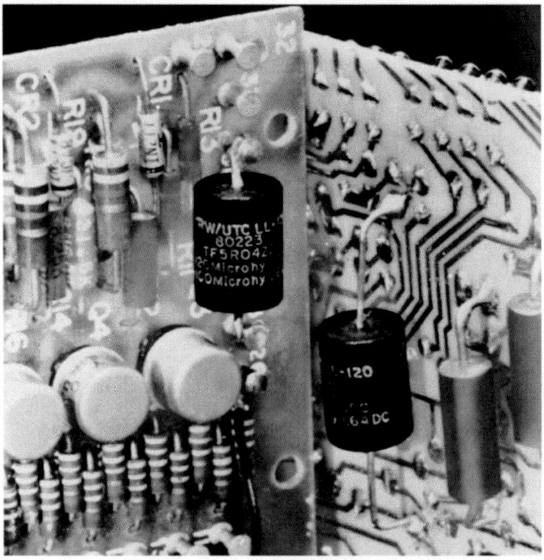

Figure E-5 Printed circuit board (PCB) (courtesy of TRW/UTC Transformers).

Figure E-6 Low-wattage soldering iron kit with several soldering tips (courtesy of The Cooper Group).

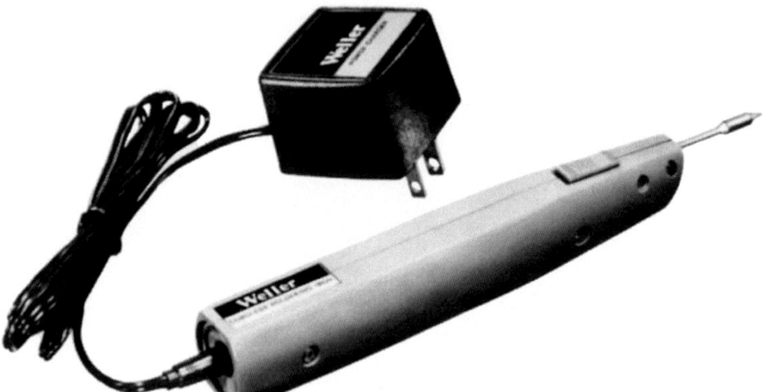

Figure E-7 Cordless soldering iron with charger (courtesy of The Cooper Group).

Figure E-8 Cordless soldering iron kit (courtesy of The Cooper Group).

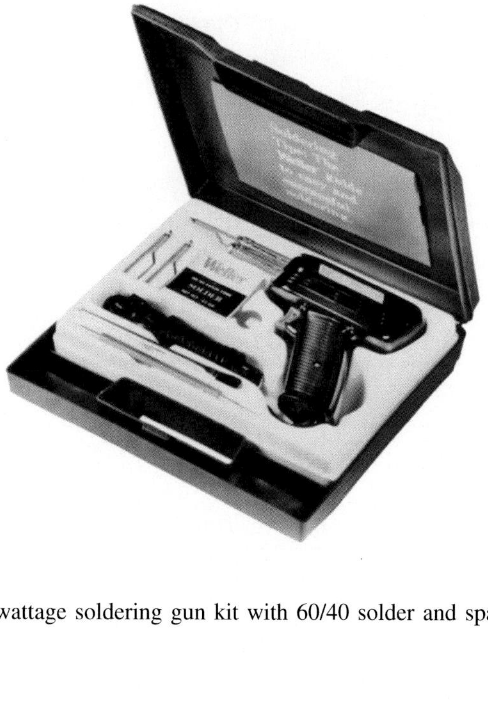

Figure E-9 High-wattage soldering gun kit with 60/40 solder and spare tips (courtesy of The Cooper Group).

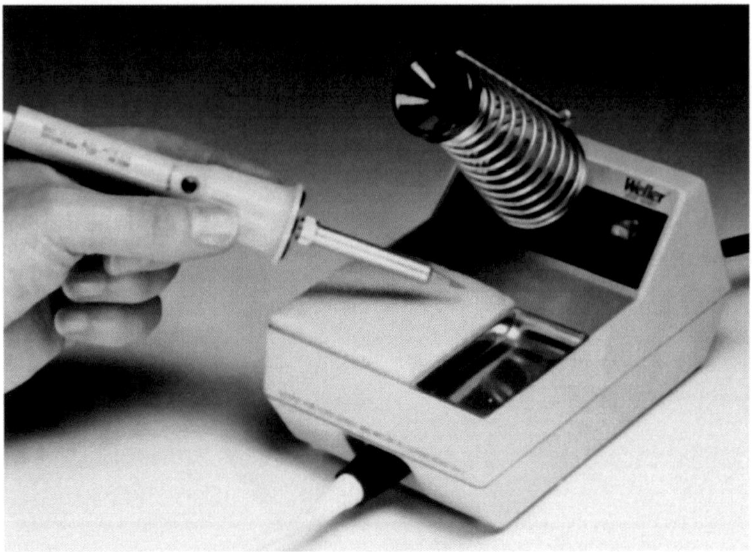

Figure E-10 High-wattage soldering gun kit (courtesy of The Cooper Group).

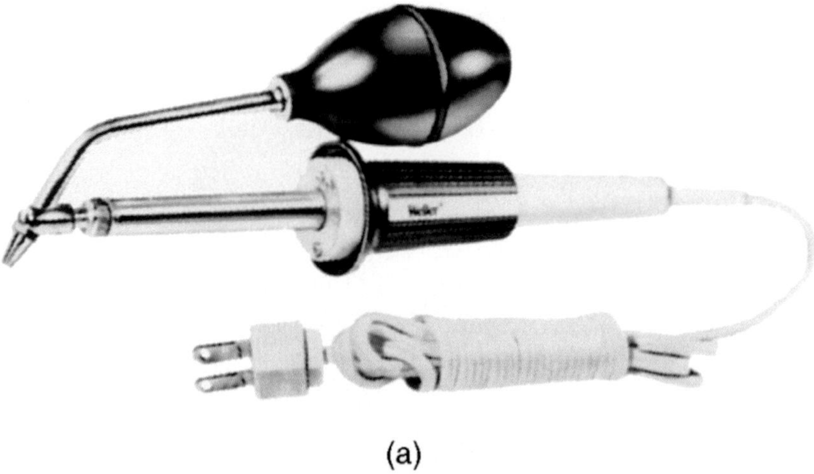

(a)

Figure E-11 (a) Desoldering tool; (b) desoldering bulb (courtesy of The Cooper Group).

Appendix F: Calculator Examples

This appendix is used to show the effective use of scientific calculators as a necessary tool for solving problems. The examples and procedures that follow illustrate how to use a typical scientific calculator. These procedures will work on most calculators with a few changes in button sequence.

OPERATING CONTROLS

[OFF ON]

Power switch

When the power switch is pushed to "ON", the calculator is ready for operation.

Automatic Power-Off Function (APO)

This calculator is automatically turned off approximately *6 minutes after the last key operation to save your batteries.* This time varies ±2 minutes depending upon the operating temperature or condition of the battery.

[ON/C]

Power on/clear key

Turns on the machine when the automatic power-off function is in effect and when pushed during operation it clears the calculator except for the memory.

[0] ~ [9] [·]

Numeral keys

Used to enter numbers.

Ex. 12.3 → [1] [2] [·] [3]

0.7 → [·] [7]

[+/−]

Change sign key

Changes the sign of the number displayed from a positive to a negative or vice versa.

[+] [−] [X] [÷]

Arithmetic function keys

Depressed for addition, subtraction, multiplication and division.

[=]

Equals key

Completes four arithmetic calculations (+, −, ×, ÷), $\sqrt[x]{y}$ and y^x calculations.

[2nd F]

2nd function designation key

Note: If pressed after the [hyp] (arc hyp) key, the function of the [hyp] (arc hyp) key will be reset and the function of the [2nd F] key will be set.

Ex. .5 [hyp] [2nd F] [sin] → $\sin^{-1} 0.5$

30 [2nd F] [hyp] [2nd F] [sin] → sin 30

[DRG]

Degree/Radian/Grad selection key

Used for calculation of trigonometric, inverse trigonometric and coordinate conversion. The [DRG] key changes the angular mode.

Figure F-1 Instructions for use of a scientific calculator (courtesy of Sharp Co.).

(When the calculator is turned on, "DEG" mode is set and the symbol "DEG" is displayed.)

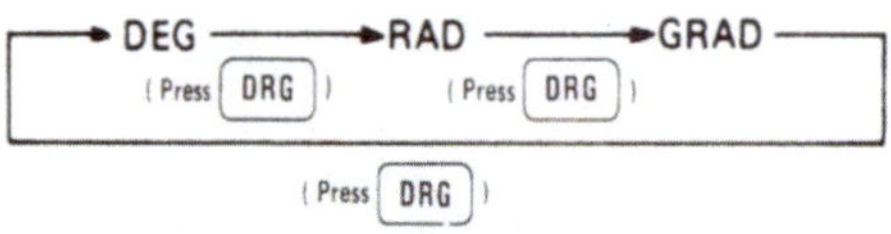

Ex. DEG → GRAD: Press the key twice (DRG DRG)

"DEG" mode – Entries and answers are in decimal degrees.

"RAD" mode – Entries and answers are in radians.

"GRAD" mode – Entries and answers are in grads. ($100^g = 90^\circ = \frac{\pi}{2}$).

TAB / F↔E

Display format exchange/tablation key

F↔E : When a calculation result is displayed in the floating decimal point system, pushing the key displays the result in the scientific notation system.

: Pushing the key once more displays the result in the floating decimal point system again.

2nd F TAB

n! / CE

Clear entry/factorial key

CE : Used to clear an incorrectly entered number.

2nd F n! : Calculates the factorial of the displayed number.

Factorial of n (n!) = n · (n − 1) · (n − 2) ····· 2 · 1

$\sin^{-1}$ / sin, $\cos^{-1}$ / cos, $\tan^{-1}$ / tan

Trigonometric/inverse trigonometric function key

→D.MS / →DEG

Degree/minute/second ↔ Decimal degrees conversion key

arc hyp / hyp

Hyperbolic/arc hyperbolic key

$^x\sqrt{y}$ / y^x

$y^x/^x\sqrt{y}$ key

y^x : Raises a number to a power.

2nd F $^x\sqrt{y}$: Calculates the Xth root of Y.

Figure F-1 (continued) Instructions for use of a scientific calculator (courtesy of Sharp Co.).

[ln]

Natural logarithm/antilogarithm key

[ln] : Used to obtain the logarithm base e (e = 2.718281828).

[2nd F] [e^x] : Calculates the antilogarithm base e of the displayed number.

10^x
[log]

Common logarithm/antilogarithm key

[log] : Used to obtain the logarithm with the base of 10.

[2nd F] [10^x] : Calculates the antilogarithm base 10 of the displayed number.

π
[EXP]

Exponent/Pi key

[EXP] : Used to enter the exponent of a number when working in scientific notation.

[2nd F] [π] : The constant π (π = 3.141592654) is entered.

$\sqrt[3]{}$
[$\sqrt{}$]

Square root/cube key

[$\sqrt{}$] : Calculates the square root of the number displayed.

[2nd F] [$\sqrt[3]{}$] : Calculates the cube root of the number displayed.

$\Delta\%$
[x^2]

Square/percent change key

[x^2] : Calculates a square of the number displayed.

[2nd F] [$\Delta\%$] : Performs the percentage or percent increase calculation.

$\rightarrow r\theta$
[$1/x$]

Reciprocal/rectangular coordinate → polar coordinate conversion key

[$1/x$] : Calculates the reciprocal of the number displayed.

[2nd F] [$\rightarrow r\theta$] : Converts rectangular coordinate into polar coordinate.

$\rightarrow xy$
[(]

Open parenthesis/polar coordinate → rectangular coordinate conversion key

[(] : Used to open parenthesis.

[2nd F] [$\rightarrow xy$] : Converts polar coordinate into rectangular coordinate.

STAT
[$]

Exchange/statistical calculation mode key

[$] : Used to exchange the number being displayed with the number stored in the working register. ($x \leftrightarrow y$)

[2nd F] [STAT] : Statistical program will be activated.

Figure F-1 (continued) Instructions for use of a scientific calculator (courtesy of Sharp Co.).

When the calculator is set to the statistical calculation mode through this key, the symbol "STAT" appears, and at the same time the entire machine is cleared. Mean-while, in the statistical calculation mode the [ )], [x→M], [RM] and [M+] keys work as the [n], [$\bar{x}$], [S] and [DATA] keys, respectively. And pushing these keys immediately after the [2nd F] key they work as the [Σx], [Σx^2], [σ] and [CD] keys.

$n\Sigma x$ [ )]

Close parenthesis/nΣx key

- Used to close parenthesis.
- When the statistical mode is set,

  [$n\Sigma x$]: Displays the number of samples entered.

  [2nd F] [$n\Sigma x$]: Used to obtain the sum of the data (Σx).

$\bar{x}\Sigma x^2$ [x→M]

Memory-in/$\bar{x}$ Σx^2 key

- Clears the number in the memory and then store the number being displayed in the memory.

  To clear the memory depress the [ON/C] key followed by the [x→M] key.

  (This means to enter zero (0) into the memory)
- When the statistical calculations mode is set,

  [$\bar{x}\Sigma x^2$]: Used to obtain the mean value of the data ($\bar{x}$).

  [2nd F] [$\bar{x}\Sigma x^2$]: Used to obtain the sum of x^2 (Σx^2).

[RM]

Recall memory/standard deviation key

- Displays the contents of the memory. The contents of the memory remain unchanged after this key operation.
- When the statistical mode is set,

  [Sσ]: Used to obtain the standard deviation of the samples (S).

  [2nd F] [Sσ]: Used to obtain the standard deviation of the population (σ).

DATA CD [M+]

Memory plus/DATA CD key

- Used to add the number being displayed or a calculated result to the contents of the memory.

  When subtracting a number from the memory, press the [+/-] and [M+] keys in this order.
- When the statistical mode is set,

  [DATA CD]: Used to enter the data (numbers).

  [2nd F] [DATA CD]: Used to correct a mis-entry. (Delete function).

Figure F-1 (continued) Instructions for use of a scientific calculator (courtesy of Sharp Co.).

DISPLAY

(1) Display format

(2) Symbols

– : **Minus symbol**
Indicates that the number in the display following the "—" is a negative.

M : **Memory symbol**
Appears when a number is stored in the memory.

E : **Error symbol**
Appears when an overflow or an error is detected.

• : **Battery indicator**
The battery indicator is a grey dot located at the left side of the display. When this dot is not on, the batteries must be replaced.

2ndF : **2nd function designation symbol**
Appears when the 2nd function is designated.

HYP : **Hyperbolic function designation symbol**
Appears when hyperbolic function is designated.

DEG : **Degree mode symbol**
Appears when the degree mode is designated.

RAD : **Radian mode symbol**
Appears when the radian mode is designated.

GRAD: **Grad mode symbol**
Appears when the grad mode is designated.

STAT : **Statistical calculation mode symbol**
Appears when statistical calculation mode is set.

( ) **Parenthesis symbol**
Appears when a calculation with parenthesis is performed by pressing the [( ] key.

(3) Display system
All answers will be displayed in either floating decimal point system or scientific notation system.

- The answer in the following area will be displayed in floating decimal point system.
  $0.000000001 \leq |x| \leq 9999999999$

Figure F-1 (continued) Instructions for use of a scientific calculator (courtesy of Sharp Co.).

Example	Operation	Readout
5 + 3 = 8	5 [+] 3 [=]	8.
2.3 + 6.8 = 9.1	2 [·] 3 [+] 6 [·] 8 [=]	9.1
10 + 14 + 5 = 29	10 [+] 14 [+] 5 [=]	29.

Figure F-2 Calculator procedures for addition.

Example	Operation	Readout
42 − 13 = 29	42 [−] 13 [=]	29.
400 − 0.25 = 399.75	400 [−] [·] 25 [=]	399.75

Figure F-3 Calculator procedures for subtraction.

Example	Operation	Readout
150 × 50 = 7500	150 [×] 50 [=]	7500.
6.5 × 86 = 559	6.5 [×] 86 [=]	559.

Figure F-4 Calculator procedures for multiplication.

Example	Operation	Readout
$\frac{600}{3} = 200$	600 [÷] 3 [=]	200.
$\frac{487}{5} = 97.4$	487 [÷] 5 [=]	97.4

Figure F-5 Calculator procedures for division.

Example	Operation	Readout
$\frac{3}{4} = 0.75$	3 [÷] 4 [=]	0.75
$\frac{32}{3} = 10.67$	32 [÷] 3 [=]	10.66666667
$\frac{3}{4} \times \frac{2}{5} = \frac{3}{10}$	3 [÷] 4 [X] 2 [÷] 5 [=]	0.3
$\frac{3}{4} \div 8 = \frac{3}{4} \times \frac{1}{8} = \frac{3}{32}$	3 [÷] 4 [X] 1 [÷] 8 [=]	0.09375
$8 \div \frac{3}{4} = 8 \times \frac{4}{3} = \frac{32}{3} = 10\frac{2}{3}$	8 [X] 4 [÷] 3 [=]	10.66666667
$\frac{3}{4} \div \frac{1}{3} = \frac{3}{4} \times \frac{3}{1} = \frac{9}{4} = 2\frac{1}{4}$	3 [÷] 4 [X] 3 [=]	2.25

Figure F-6 Calculator procedures for fractions.

Example	Operation	Readout
$3^2 = 3 \times 3 = 9$	3 [x^2]	9.
$5^5 = 5 \times 5 \times 5 \times 5 \times 5 = 3125$	5 [y^x] 5 [=]	3125.
$\sqrt{64} = 8$	64 [$\sqrt{\ }$]	8.
$64^{½} = 64^{0.5} = 8$	64 [y^x] [·] 5 [=]	8.
$\sqrt[4]{16} = 2$	16 [2nd F] [$\sqrt[x]{y}$] 4 [=]	2.

Figure F-7 Calculator procedures for powers and roots.

Example	Operation	Readout
Percent: $\frac{23}{50} \times 100 = 46\%$	[2nd F] [Δ%]	46.
Percent of: $200 \times 15\% = 30$	200 [X] 15 [2nd F] [Δ%]	30.
Percent increase: An increase from 10 volts to 12 volts $\frac{12-10}{10} \times 100 = 20\%$ increase	12 [−] 10 [2nd F] [Δ%]	20.

Figure F-8 Calculator procedures for percentage.

Example	Operation	Readout
Amperes to microamperes (1 A = 10^6 μA)		
0.0008 A × 1,000,000 μA/A = 800 μA	[·] 0008 [EXP] 6 [=]	800.
Volts to millivolts (1 V = 10^3 mV)		
0.75 V × 10^3 mV/V = 750 mV	[·] 75 [EXP] 3 [=]	750.
Watts to megawatts (1 W = 0.000001 mW)		
$\frac{50,000,000\ W}{10^6 W/mW}$ = 50 mW	50000000 [÷] [EXP] 6 [=]	50.

Figure F-10 Calculator procedures for conversion of electrical units.

Example	Operation	Readout
0.0000085 = 8.5 × 10^{-6}	[·] 0000085 [=] [F↔E]	8.5-06
52,000,000 = 5.2 × 10^7	52000000 [=] [F↔E]	5.2 07
4.698 × 10^5 = 469,800	4 [·] 698 [EXP] 5 [=]	469800.
2.32 × 10^{-4} = 0.000232	2 [·] 32 [EXP] 4 [±] [=]	0.000232
(4.8 × 10^3) × (4.5 × 10^{-5}) = 0.216	4 [·] 8 [EXP] 3 [×] 4 [·] 5 [EXP] 5 [±] [=]	0.216
(9.5 × 10^4) ÷ (8 × 10^{-4}) = 118,750,000	9 [·] 5 [EXP] 4 [÷] 8 [EXP] 4 [±] [=]	118750000
(Answer to scientific notation)	118750000	1.1875 08

Figure F-11 Calculator procedures for scientific notation.

Example	Operation	Readout
$I = \frac{V}{R} = \frac{12\ V}{5\ \Omega} = 2.4$ A	12 [÷] 5 [=]	2.4
V = IR = 5 A × 3.5 Ω = 17.5 V	5 [×] 3 [·] 5 [=]	17.5
$R = \frac{V}{I} = \frac{28\ V}{5\ A} = 5.6\ \Omega$	28 [÷] 5 [=]	5.6

Figure F-12 Calculator procedures for Ohm's law.

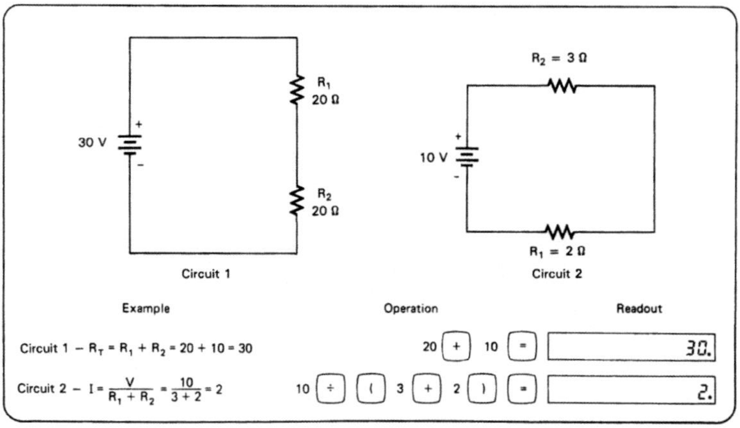

Figure F-13 Calculator procedures for series circuits.

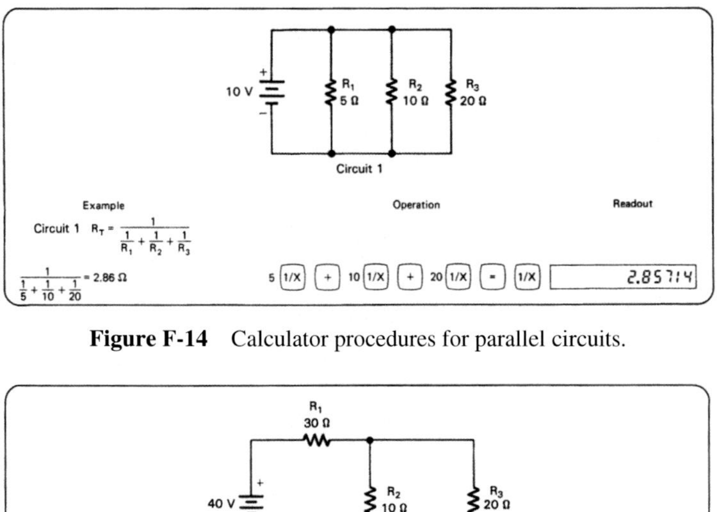

Figure F-14 Calculator procedures for parallel circuits.

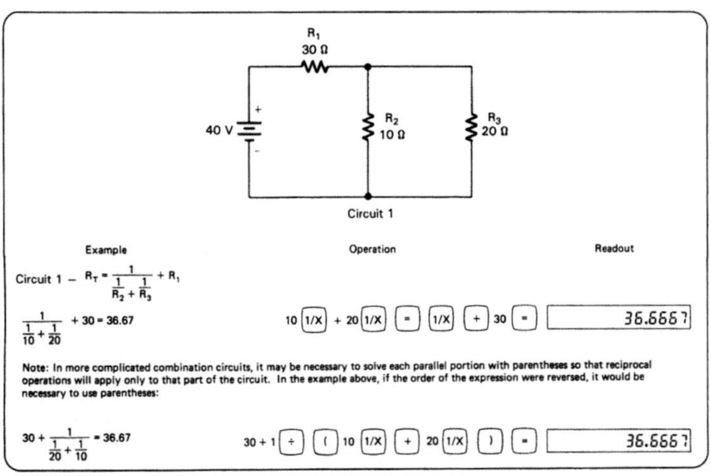

Figure F-15 Calculator procedures for combination circuits.

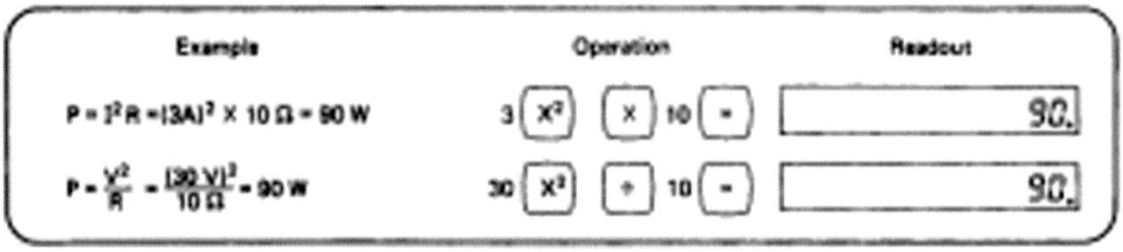

Figure F-16 Calculator procedures for electrical power.

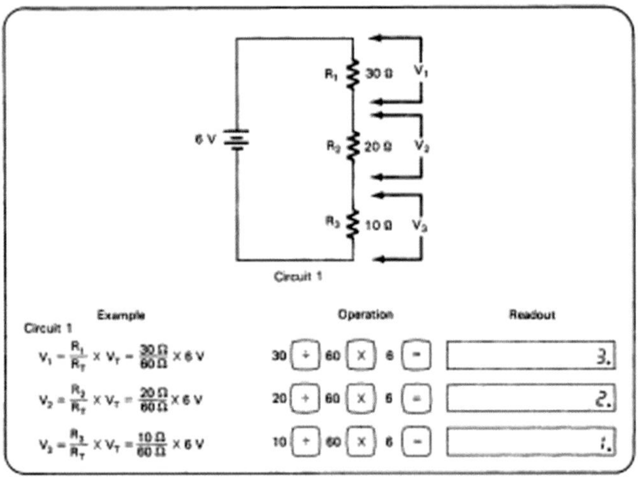

Figure F-17 Calculator procedures for voltage divider circuits.

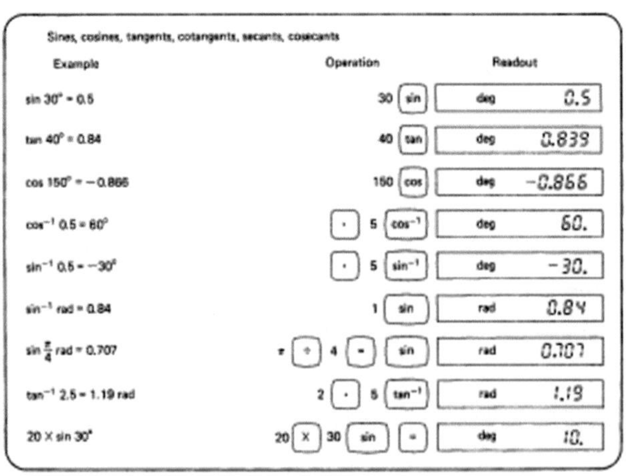

Figure F-18 Calculator procedures for trigonometric functions.

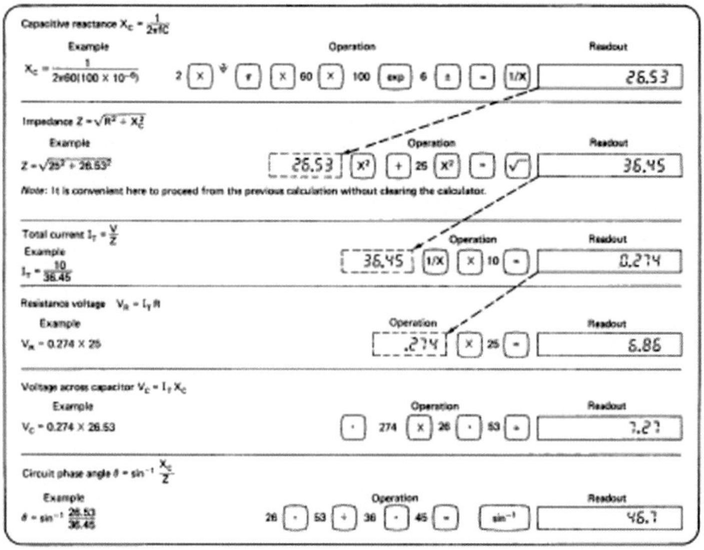

Figure F-19 Calculator procedures for series AC circuits.

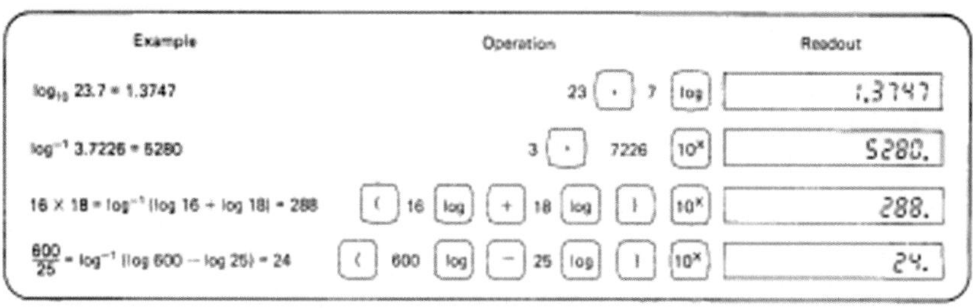

Figure F-20 Calculator procedures for logarithms.

Index

D

E

M

N

O

R

S

T

About the Authors

Dale Patrick was a Professor in the Department of Applied Engineering and Technology at Eastern Kentucky University., His experience includes teaching and organizing laboratory classes in Electrical Engineering/ Electronics Technology for many years., He completed Bachelor of Science and Graduate Degrees at Indiana State University., His experience also includes research projects, technical teacher training, and energy consulting for business and industry., He and Dr. Fardo have co-authored over 30 textbooks and laboratory manuals.

Stephen Fardo is currently a Foundation Professor in the Department of Applied Engineering and Technology at Eastern Kentucky University., His experience includes teaching and organizing laboratory classes in Electrical Engineering/Electronics Technology for many years and coordinating/ advising technical teacher programs for the State of Kentucky., He completed Bachelor of Science, Master of Science and Doctoral degrees at Eastern Kentucky University and University of Kentucky., His experience also includes research projects, technical teacher education, and energy consulting for business and industry., He and Dale Patrick have co-authored over 30 textbooks and laboratory manuals.

Ray Richardson is a Professor in the Department of Applied Engineering and Technology at Eastern Kentucky University., His experience includes teaching and organizing lecture and laboratory classes in Electrical Engineering/ Electronics Technology since 1985 as well as teaching technical seminars, technical writing, and classes in the graduate program., He completed Bachelor of Science, Master of Science degrees at Eastern Illinois University, and doctorate level at the University of Illinois., He is certified Technology Manager at the Senior level by ATMAE. Other experience includes research projects in technical education and consulting in the manufacturing, food, and electronics areas.

Vigyan (Vigs) Chandra is a professor and coordinator of the Cyber Systems Technology-related undergraduate and graduate degree programs offered by the Departments of Computer Science and Information Technology (CSIT) and Applied Engineering and Technology (AET) at Eastern Kentucky

University. He earned a doctoral degree from the University of Kentucky in Electrical and Computer Engineering and a master's degree in Career and Technical Education from Eastern. He holds certifications in various computer networking areas teaches computer systems and applications, network hardware, communication systems, and digital, analog, and machine-control electronics. He is the recipient of the 2020–21 Critical Thinking and Reading Teacher Award and the 2013 Golden Apple Award for Teaching Excellence at Eastern. His professional interests include implementing active teaching and learning strategies, metacognition, integrating open-source software/hardware with online control and computer networking technologies.